Through Mobility We Conquer

THROUGH MOBILITY WE CONQUER

The Mechanization of U.S. Cavalry

George F. Hofmann

The University Press of Kentucky

Publication of this volume was made possible in part by a grant from the National Endowment for the Humanities.

Scholarly publisher for the Commonwealth,
serving Bellarmine University, Berea College, Centre College of Kentucky, Eastern Kentucky University, The Filson Historical Society, Georgetown College, Kentucky Historical Society, Kentucky State University, Morehead State University, Murray State University, Northern Kentucky University, Transylvania University, University of Kentucky, University of Louisville, and Western Kentucky University.

Editorial and Sales Offices: The University Press of Kentucky
663 South Limestone Street, Lexington, Kentucky 40508-4008
www.kentuckypress.com

10 09 08 07 06 5 4 3 2 1

Library of Congress Cataloging-in-Publication Data

Hofmann, George F., 1935-
Through mobility we conquer : the mechanization of U.S. Cavalry / by George F. Hofmann.
p. cm.
Includes bibliographical references and index.
ISBN-13: 978-0-8131-2403-2 (hardcover : alk. paper)
ISBN-10: 0-8131-2403-4 (hardcover : alk. paper) 1. Mechanization, Military—United States—History. 2. United States. Army. Cavalry—History—20th century. I. Title.
UA30.H64 2006
357'.50973—dc22 2005037185

This book is printed on acid-free recycled paper meeting the requirements of the American National Standard for Permanence in Paper for Printed Library Materials.

Manufactured in the United States of America.

Member of the Association of American University Presses

To my children

G. Michael Hofmann

Gregory P. Hofmann

Susanne Hofmann-Linder

Contents

Preface

My interest in researching cavalry history was launched years ago at the 1971 Annual Army Armor Conference at Fort Knox, Kentucky. The late Professor Russell F. Weigley, whom I met at the Army War College at Carlisle Barracks, Pennsylvania, a few years earlier, prompted the visit to Fort Knox. At the time I met Professor Weigley, I was a graduate student conducting clumsy research for my master's thesis. One afternoon and into the evening and dinner—preceded by a few martinis—we continued our long and fascinating discussion of cavalry history. My ears were bent. Ever so gracious, Professor Weigley encouraged me to continue researching an important part of army history, especially the mechanization of U.S. Cavalry and the controversial role played by a civilian tank designer in influencing the cavalry and infantry to purchase his vehicles. This was an interest that had fascinated me earlier during my two-year tour as an armor soldier at Fort Knox.

Finally, I returned to Fort Knox, where, fortunately, I met World War II armored veterans Joseph Heard, Forrest Herbert, and Ed Reed from the Sixth Armored Division Association. This connection led to an inquiring discussion to satisfy my interest in post–World War I cavalry history and the cavalrymen's struggle with mechanization at Fort Knox. The veterans were more than kind, introducing me to their World War II divisional commander, Major General Robert W. Grow. Subsequently, General Grow introduced me to a number of key mobile-minded warriors, such as Brigadier Generals George S. Patton and Sidney R. Hinds, and Colonel H. H. D. Heiberg. My extensive interviews with these four gracious, mounted warriors centered around their knowledge of the notorious civilian tank designer J. Walter Christie, his M1928 chassis, and his four combat cars that were obtained by the mechanized cavalry at Fort Knox early in the 1930s. In addition, I was interested in how this relationship with the controversial Christie did not help the cause of mechanization but stalled a workable tank policy for years and how it affected the cavalry if not the rest of the army.

Grow, or Bob, as he encouraged his friends to call him, recalled J. Walter showing up at Fort Eustis, Virginia, in April 1931, with his M1931 tank that was undergoing government acceptance tests. "What a character who drives a V16

Cadillac about 80 miles an hour," was Grow's comment in his diary on April 26. His comment and Christie's attitude foretold what the cavalry, infantry, and Ordnance Department could expect in the years to come as the using branches and technical services debated equipment issues. The interviews that day at Fort Knox stimulated my interest even more. The following month I had the opportunity to interview Christie's adopted son, J. Edward, in Miami, Florida. What an unbelievable experience! Never did I expect to meet such a bitter man with such abhorrence. His description of the Ordnance Department and the appalling events surrounding his father's tanks is not suitable for print.

The second event in 1971 that spurred my interest was reading Timothy K. Nenninger's master's thesis, "The Development of American Armor, 1917–1940" (University of Wisconsin, 1968). What aroused my curiosity in this admirable work was the last chief of cavalry, the unyielding Major General John K. Herr, and the debate over the future of his branch on the eve of World War II. This interest over the "great debate" led to additional research. Subsequently, General Grow was kind enough to provide his personal diary for the period in question and a manuscript based on his diary, "Ten Lean Years: From the Mechanized Force (1930) to the Armored Force (1940)." As Grow recorded, "It was my good fortune to serve in positions that called upon me to play a considerable role in the development of mechanization, its application to cavalry, its acceptance, as well as a lack of acceptance, by cavalry and the eventual development of a separate Armored Force." He had close contact with many future cavalry and armor generals, as well as countless junior officers. This provided him with "an unusual opportunity to observe as well as to take part in the evolution that took place during the thirties" (2–3). Grow's diary contained valuable comments on General Herr, which he recorded while serving in the Office of the Chief of Cavalry in the late 1930s, as well as the entries that covered his experience in World War II as an armored division commander.

Recognizing the importance and extent of Grow's historical experience, first in the cavalry and then in the Armored Force, the noted historian and author of the *Patton Papers*, the late Martin Blumenson, suggested that I write a biography of the general. Grow responded in a letter to me on June 23, 1976, "It was wrong and is not worth it." As humble as the general was, Blumenson was right. Grow's historical experience with the mechanized cavalry and then armor was written down in his daily diary until he retired in 1953. It is a microcosm of important historical events experienced by a cavalry officer who was dedicated to the mounted arm and its struggle to mechanize. He was always available for straightforward discussions and most generous in sharing his diaries and records with me while I was researching the early history of mechanized cavalry and armor.

Consolidating my interest in cavalry history and research primarily based on Grow's diaries, I submitted a manuscript, "Tactics vs Technology: The U.S. Cavalry Experience," to Lieutenant Colonel Burton S. Boudinot, editor in chief of *ARMOR* (the oldest professional journal in the U.S. Army). It was subsequently published in the September–October 1973 issue. The article ended with an explanation of why General Herr could not free himself from the mesmerizing influence of the horse that precipitated the great debate, which handicapped one of the greatest transformations in the army in the twentieth century. The plan to expand this article into a book, however, was stalled for years because of increased writing and teaching obligations. Finally, after many years my interest in a mechanized cavalry history was resurrected due to two events. First, it was considered to be a follow-up to the anthology *Camp Colt to Desert Storm: A History of U.S. Armored Forces* (1999), which I coedited with General Donn A. Starry. Second, the presentation of a paper at the U.S. Military Academy at West Point in 2002 and subsequent discussions with members of the academy's history department on cavalry mechanization and the first mechanized cavalry theorist finally encouraged me to move forward. At the time a few members of the history department were working on their dissertations. Their enthusiasm over cavalry history was a catalyst to resurrect a long-delayed project. I dug into my files and restored to life sources and recorded interviews put away years ago and never used.

The framework for this history began years ago in a number of military journals and magazines for which I wrote articles on cavalry and armor history, such as *ARMOR, ARMY, Military Affairs, Journal of Military History, Journal for United Defense Studies, Journal of Slavic Military Studies*, and *Marine Corps Gazette*. I wish to thank the editors and reviewers for providing me the opportunity to publish in these outstanding journals.

Unfortunately, the far-reaching historical Grow diaries from 1930 to 1953 that were in my possession at one time are now secluded with the general's grandson and not accessible. General Grow told me on two occasions that he was considering donating the diaries to the U.S. Army Military History Institute, because he felt this was important for students and historians researching cavalry and armor history. I could not agree more.

The other unfortunate aspect in researching this history was a lack of response from the World War II Mechanized Cavalry Associations. This is understandable because this great generation is rapidly moving on to Fiddler's Green, the mythical broad meadow located halfway down the trail to hell. Here no dead trooper gets there but camps in the meadow lined with trees and many streams, exchanging reminiscences and stories with their comrades.

Acknowledgments

No historian can write a story without building on previous works. I have been fortunate during my neophyte stage to have the honor of meeting a number of military historians who were most gracious in providing information, directing and advising me as I struggled through my profession. I especially want to pay my respects and tribute to the late Robert J. Icks, Martin Blumenson, and Russell F. Weigley. Decades ago they encouraged me to write a history of U.S. Cavalry and Armor and on the controversial tank designer J. Walter Christie. Our profession will surely miss them, for they gave so much, especially to students. I will always remember their thoughtfulness and patience in addressing the many questions on military history I discussed with them over the years. I found it especially edifying to meet them and many of my colleagues at the Society for Military History and the U.S. Army Armor Conference, where ideas were always willingly exchanged.

My research on mechanized cavalry began at the National Archives at Suitland, Maryland, and continued at the National Archives II at College Park, Maryland, when I had a new research assistant, my wife, Kathy. The trips to the National Archives II were a pleasant and rewarding experience thanks to the exceptional assistance of Tim Nenninger and Mitch Yockelson, who frequently pointed us in the right direction. The U.S. Army Military History Institute was another gold mine. During my many visits over the years, Dave Keough was always available and had the knack for finding seldom-used primary sources that made this historian elated.

Fort Knox, Kentucky, with its long tradition of mounted warfare, provided a wealth of information on the history of armor and cavalry. It is a gem for conducting research. Curator Charles Lemons and Librarian/Archivist Candice Fuller at the Patton Museum of Cavalry and Armor were willing to meet my numerous requests, even at the last minute and in spite of their busy schedules. At the Armor School, Bill Hanson, Director of the Armor School Research Library, and Lorraine Allen, Technical Information Specialist, were more than generous with their time. David Manning and his team at *ARMOR* magazine, Christy Bourgeois, Vivian Oertle, and Jody Harmon, were always on call and most helpful. I spent many days in the 1990s at the Armor School and Armor School Research Library with John T. Broom, who later became a training and doctrine command instructor of the year. He read parts of the manuscript and was kind to share a number of his research efforts and unpublished manuscripts. Over the many years spent at Fort Knox, I can honestly say there was never a moment when I left the post after hours or days of research that my enthusiasm for writing armor and cavalry history was not greatly enhanced.

Burton S. Boudinot was kind to share his manuscript "The Way I Remember It," which he wrote for interested friends as a personal reminiscence of his and his father's military career. Larry I. Bland at the George C. Marshall Foundation, Lexington, Virginia, provided valuable insight on General George C. Marshall, especially regarding his relationship with Adolf von Schell and his lectures at the Infantry School. At Fort Leavenworth, Kansas, Louis A. DiMarco shared his extensive work on the history of the U.S. Cavalry and read portions of the manuscript. His Web site became a valuable source. During the bicentennial at West Point, I was fortunate to discuss ideas regarding the history of mechanized cavalry with a number of faculty members from the Department of History, especially C. J. Horn, Chris Prigge, and Matt Morton, all of whom were working on their dissertations. The most exciting visual research visit was at the Army Ordnance Museum, Aberdeen Proving Ground, where Jack Atwater spent almost a day providing a private tour of his facilities, especially letting us closely examine the only Christie tank in existence of the seven purchased in 1931. Peter S. Kindsvatter, the Command Historian at the Army Ordnance Center and Schools, was generous in supplying information on post–World War I ordnance history. John White at the Southern Historical Collection, Manuscript Department, University of North Carolina Library at Chapel Hill, was more than helpful while I was sorting through the Frank Parker Papers. In addition, my sincere thanks goes to John F. Votaw and his team at the First Division Museum in Cantigny, Illinois, for granting me access to their collection of the Parker Papers.

Sandy Slater and Curt Hanson from the Elwyn B. Robinson Department of Special Collections, Chester Fritz Library, University of North Dakota, were kind in responding to my last-minute request for archival material covering Sereno Brett's creer. Jon Hornbostel, the unofficial and competent Fort Riley historian, was generous in sharing his thoughts and photos of the important 1934 cavalry maneuvers.

Albin Irzyk, the last senior-ranking member of the U.S. Constabulary, was kind enough to read part of the manuscript and share his personal experience and insight on the post–World War II period when an unheard of and unique organization was established based on the wartime cavalry model to deal with the turmoil of the early cold war period.

I would especially like to thank the following individuals who over the years were kind enough to discuss this project and to offer advice, suggestions, documentation, and photos: Edwin Krampitz Jr., from the Society of Automotive Engineers, for providing additional information on J. Walter Christie; Randy Sowell, Archivist, and Pauline Testerman, Audiovisual Archivist, Harry S. Truman Library; Mark Taflinger, Librarian, *Louisville Courier-Journal*; M. Carl Ramga, 91st Reconnaissance Troop, 91st Infantry Division; Charles W. Barbour, 86th Reconnaissance Squadron, 6th Armored Division; William M. Tevington, U.S. Constabulary. In addition, I wish to thank Robert Bateman, Alexander M. Bielakowski, Philip L. Bolté, Robert Calabrese, Robert S. Cameron, Robert M. Citino, David Fletcher, Forrest Herbert, William O. Hickok, Richard P. Hunnicutt, Gary Nichols, Edward Reed, and Dennis Showalter for their selective input over the past years on specific historical matters regarding doctrine, leadership, and cavalry issues.

Most important, I sincerely wish to express my sincere thanks to Edward "Mac" Coffman, who read the entire manuscript. His positive attitude and approach to students and peers make him special among most military historians. A special thanks also to Lewis Sorley, who read the entire manuscript, for the advice to include the U.S. Constabulary in the mechanized cavalry story. His suggestions were valuable. And to my dear friend General Donn Starry, who more than anyone encouraged me during the frantic phase to go on and complete the cavalry history. He was always available to guide me through difficult periods, as all historians can understand during the crafting period. Finally, I would like to thank my wife, Kathy; without her help, support, and historical sense, this project would have taken much longer to finalize.

Introductory Essay

BY GENERAL DONN A. STARRY, USA, RETIRED

> The principal value of cavalry is derived from its rapidity and ease of motion. To these characteristics may be added its impetuosity; but we must be careful lest a false application be made of this last.
>
> —Baron Antoine de Jomini[1]

> This Army isn't what it used to be . . . and it never was.
>
> —General Creighton W. Abrams Jr.[2]

Through Mobility We Conquer is a chronicle of the mechanization of horse cavalry in the U.S. Army beginning after World War I (the 1914–1918 war). It includes early mechanization experiments, mechanization of horse cavalry units for service in World War II (the 1939–1945 war), and mechanization and employment of cavalry units in the years immediately following World War II. It concludes with the merger of the cavalry and armor branches in 1950.

Through Mobility We Conquer follows an earlier book entitled *Camp Colt to Desert Storm*,[3] which sets forth a history of the development of mechanized forces in the whole of the U.S. Army from the advent of tanks in battle in the 1914–1918 war to the end of Desert Storm 1991, the coalition operation to drive Iraqi armed forces out of the Kingdom of Kuwait. The chapters of *Camp Colt* do not focus directly on mechanization of horse cavalry. There has, however, been considerable demand from students of change to provide a closer examination of the mechanization of horse cavalry, for while mechanization has affected the army in all parts, from infantry, to engineers, to service and technical support branches, perhaps no branch has been as dramatically affected as the cavalry. This is so for several reasons, the most important of which have to do with the traditional functions of cavalry on the battlefield.

Without Cavalry Battles Are without Results

Attributed to Napoleon by John Ellis in *Cavalry: The History of Mounted Warfare,* the assertion that without cavalry battles are without results reflects the conviction that horse cavalry's historic primary role had been to move quickly through breaches in enemy force deployments achieved by what was frequently described as "open warfare," that being the combined action of infantry and artillery, or to go around—seeking flanks and rear of larger enemy formations. In either case it was to go deep in order to disrupt and/or destroy follow-on forces, command and control, logistics, and other support infrastructure. Thus, it turned an enemy out of his deployed battle posture by penetration, envelopment, or perhaps both, or preventing him from mounting an attack because of the disruption (in his rear) of support for the force deployed forward—in sum, collapsing the integrity of whatever the enemy operational and tactical scheme might have been, then pursuing and destroying the deployed enemy force in the field.

Seized with this traditional operational- and tactical-level doctrine and related force requirements, contending powers in the mobilization for the 1914–1918 war fielded, by one count, nearly half a million horse cavalrymen. Deployed on the western front alone were 1,254 squadrons of horse cavalry mounting more than one hundred thousand men.[4] To these numbers the United States added nominal deployments of horse cavalry from the 2nd, 3rd, 6th, and 15th U.S. Army Cavalry Regiments as the American Expeditionary Force (AEF) deployed to France in 1918.[5]

When all was done, however, no horse cavalry fought a significant, traditional-style battle on the western front in the 1914–1918 war. For demonstrated on those battlefields was the stark reality that the advent of machine guns in the hands of infantry, supported (especially after 1916) by maneuverable indirect-fire field artillery, made it virtually impossible to employ horse cavalry using traditional tactics. Neither the doctrine itself nor its attendant equipment, organization, or tactical- and operational-level training fit the changed circumstances brought on by the relentless march of Industrial Revolution technology.

This reality was of course not nearly so dramatically apparent on the eastern front, where, on famous battlefields—Gumbinnen, Tannenberg, the Masurian Lakes, and others—battles included large horse cavalry formations: German on the one side, regular Red Army as well as Cossack horse cavalry units ranging in size up to cavalry corps on the other. But since the war in the east virtually ended in 1917, with the collapse of the czar's armies, the czar's government, and the ensuing onset of revolution in Russia, there was no universal experience to parallel either the military or the national trauma of attrition warfare characterized by battles on the western front.

Some Newfangled Things

Of course, neither machine guns nor artillery, tanks, or aircraft, principal evidences of technical modernization on battlefields of the 1914–1918 war, were all that new. However, neither were their potential effects on battle tactics of the day well defined or even acknowledged. The situation represented a growing reality that all too frequently technology developments invite, even demand, change as well as offer opportunities, yet institutions and their populations are frequently all too slow in taking advantage of opportunities, as well as in comprehending the need for change. Some very early examples are instructive.

In the 1876 campaign against the Sioux, General Alfred H. Terry's Department of Dakota forces deployed into Montana included the 7th Cavalry Regiment under its field commander, Lieutenant Colonel George Armstrong Custer, and among supporting forces a battery of four Gatling guns. The Gatlings, ancestors of the machine guns of 1914–1918, were developed but not employed during the Civil War. They were big, heavy, and cumbersome, particularly when included in fast-moving cavalry columns, so much so that they were organized into (artillery-like) batteries, assigned to and employed by artillery units. However, they represented the potential for an enormous increase in firepower, especially for cavalry units, which depended primarily on speed, shock, and surprise, accompanied only by the quite modest firepower of cavalry revolvers and single-shot carbines. Nevertheless, the presence of the Gatlings on Custer Hill, late afternoon June 25, 1876, could well have made an enormous difference.[6]

The following summer saw the last army campaign against the Nee-Me-Poo—the Nez Percé Indians of the Pacific Northwest. A mixed force of cavalry, infantry, and artillery was assembled under command of several already famous Civil War era names. Beginning in the region where Idaho, Washington, and Oregon territories joined, this force pursued the Nee-Me-Poo southeastward into Yellowstone, thence north into Montana Territory as the Indians sought refuge in Canada, a campaign covering more than seventeen hundred miles and extending from June to the end of September 1877.

Then, on a cold, crisp September morning in the last battle of that campaign, one understrength battalion of the Seventh Cavalry commanded by Captain Owen Hale (also commanding K Company) attacked the entrenched Indians on the right flank of the expedition battle line at a place called Bear's Paw Mountain in north central Montana Territory. The attacking cavalry battalion lost all its officers, save one, and all its sergeants, save two. In all, seventeen were dead and thirty-six wounded, of a beginning strength not much greater than that. While this unhappiness was unfolding, there was deployed on the field one Hotchkiss 1.65-inch caliber lightweight cannon—a field piece designed to accompany and furnish fire support to fast-moving infantry and especially cavalry columns. The cannon was apparently on the field, but either out of range or out of the line of sight of the deployed Indian line, upon which its fire, for whatever reason, could not be brought to bear. Whether or not there was, earlier that morning, discussion among commanders regarding deployment of the Hotchkiss in support of the attack is not a matter of record. In any event, like the Gatlings along the Little Big Horn the previous summer, it could well have made a significant difference in the fight between Owen Hale's battalion and the entrenched Nee-Me-Poo.[7]

Owen Hale, who fell early in that fight—leading the attack—was no apprentice cavalryman. He had joined the New York cavalry at the onset of the Civil War, fought in campaigns of the New York cavalry in several grades—enlisted and officer, rising to the rank of captain by war's end. The war done, he sought and was awarded a commission in the Seventh Cavalry. Although he had fought with the Seventh in its campaigns against the Sioux, he was obviously not with Custer at the Little Big Horn. Therefore, quite possibly the significance of failure to include modern firepower—Gatlings the previous year, Hotchkiss in the ongoing campaign—simply may have escaped him.[8]

However it may have come to pass, the potential of rapid, high-volume firepower (especially from the Gatlings) was clearly not recognized, and revised tactics to take advantage of capabilities of the new weapons lagged well behind the appearance of the weapons themselves; it would happen many times over again.[9]

The incident with the Hotchkiss presaged the situation in the early years of the 1914–1918 war. The cannons of 1914 were quite reminiscent of the guns of Napoleon, in every way from appearance to tactics, techniques, and procedures—direct fire over open sights, largely directed at masses of infantry paraded to the front. By 1918, however, fewer than four short years from slow beginnings in 1914, tactics, techniques, and procedures more familiar to today's artilleryman had been set in place. These included indirect fire, massed fire, counterbattery fire, calibration, meteorological correction, and support for combined arms operations as between infantry, artillery, and supporting arms. On the German side, in 1918 in particular, one could find fire support techniques quite consistent with four principles set forth in the U.S. Army's AirLand Battle doctrine of the 1980s—initiative, agility, synchronization, and depth.[10] Integration of these improvements in fire support in the 1914–1918 war was largely the work of one remarkable German artilleryman, George Bruchmuller; he had in fact retired but had been recalled to active service and posted to a position from which he could effect infusion of modern methods into artillery fire support techniques. And his revised artillery tactics in one form or another have characterized artillery affairs ever since.[11] There again was no immediate and overwhelming stampede to take advantage of the battlefield potential offered by technology developments.

So, then, if not breakthrough and exploitation, or turning movements and envelopment, what role for cavalry? For in truth cavalry riding hard to get to the scene of battle—but once engaged dismounting to fight on foot—characterized most cavalry action in the 1914–1918 war. Cavalrymen were in truth more dragoons than cavalry. In response to this growing reality, emphasis began to shift, even in the late nineteenth century, to employment of horse cavalry on three related missions: first, as a primary gatherer of information—a reconnaissance force, finding, locating, and identifying enemy units and observing and reporting enemy activities; second, as a security force providing early warning and interdiction of enemy presence on flanks, in rear, ahead; third, as an economy force holding off or disrupting larger enemy forces to conserve the mass of friendly forces for critical missions. In sum, security for a main force much larger and slower of movement than more lightly equipped and thus more agile cavalry, but cavalry employed so as to keep the lightly armed and armored cavalryman and his mount fairly well out of the fans of fire of machine guns and increasingly accurate artillery.

Whither Cavalry?

Conceptually, then, several important questions arose. First, should cavalry be "light" enough, and therefore agile enough, to deploy with alacrity, but still "heavy" enough to fight effectively once employed? Adding weapons to satisfy the "heavy enough" requirement defied the need to be "light enough" and so questioned whether cavalry could remain agile enough to move about with sufficient alacrity. Related questions followed. Should cavalry on a reconnaissance mission be expected to fight, as well as look for information, or fight as a covering force and be heavy enough to do that (fight) routinely? Or should cavalry remain a much lighter, and so more vulnerable, force—conducting reconnaissance using "sneak and peek" tactics, evading "decisive engagement"? Against an enemy armed with modern weapons, for example, machine guns and maneuverable indirect fires of artillery—cannons, missiles and rockets, and fires from aerial platforms—would it be possible to execute security missions armed with revolvers and carbines, or would it be necessary to carry along machine guns, artillery, and other heavier weapons and to dismount and fight as infantry with heavier weapons support but much less mobility? Finally, could cavalry be successfully employed in an economy of force mission—a traditional role especially for heavy cavalry?

Those being the important questions, what, in the future, should be the function of cavalry? What should be cavalry doctrine? How should cavalry units be employed—equipped, organized, trained, commanded, supported?

Answers to these questions about cavalry of necessity sought resolution in the context of the obvious larger issues facing military establishments in all countries, reflecting lessons of the 1914–1918 war, especially those drawn from combat on the western front. So mechanization of horse cavalry came to be a lesser included event, one among many in the grander scope of mechanization of armies as a whole, not only in the two decades separating the 1914–1918 war from its successor of 1939–1945, but in years after the 1939–1945 war as well.

It is the purpose of the principal chapters of *Through Mobility We Conquer* to describe how the horse cavalry of the U.S. Army fared while the army as a whole contended with the imposing task of change—of seeking and finding at least tentative answers to the questions of mechanization.

In the Wake of the Storm

In the wake of the 1914–1918 war there was first the urgent requirement, recognized immediately (and especially) by the European powers (Germany, France, Great Britain, and Russia), to recover from the devastation of that war. This involved recovering the opportunity costs of national economic treasure given off to military forces and operations, and as well recovering from the tragic loss of generations of young men who now lay dead on the battlefields of Europe. At the same time, there was the need to decide what might happen next, especially in Europe, where war's devastation had characterized so much of the nineteenth century, and now the early years of twentieth century. Then, given newly perceived national goals and strategies, what military capabilities would be required to support the new strategies, and how could those capabilities be acquired by the increasingly rapid advance of modern technologies?

Paramount to consideration of what military capabilities might be required was the obvious need to decide how, in future armed conflict, to avoid the gross battle of attrition so characteristic of battles on the 1914–1918 western front—the destruction of whole generations of young men of combatant nations. Although no vote was taken, after the war there appeared a general consensus (the French demurring) that the best way to do that was to find means to deploy forces rapidly and maneuver quickly to employ forces expeditiously at tactical and operational levels of war. For if the ability to maneuver could not be recaptured, then future military operations were virtually foredoomed to be battles of annihilation. This dilemma brought whole armies, including all arms and services thereof, to an intellectual threshold, seeking to define how nations should design their military forces as instruments of national strategy, first of national political strategy, then of national military strategy, and against those requirements, what nations should expect their armed forces to accomplish at strategic, operational, and tactical levels of war, namely, how they should expect to fight, to be equipped, organized, commanded, trained, and supported.[12] Answers to these questions varied from nation to nation, reflecting culture, tradition, history, and other uniquely national biases, all reflected in their performance in the ultimate test, the "field trial" of the 1939–1945 war.

Developments in France, Germany, Great Britain, Russia, and the United States are subjects of individual essays in *The Challenge of Change: Military Institutions and New Realities, 1914–1918,* edited by Harold R. Winton and David R. Mets. In his introduction to this anthology, Winton notes that "military reform . . . is driven by a curious trinity of its own. The poles of this trinity are endemic strategic and technological uncertainty, the dictates of the political and social values of the state, and the defining characteristics of the military service itself." And, it might be added, the fielded operational results reflected in each case, the interstitial fiber (the culture), defining the "curious trinity."[13]

Revolution and the Red Army

In "Military Reform and the Red Army 1918–1941," another essay in *The Challenge of Change,* Jacob W. Kipp describes the political and related imperatives of the revolution, which followed on the collapse of the czar's army and government. This review sees the Soviet state as "the de facto heir of tsarist Russia"; given that, interwar reforms are seen as "the continuation of the evolution of Russian military institutions."[14]

The continuing evolution led directly to decades of development of military doctrine consistent with changing national goals, doctrine from which came requirements for equipment, force structure and organization, training of soldiers and units, and training and education of leaders. Change was not without turmoil and trauma, but change there was.

Strange as it seems, there developed between revolutionary Russia and the German Reichswehr considerable professional discussion and rather surprising collaboration in matériel, organizational development, and field experimentation, all of which laid foundations for tactical and operational schemes with which each participant went to war, the Germans in 1939, and both (against one another) in 1941. However they may have been arrived at, doctrines were different on either side, or they reflected the thinking of some of the dominating intellectual personalities in the history of the military art, from Tukachevskii and Triandafillov in the Red Army to von Seeckt, Beck, Guderian, and Rommel in Germany.

Tactical- and operational-level "field trials" in Britain, and combat operations in the Spanish civil war (Germans siding with Franco, Soviets with the Republicans), were occasions for evaluation of tentative armored force doctrine, primarily with regard to the critical question of highly mobile and concentrated mechanized elements as an independent force, or less mobile and more widely spread armor intermixed with and mainly for support of infantry. On the Mongolian-Chinese border, in battles at Kalkhin-gol and Lake Khasan, the Red Army would also evaluate developing doctrine. Both Germans and Russians benefited from joint collaboration at Kazan (in the Soviet Union), where from 1926 to 1933 the Germans operated a school and training center, literally under contract to the Soviets.

As doctrine development proceeded, enabling mechanized equipment development followed. In "Doctrine, Tank Technology, and Execution: I. A. Khalepskii and the Red Army's Fulfillment of Deep Offensive Operations," an essay in the *Journal of Slavic Military Studies*, George F. Hofmann records in detail Khalepskii's work with the controversial U.S. engineer J. Walter Christie and his tank designs, which, when adopted by the Soviets, led to fielding of the much respected Soviet T-34 tank in substantial numbers at a time when virtually no other country in the West had fielded a satisfactory tank, in design and certainly not in numbers.[15]

Then, in the mid-1930s, came Stalin's purge of Red Army leaders. While its purpose is yet a matter of debate, one accounting holds that Stalin was reacting to one or both of two fabricated conspiracies, both inspired by the NKVD (People's Commissariat of Internal Affairs). In one, Reinhard Heydrich, probably Hitler's most enthusiastic executioner, turned an NKVD "black" operation into an alleged conspiracy between senior Red Army leadership, including (then) Chief of Staff Tukachevskii, and a coterie of German general officers. At about the same time another fabricated scheme linked Tukachevskii and Red Army leaders with Trotsky. Or (it is speculated) Stalin himself may have created the situation for his own reasons. However it may have come about, one night in June 1937, Tukachevskii and seven of his principal staff were executed by the NKVD for treason without having had a trial. The event signaled the beginning of a general bloodletting in Red Army leadership, a purge that may have taken several thousand lives and cost the Soviet army dearly in years to come. Further, it stripped out former Red Army officers who were proponents of independent armored forces employed at the operational level of war, in what Tukachevskii called "deep attack," as opposed to just at the tactical level, mixing tanks with infantry, whether advancing or defending. As the latter view became dominant, the fate of literally thousands of Soviet army infantry in operations against the Wehrmacht from 1941 to 1945 was sealed. But it was, after all, the people's war, and they were the people.[16]

From Reichswehr to Wehrmacht

The 1914–1918 war ended with Germany's surrender, but with German forces yet on French soil and many Germans (soldiers for certain) believing that, however it had come about, their political masters had lost the war. Further, they were aghast at the severe limitations imposed on Germany by the Treaty of Versailles.

It was against this foreboding background that the German Reichswehr, the one hundred thousand man force permitted by the Treaty of Versailles, came into being. At its head was General Hans von Seeckt. A conservative Prussian soldier-aristocrat, von Seeckt was, and remains, the epitome of the German military culture that produced what Trevor Dupuy once called "a genius for war."[17] Von Seeckt foresaw that Germany's struggle for survival in central Europe was not at an end—there would be more war. Germany needed to be able to defend itself against predators attacking from either the west or the east. Germany needed to rearm. But how to rearm in the face of the stringent armament (and other) limitations imposed on Germany by the Versailles treaty?

Von Seeckt's strategy was to somehow circumvent, without appearing to do so, the disarmament provisions of the Versailles treaty and build the Reichswehr into a modern military force fully equipped, organized, and trained for mobile warfare, but with modern equipment, new force structures and organizations, new operational and tactical schemes, and revised training and education for soldiers and leaders. To do this, it would be necessary to avoid war at all costs until the results of reform were in place.

Further, following long-standing traditional relationships between political bodies and the military in Germany at the time, von Seeckt took charge of formulating concepts for a new national strategy. In it von Seeckt considered it likely that in the next war, as in the last, Germany might need to fight on two fronts—to the west and to the east—but also recognized that simultaneity of those fights would be most difficult. Therefore, sufficient force mobility—deployability—needed to be achieved should the two-front problem arise.

In *Panzer: A Revolution in Warfare, 1939–1945*, Roger Edwards chronicles development of the concept of panzer forces as spearheads leading German infantry, artillery, and other forces, nearly all afoot and all quite heavily horse-drawn, moving to exploit the panzer force breakthrough. This formulation was at the core of the panzer concept. Outside Germany the concept came to be called "blitzkrieg"—lightning war, a term often said (true or not) to have been invented by *Time* magazine in the United States. Edwards's book unfolds the idea, its history, leaders, the contribution of each to the process, and a detailed accounting of Wehrmacht panzer units and their battle history. Early chapters summarize the German experience from 1919 to 1939; and later chapters set forth details of panzer force deployments and employments from 1939 to 1945.[18]

Characteristic of the concept was the Wehrmacht order of battle—force deployments on the eastern front on the eve of Operation Barbarossa, the 1941 German invasion of the Soviet Union. In a typical organizational snapshot, Army Group Center (von Bock) consisted of two armies—Fourth (von Kluge) and Ninth (Strauss), in all twenty infantry divisions; two *Panzergruppe*—Panzergruppe Two (Guderian) and Panzergruppe Three (Hoth). Panzergruppe Guderian included four corps: XII (Schroth) with three infantry divisions; XXIV (von Schweppenburg) with one motorized infantry division, two panzer divisions, and the First (horse) Cavalry division; XLVI Panzer Corps (von Vietinghoff-Scheel) with one panzer division, the motorized SS-Division Das Reich, and the Grossdeutschland Infantry Regiment; XLVII Panzer Corps (Lemelsen) with one motorized infantry division and two panzer divisions. Combined arms formations all, each configured on evaluation of mission, enemy, terrain, and troops available. Operationally, *Panzergruppe* spearheaded a penetration, followed by motorized infantry divisions, supported by artillery, largely horse-drawn, and by wagon trains, also largely horse-drawn. Parenthetically, total numbers of horses (and mules) employed by the Wehrmacht in the 1939–1945 war stood at about 2,750,000. These numbers illustrate the reality that the Wehrmacht was, except for the panzer spearheads with their accompanying motorized infantry divisions, by no means a mechanized (or an armored) army—contrary to popular belief.[19]

Further is a new translation of *Die Truppenfuhrung*, a 1933–1934 edition of what many armies call "field service regulations" or "field manuals." In the 1914–1918 war the German army fought with a field service regulation issued in 1905. That doctrine reflected earlier writing of Carl von Clausewitz and the thinking of Field Marshal Helmuth von Moltke the elder, all reworked at the hand of General Alfred von Schlieffen, chief of the Great General Staff from 1891 to 1906.[20]

Among von Seeckt's early reforms was a revised field service regulation *Herresdienstvorshrift 487, Fuhrung und Gefecht der verbundenen Waffen (Command and Combat of the Combined Arms),* foreshortened to H.Dv. 487—popularly known as *Das FuG* (pronounced "Das FooGay"). Published between 1921 and 1925, *Das FuG,* like postwar doctrinal publications of other armies of the period, rejected position (read trench) warfare, emphasized the need for maneuver—for offensive action tactically and operationally, for holding attacks to enable encircling maneuver—and the notion that defense should be but a temporary measure, adopted in order to enable attack elsewhere. *Die Truppenfuhrung (Unit Command) H.Dv.300,* written largely by Colonel General Ludwig Beck, Werner von Fritsch, and Otto von Stulpnaegel beginning in about 1925, leaned heavily on von Seeckt's *Das FuG.* Part 1 was published in 1933 and part 2 in 1934.[21]

Die Truppenfuhrung is quite likely one of the most significant expositions of doctrine in the history of the art of war.[22] It was the basis for Guderian's demonstration for Hitler at Kummersdorf in 1934, following which Hitler himself "signed on," enabling development of doctrine, equipment, and forces needed to rearm Germany, an undertaking Hitler commenced in 1935, and which included creation of the Wehrmacht from the Reichswehr.[23]

Perhaps most interesting in all this is the fact that, from the beginning, doctrine developers were instructed by von Seeckt to ignore resource constraints—budgetary, matériel, and manpower boundaries of the Versailles treaty; in other words, "just say what needs to be done, what is required to do it, and assume the wherewithal will be made available." Further, it is relevant that from the beginning of serious doctrine development in 1925 it took nine years to produce the 1933–1934 edition of *Die Truppenfuhrung*, and beginning in 1935, another five years to organize, equip, and train the early panzer spearheads—the forces that led the attacks into Poland and France in 1939 and 1940. Those fourteen years were full of hard thinking, rigorous professional debate, intense analysis, and lively discussion, all followed by field trials in many venues—in Britain, at Khalkin-gol and Lake Khasan, at the joint training center at Kazan in Russia, in the Spanish civil war, and by an urgent and frustrating search for, and eventual accelerated equipment development and production of, mobile armored equipment—especially tanks. Busy times, indeed, but Hitler's endorsement was instrumental, as was his decision to set aside restrictions of the Versailles treaty and rearm Germany. Resource priorities were rearranged in order to make something like what was needed come to pass, all as von Seeckt had foreseen.

The lesson for those who would institute change—those who seek new doctrine to act as the engine of change—is that it cannot be done without considerable clear thinking, honest and rigorous evaluation in both analysis and field trial, and extensive and consistent marketing to the ultimate user. Most important, it cannot be done overnight.

The debate over mechanization notwithstanding, horse cavalry survived in the post-1918 Reichswehr and in its successor, the Wehrmacht. The structure and organization of Germany's armed forces stipulated by the Treaty of Versailles called for seven infantry and three (horse) cavalry divisions. After the Reichswehr became the Wehrmacht in 1935, the number of horse cavalry divisions in the force structure rose to a total of eight. These, plus a Cossack cavalry division (later expanded to a Cossack Cavalry Corps), fought in the 1939–1945 war, largely on the eastern front. Of note is the fact that Germany in the 1914–1918 war deployed eleven cavalry divisions, only three more than the number in the Wehrmacht structure in the 1939–1945 war.[24]

Of all participants in the 1914–1918 war who (after the war) sought answers to the obvious need to mechanize, none retained the numbers of horse cavalry found in the Wehrmacht. Increasingly, Wehrmacht horse cavalry units found themselves employed either as dragoons who dismounted to fight or as reconnaissance units, which were found to be inadequately armed and equipped to fight for information, rather than reconnoiter by "sneak and peek," a dilemma highlighted earlier. Hence, as experience dictated, the horse cavalry found itself more and more encumbered by equipage—both weapons and vehicles—which inhibited its "rapidity and ease of motion," so touted in Jomini's observation on cavalry cited earlier. First there were added bicycles, later motorcycles, then motorcars, then armored motorcars, then light tanks. First there were machine guns, then mortars, then light artillery, then air defense weapons. First there were small trucks in the trains, then as the need for more transport for weapons and ammunition grew, there were larger and larger trucks, but never enough of any. Illustrative of this experience is that of the First Cavalry Division, observed earlier as part of XXIV Corps in Guderian's Panzergruppe Two, part of Army Group Centre in the lineup for Barbarossa. After a few months of combat, struggling for sufficient mobility to keep up with the panzers, and with the need to heavy-up the cavalry to enable it to fight if need be (and it usually did need be), the division commander (Feldt) petitioned Guderian to mechanize his division. Thus First Cavalry Division became Twenty-fourth Panzer Division, rearmed, reequipped, and retrained in remarkably short order to fight the remainder of the war as a panzer division.

So as the Wehrmacht, thus far strikingly invincible, moved eastward against the Soviet Union in 1941, two competing operational concepts came into conflict, with the Wehrmacht's panzer spearheads very nearly prevailing, but in the end failing against the vastness of Russian geography, of numbers of soldiers and equipment, of winter, and of systemic flaws in Wehrmacht support capabilities for operations on such a grand scale—especially shortages in replacement personnel and equipment, and in the means to resupply rapidly moving mechanized forces over enormously long distances. What happened, of course, was that Hitler forced the course of action that von Seeckt had feared most at the outset—and encountered the extreme difficulty of fighting on two fronts simultaneously.

The Admiralty Landships Committee

In his introduction to *The Challenge of Change: Military Institutions and New Realities, 1918–1941,* editor Harold Winton asks why the British army, which invented the tank and was home to at least the most quoted advocates of mechanization, especially of Britain's armed forces after 1914–1918, was never able to set forth and institutionalize a relevant doctrine for armored warfare.[25] It is a question raised a decade earlier by Field Marshal Lord Carver, who asked, in his 1979 Lees Knowles lecture, "why the British Army had done so little, by the outbreak of the Second World War in 1939, to transform itself into the mechanized and armoured army that Fuller had foreseen twenty years before."[26]

In *Men, Ideas and Tanks: British Military Thought and Armoured Forces, 1903–1939,* J. P. Harris reevaluates (debunks) much conventional wisdom about icons of mechanization from Swinton to Fuller to Liddell Hart, based on performance of British army armored units in the 1939–1945 war, especially in Europe and North Africa. He ends by asking the same question set forth by Harold Winton and Lord Carver but shifts a good part of the blame onto a different subset of the constellation of actors on the mechanization stage.[27]

Then, in *Tank Tactics from Normandy to Lorraine*, Canadian historian Roman Johann Jarymowycz cites, with few exceptions, commanders of Allied forces at all levels in the campaigns in northwest Europe in 1944–1945 as being delinquent in their doctrinal understanding of mobile warfare, as well as their operational and tactical skills in the employment of armored forces in their battles and campaigns.[28]

Finally, at the fighting vehicle level of matters, G. McLeod Ross, in a book titled *The Business of Tanks*, asks why, if "British brains and effort had created the first tank ever to take the field in 1916; yet in 1940, twenty-four years later, the only British Armoured Division was sent hastily to fight in France, not only with a numerical deficiency in armour, but with the majority of it inferior to that of the enemy."[29]

In the search for answers to the question, Harold Winton, in an essay titled "Tanks, Votes and Budgets: The Politics of Mechanization and Armored Warfare in Britain, 1919–1939," part of Winton and Mets, *The Challenge of Change,* cites two factors: first, the influence of external circumstances, such as foreign and domestic political and economic conditions; second, actions taken or not taken by military advocates of change to gain support for their reforms from constituencies external to the uniformed British army (he might have added internal to the army as well).

Lord Carver reminds us of several outside factors that affected Britain's search for a national strategy in the interwar years: the world economic crisis of 1929–1933; the Italian incursion into Abyssinia; German and Soviet intervention in the civil war in Spain; the Japanese invasion of China; and the growing threat from Germany, especially after Hitler commenced rearmament in 1935. Combined, all these factors created in Britain substantial disagreement in the search to define national strategy, as well as to define what should be required of the nation's military forces. The War Office is cited for not being bold enough in effecting radical changes in army organization in favor of mechanization, the principal justification being that much infantry and horse cavalry would be needed to defend the empire, which defense was after all the primary task of the army, most especially in India, where internal security was the preeminent challenge. Contrariwise, there was an obvious need to contend with the growing threat on the Continent. So proceeded the dialogue: the empire? The Continent? Both? The compromise struck was to wait a bit and see, in the meantime turning the funding—the whole a trickle at best—to the Royal Navy and the Royal Air Force (RAF) until a reasonable land force strategy could be perceived. Hopefully the world economic situation would improve, and a land force funding strategy set in place. This euphoric rationalization lasted until late in 1939, when, following events in Munich in 1938, the need for focus on Continental strategy became both obvious and imperative. Indeed, not until it was acknowledged that the RAF could not range deep into Germany unless operating from bases in the Low Countries, was it decided to plan for commitment of six divisions (one of them armored) to secure an operational foothold (for air bases) on the Continent.

Into this lacuna in strategy and operational concepts for providing the means for mobile warfighting at operational and tactical levels, the cohort of those whom Kenneth Macksey calls "the Tank Pioneers" moved with alacrity.[30] At the end of the 1914–1918 war, tanks were enormous (British Marks I through VIII) with large crews of up to twelve soldiers; in contrast, there were also smaller tanks (French FT) with but two crewmen. All were to cross trenches, crush wire, move quickly, protect crews, and destroy machine gun emplacements and artillery positions, disrupting the integrity of enemy defenses. The "Pioneers" were, in Macksey's words, a "highly articulate and influential group of men who were both intimately connected with tank development and the reporting of contemporary events."[31] The latter of course referred to B. H. Liddell Hart, who, having been gassed at the first battle of the Somme, gave up his military commission and for decades after wrote as a commentator on military affairs, served as adviser to members of the Cabinet, and became general kibitzer to military persons who were trying to command units, conduct trials, devise tactical- and operational-level doctrines, and in general decide what to do next, but from the purely military aspect of the matter.

The tank idea—the machine itself—is generally considered to have been a product of the fertile imagination of (at the time) Major Ernest Swinton, an officer of Royal Engineers. Out of line with army thinking, as well as funding, the first tanks produced from Swinton's ideas came to be, of necessity, funded and built under the auspices of (then First Lord of the Admiralty) Winston Churchill and his Admiralty Landships Committee, which sought to devise a machine capable of coping with trenches, wire, machine guns, and mud in the assault to break through enemy lines on the western front.

Ideas for employment of machines such as Swinton visualized had originated quite early in the 1914–1918 war, by Swinton himself hypothesizing tactics, techniques, and procedures for employment.[32] By 1917 the Tank Corps had been established, and Swinton had moved on to another assignment; his work in fielding the tanks had been taken up by another Royal Engineer officer, (then lieutenant colonel) Hugh Elles, who would in fact for a time command all the British tanks then in France. With Elles went (then captain) Giffard le Quesne Martel, himself destined for long tenure as one of the "Apostles of Mobility." By and large the role of spokesman for operational- and tactical-level ideas devolved on J. F. C. Fuller. An infantryman, a veteran of the Boer War, and an officer of towering intellectual capacity, "Boney" Fuller would become a most vocal and persuasive advocate of mechanization—concepts for mobile and armored warfare—for nearly fifty years of extensive thinking and writing. "Other things being equal," he said frequently, "the most mobile side must win."

In the decades from 1919 to 1939, during which the Germans produced *Truppenfuhrung* and the panzer concept, Swinton, Elles, Martel, "Boney" Fuller, Liddell Hart, and others in Britain tried persistently to get the British army to abandon the horse and infantry afoot in favor of machine power and to accept the fact that the Industrial Revolution had invoked a new requirement for mobility in war, as well as the technical opportunity to provide that mobility. The history of this matter has been in many ways virtually overwhelmed by the writings of Fuller and Liddell Hart, especially the latter, whose interpretations were for the most part set forth in his landmark history of the Royal Tank Regiment, *The Tanks.*[33]

In *Men, Ideas, and Tanks,* Harris sets out what he calls the mainstream of British military thought about armored warfare and armored forces to fight armored warfare, from 1903 to 1939. Harris writes that at the Cabinet level in the mid-1930s—critical years—the primary effort of senior General Staff members seems to have been devoted to battles with the Treasury and Cabinet to secure acknowledgment of the need for a field force, well funded, well armed, well organized, well trained, to contend with the growing German threat on the Continent. For faced with that threat, it did seem foolish to be theorizing about empire versus Continental strategies at the Cabinet level of involvement.[34]

But it was an issue with a long and divisive history, to include divisiveness in the Royal Tank Corps itself, as well as among those (primarily Liddell Hart) whose gratis comments on its activities frequently got wider audience than likely was warranted. There seems to have been no central core of responsibility for doctrine development—everyone held his own opinion about tactics, about operational-level matters, about equipment, about organization, about training and education. It could be said that no one was in charge, but it could also be said that everyone was in charge. This inbred dysfunctionality erupted in the wake of the 1934 field trials. Harris takes to task Royal Tank Corps commanders of the interwar years for being excessively "tank-centred," underplaying or ignoring the need for all-arms cooperation. Further, he describes the most unfortunate outcome of the 1934 Mobile Force exercise in which George Lindsay, Royal Tank Corps—the force commander—was wedged between an insubordinate subordinate, Percy Hobart, and the exercise director (and 3 Division Commander) General Sir John Burnett-Stuart. The unfortunate outcome brought Hobart to command of the Royal Tank Corps; in Harris's words, "He [Hobart] had a narrow, intolerant mind, was a poor ambassador for the RTC and held tactical ideas which were dangerously misconceived."[35]

The 1934 Mobile Force exercise was a watershed in many ways. In preparation for the exercise, Fuller (designated, and the most natural to command) became disenchanted with command arrangements to support the trial. For in addition to command of the Experimental Force (a mixed battalion of armored cars and tracked machine gun carriers for reconnaissance, a battalion of medium tanks, a motorized machine gun battalion, and detachments of mechanized artillery and engineers), he was at the same time to command a conventional infantry brigade, and, in addition, serve as garrison commander for the entire force. Fuller protested vigorously to both the British CIGS Sir George Milne and the exercise director (and Third Division commander) General Sir John Burnett-Stuart. After much acrimonious debate, Fuller elected to retire, and from that point forward his influence within (and on) the army was on the wane. Liddell Hart largely took up his role as spokesman for mechanization in the pages of the press.

After-action reviews of the 1934 exercise resulted some two years after the fact (1936) in the first serious attempts to mechanize the cavalry arm of the British army. The cavalry had stood back from both the general debate about mechanization of the whole force and the debate about tank development and employment. As cavalry mechanization unfolded, its development encountered all the difficult questions set forth earlier, and fairly well followed the experience of the Wehrmacht, also cited earlier. Thus the British Army Cavalry arm emerged as a lightly armored force whose doctrinal missions were reconnaissance, security, and economy of force, but which was, all done, not heavy enough to fight effectively—in reconnaissance, security, or in offense or defense against heavier forces when (all too frequently) required to do so.

Finally, in the wake of the 1934 exercise, with Fuller's influence diminished within the army, General Sir John Burnett-Stuart became the quiet champion of change. His influence in this role is recounted quite well in Harold Winton's excellent biography, *To Change an Army.* Not an outspoken commentator, either in the tank debate or in the general conversation about mechanization or motorization of the army as a whole, General Burnett-Stuart became the apostle of all-arms mechanized force doctrine—operational concepts, equipment, organization, and training and education of all ranks, all separate from the tank debate, but all necessary to change an army. Burnett-Stuart left active service in 1938, but in a few brief years he had led the expedition "to change an army."[36]

All of this caused Lord Carver to observe that the "Apostles of Mobility" (among whose names he does not list Burnett-Stuart) "had to fight conservatism, self-interest, and sheer stupidity" to bring about mechanization of the British army.[37] But in this they were not alone, as the chapters here will seek to illuminate.

Le Jour de Gloire est Arrivé

July 14, 1919. It seemed that almost all Paris had assembled on either side of a five-mile-long stretch of the Champs-Élysées to celebrate *le jour de gloire*—the day of glory—with a victory parade. France had won at last against the kaiser's Germany. Forty-eight years earlier the Prussians had insisted on staging a victory parade (theirs) along this same avenue, following which the French had encircled the Arc de Triomphe with a heavy link chain—a chain of bondage. Now the bondage was removed, Alsace-Lorraine would be returned to French control, and the Germans would be bound by a different kind of chain in the strictures of the Versailles treaty.

Leading the way through the now opened Arc de Triomphe was a remarkable group. No cavalry resplendent, no Zouaves colorful, no poilus celebrant—just three young men, or what remained of each of them, trundled past in chariot-like conveyances. Three young men horribly and forever crippled in action in the war just concluded. They were followed by a detachment of *grands mutiles*—all ranks, in all sorts of attire, hobbling on crutches or canes, arms in slings, heads in bandages, a dozen abreast but otherwise in no semblance of marching order. As they passed, there was a nervous quiet along the street, as if the watchers should be doing more for these tragic creatures than just looking.[38]

Then followed the grand parade. But the lead contingent just described bespoke silently but dramatically a significant (and not so grand) story—one that would have unanticipated consequences for years to come. For they represented the living remains of possibly as many as a million and a half French soldiers who had died in the cause of victory in the war just concluded. That number represented 27 percent of all fit French males of metropolitan France between eighteen and twenty-seven years of age. No combatant in the 1914–1918 war, save little Serbia, could claim a higher mortality ratio. For France, then, those dead soldiers represented lost treasure that simply could not be replaced, nor could their numbers be replaced in the next generation. And that reality created a military-political dilemma, solutions to which, among other factors, led to defeat of French forces in the field in the summer of 1940.[39]

Now it is June 14, 1940; German armies enter Paris. France, its armies defeated in the field, is split asunder, sues for termination of hostilities, is divided into an occupied France, which includes Paris, and a pitiful remnant whose government sits at the beautiful little spa town of Vichy. Four days later a refugee brigadier named Charles de Gaulle, recently escaped to London with other French refugees, hoists a flag there, declares himself head of a Free French government, and begins a tenure in French military and political affairs destined to last well into the twentieth century and to have a profound impact on his country, both in its relations with its European neighbors and most especially in its colonial affairs in Indochina and Algeria.[40]

But what happened (or did not happen) in the years between the *jour de gloire* of 1919 and the *jour de catastrophe* of 1940?

Answers to both questions reveal a profusion of political-military complexities. In *To Lose a Battle: France 1940*, Alistair Horne seeks to explain why the French army "fought as it did, so very differently, on four various occasions: in 1870, with almost complete ineptitude; in 1914, with heroic fortitude, final success, but colossal losses; in 1940 with the most cataclysmic disaster; in the post-1945 colonial wars of Indochina and Algeria, where various blends of military competence found themselves in grave discord with the political will." Our focus here is, of course, on what happened or failed to happen between the second and third of those episodes.[41] Were the great destructive wars for control of central Europe done with, or would there be more of the same?

As for national strategy, there remained a consensus on the imperative of defending France (French soil) against future incursions by armies—read German armed forces, from the east. An overall strategy of defense seemed the order of the day. This perception led to the construction of a series of fortifications commencing at Basle on the Swiss border in the southeast, extending to Longwy, close to where the borders of France, Belgium, and Luxembourg meet. Thus the famous Maginot Line. In perception impenetrable, but not truly so, it was nonetheless an imposing barrier designed to at least disrupt a potential assault toward Metz and Nancy to the south, and in the north to guard against an intrusion southward from the Lower Alsatian plain. It would cover a distance of nearly ninety miles, cost several times the original engineering estimate, and need to be completed by 1935 when French forces were by treaty required to remove from the Rhineland.

Turning westward from the northern terminus of the Maginot Line, there would be a series of fortified strongpoints extending toward the Atlantic coast, but not nearly so impressive and well integrated as those of the Maginot complex. Through this less well-defended zone, French forces would debouche northward to meet and defeat an intrusion from the east, and do that on the Belgian Plain, thus defeating attackers from the east before they could occupy even one square millimeter of French soil—ever an emotional and so political issue with the French.

The decision to embrace a strategy of defense required supporting decisions for combat at operational and tactical levels of war. Of long standing was the French tradition that the offense should dominate; that success would accrue from large numbers of well-led infantry units, the élan of their soldiers, and the cold steel of their polished bayonets. In *The Seeds of Disaster: The Development of French Army Doctrine 1919–1939*, Robert Allan Doughty observes that a dramatic transition from an offense doctrine to one of defense at operational and tactical levels took place about midway in the two decades from 1919 to 1939.

Characteristic of the newly adopted defense doctrine was the operational-cum-tactical concept of the methodical battle—*la bataille méthodique.*[42] Here, infantry protected by literally curtains of massed artillery fires would move to defend against hostile attack. Strangely enough, infantry was yet touted as the "queen of battle"; under the new doctrine, however, the rate of movement of the force as a whole would be determined by the rate at which the shroud of artillery fire could be moved about to protect the infantry. This reality, it was considered, called for close and centralized control of the battle—from a high level. Thus the entire operational-level battle dynamic—the pace of battle—would of necessity be rather rigorously constrained (read slowed) by the need to coordinate the whole, and by the metric of sufficient and sufficiently maneuverable artillery fire to protect the infantry. It was, just conceptually, quite a cumbersome system, for it would be possible only if massed artillery fires could be made both sufficiently accurate and maneuverable to support a reasonable rate of advance for infantry. The whole concept did not augur well for mobility at tactical and operational levels.

Further, *la bataille méthodique* could be made possible only if supported by a mobilization system designed to engage the full resources of the nation in an intense effort to support the nation's armed forces. Conceptually this idea had basis in the wars of Napoleon, reinforced by lessons from both the Franco-Prussian War and the 1914–1918 war. But in the context of readying for what might happen next, the French bumped head-on into legacies from the past. The most significant of these was the problem of sufficient manpower to fully mobilize for war. This, too, was a legacy from the past, again beginning with the wars of Napoleon, and more recently the near million and a half dead in the military-age generation lost in the 1914–1918 war. Because of insufficient numbers in relevant age-groups, conscription was enacted in the absence of an ability to attract sufficient volunteers. Terms of conscription varied. Absent an immediate and obvious threat, conscription for military service became unpopular and thus a political issue. Laws of the 1920s called for one-year conscript terms, reduced from two or three years under earlier laws. The law of 1927, which established one-year conscript service, clearly established that without mobilization the French army could not go to war, but that should it be required to do so, the only reasonable operational strategy was one of defense.

So by 1935 the active-duty French army, once some six hundred thousand strong, with professional leadership and reasonably well-trained soldiers, was barely twice the size of the hundred-thousand-man force allowed the German Reichswehr under the Versailles treaty; further, it consisted of not so well trained conscripts led by equally not so well trained and educated reservist leaders.

Further was a self-inflicted leadership lacuna brought on by a decision to reduce the officer strength of the army based on the presumption that Maginot Line manning would require fewer officers than an army maneuvering in the field. So in an army already suffering significant leadership problems, there would subsequently be five thousand fewer officer leaders.

Among other legacies from the 1914–1918 war was a veritable surfeit of equipment; especially tanks and artillery. It is frequently considered that the British were first to acknowledge the need for and so invented the tank. That is not really the case. As early as 1915, a French lieutenant colonel of artillery named John-Paul Estienne described requirements for a fighting vehicle, which would become the tank. Estienne was promoted twice in quick succession for his initiative, and as General Estienne appointed "Commandant de l'Artillerie d'assaut des armées." Note from the title, as well as from his branch affiliation, that tanks were perceived of more as mobile artillery than as anything else. Two French firms, Schneider and Saint Chamond, were contracted to produce several hundred machines. They were big and heavy (13 tons or more), and both mounted large-caliber cannons (75mm) as well as two machine guns. While production was under way, Estienne visited the United Kingdom, saw the British Mark I, also big, heavy (28 tons), and cannon mounting (two 6-pounders), concluding from his observations that a much lighter machine was needed. Thus was begun large-scale production of the famous Renault tank. By May 1917, some thirty-five hundred Renaults had been ordered; production was under way despite lingering objections of the Ministry of Munitions. After-action review of the British tank attack at Cambrai in November of that year overwhelmed ministry objections, and tank production was further accelerated. There would be thirty tank battalions of seventy-two tanks each plus three signal tanks per battalion. The Renault, at 6.5 tons, armed with only a machine gun, protected by thin plates of fairly ductile steel, manned by a crew of two—driver and commander—with no intercommunications save voice or foot and hand signals inside the tank, or by hand or flag signal, or (yes) carrier pigeons outside the vehicle, hardly represented a major contribution either to trench and wire smashing or to fire support for infantry; but it was light and so somewhat more agile than its heavier cousins.[43]

Shortly after the war, (then) Colonel J. F. C. Fuller provided a summary of the French Tank Corps as of the end of August 1918. In it Fuller noted the personnel strength of the French Tank Corps stood at just fewer than fifteen thousand of all ranks. There had been in that year nearly 4,000 combat engagements that included tanks: 3,140 by Renault, 473 by Schneider, and 373 by Saint Chamond tanks. French and Germans alike conceded that the Battle of Soissons (July 1918) was a watershed engagement on the western front and could not have resulted in French victory had it not been for the number and effectiveness of the French tank forces committed.[44]

The French made significant numbers of Renaults available to the newly created Tank Corps of the American Expeditionary Force, which deployed to France in 1918. U.S. Army tank soldiers took basic training (without tanks) at several posts in the United States, then deployed either to France or to the United Kingdom. The most famous of tank training centers in the United States was at Camp Colt, in Gettysburg, Pennsylvania; there the commandant was Captain (later in that war lieutenant colonel) Dwight D. Eisenhower, an infantry officer of whom much more would be heard in the 1939–1945 war—and after. Trainees deploying to France went first to a school at Langres; there the commandant was Lieutenant Colonel (later in that war colonel) George S. Patton Jr., a distinguished horse cavalryman. Colonel Patton would lead a brigade-sized tank force of the AEF into the Saint-Mihiel salient, then again into the Argonne in

September 1918. In both operations he was afoot—reflecting the need for command control (communications) between vehicles in the formation, and for command from well to the front. From that vantage point on September 26 in the Argonne, he took a small arms round through the buttocks, an event of which he was ever quite proud, but which reflected a primitive state of command control affairs in tank formations, a problem he sought to rectify by commanding afoot in and among his tanks.[45]

So it was that in November 1918 there were about four thousand tanks of French manufacture in the French army inventory, all of which now were legacies from the past. Such an impressive inventory represented both a blessing and a burden. It was a sunk cost in modern industrial terms; but in times of limited budgets it would be quite difficult to secure new investment in machines of war with such a large inventory already in the vehicle parks.

Nonetheless, the French army pursued new operational concepts—new ideas, and new equipment to provide new capabilities. By 1926 the War Ministry had set forth a new program calling for three tank types: a light tank, a battle tank, and a heavy tank. By 1930 the heavy tank program had been sacrificed on the altar of the Geneva Disarmament Conference. By 1934 Charles de Gaulle had advocated in his controversial book *Vers l'Armée de métier* the creation of a professional army instead of a conscript army; it would be a force of mechanized shock troops. When this recommendation was put forth by Premier Paul Reynaud, the political opposition remonstrated not only that a professional army was antirepublican but also that armored divisions such as those proposed by de Gaulle were offensive weapons and France had embraced a defensive national security strategy.[46]

Undaunted, the innovators persisted. In 1931 the cavalry introduced a proposal for mixed cavalry division—a combination (in the same division) of mechanized units, motorized support units, and horse cavalry units. By 1935 this proposal had become a light mechanized division (*division légère méchanique*, or DLM), actually a light armored division. As observed earlier, the light units became heavier as antiarmor weapons proliferated.

So it was not so much that the French had not realized the need for, and the opportunities offered by, mechanization; and not that they had not responded to the perceived need. It was rather that, as with so many other issues, they tended to simply insert that which was new into an old operational paradigm; tanks (conceptually) offered no operational advantage other than for support of infantry by crossing trenches, smashing wire and other obstacles, protecting crews from hostile fire, and in so doing making battle life simpler for the poilu afoot. Thus the built-in dichotomy that the ability to maneuver would be determined by the rate at which artillery fires could be moved about to support the advance of infantry (now infantry with tanks). Cognitively, the whole concept was frustrated by internal inconsistencies. In 1940 the left wing of the French army did deploy northward toward the Belgian Plain, anticipating that according to plan (French plan) they would there encounter the Wehrmacht's attacking forces. However, Guderian's panzers, pursuant to von Manstein's plan, moved quickly and with virtually complete surprise through the Ardennes. The aggressive panzer leader moved his force behind (south of) deployed French and Belgian forces, at the hinge in the line between the Maginot fortifications and those of the northeastern frontier, forcing French, Belgians, British, and whomever else may have been deployed there northward into a surrounded cul-de-sac on the Channel coast at Dunkirk. There the British began the heroic and miraculous evacuation of more than 337,000 British and French soldiers from that Dunkirk cul-de-sac to the United Kingdom.[47]

The French sued for peace. Eugenia C. Kiesling, in her essay "Resting Comfortably on Its Laurels: The Army of Interwar France," characterized French mechanization development, particularly in the 1930s, in terms of three themes: development of an effective cavalry tank with which to equip the DLMs; the difficulty of integrating any tank into the operational concept of *la bataille méthodique*; and development of a breakthrough (fast, medium) tank for exploitation (doctrine for exploitation and pursuit not further specified). In the first instance, doctrine, organization, and technology seemed compatible, despite the absence of an operational-level concept for employment of the DLMs. In the second instance, a new technology intruded into an already carefully thought out concept of operations with which it was basically incompatible: infantry at the pace of maneuvering artillery—so why tanks? In the third instance, General Estienne and his supporters forced development of a machine (the medium tank) for which there was no operational doctrine, and so no user support for, or statement of requirement for, a capability that could be provided by a medium tank. In sum, then, carefully thought out operational concepts—doctrine—must inevitably become the engine of change. An engine that forces user requirements for everything else—equipment capabilities, force structure and organizational design, training and education of all ranks. Absent any of those elements, the unhappy history of such matters testifies that the outcome of change fares not at all well. And so it was that *le jour de gloire* disintegrated into *le jour catastrophique*.[48]

Bring the Boys Home

In remarkably few months, beginning in April 1917, the United States mobilized for war. More than three million men were conscripted or enlisted. The limited industrial capabilities of the nation were set to work producing the weapons and equipage necessary for war. A supply system was created to deliver weapons, equipment, and supplies to mobilized forces, both those training in the United States and those deployed to Europe. For operations in France, an American Expeditionary Force (AEF) was created; it included a Tank Corps. As described earlier, soldiers of the Tank Corps took basic training in the United States, trained in the United Kingdom or in France, and fought in France in various models of British or French tanks. By November 11, 1918, there were no more than a dozen tanks in France that had been manufactured in the United States—and all were the French Renault design. Leadership of the Tank Corps was provided by officers of several branches. Many, but not all, were cavalrymen. Also noted earlier, the AEF included horse cavalry elements of several U.S. Army regiments. Deployed without horses, they carried along tack to mount once horses had been procured (from the French Cavalry School) in Europe after deployment.[49] One year after the Armistice of November 11, 1918, more than three million men had been mustered out of service, and by end of June 1920, the army was scarcely larger than it had been in April 1917 when mobilizatiion began. Compared to previous mobilization and demobilization experiences it was a truly stunning performance.[50]

Absent from the ranks of those returning to be mustered out of service were some 48,000 young Americans who had died on the battlefields of Europe, in the "war to end all wars." While the loss of any number of young lives in battle is difficult to accept, the numbers lost by other major contenders considerably overshadowed those of the Americans. In all the Central Powers (who lost) suffered more than 3.5 million dead, the Allies (who won) lost 5.1 million dead; each day of the war, 5,600 soldiers died in battle; British losses alone in the first day of the Somme battle numbered 20,000 dead.[51]

Back home, the U.S. Army was confronted by several challenges. First there was demobilization of the more than 3 million men who either volunteered or were summoned to the colors as the army mobilized; in fact, the mobilization system, yet at work as the war ended, needed to shift to reverse gear and with the utmost alacrity commence demobilization. Second, the system devised to manufacture or otherwise acquire and supply the army with everything from aircraft, tanks, rifles, artillery pieces and other weapons to tents, mess kits, shoes, socks, and other soldier gear needed to be shut down or dramatically reduced in output to meet requirements of the postwar army. What requirements?

That question brought to focus yet a third (but really the first-order) problem. Just how large an army did the nation need now that all wars had been ended with the close of the war that was said to be the end of all wars? Answers to that question needed to be established in order to postulate numbers (manpower strength—officer and enlisted), structure and organization (numbers of armies, corps, divisions, brigades or regiments, and other units). Then, on the basis of those determinations, how much equipment would be needed—weapons, vehicles, supplies, and support of all kinds. Finally, how much was all that expected to cost? Compared with how much Congress was likely to be willing to spend? In all, answers to such questions depended as a first order of business on agreement with regard to a national strategy—what were likely to be threats to national security and how best to counter those threats. The calculus of numbers would be much easier to handle than would be agreement on a national strategy. Absent an immediately apparent threat to national security from abroad, defense of the homeland seems to have been, sort of by consensus but without official definition, perceived as a necessary national strategy.

Deciding what might need to be done next, then spelling out requirements for mobilization, organization, and training of manpower, providing weapons and equipment, and supplying, deploying, and sustaining units in the field were not new problems. Indeed, they had plagued soldiers and public officials since the beginning of the Republic. General George Washington's opinions on these matters are well preserved in Brevet Major General Emory Upton's classic *The Military Policy of the United States.*[52]

Although General Upton's commentary on the military policy of the nation ends with the close of the Civil War, the problems he cited remained unresolved, for after the Civil War there was no obvious urgent need for resolution—relatively, the army was much too small, and there was no foreseeable circumstance that might call for another mobilization.

However, shortcomings in all functions of mobilization for war were brought into sharper focus and made more public beginning in 1898 with declaration of war against Spain over the issue of independence for Cuba. While soldiers and units performed rather well tactically in the short span (four months) of that war, planning for the war, transportation to embarkation sites, deployment to the combat theaters (both Cuba and the Philippines), equipage—from modern weapons to uniforms suitable for operations in hot climates, to food (quality and quantity) and medical treatment, especially for the unanticipated mass outbreak of yellow fever in the force deployed to Cuba—created what seemed to be one crisis after another.

The Spanish-American War was a watershed event for the nation and the nation's armed forces, especially ground forces, for it required a dramatic change in national strategy and attendant requirements for forces and their support: change from a small army deployed over vast distances against North American Indian tribes, the first of many insurgent forces with which the nation would need to cope in future years, to requirements for expeditionary forces deployed overseas to conduct operations against military forces of foreign countries whose ambitions were considered inimical to the security of the United States.

The second national-level dilemma highlighted by the war with Spain was the almost total absence of planning capabilities at the strategic and operational levels of war. The ability to reach agreement on national goals and aims, forces required to attain those goals and aims, and a general framework of strategic and operational-level strategies consistent with achieving those goals and aims simply did not exist. For example, although there was a commanding general of the army, he had no real command authority or responsibility, and no staff to define goals and aims, ways of getting from here to there, strategies consistent with those goals and aims. Functional requirements for everything from weapons design and procurement, to procurement and issue of food and clothing, provision of medical support of all kinds at all levels, to transportation and deployment of units and equipment, were firmly in the hands of eight departments: Adjutant General, Inspector General, Judge Advocate General, Quartermaster, Subsistence, Medical, Pay, Ordnance; two bureaus—Provost Marshal General and Refugees, Freedmen and Abandoned Lands; two commissary generals—of musters and of prisoners; and two corps—Signal and Engineer. Chiefs of each of these groups, regardless of title, reported directly to the secretary of war (if and when they chose to do so), who also had no staff to organize his work, rationalize differences, or coordinate and direct effective support in any or all categories. Some chiefs of departments, bureaus, commissariats, and corps established quite close relations with members of Congress, especially those members on committees responsible for anything requiring large amounts of money—money for procurement of weapons and other equipage, for supplies of all kinds, for transportation, for services in the United States as well as in the field (medical, for example), or for base infrastructure in the homeland. And in time of crisis, apparent or suspected delays or mismanagement in one or more functional areas invited, if not demanded, immediate intrusive assistance by Congress. These functional groups, while sometimes referred to as the "General Staff," were not by definition a General Staff. For there was no chief of staff charged with directing and coordinating their work—neither in the Department of War, nor in the office of the commanding general, nor that of the secretary of war. They were virtually autonomous entities, answering only to their chiefs and with no requirement for either coordination or cooperation with other functional groups. And therein lay the heart of the problem.[53]

The unintended mishaps of the four-month war in 1898 were for the most part laid by press and public (informed by the press) on the head of Secretary of War Russell Alger, whether or not he had either knowledge of what may have gone wrong, or, had he possessed such knowledge, the ability to do anything meaningful about the crisis.[54]

Almost in frustration, President McKinley took on many, if not most, tasks normally the function of the secretary of war. And the president clearly had other work to do. Finally, he was able to get Secretary Alger out of office, replacing him with a lawyer with no military background, no particular political ambition, but with a long and successful career as an administrator in the legal profession—Elihu Root. For the significant and enormously important changes required in the army, President McKinley had brought in just the right man. Root was secretary of war from August 1, 1899, until January 31, 1904. In that all too brief time, he set in train reforms in command, in organization, in education, and in the management of the U.S. Army that would last through the army's participation in the 1914–1918 war (for the United States, 1917–1918) and well beyond. Secretary Root had studied Emory Upton's book *The Military Policy of the United States.* When General Upton died in 1881, he left a manuscript of *The Military Policy,* which Secretary Root, after having studied it thoroughly, caused to be published. In it were recommendations about mobilization—raising an army, drawn from testimony beginning with George Washington and including the mobilization experience of the War of 1812 and the Civil War. As well, Root seized on concepts concerning the need for and work of a General Staff taken from British historian Spenser Wilkinson's *Brain of an Army.* First published in 1895. Wilkinson's book describes the functioning of a General Staff system (on the German model), together with commentary notes by Count Moltke, a designer of the German system and a persuasive spokesman about the need for and proper functioning of a relevant staff and command system. Included in *The Brain of an Army* are relevant commentaries about an education system designed to bring officers with relevant staff training to appropriate postings in army commands.[55]

Secretary Root crafted necessary changes in the mobilization system, the education system, and the "bureau system," cited earlier, into binding legislation. It would obviously be, and was, strongly resisted (indeed lobbied against) by incumbent heads of the fourteen departments, bureaus, commissary-generalships, and corps cited earlier. So strong was opposition to the legislation that it required submission to successive sessions of Congress, but it was at last enacted into law as the Dick Act of 1903. It was a landmark piece of work that would, however, continue to be challenged even as the United States began to gear up for its part in the 1914–1918 war.

Under provisions of the Dick Act, a chief of staff was appointed (the position of commanding general was abolished). The first chief of staff was Lieutenant General Samuel B. M. Young, who served for a brief three and a half months.

Lieutenant General Adna R. Chaffee, who served for two years, followed him. The reader will in later chapters meet his son, a younger General Chaffee, who was a prime mover of mechanization in the U.S. Army and first commander of the army's Armored Force in the run-up to the 1939–1945 war.

Subsequently six officers would hold the post of chief of staff until appointment of Major General Peyton C. March, the primary wartime chief, in May 1918. Among the longest serving was Major General Leonard Wood (April 1910–April 1914), who became an important actor in implementation of the Root reforms, for by the time General Wood took office, vocal and institutional resistance to changes embodied in the Dick Act had stiffened considerably.

One of Secretary Root's first acts, once the Dick Act was law, was to create an Army War College. Root spoke at the laying of the cornerstone of the War College building in Washington, D.C., on February 21, 1903. In his remarks he lionized Emory Upton, citing the need for a General Staff as set forth in General Upton's *Military Policy of the United States.*[56] Root then established the War College Division of the Army General Staff, responsible for providing the staff planning and coordinating functions whose separation into the various bureaus and departments was at the heart of the army's long history of difficulty with planning for, directing, and supporting the military affairs of the nation.

The General Staff thus formed immediately came to cross-purposes with chiefs of the bureaus, departments, commissariats general, and corps. Leading the resistance was the adjutant general, Major General Fred C. Ainsworth. Both Generals Wood and Ainsworth were medical officers. Then Colonel Wood may be remembered as commander of the 1st U.S. Volunteer Cavalry Regiment in the Spanish-American War; General Ainsworth had made himself indispensable to members of Congress as custodian of records from which Civil War pension claims were validated. Further, since it was the role of the adjutant general to sign all official instructions to the army at large, Ainsworth concluded that he, not the chief of staff, was in charge of the staff. The two men occupied adjacent offices but came to such an impasse that they would not speak to one another, communicating only in writing. That correspondence grew to famous (or infamous) proportions—some seventeen endorsements back and forth. Finally General Wood appealed to Secretary of War Henry L. Stimson, who, reaffirming the intent of Elihu Root's reforms and implementing provisions of the Dick Act, forced Ainsworth into retirement. Undaunted, Ainsworth then went to work as a congressional staff person, from which vantage point he continued his antireform campaign against the General Staff and whomever might be its chief. The adverse effect of Ainsworth's campaign, both in uniform and out, is difficult to estimate. But he takes his place as one of at least two officers in army history who were primary (to the point of insubordination) obstacles to change. The other was Major General James Wolfe Ripley, Civil War chief of ordinance, who literally blockaded modernization of army weaponry for nearly fifty years.[57]

The U.S. Army that went to war in 1917–1918 had its functional heritage in the reforms instituted by Secretary of War Elihu Root after the Spanish-American War, later revalidated by Secretary of War Henry L. Stimson. As war clouds roiled over central Europe in 1914, there were concerns in North America—would the United States become involved? If so, how, when, and where? These concerns generated the impetus for passage of the National Defense Act of 1916, which included echoes of the Wood-Ainsworth controversy over responsibility for war planning, especially with regard to manpower mobilization, the role of the National Guard, and need for conscription—all heard before. But the army's performance—mobilization, training, deployment, support (sustainment) of forces once deployed—was in the largest sense a test of the reforms set forth in the Dick Act of 1903 and amended by the National Defense Act of 1916.[58]

Deciding What Needs to Be Done Next

The attempt to decide what to get ready for next came to be (unintentionally perhaps) divided into two parts—operational and related lessons from the combat experience of the AEF, and lessons from preparation for and support of the war at the national level—the War Department and other departments of the executive branch, and Congress. This division of interest and action would continue through the two decades from 1919 to 1939, much as it had in other major powers in the 1914–1918 war. As noted earlier, however, the background experience of each nation was quite different; therefore, each nation proceeded differently when seeking answers to many of the same questions. The United States was also unique in this regard for two reasons. First, the country had come late to the war and had suffered far fewer casualties than did our allies. And second, the United States military had struggled for seventeen years with doctrine, strategy, leadership, and administration, but as the need to mobilize for war came upon us, the Dick Act and the National Defense Act of 1916 were implemented. Some of the significant controversy with regard to those reforms had not been resolved; it would not be satisfactorily resolved in the 1919–1939 decades and would continue to haunt the army through and after the 1939–1945 war. Some would in fact be brought under scrutiny in the administration of Robert Strange McNamara (secretary of defense, 1961–1968), who opened once again the issue of how much central control was required over the varied functions of the armed forces of the nation, and how and by whom that control should be exercised.[59] Therefore, in considering the U.S. Army in the 1919–1939 period, *Through Mobility We Conquer* will focus primarily on the search for doctrine at the operational (theater) and tactical levels of war. For it was there that the effects of the Industrial Revolution and rapid advances in technology in many areas posed the greatest challenges to the warfighting capabilities of the armed services—especially the army and later the air force.

Back to 1919. About operational- and tactical-level matters, General of the Armies John J. Pershing, much-heralded commander of the AEF, testified before Congress that he considered a doctrine of the "open battle" to have been confirmed by the recent war and to be an acceptable basis for preparation for future wars. From his own description, "open battle" seems to have been massed infantry supported by massed artillery, employed so as to penetrate enemy defensive deployments, opening the way for pursuit by follow-on forces, which might or might not include cavalry. He decried French and British fixations on trench warfare. In his way of thinking, "open battle" was an Americanized version of French doctrine, but he failed to note that it was doctrine the French themselves had set aside as they went to the trenches in order to escape devastation from machine guns, artillery, tanks, and aircraft. Mechanization, in whatever form—tanks, aircraft, indirect-fire artillery, machine guns—fit neatly into an established paradigm of battle. All were simply support for masses of infantry. All reflected recommendations of the Superior Board convened by General Pershing to analyze and report on the wartime experience of the AEF. It was largely on the basis of the Superior Board report, and General Pershing's testimony, that Congress enacted the National Defense Act of 1920. It was fundamentally a revised edition of the act of 1916, although with important changes. First, and most noteworthy, was abolition of the Tank Corps as a separate branch. Second was the assignment of everything relating to tanks to the Infantry Branch. Leaders who had become tank enthusiasts during the war, especially those who saw in the fledgling tank forces a way of restoring mobility to the battlefield, and a role for mechanized forces other than simply support of infantry, reverted to their parent branches. Implementing provisions of the 1920 National Defense Act were set forth in *Field Service Regulations 1923*, enacted into regulation under the watchful eye of General Pershing, who came to office as chief of staff in 1921. In *FSR 1923* the combatant arms were defined as "the infantry, the artillery, the cavalry, the signal corps, the engineers, and the air service." The document asserted, "No one arm wins battles. The combined employment of all arms is essential to success." The *FSR* crowned those pronouncements with this statement: "The coordinating principle which underlies the employment of the combined arms is that the mission of the infantry is the general mission of the entire force."[60]

In paragraphs describing cavalry, the 1923 regulation ignored all the battlefield realities of the war recently ended, realities that brought into question the traditional functions of cavalry on a battlefield now dominated by the fires of modern weapons—for example, machine guns and artillery.

Another icon of the Great War, General Douglas MacArthur, would later add to the mechanization dilemma in the U.S. Army by declaiming in 1931, early on his watch as chief of staff, that chiefs would be responsible for mechanization as it applied to their branch.

FSR 1923 would govern until the edition of 1939, which is sometimes said to reflect a modest evolution. In the 1939 edition there was more emphasis on "combined arms" and the idea that "no one arm wins battles." Even though infantry was cited as "charged with the principal mission," tanks would "as a rule be employed to assist the advance of foot troops," despite recognition that they could be a "powerful maneuver force in the hands of a higher commander with which to influence the course of combat." However, the tank battalion was the largest tank force visualized. Finally, despite much visible furor over the role of airpower, and some increased emphasis thereon in *FSR 1939*, it still cited airpower as being for support of the ground force—the debate over a separate air force and strategic bombardment was simply set aside for another forum.[61]

George F. Hofmann's chapters in this book deal with two primary subjects. First is the phenomenology of change, that is to say, the search for new doctrine at tactical and operational (theater) levels of war, hopefully informed by lessons of the past war but also by changing perceptions of just what the armed forces of the nation might be required to do for their country in the years that lay ahead as international dynamics became more persuasive due to the rise of fascism. Second, they present the history of changing doctrine and requirements for weapons, equipment, organization and force structure, training, and education, which is the saga of people, the key personalities leading the search for new doctrines. For it would be those personalities, their experience and cognitive biases, however derived, that would provide the armed capabilities with which the nation would begin its next military endeavor, whatever it might be.

In chapter one we meet perhaps the first of the mechanization enthusiasts in the U.S. Army—Frank Parker. Often described as a "flamboyant, ambitious and energetic cavalry officer," he was fluent in the French language, had graduated from the French Cavalry School at Saumur, and later the France École superieur de guerre. This background more than qualified him for assignment as head of the American military mission to France as the AEF mobilized and deployed. On deployment, General Pershing posted Parker to a small group to advise him (Pershing) on Allied conduct of the war thus far, on tactical doctrine in trench warfare, and on early mechanization efforts by both French and British armies—especially on tank development. Parker was also instrumental in defining a staff organization for the AEF following, in large part, French army staff concepts. Later in the war he would command the 18th Infantry Regiment of the 1st Infantry Division and, for a brief few weeks at war's end, the 1st Division itself. General Parker was among the first to recognize that machine guns, indirect-fire artillery, tanks, and perhaps aircraft as well raised important questions about the utility of horse cavalry in a traditional role on a modern mechanized battlefield. He also saw in mechanized forces the need for vastly different doctrine for war at the operational level. The chapter ends with a look at the last horse-mounted operation in Europe at the end of the war. It foretold the end of the traditional mount and the upcoming transformation of the cavalry branch due to the immerging chaos of the interwar period.

In chapter two is a detailed survey of the work of General Pershing's Superior Board, Pershing's own influence on and development of *FSR 1923*, details of *FSR 1923*, and the reactions of officers who would be principals in the growing debate about mechanization of the army as a whole, the role of cavalry, and mechanization of cavalry.

Chapter three recounts the "struggle for an innovative doctrine," primarily in the first decade after 1919. In it the reader will find the debate over mechanization of the force as a whole—what kind(s) of tanks and other armored vehicles should there be, what should be tactical- and operational-level doctrine for their employment, and what about cavalry? Here it becomes apparent that most of the mechanization experimentation was being conducted by cavalrymen—men who were convinced that, alas, the horse cavalry may be about to become history, but what about the role of cavalry in battle, and if such a role remains, how should the force that conducts those operations be equipped, organized, trained, and employed?

Chapter four finds mechanization experiments in full swing, centered around an experimental force at Fort Knox, commanded and staffed by men whose names would become legend in the run-up to and conduct of the next war—Daniel Van Voorhis, Adna R. Chaffee (whose father was the second chief of staff of the army), Robert W. Grow, I. D. White, and many others—many in grades from lieutenant to major or lieutenant colonel. Experimenters at this point were focused largely on determining how much mobility and agility could be had from the (primitive) machines with which they were experimenting, for answers to that question would determine what they could expect to demand of forces equipped with the machines available. The great Christie tank debate comes to the fore in these pages. It was a time of scarce resources—the economic crisis of 1920 had advanced into the Great Depression of the 1930s. In the face of what he saw as virtually overwhelming resource limitations, Chief of Staff General MacArthur demurred on proposals for centralization of mechanization by simply passing responsibility for mechanization to the branch chiefs—each for his own branch. It was a decision that would have far-reaching consequences—to, through, and well beyond the next war. The chapter also examines the impact of two German staff officers who were guests of the U.S. Army.

Chapter five looks at a number of controversial historical issues involving intense debates between cavalrymen over their future course in light of mounting international events, especially from attachés' reports from Spain. In 1937 a German staff officer and a friend of future general George C. Marshall visited Fort Knox and other military bases. His discussions with cavalrymen and infantrymen are analyzed. A future president of the United States, then Senator Harry Truman, also visited Fort Knox to inspect the status of the mechanized cavalry. The chapter also examines the important mechanized cavalry maneuvers at Fort Knox that rapidly changed the tempo of operations for a future mechanized force. It was a time that seriously questioned the future role of cavalry as a combat arm. The problems were accentuated by the debate over three types of cavalry: all horse, horse-mechanized, or all mechanized. Was it to be a specialized reconnaissance force or an all-purpose mobile combat force capable of carrying out a variety of missions?

In chapter six we find panzer forces of the Wehrmacht devastating Poland at the same time the U.S. Army published *FSR 1939 (Tentative).* In *FSR 1939* one again finds General Pershing's formula for "open warfare"—infantry and artillery supported by all other arms. Chief of Cavalry John K. Herr published *Field Manual 2-15 Cavalry* (1941), reaffirming the traditional role of his branch but adding some words about mechanization, almost in an "oh, by the way" tone, thus raising for the U.S. Army the dilemma already confronted by other armies of the world. Adding firepower and motor vehicles to cavalry, making it heavy enough to fight in its traditional roles, threatened to make cavalry so much less mobile and agile that it simply could not effectively accomplish the mission.

Chapter seven lays out in tactical-level detail the dilemma between arming and equipping the cavalry to fight in order to perform its reconnaissance and security missions, or keeping it light, ill protected, underarmed, and forced to avoid combat, and so much less effective. In the North African theater and during the Italian campaign the battles brought to the forefront that precombat training, field manuals, organization, and equipment did not prepare the cavalry to meet the demands of the battlefield. This chapter also looks at leadership and its view of cavalry as an economy of force.

Chapter eight sets forth a series of battle summaries describing the employment of mechanized cavalry units in campaigns in Europe from Normandy into the German heartland. It is taken largely from after-action reviews conducted by one or more of a series of boards or committees assembled in Europe for that purpose at the end of the war. Participants reported and reflected on the employment of armored cavalry units in or under their command. Their significant value is that testimonies of the participants are in many cases strikingly different. Teamed with tank and artillery units, mechanized cavalry units performed as part of combined arms teams, adding their unique cavalry skills to the capabilities of the team as a whole. In one unique case a mechanized cavalry squadron (the 6th Cavalry Reconnaissance Squadron, 6th Cavalry Group) saw service as an information link between armor task forces leading the attack of an army (Patton's Third Army) and the army headquarters, where the army commander and his staff, because of that link, knew more about actions of forward elements than did the parent headquarters of those elements. General Patton called it the Army Information Service. It was one of a kind, and contributed enormously to Third Army's legendary agile and quick response to crises.

Chapter nine and the conclusion bring the reader to postwar Germany and the need for a force to perform security functions of all kinds. Given that mission, the U.S. Army in Europe created the U.S. Constabulary, whose function was to maintain order, provide security within the U.S. Zone of Occupation, and maintain border security along the geographic boundary between U.S.-occupied Germany and the German Democratic Republic as that political entity was being established by the Soviets in East Germany. That border was of course not well marked, if marked at all, and would remain substantially ill defined until erection of the Berlin Wall and its extension along a series of border fences following erection of the wall itself. Reconnaissance squadrons of the Constabulary performed the border mission until the 1950s, when formed-up the Armored Cavalry Regiments (2nd, 6th, 11th, 14th) would at one time or another perform that mission. This situation continued until 1993 when, after the collapse of the Soviet Union, the fall of the Berlin Wall, and disassembly of the border fences, the last U.S. Armored Cavalry Regiment (the 11th) stood down, folded colors, and redeployed to the United States to become the Opposing Force at the National Training Center, Fort Irwin, California. This chapter is quite likely the most comprehensive account yet of the formation and service of the Constabulary. The mission was not exactly a cavalry mission. The organization was formed out of remnants of the 4th Armored Division at war's end. But, equipped and organized as had been cavalry reconnaissance squadrons during the war, it took on all the esprit and dash of the cavalry.

Organized and Equipped to Perform Any Mission

"Remember, gentlemen, that the ground reconnaissance units of the U.S. Army are organized and equipped to perform any mission for which they are organized and equipped." So spoke the chief of cavalry branch of the Command and Staff Department at the army's Armor School at Fort Knox, Kentucky, in the spring of 1949, to a class of second lieutenants, newly commissioned, all en route to join armor and cavalry units of their army scattered from Germany to Japan.

The lecturer was a quite young (in his thirties), well-turned-out, well-spoken, indeed eloquent, lieutenant colonel of cavalry. Commissioned out of West Point in 1940, he had joined a cavalry branch that was at the time in a somewhat confused transition from horses to jeeps, wheeled scout cars, and other mechanized paraphernalia. And with that change in technology there was further confusion in regard to fundamental military matters such as tactics, techniques, and procedures (doctrine) at tactical and operational levels of war, what equipment was available or required to enable execution of the doctrine, how best forces should be structured and organized to fight battles yet to be fought, in campaigns yet to be planned, against enemies yet to be very clearly identified (at least as to military capabilities), and how best to train soldiers and train and educate leaders to command and control the new force. In the course of his participation in the war just finished, the lecturer had commanded a mechanized cavalry squadron in the Sixth Cavalry Group in the European theater of operations and for a brief time had commanded the group itself.

The war finished, he and a select group of his peers had testified before one or more boards convened (in Europe) to digest the lessons of the war. Subsequently, some of that group had been posted to the army's Armor School at Fort Knox, where they were to record and analyze lessons and develop and teach new doctrine for armor and mechanized units—including, of course, armored cavalry units. For during World War II, almost universally without approved and published doctrine for employment of armored and mechanized forces, and the equipment, organization, training, and education to match the doctrine, commanders of those forces simply put together on their own initiative, in training as well as combat, whatever tactics, techniques, and procedures seemed appropriate to the situations with which they were confronted—at tactical as well as at operational levels—and trained their units as best they could in the time available to ready themselves for battle. This was true of most armor units, but especially so of cavalry units; the change from horse to horsepower had been rapid, and there had been little enough time to design, test, and evaluate new doctrine, equipment, structure, and organization. So they learned, evaluated, and improvised as they went.

Lieutenant Colonel Samuel McC. Goodwin closed every lecture with the "every mission" statement just quoted. His students could never quite determine if he could not decide what role cavalry was to play on a modern battlefield, and so was begging the issue; or if he simply wanted to challenge his students to think it through for themselves; or perhaps he was just reflecting, for whatever it might be worth, what his students could expect in the units they were about to join, and, in the latter case, trying to encourage the next generation to think things through, try it out, and take time to evaluate the results of trials.[62]

At the head of the Command and Staff Department at the Armor School during this period stood a lieutenant colonel named Creighton W. Abrams Jr. Commissioned from West Point in 1936, he served briefly in horse cavalry units, then joined the 4th Armored Division, then forming up under command of the legendary Major General John Shirley Wood, where he (Abrams) commanded the famous 37th Tank Battalion, a unit that, among other daring deeds, relieved the besieged garrison at Bastogne in the counteroffensive against the Wehrmacht winter offensive through the Ardennes in December 1944. Later he would command a Combat Command in, and become chief of staff of, the 4th Armored Division.[63] He was thus a natural choice to head armor and cavalry doctrinal revisions after war. A few short months later he would take command of the 63rd Heavy Tank Battalion, 1st Infantry Division, in the U.S. Army Europe—the only tank battalion in the only U.S. Army division in the European theater at the time. There he would find several lieutenants, by now platoon leaders in that battalion, who had been students in Lieutenant Colonel Sam Goodwin's cavalry classes a few brief months earlier. As was ever the case in units commanded by Abrams, everyone assigned would learn firsthand that doctrine development is an ongoing, almost everlasting process, intellectual as well as physical—"think it out—try it out—see if it works," he would say. We would also learn that technique was something he (Colonel Abrams) himself had learned from General Wood's command of the 4th Armored—always looking for a better way, checking if what was being done was right, changing if it made sense, all of which teaches that in the development of military forces to support the national goals, military folk of all services are ever confronted with questions: What is to be done? How best to do it? What resources are needed to get it done—equipment, force structure, organizations, trained units and leaders? How best to put it all together?[64]

Indeed, one clear thinker about these matters once opined that doctrine is what about 51 percent of the people are willing to do on any given day in any given circumstance. Early on, the doctrine developer is confronted by this reality. If it does not make sense to those who are to implement it—live with it, perhaps die with it—they will do whatever seems to them to be the best thing under the circumstances with which they are confronted.[65]

For, as has ever been true, battles will be won by three inevitable constants: the courage of soldiers, the character of leaders, and the excellence of unit training before the onset of battle, training in the units to which the soldiers and leaders are assigned, training with those in whose company they will fight.

Introduction

Yellow Legs

Sound the call to horse for the yellow legs,
Throw down your whiskey to the dregs,
Keep on your boots, you Cavalry,
And spur your horse to charge with me.

Mechanized wars are now the style,
Parachute troops drop in a pile,
Some yellow legs are going far,
By fighting from a Combat Car.

This is a history of mechanization of the U.S. Cavalry during the first half of the twentieth century. The success of mechanization during this period flows from a sum total of measures not only in cavalry doctrine but also in politics, economics, budgets, and domestic and international dynamics, which in their entirety decided the future of the U.S Cavalry until 1950, when Congress passed the Army Organization Act. That law ordained that the cavalry be absorbed by the armor branch. This book is also a history of Fort Knox, Kentucky, where new ideas had germinated at times and been discussed with German staff officers who visited the base in the 1930s. More so, it is a reassessment of the major cavalry leaders at Fort Knox who made mechanized cavalry history.

The history begins with cavalrymen called marketeers or innovators who represented the voices for change that started in World War I with the untold story of the first mechanized cavalry theorist. A lack of major American cavalry deployment to western Europe during World War I and then during the occupation of the Rhineland foretold that fate was beginning to dictate significant changes in decades of mounted warfare based on the horse. After the war, more voices for change began to emerge, calling for the adoption of mechanized warfare. However, it was a period in American history that was pervaded with the domestic attitude that World War I was the "war to end all wars." Concurrently, it was a period of rising political and economic international instability in spite of collective security efforts by the League of Nations and disarmament agreements among the major powers. Unfortunately, this stoic period in American history sustained intellectual military conservatism, which caused hesitation for an engine of change. Nevertheless, the emerging hostile international environment arising out of the unresolved issues over the effects of World War I, bolshevism, and the Versailles treaty of 1919 began to unfold new approaches to warfighting. The postwar chaos fostered the rise of communism, fascism, and civil wars in Europe. However, as these international events unfolded, the United States resorted to hemisphere defense and embraced a form of political isolationism. The cavalry, as before the war, again expected to find itself on the Mexican border. Meanwhile, Congress was not willing to fund military preparedness and became more adamant on this issue during the Depression until war clouds again threatened the country. This background had an impact on the future of U.S. Cavalry, which struggled for a reason to be—horse-mounted or mechanized or later horse-mechanized mounted.

By the late 1930s, American military conservatism was secured and dominated by the infantry and its supporting combat arms branches, the cavalry and field artillery. The combat arms branch chiefs, by orienting themselves to tradition and congressional dictates, failed to produce an engine of change, a coordinated all-arms mechanized doctrine. More so, they viewed the next war in terms of World War I with leftover, obsolete equipment. The compartmentalization of their autonomous institutions also depreciated the value of change necessary to prepare and fight on a mechanized-driven battlefield at an operational level. For the cavalry this period brought into question issues of doctrine between the horse traditionalists and a new breed of mechanized mounted warriors, the marketeers. It became a crucial debate that had a negative effect on formulating a new warfighting doctrine.

The Reichswehr and then Wehrmacht staff officers who visited U.S. military installations, including Fort Knox, Kentucky, during the 1930s, plus numerous and comprehensive reports of American military attachés, provided information on important voices and engines of change going on in Germany and even in the Soviet Union. The U.S. military had the opportunity during this amiable period to absorb the adage that almost always in major wars the losing side learns the most. The Germans saw World War I as a consequence of faulty actions and presented these to their hosts. The Germany military was moving to a new form of operational art where campaigns and major operations are fought with flexible combined arms elements to accomplish strategic objectives in an expanded mechanized battlefield. The marketeers at Fort Knox also understood these ideas. However, many of their ideas on mechanized warfare never were implemented by army elites or achieved an operational level of warfare. This was due to the self-serving restrictions of the branch system controlled by the combat arms chiefs. They were more concerned over their own tactical function than moving to all-arms cross-cooperation at an operational level where offensive operations directly or indirectly are conducted against the enemy's center of gravity.

The impact of the Spanish civil war also influenced U.S. Cavalry thinking and the army as a whole, creating a break between the marketeers and the traditionalists. Reports from the American attaché in Republican Spain, who was a former chief of infantry, questioned the future role of tanks due to the extensive use of antitank guns. The marketeers at Fort Knox considered these reports flawed lessons and moved on a path of preparedness and mobile relevance by increasing the depth of mechanized combat in time and space.

The function of mechanized cavalry at an operational level was to deny the enemy freedom of action by finding and fixing the enemy forces and disrupting his sources of strength by fighting mounted or dismounted so as to create the conditions for operational and tactical success by high commands. The last chief of cavalry seriously believed this could be accomplished only on a mechanized battlefield with his horse divisions or horse-mechanized divisions. As a result, his unyielding attitude made the cavalry an obsolete arm and stifled the creation of mechanized cavalry divisions. The army chief of staff did not consider him a loyal team player and subsequently had him removed from office. Consequently, he lost his mechanized cavalry divisions, and an Armored Force was created.

In addition, during the interwar period and until the eve of World War II, the cavalry, like the army and government, had to deal with a cantankerous civilian tank designer whose unsavory reputation is still emotionally discussed today within the military and by military historians. For more than a decade these notorious debates hampered the development of a workable tank policy. Instead, emphasis was placed on lightness and speed at the expense of armor protection and greater firepower. Accordingly, this policy had an affect on preparedness, making it difficult to provide a suitable combat car for the cavalry and a tank for the infantry by the time American ground forces were preparing to go to war.

After the Armored Force was created, the remaining horse and horse-mechanized cavalry units set out on the road to mechanize for stealth reconnaissance. The Armored Force was now taking over the cavalry's traditional tactical mission of exploitation and pursuit. In spite of this reorganization process, the army was far from being prepared for war. It had neither suitable equipment nor a doctrine until 1941 to provide guidance on how to fight a modern mechanized war. The cavalry not only was influenced by the horse mentality but also was left with the mission of stealth reconnaissance that had not matched the realities of the battlefields as they unfolded in the Mediterranean theater and then in western Europe. Mechanized cavalry learned to assume a number of missions that had to be reinforced by other combat elements and at times fight as infantry. After the war, cavalry elements were again influenced by a whole new set of demands, the dynamics of the early cold war period. This period produced a new breed of mounted warriors, the U.S. Constabulary. By 1948, Constabulary units were being reorganized into armored cavalry regiments, an organizational structure that is in essence part of the U.S. military today.

It would be a monumental task to cover all the cavalry unit histories during this period. Due to time and space, efforts were made to select meaningful stories with an emphasis on the human element and how domestic and international events affected the cavalry branch, its troopers, equipment, organization, doctrine, and leadership. With the use of archival materials and primary and secondary sources, a story is told herein of cavalrymen who were confronted with an intense debate over the ever-present issues of change and innovation. It became a struggle between the voices and engines of change that were influenced by domestic and international dynamics, all within the confines of command leadership.

1

Frank Parker

Early Mechanized Cavalry Theorist

In September 1939, Adolf Hitler launched what became known as the "blitzkrieg," or lightning war. This German word was Americanized after the Wehrmacht's rapid victory over Poland. The Wehrmacht employed a mechanized force where panzers were deployed as the main maneuver element supported by tactical airpower and infiltration tactics. Since then the word "blitzkrieg" has become a debatable legend over its origins as new doctrine on mechanized warfare. The German idea for a short, lightning war came out of World War I as an answer to the impasse caused by the harsh subterranean trench environment and the unfathomable slaughter that existed on the western front. This stalemate led to a dispute over what mechanical means to employ to break through the long, continuous front and reduce meat-grinding infantry losses. Arising from this dilemma were early concepts on what constituted a mechanized force of combined arms armored elements capable of relatively deep independent action or an armored force consisting chiefly of tanks where personnel fight mounted. Emerging as one of the foremost theorists on armored warfare was J. F. C. Fuller, a staff officer posted to the British Tank Corps in 1916. The following year he was advocating a new model tank force and other modern equipment for mobile operations designed to strike deep into the German rear, thus neutralizing the reserves and supporting communication systems.

Although military historians have devoted considerable time and effort to analyzing the blitzkrieg and Fuller's theories on armored warfare, little or no attention has been paid to concepts of mechanized warfare advanced by certain U.S. Army personnel in World War I. Much has been written about the exploits of Colonel George S. Patton Jr., who deployed the tank as a tactical offensive weapon. Only later in World War II did this dynamic cavalry officer unveil concepts of operational depth employing a tank-dominated combined arms mechanized force supported by tactical airpower. The first concept of a combined arms mechanized force, however, belongs to another cavalry officer, Frank Parker. During World War I he understood the idea of operational depth (campaigns and major operations required to sustain strategic objectives) of mechanized warfare supported by tactical aviation as a solution to trench warfare. Like Fuller, Parker became a visionary after he witnessed the introduction of two of the war's most revolutionary weapons, the airplane and the tank. One could argue that Parker became the U.S. Army's first "blitzkrieger." Parker's ideas challenged the U.S. Army's primary weapon system on the battlefield, the infantry that adhered to the traditional doctrine of open warfare or a war of movement with looser formations based on maneuver and infantry firepower.[1]

Frank Parker was born in Georgetown County, South Carolina, in 1872, and in 1894 he graduated and was commissioned in the cavalry from the U.S. Military Academy at West Point, where he developed a fluency in French and Spanish.[2] His career was marked by a strong sense of vision and purpose. Parker became a flamboyant, ambitious, and energetic cavalry officer with a strong ego. More so, he became a trendsetter of new ideas for a future U.S. Cavalry. During the Spanish-American War, he served at Tampa, Florida, the launch site for the U.S. expedition to Cuba. Early in the century, he was an instructor at the Military Academy and in 1903–1904 a student at the École d'application de cavalerie at Saumur, France. In 1912, he attended École superieure de guerre, returning in 1914 to complete the course. However, with the war heating up, he was ordered back to the United States. No sooner had he returned than he was ordered in December 1915 back to France as a military observer until the early part of 1917. By then the war on the western front had become a quagmire of blood, mud, craters, and long trenches. France was rapidly losing a generation of men to poor generalship. Against this background, Parker became chief of the American Military Mission. His role was invaluable, since he was well schooled and knew the French language intimately. This helped to establish the groundwork that eventually led to coalition warfare. As the war progressed, Parker's diplomatic demeanor greatly enhanced the relationship between the future American Expeditionary Force (AEF) and the French military when the United States entered the war. Most important, Parker became one of the first U.S. Army observers to critically view military operations, employing mechanized forces in an attempt to overcome trench deadlock and ravaging infantry losses.

When World War I broke out in August 1914, European military indoctrination differed little since the Wars of German Unification, 1864–1871. Even with the introduction of new weapons technology capable of devastating firepower, the strategy of major European powers such as France, Germany, and Russia was based on the nation-in-arms offensive spirit. This grandiose national military strategy was designed to rapidly deploy large conscript armies dependent upon railway timetables and strict schedules to quickly win the big battle of annihilation. One U.S. Marine Corps officer described the four years of bloodletting "as the demise of the Frederickan legacy of war-by-the-numbers in the blast furnace of the Industrial revolution."[3] Only Britain and the United States had not maintained a nation-in-arms. By December 1914, the race to the English Channel with French and British armies competing against the invading Germans had ended their vision of a short war of annihilation. The era of stabilized trench warfare had begun, inaugurating a war of attrition characterized by the wholesale use of large numbers of machine guns, rapid-firing artillery, and extensive barbed wire entanglements. The dreams of the offensive spirit were relinquished, including the traditional role of cavalry. It now became vital to break the impasse of trench warfare and return to the offensive. This endeavor became an ongoing process to find some means of defeating the growing elastic defense doctrine in depth the Germans were beginning to perfect.

Attempts at dealing with the continuous front impasse and the absence of exposed flanks led at first to the use of the motorized armored car and then the caterpillar tractor. First the British and later the French deployed the new weapon after developing it from a farm tractor to the first clumsy tracked armored fighting vehicle. The idea, however, for using it for warfare preceded the Great War. In November 1913, Benjamin Holt of Holt Manufacturing Company of Stockton, California, wrote the chairman of the Senate Committee on Expenditures in the War Department, Miles Poindexter, that the company's tractors reduced the use of horses in wartime. Holt's idea of mechanization had not created an immediate interest with potential military users in his own country. Instead, the British and French adopted Holt's tractor designs during the war to move artillery and pull supplies, and eventually tanks, over the war-torn landscape.[4] In addition, consideration was given to armored fighting vehicles as replacements for traditional horse cavalry.

While the British and French were applying Holt's designs to deal with problems of trench warfare, the U.S. military began deploying Holt track-type tractors during the 1916 invasion of Mexico in an attempt to capture the bandit Pancho Villa. The action was precipitated when Villa and his bandits plundered Columbus, New Mexico, killing numerous soldiers and citizens. As the punitive expedition—led by Brigadier General John J. Pershing—deployed to Mexico, it experimented with flimsy, unreliable airplanes from the Signal Corps for observation and reconnaissance. The idea of using aviation germinated as early as 1909 in an article in the *Journal of the United States Infantry Association* suggesting the airplane had potential for a new cavalry by providing air reconnaissance, thus extending operational mobility. The author stated that traditional cavalry

Motorized convoy moves supplies to General Pershing's troops in Mexico, 1916. Courtesy of the Patton Museum.

had two functions, screening and covering a position. If necessary, it also could occupy a territory not reached by the infantry due to its lack of mobility. The airplane, he argued, could change the traditional role of cavalry. He even went on to say that in the future, armored automobiles would lead the vanguard of an attack with dragoons fighting mounted and dismounted as infantry. When this happens, he maintained, the traditional cavalry role of screening may have ended.[5]

For logistical support, Pershing's expedition relied on tractors and motor transportation. Though limited, the motorized expedition became the first for the army. A few commercially built armored cars served with the army but never saw combat. However, the desert and mountains of Chihuahua were rough, so the expedition became more of a horse cavalry expedition with supporting infantry and artillery. The cavalry played an important part in its traditional role of reconnaissance, pursuit, and fighting mounted and dismounted when

The U.S. Army Ordnance Department in 1916 authorized Holt Manufacturing Company to build a self-propelled track-laying carriage to mount a 3-inch antiaircraft gun. Courtesy of the U.S. Army Ordnance Museum.

necessary. The attempt to capture Villa, nonetheless, failed. However, Pershing's expedition was the last large-scale cavalry operation on horseback.

That same year Holt began designing the first self-propelled track-type gun mount. The vehicle was eventually delivered to the U.S. Army's Aberdeen Proving Ground in Maryland.[6] This event changed the whole direction of the role of field artillery in mechanized warfare, since it raised the question of how to mold a combined arms warfighting doctrine to the ever-increasing application of scientific means brought about by the war. The Holt-type track-laying fighting vehicle led to a disputatious reassessment of battlefield doctrine by the U.S. Cavalry, field artillery, and infantry, which became even more intense after the war.

Meanwhile, stabilized trench warfare caused a reevaluation of old ideas and doctrine in the U.S. Army. Before the advent of the automobile, the U.S. Cavalry provided the greatest operational mobility in reconnaissance, flank security, and pursuit. In addition, American horse soldiers tactically functioned as mobile infantry to fight dismounted. This military tradition was abundantly demonstrated during the Civil War, on the frontier with the various Indian tribes, and in wars beyond U.S. borders. However, with the introduction in World War I of massive and combined defensive power of machine guns and rapid-firing artillery, the U.S. Cavalry looked for new ways to retain its operational mobility.

While Frank Parker was appraising French army operations, a cavalry officer stationed at the U.S. embassy in Paris, Lieutenant Carl Boyd (later to become an aide to General Pershing, the commanding general of the American Expeditionary Force), submitted an interesting report in February 1916. Apparently, Boyd was concerned with the traditional role of European cavalry. Not until the realities of the war set in did European cavalry learn to fight dismounted. By then military technology far outstripped traditional cavalry tactics. It became evident that the practicality of the *arme blanche* as a cavalry theory was in doubt. The shock value of the horse cavalry ended on the fields of western Europe. Boyd noted in his report that a military and civilian attitude had emerged on the western front, believing that rapid-firing machine guns and artillery along with conditions of siege warfare made horse cavalry obsolete. Airplanes were now viewed as the means for long-distance reconnaissance. Boyd was concerned that the infantry and artillery had eclipsed the cavalry, thus relegating the horse soldiers to history. He argued that the cavalry's future was with armored cars, which could be more useful in open warfare. The young American officer believed the branch's strength lay in partial mechanization by adding armored cars to horse regiments to enhance mobility. More so, he argued that motorized combat vehicles were necessary because in the near future American horse soldiers surely had to fight an enemy with armored cars. Therefore, the cavalry needed to improve its tactical ability for pursuit.[7]

Boyd's report was forwarded to Parker for endorsement before it was sent to the War Department and then on to the War College Division, the Cavalry Board, the chief of ordnance, and the quartermaster general. Parker found the report interesting. At the time, he had a discussion with a French cavalry officer regarding his experiences since the war started. The French officer suggested cavalry be equipped with automatic weapons and sidecar motorcycles, adding that the bayonet was still indispensable. He believed the U.S. Cavalry could be more useful in America than on the western front. Parker, nevertheless, agreed with Boyd's concerns. He added that the U.S. Cavalry must present the same fighting features as the infantry while adding an element of increased mobility that motorized armored cars could provide. The cavalry must be armored but also equipped as foot soldiers in order to fight a war of movement, he argued, concluding that this increase in firepower and mobility for the cavalry could reduce the infantry's bloodletting in France.[8] Thus, early in 1916 Boyd, with Parker's strong support, had presented to the War Department the first ideas on the future role of the cavalry in mechanized warfare. At the time, Boyd and Parker understood the problem and were looking for a way to overcome the defensive problem and restore mobility on the western front. They were looking at armored cars to increase operational mobility for cavalry.

As noted, the race to the Channel led to linear immobility and human attrition, which caused the invention of the tank, though not before other tactical means were tried. To overcome the deadlock in 1915, a sharp increase in intense artillery preparation was used to support sudden infantry attacks. Still, infantry breakthrough tactics were unsuccessful. The attacking force lacked the element of surprise, speed, and the energy to pursue the enemy. Warfare on the western front now changed to methodically prepared blasting tactics where the infantry attempted to occupy ground conquered by massive artillery concentrations. Again, the attempt at a breakthrough was unsuccessful. As the infantry advanced, it left the protection of its supporting artillery, leaving the attackers in German artillery range. By 1916, the principles of siege warfare were applied at Verdun, where the tactical objective was even more limited. The next tactical consideration of importance was the Battle of the Somme in 1916, where the first tanks were introduced in September. Yet artillery-blasting tactics still dominated the battlefield. Again, the Germans' growing use of elastic defenses-in-depth and the tattered ground once again limited objectives, making infantry breakthrough and cavalry exploitation and pursuit impossible. The excessive use of a large number of artillery shells into a given area for the attacking infantry to traverse did no more than swallow them into thousands of graves. The infantry had succumbed to the doom of artillery craters that inundated the battlefield. By now the western front had taken on an intense atmosphere of a decaying landscape that was rapidly absorbing generations of combatants from both sides of the trenches. The fields of western Europe were no place for horse cavalry.

Meanwhile, a U.S. infantry observer with the British armies in France learned in August 1916 of a new 28-ton heavy armored motorcar. The new invention was expected to provide firepower, protection, crushing power, and shock action for assaulting infantry. In his report to the War College Division, the observer humorously added that in one test of the heavy armored car fourteen sheep were secured inside. Subsequently, they were subjected to a storm of rifle and machine gun fire and shrapnel. The armor held up. The sheep were still alive and unhurt; however, they had defecated all over the inside of the vehicle. The

London Times referred to the heavy armored motorcars as mysterious "tanks," a blind name used to conceal their real purpose.[9] Along with the *Times* article, a current issue of *Iron Age* was sent to the chief of ordnance that reported on the wonderful success of an American invention, the armored tractor.[10]

True armored warfare, as noted, began during the battles on the Somme. The British Mark I tanks employed tracks that completely encircled each side of the chassis so obstacles could be crushed and trenches crossed. The proposed utility of the armored tractor was to put the offensive back in the hands of the infantry by overcoming the shell-pitted ground and trench system, neutralizing wire entanglements and enfilading German machine gun nests. Other than surprising the Germans, tanks were a failure because of numerous mechanical breakdowns and inexperience in employment. Nevertheless, this first test of armored warfare had demonstrated a new method of mobility combined with firepower, shock action, and armor protection. The tank idea began to gain prominence, resulting in adjustments being made for future deployment.

Just how did U.S. Army observers react to this new method of warfare?

One army ordnance officer described them as heavy armored cars, alligators, and land cruisers. His reports were submitted to the chief of ordnance through the chief of the War College Division for evaluation. Initially, the reports were optimistic regarding the first employment of tanks: "They have proven extremely efficient . . . as they cannot be put out of action except by direct hits from a field gun." Later he modified his report, noting that a number of tanks used during the battle stalled due to cutting into the shell-torn ground. This situation caused a number to be damaged and destroyed by direct artillery fire. By the end of 1916, the ordnance officer made a more detailed assessment by noting that the Mark Is had been overrated in the public press. Nevertheless, he believed tanks had an important effect on the morale of the attacking infantry. Conversely, he noted, the chief weaknesses were their vulnerability to artillery fire and the effect of heat and noise on the crew. He concluded that the French were also building tanks, but different from the British design.[11]

The driving force behind French tank development or the assault artillery program was Colonel Jean-Baptiste Estienne, an artilleryman who looked for solutions to deal with the impasse on the western front. Estienne, avoiding bureaucratic channels, submitted to Schneider and Company in 1915 a design for an artillery assault vehicle based on the Holt tractor. By January, trials were completed and four hundred Schneiders were ordered, thus becoming France's first mobile artillery assault tank. Estienne, regarded in France as the "father of the tanks," by the end of the war advocated light tanks for close infantry support, heavy tanks for breakthrough, and medium tanks for pursuit. In 1916, Estienne anticipated the Schneiders and their rival, the Saint Chamonds, could

(Above) French assault artillery, also called tanks, circa 1917. Saint Chamond. Courtesy of the Patton Museum. *(Below)* French assault artillery, also called tanks, circa 1917. Schneider. Courtesy of the Patton Museum.

provide continuous mobile armored artillery support for the attacking infantry. Later this tactical approach did not escape the attention of the chief of the AEF Tank Corps.

Meanwhile, at the Saint Chamond Company another army observer from Parker's mission reported on French attempts at armored warfare with what he called mobile steel boxes. The Saint Chamond was a heavy assault vehicle designed by Service Technique Automobile, which Estienne had bypassed. Like the Schneider, it was based on Holt's tractor designs. These vehicles were propelled by self-contained caterpillar tracks and armed with cannons and machine guns. In April 1916, four hundred were ordered. The French were extremely optimistic about the tactical merits of their two-tracked armored assault vehicles. Potentially, they were at first to be tactically positioned to follow the assaulting infantry, destroying enemy positions not eliminated by preparatory indirect artillery fire with direct artillery fire. The American observer noted that the steel boxes were called *artillerie d'assaut*. Candidly, he admitted that the French told the most fantastic tales about the merits of their new tracked vehicles. He added that the French and British were using design principles appropriated from an American firm, the Holt Manufacturing Company. Of interest, the observer added that French military officials complained that the use of British tanks on the Somme was premature.[12] This prevented surprise coordination, because the French were not ready to deploy their Schneiders and Saint Chamonds.

Back at French General Headquarters, Parker's mission was able to acquire additional information on armored tactics. The French maintained that open warfare was a failure, causing a change from infantry action supported by artillery to an artillery battle in which the infantry occupied the ground conquered by the barrage. This change produced a system of siege warfare dominated by artillery battles, which—because of the enemy's elastic defenses—was no longer suitable for the doctrine of open warfare that depended on maneuver and infantry firepower. The solution, according to the French, was mobile artillery. This tactic required two conditions: a method of traction to maneuver artillery over the rough terrain and trenches, and armor protection for the operating personnel. Out of this tactical consideration, however, emerged another opinion over the employment of mobile artillery advanced by the French cavalry, whose adherents believed in the traditional shock en masse. This cavalry tactic called for mobile artillery to open a surprise assault without traditional artillery preparation followed by infantry, cavalry, and mechanized artillery for pursuit. This offensive technique was expected to complete the assault.[13] Though information was available on British experience with their heavy Mark Is on the Somme, the French made no use of their mobile steel boxes until the spring of

1917. It was reported a new French light tank, the Renault, was being built as a possible cavalry tank.

Before General Pershing and his small staff sailed for France at the end of May 1917 to lay out basic war plans for the AEF, the chief of the War College Division directed Parker's mission to acquire more specific information on the "tactical employment and distribution of tanks." Consequently, Parker inquired about the first use of French tanks in the Battle of Chemin des Dames on April 16, 1917. The French military, he noted, considered that on occasion their Schneiders produced consternation among some of the Germans. There were, nevertheless, marked criticisms. First, there was faulty liaison between the Schneiders and infantry skirmishers. Consequently, many were destroyed for a lack of infantry support. Second, they provided insufficient protection against internal fire because of thin armor and a lack of extinguishing equipment. The following month the Saint Chamonds received their baptism of fire. They also experienced a number of problems, mainly that they were too top heavy and nose-heavy for the tracks. Both armored vehicles lacked breakthrough capabilities.[14] The French offensive with tanks was unfortunate. Heavy infantry losses with virtually no gains produced a major morale crisis in the French army.

While French tanks and tactics were being assessed, another member of the American Military Mission visited Tank Headquarters in London requesting guidance regarding the manufacture of tanks. He was advised that if the U.S. Army expected to fight in Europe it needed tanks, meaning tanks manufactured in the United States. Unfortunately, no industries were yet devoted to tank production. To speed results, the British hosts suggested Americans send mechanical engineers to Europe to acquaint themselves with the refinements of tank construction. The British were more than cooperative, offering facilities and drawings of their latest tanks. The U.S. Army representative was also advised that the British viewed armored cars as of little value because they had not been employed to any extent. It became apparent Boyd's report and Parker's interest in the use of wheeled armored cars for the cavalry were impractical due to the preponderance of artillery blasting that dominated the terrain, making off-road deployment almost impossible. The British, nevertheless, still believed there was a need for a wheeled armored cavalry-fighting vehicle.[15]

On the eve of America's entry into the war, the American Military Mission had also acquired considerable information on British experiences with their heavy rhomboidal tanks that looked like giant legless centipedes. In the summer of 1917, the British Tank Corps, previously the Heavy Branch Machine Gun Corps, came into being. Generally, the British initially preferred using heavy tanks alone and forward of the infantry to crush the barbed wire for the attacking infantry, cross the trenches, and then deploy behind the German lines and exploit the breakthrough. On fronts where tanks were unsuccessful, British opinions were generally adverse, a luxury many "Tommies" felt was useless. Another school of thought at British Tank Headquarters and Training Center in France found tanks indispensable. Also reported were reworked British tank tactics, especially those expressed by Colonel Fuller and based on the Somme experience in September 1916.

At Tank Corps Headquarters lessons learned determined that tanks needed to be deployed en masse to open the way for the infantry, disorganize the enemy's infantry, reorganize, and then prevent the Germans from throwing up new defenses. This later phase of tank tactics required deep mobile operations to neutralize German second and subsequent lines of their elastic defenses-in-depth. The British also emphasized tank cooperation with the infantry, artillery, and cavalry. It was suggested, however, that tanks for deep operations could replace cavalry unsupported by infantry. The American Military Mission also reported that the British proposed that tanks deployed with horse cavalry could be used as a combined mobile force.[16]

By the time an unprepared America entered the war, Parker had digested numerous ideas on tactical solutions for breaking the stalemate with the use of Holt track-type tractors as some form of movable artillery to support the attacking infantry. It was also expected that French tanks rather than cavalry armored cars be eventually employed as artillery of pursuit. Parker's mission, however, was not impressed with designs and operational problems experienced with the Schneiders and the long-hull Saint Chamonds. It believed that the lightly armored artillery carriers—now referred to as *char d'assaut*—were not suitable for the AEF because the armed steel boxes were easily impaled and derailed. It was determined that maneuver was critical for the track-type tractor's protection; however, as the Saint Chamonds and Schneiders stalled, they became easily vulnerable to hostile fire and German counterattacks.

No sooner had General Pershing landed in Europe than he reviewed the reports that were submitted to the Army War College by the American Military Mission regarding British and French tank tactics and operations. The military missions now became permanent due to the suggestion of General Pershing's operations chief, Lieutenant Colonel John McAuley Palmer. With this arrangement, Palmer believed technical and operational information was more valuable.[17] Again, Parker, the French-speaking soldier-diplomat, was considered critical in maintaining the continuity that led to coalition warfare.

Conceptual doctrines were now beginning to have an effect on how new war technology was to be applied to traditional army doctrine that emphasized open warfare, rifle marksmanship, and the time-honored cult of the bayonet.

It became a question of assimilation when the AEF entered the conflict. Not accepted were years of British and French experience in dealing with the problem of penetration associated with the defensive nature of trench warfare. Their solution was focused on massed artillery, which was not suited to the American way of warfighting. The question was which tactics were suitable for the AEF: French tactics of using lightly armored track-propelled movable artillery following the attacking infantry, or the British armored breakthrough tactics of firepower, protection, crushing power, and shock action? In July, General Pershing requested the army chief of staff to detail a board of select officers to study Allied combat operations and new weapons of war, including a careful and confidential study of Allied tanks. Four members were detailed to the Board of Officers: Parker, along with Colonel Fox Conner (field artillery), Lieutenant Colonel Clarence Williams (ordnance), and Major N. E. Margetts (field artillery).

Meanwhile, Parker presented an interesting report on the role of airpower at another meeting of the AEF Board of Officers. Parker was suitable for these proceedings because he had married into a family of early aviation enthusiasts. Apparently this report was made without contributions from Lieutenant Colonel William "Billy" Mitchell, who was rapidly becoming a dominant figure in the newly formed AEF Air Service. Discussing combat operations, Parker suggested two aeronautical techniques for aviation, tactical and strategic. He proposed the former to be directly attached to ground forces in direct support, while the latter operated deep and independent of ground forces. Parker's report was interesting for the early ideas he expressed on tactical and strategic airpower in support of ground operations.[18]

That hectic July, Conner—a member of the AEF advance staff's Operations Section—was the first to witness trials of a pilot model French 6-ton Renault tank. Conner, like Parker, spoke fluent French. Before the war, he had served a tour with a French artillery regiment. The industrialist Louis Renault, the Henry Ford of France, was persuaded by Estienne to build a light, small, and inexpensive tank. By October 1916, a wooden mock-up was completed. Conner believed that the Renault was superior to the Schneider and the Saint Chamond, and to the British rhomboidal tanks. Its size reduced visibility and vulnerability, and it carried a crew of two. The early British heavy Mark Is were clumsy, difficult to maneuver, and offered poor crew visibility. Each rhomboidal tank required four of the eight men for driving: a coordinator, a main gear man, and two gear men who controlled the tracks, all operating in an oppressive environment. The Renault, according to Conner, had a lower silhouette, thus reducing its vulnerability to artillery fire. The tank, he noted, required only a crew of two, a driver and a gunner, who—unlike the British gunner—had the ability to lay

Louis Renault (straw hat) and Col. Frank Parker (smiling) after viewing test of a Renault Char FT light tank, August 1917. Courtesy of Renault Communications.

his weapon from a 360-degree traverse turret. Conner recommended further investigation of the Renault development.[19]

The following month the busy Parker, along with Colonel Williams, pursued Conner's recommendation and visited with Louis Renault. Parker and Williams were given an extensive tour of Renault's automobile factory at Billancourt, where the new 6-ton tanks were being built. Unlike the Schneider and Saint Chamond, the Renault was designed as an infantry-accompanying tank, and possibly was considered as a cavalry tank for pursuit. Parker, the cavalryman, was impressed with the demonstration. He also had an opportunity to ride on a new caterpillar truck designed to transport the Renaults to the battlefield.[20] Consequently, he came away with a number of positive conclusions regarding the Renault, considering the tank suitable for the AEF's idea of open and maneuver warfare. He then consolidated this information with reports from Conner and other army officers who had consulted with the British regarding tank production, technology, tactics, and doctrine. The purpose of this consolidation was to outline a new organizational plan for armored warfare for the arriving AEF that could be integrated into Pershing's doctrine of open warfare, which emphasized maneuver and infantry firepower. The Board of Officers detailed to look at allied tanks summarized its observations and submitted "Report on Tanks" to General Pershing. This report analyzed the classes and tactical employment of heavy British and light French tanks, suggesting their use with the AEF as infantry-accompanying weapons. This was not the tactical position Parker had in mind, as we shall see.

It was also suggested that immediate steps be taken to produce the heavy British rhomboidal tank and the 6-ton French Renault in the United States.[21] Due to Parker and Conner's influence, the AEF rejected the Saint Chamond and Schneider.

Meanwhile, the British were also analyzing the use of tanks on a large scale. Colonel J. F. C. Fuller, a staff officer at the Tank Headquarters at the time, wrote that the solution to static warfare was determined by two operational functions of battle, penetration and envelopment supported by aviation. He claimed the tank was ideal to carry out these operations. Regarding the tactics of penetration, he recommended eliminating the use of heavy artillery fire on the ground to be traversed by the infantry except for counterbattery fire. A continued advance not only during the heat of battle but deep behind the battlefront could overcome the defense-in-depth, Fuller argued. He proposed a Mechanical Striking Force consisting of breaking and exploiting components. The breaking component, composed of heavy tanks with infantry and mechanically transported field artillery, would penetrate the first two lines of the enemy's defenses. Light tanks in turn followed this action with infantry and field artillery to disorganize the enemy, forcing them farther to the rear. This exploiting component consisted of light tanks followed by infantry, cavalry, and artillery moved by mechanical carriers. It was expected that tactical airpower could support the ground forces.

Colonel Fuller prepared a most innovative addition to the AEF "Report on Tanks," which was attached to the appendix on September 1, 1917, before the Battle of Cambrai, entitled, "The Tactical Employment of Tanks in 1918."[22] The plan was one of Fuller's essays on tank tactics, which eventually led to the first large-scale deployment of tanks during the Battle of Cambrai in September 1917. What had become evident, he wrote, was that mechanical warfare (the term used at the time) replaced horses and men's legs. This new mechanical approach provided a potential solution to trench warfare. The Allies, Fuller asserted, should raise mechanical armies to expand the base of attack and restore mobility to the battlefield. The projected main maneuver element, the tank, was considered a time- and life-saving machine that could alter infantry tactics. The main problem, he explained, was that tanks currently in use could not keep pace with tactical conditions caused by the German's defense-in-depth and ground obliteration due to extensive artillery bombardment. Fuller criticized these artillery-blasting tactics as removing one obstacle and replacing it with another.

The most innovative suggestion in Fuller's Mechanical Striking Force was an organization built on a combined arms grouping with tanks forming the basic maneuver element. He realized that this tactic was a major problem, chiefly because of a lack of experience and training. In spite of these problems, Fuller believed that once enemy lines were ruptured and the mechanical force moved to battle depth, the tactics of envelopment could act to degrade morale and cause bewilderment by neutralizing the regrouping German reserves. Therefore, he argued, the idea of mechanizing the infantry and horse-drawn guns on the same footing as tanks should not be abandoned but pursued.

Fuller's "The Tactical Employment of Tanks in 1918," which he discussed with Parker in August, emphasized a combined arms tank raid. This innovative concept was a forerunner of mechanized or armored cavalry designed to demoralize, disorganize, confuse, find, and fix the enemy through unexpected quick mobile operations.

By August 1917, Fuller had presented to the AEF Board, especially Parker, a new method employing a combined arms tank-mechanical force capable of rapidly defeating the German's defense-in-depth. He emphasized that for tanks to be tactically successful they had to be employed in large numbers. Fuller's mechanical force was designed not only to return mobility to the battlefield but also to create psychological panic through deep penetrations at an operational level of war. This approach meant that the war could be brought to a quick conclusion without heavy casualties and without trying to fight the infantry-artillery attrition warfare with a series of accumulating tactical victories. Fuller said that success in war greatly depended on timely mechanized mobility.

What Fuller was presenting to AEF representatives were elements of his plan using massed assaulting armor for the upcoming offensive at Cambrai in November 1917. These later became the essence of his famous armored warfare plan, "The Tactics of the Attack as Affected by the Speed and Circuit of the Medium D Tank," or, more concisely, "Plan 1919." The war, however, ended before Fuller's radical "Plan 1919" could be executed using newly designed medium tanks for rapid and deep penetration. The Medium D was the last British design of the war and influenced the U.S. Army's postwar medium tank program. The British War Office also entertained the idea that this tank could eventually replace the horse cavalry.

Colonel Fuller recalled meeting a U.S. Army officer in August 1917, who turned out to be Parker. They exchanged ideas for deep operations with a mechanical force and tactical airpower. Fuller claimed Parker held ultramodern views. Since arriving in France, Parker had questioned the necessity of frontal attacks so common on the western front, suggesting instead some offensive means other than dismounted infantry. According to Fuller, Parker showed him a paper marked "Very Secret," which set forth his tactical concepts. Later Fuller claimed Parker's paper found its way to Tank Corps Headquarters.[23]

In the paper, according to Fuller, Parker outlined the role of aviation in clearing the air of enemy aircraft and tactically supporting the ground forces. He believed that once airpower prepared the battlefield, tanks, motorized infantry and artillery could all drive "Hell Bent" into the enemy's territory, widening the breach and swinging deep around the enemy's defense-in-depth. The Germans did not have the ability or the resources to deal with such a plan, Parker argued. He ended his paper with the comment, "Plunge on the idea." This was the earliest presentation of a combined arms concept for deep operations with a mechanized force presented by an American officer. Parker's new tactical approach challenged the AEF's infantry-driven doctrine of open warfare and firepower. Fuller and Parker at the time rejected the acceptable exchange rate of manpower destruction characterized by attrition warfare. They proposed a mechanized force capable of collapsing the Germans by disrupting and shocking their forces through high-tempo operations.

After the war, Fuller wrote about that meeting in August and referred to the American officer as a "veritable he-man," one whose proposed tactical approach to mechanized warfare was a remarkable concept for ending the war on the western front.[24] No doubt, Fuller gave credit to Parker's views because they agreed with his own. Later, in reference to their views and credits, Fuller wrote that the Germans had not executed theories of mechanized warfare until 1939 against Poland.[25] It then became known as the blitzkrieg, a term coined in the United States by a popular magazine, *Time*.[26]

Because of information provided by the Military Mission regarding British and French tactical tank deployment, an organization for AEF tanks was established on September 14. Using data primarily supplied by Parker's Military Mission, general principles were outlined. General Headquarters should only use tanks in large numbers according to the nature of the front. Tanks were expected to play an important role in the AEF as infantry-accompanying weapons. For this to be possible, the report noted, a separate tank force had to be created with a single chief reporting to General Pershing.

This recommendation led to the creation of the AEF Tank Corps. At the end of December 1917, General Pershing assigned Colonel Samuel D. Rockenbach (cavalry) to duty as chief of AEF Tank Service. The following month, on January 26, the U.S. Army Tank Corps was created, and Rockenbach was assigned as its chief. The Tank Corps was created like the AEF Air Service on a temporary basis under wartime legislation. Personnel for the Tank Corps received training at British and French training centers. Such personnel needed to be familiar with gasoline motors and accustomed to handling tractors, heavy trucks, and farm machinery. Officers were expected to be young, capable of exhibiting audacity and initiative.[27]

Only the 2nd, 3rd, 6th, and 15th Cavalry Regiments were sent overseas, detached, and scattered over France. They left the United States without their horses, which were to follow later as shipping space became available. In spite of this situation, some cavalrymen made light of their voyage. Just before a German submarine alert, one trooper from the 2nd Cavalry filled his mess kit with a generous portion of slumgullion, a meat stew of various sorts called "slime." At the same time, he was holding his hat in the other hand. Suddenly the ship fired at the submarine, and as the trooper raced to his lifeboat station, "he cast his hat overboard and hastily put the mess kit on his head." He appeared to his fellow cavalrymen as "a cake well pasted with caramel." Another trooper was too close to the ships guns when a blast sent him flying down the hatchway. His overcoat was ripped off by the blast. While this alert and confusion were occurring, a few enterprising cavalrymen liberated all the pies that were to be served for dinner.[28]

Once the cavalry reached France, the troops were organized for various missions, such as maintaining and grooming horses at remount stations, liaison between units, police duties, traffic control, reconnoitering ahead of the infantry, night patrolling, dismounted action, and, toward the end of the war, combing the battle areas for stragglers and prisoners. Pershing later believed that if the cavalry had organized in its traditional role as an economy of force, it could have been effective in pursuing the retreating Germans toward the last weeks of the war. In this case, he noted, American cavalry tactics were suitable for his tactical concept of open warfare.

Boyd and Parker's early desire to add wheeled armored cars to the cavalry was not realized until years after the war. No armored cars were employed by the AEF during the war, and no cavalry divisions were sent overseas other than four regiments previously mentioned. The most notable to see action against the Germans was the Provisional Squadron of the 2nd Cavalry (Second Dragoons). After remount duty, 2nd Cavalry (Provincial) was fragmented and finally committed to combat, seeing some action in the battles of Aisne-Marne, Saint-Mihiel, and Meuse-Argonne, where it was attached to the 35th Division and provided left flank protection in an eight-division attack during the Allied final offensive. The Second Dragoons, however, were not deployed as an economy of force but instead, as with the other cavalry units, detached a num-

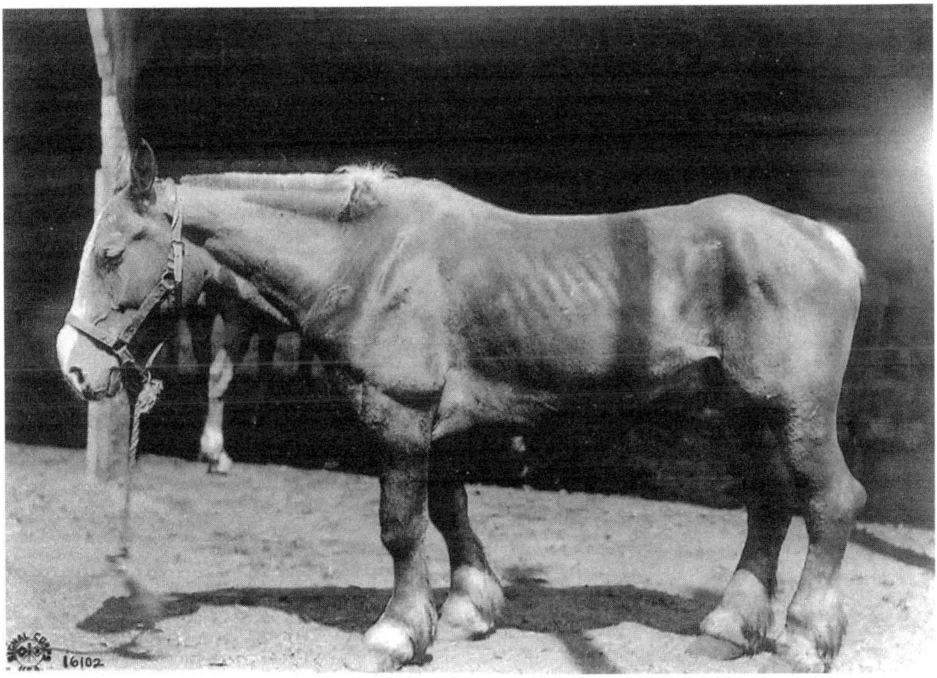

Pectoral influenza, a typical case U.S. Cavalry elements had to deal with at the AEF Remount Depot, St. Nazaire, France, August 1918. Courtesy of the National Archives and Records Administration.

ber of times during tours with the AEF. As one officer from the 2nd Cavalry noted, "The Regiment was sent practically to the four corners of France in April 1918."[29]

Captain Ernest N. Harmon, a troop commander and a future accomplished armored division commander in World War II, recalled that the Provisional Squadron, 2nd Cavalry, was the largest cavalry force in the AEF, but it was young and inexperienced. The historian of the 2nd Cavalry wrote after the war that the regiment "was the only cavalry regiment that saw any action as cavalry during the World War."[30] The French horses Harmon's units inherited in August 1918 were in deplorable condition, having just arrived after treatment and recuperation in veterinary hospitals. These horses were untrained as cavalry mounts. Once they were trained, Harmon questioned their use in cavalry operations. His reasoning was influenced by enemy control of the high ground in a combat zone, where lines shifted a few thousands yards a day in a defense-in-depth terrain covered with machine guns, barbed wire, and artillery. German reconnaissance balloons easily spotted the horses, which became artillery targets. On one occasion, Harmon recalled, his troop stopped to water the horses when a salvo of high-explosive shells landed short of a water trough. Then a second salvo landed beyond the trough. Immediately the troopers realized that this was a bracket in artillery ranging and left at a gallop. Subsequently, a third salvo hit the water trough. Muddy trenches and shell holes also became a problem for the cavalrymen, as did barbed wire, which the Germans placed high and low. Harmon noted that the barbed wire was too high for the horse to jump, and the low-strung barbed wire was at a level to trip the horses. By the end of the conflict, when the squadron was attached to the 1st Division, the horses were decimated and in appalling condition. No provision was made to assign veterinarians to treat the horses who were poor in flesh, according to Harmon.[31]

Harmon also made a comment on cavalry leadership. He acknowledged as a young troop commander he learned from some of the problems in deploying his cavalrymen, especially when it came to patrolling both mounted and dismounted in a terranean environment covered with countless shell holes. Harmon respected Lieutenant Colonel O. P. M. "Happy" Hazzard, the commander of the 2nd Cavalry (Provisional). However, Hazzard was plagued with recurring tropical dysentery that he had acquired in the Philippines. This recurrent microbic event kept him in bed more often than leading his cavalrymen. Harmon felt Hazzard was not fit to lead because the miseries of his disease compromised unit integrity and the welfare of the cavalrymen under his command.[32]

Two cavalrymen, Lieutenant Colonel George S. Patton Jr. and Major Ralph I. Sasse, chose not to stay with their branch. An infantryman, Major Sereno Brett, joined the cavalrymen in their move to tanks. They became leaders who displayed audacity and initiative while leading tanks against the Germans. Before the war ended, Patton suggested a perfect tank—light and capable of traveling on its own wheels over roads and converting to tracks for rough terrain. Patton was more interested in a light tank than a heavy tank because it provided more speed and mobility than the heavy British rhomboidal tanks. Light tanks, he argued, could get to the battle faster, thus providing both a strategic and a tactical advantage. His idea was visionary and farsighted.[33]

In the United States a civilian tank designer, J. Walter Christie, who eventually befriended Patton, had just completed an 8-inch self-propelled howitzer for the engineers to weight-test self-propelled vehicles on domestic bridges. This was a period early in America's involvement in the war when confusing executive control over the numerous bureaus caused overlapping and duplication of functions. The 8-inch self-propelled howitzer was based on an innovative convertible design, meaning it could travel on wheels or tracks. The piece went through a few modifications but was never fully developed or used by the field artillery. At the same time, Christie was also in the process of designing a convertible tank. Like Patton, Christie looked at the convertible principle as a means to enhance strategic mobility by deploying on wheels to the battlefield and tactical employment in the tracked mode on the battlefield.[34]

Due to the immediate demand for manpower and serious problems with industrial mobilization in the United States, Pershing decided to use British and French tanks. Other than this action, Pershing was not imaginative regarding the use of mechanized warfare as a revolutionary way to deal with trench warfare and the German's defense-in-depth. Parker's visionary ideas were ignored. Instead, Pershing and his staff concluded that the French did not know how to use the rifle, and the British infantry lacked initiative and resourcefulness. To the end, he emphasized open warfare that meant maneuver and infantry firepower. After the war, he argued that rifle firepower and the bayonet should remain as dominant elements for the infantry.[35] The American doctrine of open warfare with rifle firepower and looser formations, however, was not realized in the Meuse-Argonne counteroffensive beginning in late September 1918. It was the use of massive manpower in offensive warfare by the recently designated commander in chief of the Allied armies in France, Marshal Ferdinand Foch, that led to attrition warfare. This was not General Pershing's tactical idea for the AEF, which was to engage in open warfare. This, he believed, was the key to victory over the Germans.[36] Instead, the American method of warfighting gave way to the Civil War strategy of attritional warfare that was characterized by inflicting an acceptable exchange rate of destruction of enemy through the direct approach using manpower and firepower. This World War I legacy became the example for many future regular officers that attrition prevailed in combat.

Because of unpreparedness, U.S. industry was never able to contribute to the production of tanks for the war effort. Instead, the AEF relied primarily on French Renaults and British Mark IVs, Mark Vs, and Mark V Stars. By the armistice, the U.S. War Department had ordered 4,440 domestic-built Renaults, 15,015 Ford 3-ton tanks, and 1,450 35-ton Mark VIII Liberty tanks. The Mark VIII—originally an American and British joint development program—was expected to have the offensive power of one hundred infantrymen. With the end of hostilities, orders were canceled. Eventually, the U.S. Army accepted 950 M1917 6-tonners, 15 Ford tanks, and 100 Mark VIIIs.[37] The Mark VIIIs and M1917s were to serve the infantry during the postwar period. The Ford was greeted in the press as the two-man cavalry tank that could revolutionize the battlefield. The Ford, however, was tested by the AEF and the Ordnance Department and rejected as unsuitable.[38]

By the end of the war, the slow Allied tank had come into its own as an infantry-accompanying assault weapon with limited exploitation capabilities. The heavy tanks were used to break the way for the attacking infantry, and light tanks were to accompany the attackers. The tanks finally adopted by General Headquarters, AEF, were based on British heavy and French light models, tactically positioning them as infantry auxiliaries or accompanying weapons.

Also, by November 1918, the AEF Tank Corps had to resolve the AEF tank issue and how it applied to Pershing's idea of open warfare. It caused a struggle over tactical designs for different tank models, a heavy, a light, and a medium or fast raider tank. Heavy tanks were to go in front of the leading infantry, creating passages through wire entanglements and over trenches. This was the breakthrough engagement. Light tanks, in turn, were to operate outside the trench system along with the assaulting infantry, reducing remaining strongpoints. On occasion, it was suggested raider tanks be used independently by the infantry for pursuit. This employment was traditionally a cavalry operation attacking a disorganized retreating enemy or causing confusion. What influenced this tactical design was the British Medium A, or the "Whippet." Introduced to combat in March 1918 by the British, this tank was developed as a fast cavalry tank armed with machine guns and designed to exploit breakthroughs created by heavy tanks. There were problems, however, with resolving the issue of developing a raider or cavalry tank.[39]

This was, in part, due to British attempts to combine Whippets with their horse cavalry during the Battle of Amiens in August 1918. The tactical experiment proved a failure, resulting in disappointment to the cavalry. The Whippets were designed to push forward as fast tanks for use against German machine gunners and to keep the British horse cavalry mounted. The effort of coordinating the machine with the horse resulted in a lack of success in attempting to execute a deep operation. It was subsequently determined that the Whippets operating independently could penetrate the defense-covering machine guns. The remaining German machine guns and reserves, however, easily neutralized the horse troops.[40]

Shortly before the war ended, Brigadier General Samuel D. Rockenbach, the chief of the AEF Tank Corps and a proponent of three distinct types of tanks, was impressed with the speed of the Whippet and its radius of action. He was an adamant supporter of tanks as infantry-accompanying weapons, emphasizing they were not mechanized cavalry, field artillery, or armored infantry. He argued that a high-speed tank like the Whippet was more of an infantry raider type that caused confusion and destruction behind enemy lines under the same conditions that justified a cavalry raid.[41]

It was quite apparent the mounted cavalry met its twilight due to the introduction of rapid-firing weapons, barbed wire, and the subterranean nature of extensive trench warfare. Infantry and tank proponents were looking at the tank to assume many of the traditional cavalry roles of exploitation, reconnaissance, shock action, and pursuit. Army chief of staff, General Peyton C. March, concluded the use of cavalry on a large scale was not useful on the western front due to the extensive use of wire entanglements. He made the decision to transfer five regiments of cavalry into field artillery units for service with the AEF in France. He reasoned that cavalrymen already knew how to take care of horses. All that was now necessary was to train them in the use of field guns once they arrived in France. More so, he argued, shipping American horses was a serious problem because the insides of transports had to be rebuilt with stalls and room for large quantities of forage for horses along with numerous saddles and cavalry equipage. March preferred not to tear out the inside of transports at the expense of sending fighting men and their supplies to France. Yet Pershing still called for eight corps cavalry regiments for the Meuse-Argonne offensive. Years later Lieutenant General Liggett Hunter, who commanded the AEF First Army during the last months of the war, lamented that he wished for a couple divisions of cavalry. "Had these been available on November 2nd," he said, no Germans would have crossed the Meuse River, and the capture of their war materials would have been much greater. Nevertheless, March thought Pershing's request for cavalry units was ludicrous because the terrain and shipping capabilities precluded the use of large cavalry units. More important was the chief of staff's desire to increase infantry-artillery capabilities rather than having large bodies of horses "simply eating their heads off and serving no useful purpose," as he noted from the British and French experience during the war.[42]

Other U.S. Cavalry regiments were used mainly in patrolling the Mexican border because of the revelations of the Zimmermann telegram. Supposedly, German foreign minister Alfred Zimmermann attempted to embroil the United States in a war with Mexico, promising Mexico the states of Texas, Arizona, and New Mexico if that country joined imperial Germany in war against the United States. This situation made cavalry patrolling more imperative along the borders between the two countries.[43]

Because of years of warfare in France, it was evident that traditional cavalry doctrine was in serious jeopardy, as Boyd and Parker had predicted in 1916. The horse soldiers were unable to change doctrine to adapt to new weapons and had conflicting ideas regarding what constituted a new mount. Parker—like a number of cavalry officers during the war—saw no future for his branch. In part, they sensed traditional cavalry had become an anachronism. Parker finally gained combat commands in 1918 in the distinguished 1st Division in

Major General Charles P. Summerall's V Corps. First, he commanded the 18th Infantry, then the 1st Infantry Brigade, and finally the division. Parker ended his wartime career, however, in controversy when units of the 1st Division crossed over elements of the 77th and 42nd Divisions in Major General Joseph T. Dickman's I Corps on the road to the historic and important city of Sedan during the last stage of combat in the Meuse-Argonne offensive. Parker and his V Corps commander, the eager General Summerall, were as desirous of beating the rest of the AEF as Pershing was desirous to beat the French to the city. This attitude compromised military requirements for the sake of, as one officer called it, "glory chasing." The First Army commander, Major General Hunter Liggett, who, like Dickman, was a West Point graduate and was commissioned in the infantry, called the action a military disgrace. Dickman never forgot this episode, which added to an already strained relationship between the I Corps and the V Corps commanders.[44] Pershing, on the other hand, gave the impression it was not so serious and credited Parker's tired division with spirit.[45]

General Summerall, who had wartime experience and knowledge of tank tactics, complained about the lack of liaison between the infantry and tanks. He felt the doughboys were disorganized and without apparent leadership. In addition, he criticized the artillery for not providing enough assistance to the tankers and the engineers or pioneers for not properly preparing the way for tanks.[46] Summerall, however, considered Parker one of his most skillful leaders. Parker could be relied on to take his objectives by sending his men against the Germans with unflinching determination; he attacked with "perfect stoicism," recalled Summerall.[47] Pershing also praised Parker's leadership when commanding the 1st Division during the last month of the war.[48] Nevertheless, after the war Parker became the scapegoat for the Sedan debacle.

When the armistice was signed on November 11, 1918, it was apparent General Pershing's open warfare had proved questionable. It became infantry–artillery-centered attrition warfare. It was not how Pershing wanted the AEF to end the war. He did admit that the war provided little opportunity to employ cavalry due to new weapons, such as tanks, airplanes, and the increased use of machine guns and artillery. He nevertheless believed the U.S. Cavalry was better trained, because its mobility was strongly emphasized. He visualized horse cavalry playing a more important role in the defense of the American continent, where he felt the army could find its future mission, especially on the borders with Mexico.[49] Pershing, however, had not understood that the war gave a new meaning to cavalry. Meanwhile, the doctrine acknowledged by the victorious Allies was that tanks were close support weapons for use in conjunction with the infantry. Air reconnaissance, light tanks, and armored cars were used to perform the role of cavalry more as the war ended. For the next two decades, this doctrine of in-

Brigadier General Frank Parker, Commanding General 1st Division with admiring dog. Courtesy of the National Archives and Records Administration.

fantry supporting tanks created a major debate between the U.S. Army combat branches, especially the cavalry, infantry, and field artillery.

No sooner had the war ended than the huge American army began to demobilize. A surge to the ports of embarkation began. By November 1919, fifty-five of the army's sixty-two divisions had demobilized. In spite of this rapid downsizing, elements of the AEF still had a mission. As a prelude to a guaranteed peace treaty after the war, Marshal Foch was determined to enforce the terms of the armistice, which he dictated. The terms charged French, Belgiam, British, and American divisions to occupy positions along the western bank of the Rhine River and establish bridgeheads on the eastern side of the river for possible future military operations toward Weimar and Berlin if the treaty with the new German government failed. The Allied forces were to secure bridgeheads, in an eighteen mile radius east of the Rhine—the French at Mainz, the Americans at Koblenz, and the British at Cologne. Beyond the bridgeheads a six-mile-wide neutral zone was established. The Belgian forces were to advance to Aachen.

The American responsibility fell on the recently and hastily assembled Third Army on November 7, 1918, under the command of General Dickman, a serious officer who still fumed over the Sedan affair. He had been assigned to the 3rd Cavalry after graduating from West Point. Later he became an instructor at the Cavalry and Light Artillery School. During World War I, he commanded the 3rd Division, which became known as the "Rock of the Marne" for its aggressive offensive action after two French divisions on either side retreated. In August 1918, Dickman took command of the IV Corps, and in October he commanded the I Corps during the Meuse-Argonne offensive. His Third Army chief of staff was another cavalry officer, Brigadier General Malin Craig, also a West Point graduate who shortly after the war became a chief of cavalry and then an army chief of staff during the 1930s. The 2nd Cavalry at the end of the war was scattered throughout France. But now the regiment was given orders to form up the 1st Squadron and act as advance cavalry on the Third Army's march into Germany. Troops A and I were to operate with Major General John A. Lejeune's 2nd Division, while Troop B operated with the 5th Marines, 2nd Division. Troop C was to act as advance cavalry with the 32nd Division, and Troop M with the 3rd Division. Troop D operated with the 64th Brigade during the advance into Luxembourg.[50] In all approximately two hundred thousand men were employed in the Third Army, divided into three corps and eight divisions as the main units of the American Army of Occupation.

General Headquarters, AEF, also created a mechanism, the Office of Civil Affairs, under the command of Brigadier General Harry A. Smith. He became the deputy chief of staff, civil affairs, Third Army, to deal with the various contingencies of civil control, management of utilities and public works, sanitation, schools, and legal matters growing out of relationships between German civilians and American soldiers. As with all units of the Third Army, the mission was to advance into Luxembourg, cross the German border, move through the Moselle Valley, and occupy the area around Koblenz. The American Army of Occupation Headquarters was to be established at Fort Ehrenbreitstein across the Rhine River at Koblenz, occupying an eighteen-mile semicircular bridgehead on the east side of the river. In addition, one division was deployed to guard the army's zone of communication. The mission of the American Army of Occupation, along with its allies, was to prevent renewed German military aggression. Critical to this major operation was securing bridges over Germany's economic and defensive artery, the Rhine River.[51] At the same time, the American Army of Occupation was to manage civil affairs and maintain order in its zone. By the end of the year, the Third Army strength was at approximately 250,000 and prepared for any contingency should armistice agreements fail. However, the desire to return home caused personnel turbulence that made it almost impossible to retain experienced soldiers for occupation duties.

As in almost all Third Army units, cavalry elements from the 2nd Cavalry that moved through Luxembourg the first few days after the armistice encountered marauding bands of German soldiers and mutineers from the German naval base at Kiel. Many encountered were wearing red armbands of Bolshevist revolutionaries and carrying red flags. Soldiers' councils (*Soldatenrat*) were established and attempted to infiltrate a number of German regiments that were retiring from the front. An intelligence report from General Craig's staff noted that most of the German soldiers' insubordination came about from rear-echelon communication and supply troop agitators. In almost all cases, the Bolshevist revolutionaries when encountered gave the Americans no trouble. Nevertheless, there was some concern. A Third Army report noted that German morale suffered the most during the days after the armistice. By December 1, Craig's staff reported many units returning to Germany began to exhibit good road discipline, yet he still believed that German supply troops would continue to engage in disorderly and insubordinate conduct to impregnate retiring frontline troops with Bolshevist doctrines. Meanwhile, the 1st Squadron, 2nd Cavalry, had to deal with maintaining healthy horses during adverse weather because many of the German horses encountered in the withdrawal were in poor condition and had hoof-and-mouth disease.

While Craig's staff was concerned with the Bolshevists and preparing for the move into Germany, the III Corps assistant chief of staff, G-3, Colonel Adna R. Chaffee Jr., gave Troop I a mission-type order. Chaffee had been commissioned in the cavalry after graduating from West Point in 1906. In the academy's

Howitzer, Chaffee was referred to as the "King," whose ambition was "to shut up in a room full of 'Bull' with orders to smoke his way out." He was considered to have a "blasé manner, but this same manner has carried his bluff through many a tight squeeze and has been his stock in trade."[52] Like Parker, he became an unwavering proponent for mechanization of the army during the interwar period.

When the Third Army crossed the border into Germany on December 1, 1919, the 2nd Cavalry Headquarters and 1st Squadron were to be deployed as an advance guard for the III Corps. Chaffee told Captain Stephen H. Sherrill, commander of Troop I, to advance on the corps' left and patrol, reconnoiter towns, and maintain liaison with a Canadian corps along the route into Germany.[53] This deployment of U.S. Cavalry was the last horse-mounted operation of that arm in Europe.

Sherrill's troop was to head for Ehrang and then reconnoiter toward the Ahr River and then to the town of Remagen on the Rhine River. The Third Army was not expected to reach the Rhine until December 11 to give time and space for the retiring German military to cross the river. However, intelligence reported that the citizens of Koblenz were afraid of the Bolshevist revolutionaries, especially when a Soldiers and Workmen Council was formed after the departure of the German army. The city's βürgermeister requested that as soon as possible units of the Third Army be sent to Koblenz to perform police duty. On December 7, a battalion of the 39th Infantry, 4th Division, was dispatched to maintain order before the main force arrived.[54] Meanwhile, on December 8, patrols from I Troop reached the Rhine, secured the Remagen railroad bridge, and reported it suitable for crossing once the congestion was cleared. The bridge was blocked with coal and rocks and subsequently cleared. According to Sherrill, I Troop was "the first American organization to pass along the route to and into Germany." He also claimed his troop were the first Americans to reach the Rhine. A guard was posted at the bridge until units from the 2nd Division arrived. Meanwhile, I Troop secured the town of Remagen. After the horses were groomed, the cavalrymen pushed patrols across the river and reconnoitered along the bridgehead area. After reporting to corps headquarters, I Troop moved to Ehrenbreitstein. Early in January 1919, Chaffee informed Sherrill that I Troop was to be drilled for escort duty. Just over twelve hundred cavalrymen were left.[55]

Troop I's mission was tactically important because the Remagen bridge was a vital communication and supply artery for the German military and intended for possible future operations if the armistice failed. Thus, the cavalrymen completed a portion of Foch's strategy to blunt renewed German military aggression by securing a key bridge for an Allied move to the east bank of the river at Remagen and deployment into the heart of Germany if necessary. In April 1919, the 1st Squadron, 2nd Cavalry, was relieved and in June ended its overseas assignment. Elements of the 3rd, 6th, and 15th Cavalry Regiments also completed their assignments in Europe, and by the end of June 1919, all cavalry elements had embarked for the United States.[56]

Marshal Foch was not at all happy with the decision to muster so many doughboys and send the few cavalrymen home before a treaty was signed with Germany. As the negotiations stalled on June 17, 1919, Foch attached the French 2nd Cavalry Division, along with its complement of armored cars, motorcycles, and artillery, to the American III Corps for a possible Allied exploitation beyond the Rhine and into eastern Germany.[57] The American Army of Occupation no longer had any cavalry units left in Germany or enough doughboys for such a military adventure. Finally, on June 28, 1919, the Treaty of Versailles was signed, and the French cavalry returned to its zone of occupation. After months of heated debate, Congress in a joint resolution terminated the war with Germany and Austria-Hungary in July 1921.

Nonetheless, there were ongoing and mounting problems with the former Allies after the war. By April 30, 1919, General Pershing was not happy with the pretentious Dickman and replaced him with General Liggett, who commanded the AEF First Army. No sooner had Liggett taken command than he inherited French agitation for a Rhenish separatist movement. The commander of the French Tenth Army on the right flank of the American Third Army, General Charles Mangin, like Foch, was determined to encourage a separatist movement and political autonomy for the Rhineland, including the American zone. The idea of a buffer state was important for France's future security. The movement, however, was not acceptable to England and the United States, since it was perceived as a traditional political and military imperialistic effort to push the German frontier back to the Rhine from Alsace-Lorraine and the Dutch border. The area encompassed about ten thousand square miles with more than five million German-speaking inhabitants.[58] Mangin requested Liggett's assistance in France's efforts. Dickman and then Liggett, along with President Woodrow Wilson, strongly protested efforts to establish a Rhenish separatist movement in the American zone. Mangin's representative was informed that the United States refused "to recognize revolutionary movements of any character, and that if agitators for a Rhenish republic entered the American area, no matter whether of French or German nationality, they would be treated on the same basis as other agitators." Liggett informed Mangin he would not deviate from this policy.[59]

On June 1, 1919, the American Office of Civil Affairs in the American Zone was abolished. The powers of governmental administration were transferred to Major General Henry T. Allen, who replaced Liggett in July. Subsequently, the

Third Army was dissolved. By now, the majority of the Army of Occupation was gone. Much to Foch's displeasure, President Wilson desired to have a small detachment known as the American Forces in Germany, consisting of about sixty-eight hundred men. Wilson believed that peace required reconciliation rather than the revenge that historically possessed the French. This had not set well with Foch and his ambitions to shed some of Germany's territories and militarism for France's future security. He wanted a larger representation of American military for continued occupation and to deal with any contingencies. Sarcastically the French commented on Wilson's decision that he wanted "merely to keep the American flag in the Rhine."[60] The United States began to embrace normalcy and isolationism as a means to avoid future conflicts.

Two cavalry officers who made a major impact on American military history served as assistant chiefs of staff, G-3, under Allen. Both graduated from West Point and were commissioned in the cavalry: Lieutenant Colonel Jonathan M. Wainwright and Major John K. Herr. The former surrendered the Philippines to the Japanese in 1942, and the latter, as the last chief of cavalry, became a stubborn obstructionist by impeding the mechanization of his arm in the late 1930s and early in World War II. Both were cavalry traditionalists, believing in the utility of the horse in modern warfare. They failed to understand that because of the war, technology provided new opportunities for the cavalry to reconsider its future doctrine.

For the next three years, a token contingent of American Forces in Germany remained on the Rhine. During this period, Allen became increasingly frustrated in his attempt to restrain French ambitions in the Rhineland. Early in January 1923, Germany was held in default of reparation payments as stipulated in the Versailles treaty. Consequently, French and Belgian forces again occupied towns in the Ruhr District. At the end of the month, Allen turned over his command to the French army, which now replaced the American Army of Occupation.[61] The War Department ordered the occupation terminated and American forces returned to United States. In June 1930, the French finally left the Rhineland.

What were some of the issues that caused a relapse in American and French relations after the war?

As the American Army of Occupation crossed the German border and entered its zone of occupation, the German population displayed a mixture of suspicion and inquisitiveness. For the time being, static occupation by the Americans did quell any potential hostility by the civilian population, in spite of the hasty redeployment of American forces to the United States. Regarding the remaining troops, various programs were established to help relieve the anticipation and irritation of shortly mustering out of the army. For example, in nearly every unit, well-organized sports and games, schooling, and especially horse shows were designed to lessen the doughboys' anticipation of going home.

The rapid deployment to the United States became problematic because the heavy AEF organizational structure was now being depleted and was neither prepared nor trained to stabilize a civil government in an occupied territory during the turmoil of an immediate postwar period. Once the war ended, almost 2 million doughboys wanted out of France and returned home so they could reenter civilian life. Only 5,770 of that number were classified as cavalrymen. Due to the rapid deployment to the States, Dickman's recently created Third Army at first lacked the expertise required to restore public order necessary for a uniform civil affairs administration based on the traditional German hierarchical bureaucracy. There were no concrete plans for moving from a tactical role to a static occupying role once the armistice was signed. The strategic contingency was paramount over occupation. The problem was the use of tactical formations as military government units during rapid manpower demobilization while at the same time the German military was disintegrating. The occupiers had not understood at first that the German political system was a more minutely and carefully organized bureaucracy than the American democratic system. Soldiers also had not understood traditional German civil hierarchy with a kaiser as a monarchical head of the country. Accordingly, civil affairs personnel drawn from the Third Army had to adjust to traditional German bureaucratic particularism. Nonetheless, in spite of this situation, the American Army of Occupation adjusted and eventually prevailed in maintaining security and civil order in its zone in spite of French ambitions to continue the war and severely punish Germany. The Germans in the American Zone were generally friendly, causing considerable friction between the Americans and the punitive French. The late Pershing biographer Donald Smythe wrote that the American occupiers were friendly and the Germans acted in kind. This had not set well with the French, who were sterner in their zone in dealing with the German population, demanding huge reparations and Rhenish separatism. As Smythe noted, the Americans considered French actions "unnecessarily harsh."[62]

The Paris Peace Conference dominated by France and England produced the Versailles treaty of 1919. Though Wilson's coveted League of Nations was established, much of his idealism and efforts for an honorable peace were rejected. The treaty did produce a new world order but gave rise to new grievances, which were hardly a manageable job for an occupation force. Decades later occupation of Germany took on a whole new historical meaning. After World War II, the management of police-security action in occupied Germany was more thought out. The cavalry had exchanged their horse mounts for mechanical mounts. The horses had been stabled, and U.S. mechanized cavalry elements would take on a much greater role, becoming a unique constabulary force to maintain order and security in a defeated country during a period when the United States, unlike after World War I, became a major player in international affairs.

After the Imperial German Army began to dissolve and the Versailles treaty was forced upon Germany, civil disorder rained upon the country as it endured political collapse, preparing the way for the Free Corps Movement led by bands of disgruntled military volunteers known for their savage actions and warped ideology. One noted historian called them the "vanguard of Nazism."[63] In numerous operational reports, General Craig and his staff noted the activities of Bolshevist elements as the Third Army started its advance toward the Rhine River. These elements were determined to establish a new Germany based on the communist model, which through a violent revolution deposed czarist Russia. Their activities and the separatists, however, were subdued in the American zone due to the repudiation of the Rhinelanders' reaction to the French policy of separation, and the civil order and friendliness established by the American occupation troops.[64] The 1st Squadron, 2nd Cavalry, contributed to the American attitude in the Rhineland that the war was over, and it was not a time to act as overbearing conquerors. Nevertheless, in other parts of Germany, Free Corps units were used by the newly established German Weimer Republic to fight the encroachment of communism, which was avoided in the American Zone of occupation.

After the war, Major Guy V. Henry Jr. made some scathing remarks about the cavalry. He had graduated early from West Point during the Spanish-American War due to wartime demands and a few years later from the U.S. Cavalry School. In 1912, he was a member of the American riding team in the Olympics and became a renowned lifetime equestrian. Much to his displeasure, he was first commissioned in the infantry. However, given that his father was a cavalry officer, young Henry spent most of his early years with horse soldiers in various frontier posts.[65] Right after World War I ended, he indicated that he regretted that the war had taken most cavalry officers away from their arm. As a result, the branch could only rely on its future as a deployable combat arm with little wartime experience or limited postwar occupation of the Rhineland. Henry wrote that many officers serving in other combat branches were now returning to their arm. He believed the war demonstrated that American theories for training and the employment of cavalry were still sound. It was evident, however, that heavy firepower and mobility were now essential features of warfare, which he said the cavalry should take into consideration. Regarding mobility, Henry was critical of cavalrymen because their horses lacked conditioning and their officers lacked the will and desire to ride. He also cited indifferent discipline, poor training of remounts, lack of attention to feeding and watering the horses in the field, and an absence of both men and their horses from mounted

Captain Guy V. Henry, Cavalry, 1912. Courtesy of the Patton Museum.

work. His solution for the postwar period was to improve education for officers at the Cavalry School in every subject relating to the mounted branch. Furthermore, he suggested cavalrymen needed to become more proficient in riding to increase mobility and arouse interest in mounted sports and contests. His most important suggestion was to appoint a chief of cavalry to direct postwar doctrine, training and efficiency, and horse mobility.[66] Shortly this came to pass when Congress passed the 1920 National Defense Act. In ten years, Henry was appointed the chief of cavalry.

Meanwhile, in the United States before the new defense act was passed, the country's politicians began arguing over a postwar military policy while army leaders debated a future warfighting doctrine. While the army was rapidly demobilizing and readjusting from what the nation and politicians believed was the war to end all wars, Parker was off to school in France. He did not rejoin the cavalry. Instead, he returned to the École superieure de guerre, finally graduating in 1920. He then became an assistant professor and student at the Centre des hautes études, the highest military school in France.

2

A Reason to Be!

When the war ended, a conflict arose over the transition to a peacetime military policy and doctrine, meaning how the United States would prepare for the next conflict if there was to be one. After the armistice, the army endeavored to develop in conformity with the lessons of the Great War, especially with the creation of the Tank Corps and Air Service on a temporary basis under wartime legislation. What were the visions for future combat missions for each combatant arm, especially the cavalry? What were the problems that muted cavalry innovation and leadership during the 1920s? These questions will be examined in this chapter.

Undoubtedly, the war seriously brought into question the mobility of horse cavalry executing missions on the fringe of battle. Were the days of "hell bent for leather" cavalry charges with a spark of sunlight on drawn pistols or sabers over? At issue was the diminished use of cavalry during the war because of the introduction of machine guns, a preponderance of rapid-firing field artillery, and airplanes for deep reconnaissance. More so, the cavalry lacked experience in static warfare and combined operations with the infantry. In various army circles tanks were considered for taking over the traditional mission of cavalry. The question was what would be the value of cavalry in the future. The chief of the Tank Corps, Brigadier General Samuel D. Rockenbach, viewed tanks as weapons restoring mobility to the battlefield once provided by the cavalry. Weeks before the armistice, he presented at an Allied Expeditionary Force conference a lecture entitled "The Role of Tanks in Modern Warfare." Rockenbach stated that tanks were not mechanized cavalry, artillery, or armored infantry; they were separate weapons to aid the infantry. He rejected a pure tank attack. He supported, however, high-speed infantry raider tanks to carry out the traditional cavalry raid of creating havoc behind the enemy's front lines. Heavy tanks facilitate the breakthrough, and medium and light tanks exploit the success, he argued.[1]

Even toward the end of the war, the Germans reported that tanks completely changed tactics. One influential article in the *Neue Freie Presse,* which found its way into an important postwar U.S. Army Ordnance report on future tank development, admitted they were more suitable for effecting a breakthrough. Tanks were the best means for exploiting success, a role formally allocated to cavalry, the *Presse* reported. Continuing, this opened a new field of study for the tactician in which tanks were a new arm requiring cooperation with other arms, such as motorized infantry. The best means for fighting tanks were undoubtedly tanks, the article predicted.[2]

To deal with these controversial issues and analyze the experiences of the AEF during the war, a number of studies and boards were established during a period when the army was in the process of convulsive downsizing. This resulted in a serious debate within the army over a postwar doctrine adaptable to the new weapons of war.

The most controversial board convened to deal with lessons learned, doctrine, and the future of warfare was the Westervelt or Caliber Board appointed in December 1918 by the chief of staff, General Peyton C. March, who one young officer called "a grim, cold-blooded driver," detested by everyone. Yet he "built a fire under foot draggers." The board of officers studied and then made recommendations for the future use of field artillery. Its conclusion had a major impact on artillery as well as on the cavalry and infantry, declaring that mechanical transport was the prime mover of the future.[3] Brigadier General William Westervelt, who headed the board, stated its report produced a noticeable look of amazement on the faces of ranking officers, especially those among the field artillery who still relied on the future use of horses.[4]

In addition to the Westervelt Board was the influential Superior Board convened by General Headquarters (GHQ) AEF, France, in April 1919 to consider the lessons as they affected the organization and tactics of all combat arms, especially the infantry. Major General Joseph T. Dickman, whom Pershing had just relieved as commander of the Third Army in Germany, chaired the review board, which was composed of senior AEF officers. The Superior Board examined the combat results of heavy infantry divisions during the war and their future organization. It was determined that heavy square divisions of twenty-nine thousand officers and men should be sanctioned. Motorized artillery, along with tractor-drawn artillery, was considered appropriate providing there were suitable roads. For flexibility, the board suggested one regiment of horse-drawn 75mm guns for each infantry regiment. It was also recommended that the other new weapon, the airplane, be employed as a squadron by the division for distant reconnaissance, photography, and observation and registration of artillery.[5]

The Superior Board endorsed the Civil War doctrine of cavalry based on the mobility of the horse and the breech-loading rifle as the principal weapon.

Though the war had no influence on cavalry operations because of a lack of combat employment, traditional attitudes persisted. The war, the board noted, furnished but few reasons to change cavalry doctrine. However, it was decided deep reconnaissance was now more obtainable by aircraft. Thus, the strategic employment of cavalry again was directed to defeat the enemy's cavalry, break up his communications, provide flank and rearguard security, employ shock action, carry out pursuit and harassment, and engage in tactical reconnaissance. The board concluded that the conditions of northern France probably could not be reproduced in America. As a result, U.S. Cavalry was expected to find useful employment as in the past. No consideration was given to wheel or track mounts to supplement the horse.[6]

New weapons such as the tank were specifically seen as infantry weapons held at army level to be allocated to divisions as needed. The Superior Board recommended that current light and heavy tanks be recognized as supporting and accompanying weapons and organized for association with the dominant fighting arm, the infantry. The board emphasized that there was no such thing as an independent tank attack. It was, however, conceded with Rockenbach that high-speed medium tanks needed to be developed for raiding behind enemy lines. General Pershing's endorsement of the Superior Board report supported the conclusions, adding that tanks were best employed with the infantry.[7]

In April 1919, another board convened in France by Special Orders of GHQ, AEF, was the Lewis Board, or Infantry Board. Its purpose was to consider infantry tactics and organizational lessons by formulating a series of questions to be discussed with a number of key officers regarding their experiences gained during the war. There was a general indifference regarding the employment of cavalry with infantry because of a predominance of machine guns on the battlefield. Only two questions dealt with the use of tanks: should they be assigned to division control or to the chief of the Tank Corps for particular missions, and should there be a cargo-carrying tank? Most officers interviewed had no tactical experience with the deployment of tanks. One general officer who did was Lieutenant General Hunter Liggett, I Corps commander and then First Army commander during the Meuse-Argonne offensive. Pershing considered Liggett one of his best and most effective corps commanders. Liggett alleged infantry and tank cooperation was poor, emphasizing that tanks were infantry auxiliaries and the idea that they could be used for independent operations was "sheer folly." He strongly believed the main value of tanks was as artillery accompanying guns.[8] When Major General Frank Parker, the commander of the 1st Infantry Division, was interviewed, he stated that tanks and airplanes should retain their own tactical control and direction, suggesting more cooperation between the infantry and its supporting arms.[9] It was apparent that Parker was not convinced by the infantry solution of controlling tanks as adjuncts for Pershing's doctrine of open warfare. General Charles P. Summerall, Parker's corps commander during the last part of the war, took a different view, maintaining tanks and airplanes should constitute organic parts of the division along with substantial supporting artillery.[10]

When Colonel Adna R. Chaffee Jr. was interviewed, he supported the cooperation of tanks and airplanes as corps troops and viewed cargo-carrying tanks as useful. Chaffee also supported a horse cavalry regiment for each infantry division. Like Frank Parker, Chaffee had graduated from West Point, was commissioned in the cavalry, and attended the École d'application de cavalerie at Saumur, France, before the war. During the war, he served primarily as a general staff operations officer.[11] Later he became one of the instrumental officers in changing the meaning of mechanized cavalry while serving under Parker, who was then assistant chief of staff, G-3 (Operations). One of Parker and Chaffee's future antagonists, Colonel Steven O. Fuqua, who later became a chief of infantry, supported Liggett's position. After the war, Fuqua maintained tanks were an essential part of the infantry and should be placed under the control of the divisional commander. He supported the use of cargo-carrying tanks along with the wartime caisson companies of horse units and ammunition trains, providing they were motorized.[12]

The consensus of those interviewed supported the doctrine of open warfare General Pershing brought to France. He firmly believed that infantry trained in rifle practice and maneuver was always equal to the conditions of trench warfare, adding the importance of close cooperation with the artillery. Furthermore, the infantry now had the Browning automatic rifle (BAR), a squad support weapon first developed in 1917. However, Pershing's opinion was that men rather than armament were the most important factor in war. Regarding the employment of tanks as infantry-accompanying weapons, the consensus can best be identified with Liggett and Fuqua's strong beliefs along with Pershing's support. Virtually no suggestions were made on the role of cavalry operating with infantry divisions.[13]

The view of the AEF Cavalry Board was that the recent conflict had changed little regarding a war of movement. It was admitted, however, that mounted combat with large bodies of cavalry was outdated. Regarding future organization, the board objected to cavalry being assigned as organic elements with the infantry, arguing that it should be attached only for operations as needed.[14]

While the various boards in France were studying the lessons of the world war, Major General Clarence C. Williams, chief of ordnance since April 1918, established an organizational plan in early 1919 to cope with new developments in war technology. As a result, the Technical Staff was created to serve as a staff organization, providing technical advice to the chief of ordnance. The true innovation was the Ordnance Committee, composed of representatives from the using arms and services, to serve in an advisory function to the Technical Staff. This committee gave the using arms and services more input on establishing type, design, tests, and final judgment on army ordnance. Thus the using arms, such as the cavalry, infantry, and field artillery, now had a little more influence for specifying their needs. On many occasions, however, the Ordnance Committee controlled the formal seal of approval or rejection on military equipment. The "Minutes of Ordnance Committee Meetings" became the standard for specifications and design of ordnance to be used by the army.[15]

To keep posted on industrial developments in automotive equipment, the Ordnance Department organized—with the Society of Automotive Engineers—the Ordnance Advisory Committee. At the first joint meeting, held in September 1919, General Williams stated that the introduction of automotive equipment during the war was a "matter of futurity." He predicted that for some time to come the conflict between the horse and track-laying vehicles would produce uncertainty in the tactical use of these new weapons of war. The military mind, he stated, is a rigid one and does not lend itself to keeping pace with advanced technology and its impact on tactics. For future meetings, it was planned that all Ordnance Department projects be submitted to its members.[16]

Almost a year after the war ended, the Army General Staff was still baffled over a future tank development policy. To facilitate a postwar program for the Ordnance Department, Major Raymond E. Carlson, a member of the wartime Anglo-American Tank Commission that developed the heavy Mark VIII, made a complete review of the tank situation. He then submitted recommendations on future tank types for the army. His "Memorandum on Development of Tanks" was farsighted because it attempted to establish a new warfighting doctrine to conform to new weapons. Like Parker's earlier ideas on a combined arms mechanized force, Carlson's "Memorandum" offered a similar approach to combined arms based on the tank as the main maneuver element.[17]

From a tactical view, Carlson wrote that the infantry needed to be armored. Armored infantry, he suggested, required a small tank with extreme mobility. This, he argued, would eliminate human fatigue caused by negotiating a shell-torn battlefield. In addition, Carlson proposed an armored cavalry vehicle capable of speeds of twenty to twenty-five miles per hour. The cavalry tank he had in mind was similar to the British Medium D, the key maneuver element in Fuller's unused and potentially hazardous "Plan 1919." This new and untried tank was expected to have an operating radius of some two hundred miles and a speed of eighteen miles per hour, its greatest asset. As a fast-moving tank, it was

(Above) The M1919 was Christie's first tank endeavor with the U.S. Army, but it failed acceptance tests and was rebuilt to the M1921 as a turretless tank. It was not purchased by the army due to mechanical problems and funding. Courtesy of the National Archives and Records Administration. *(Below)* The M1921. Courtesy of the National Archives and Records Administration.

also expected to deploy independently as a disorganizing and pursuit force, extending far into the rear of the enemy's lines. Carlson predicted airpower could work in close cooperation with tanks. He also prepared a document entitled "Paper on the Development of Tanks," which contained a number of interesting articles on wartime tank experiences from various participants.[18]

In August 1919, Carlson met with General Rockenbach to discuss the medium tank program, which called for two tanks, an ordnance design and a Christie wheel and track convertible. The latter, designed by the eccentric J. Walter Christie, gave promise of revolutionizing cross-country transportation and fighting by eliminating tank carriers. The idea was to strategically deploy convertible tanks for combat in the wheel mode and then employ them in the track mode for the tactical battle.

Before concentrating his efforts and resources on armored fighting vehicles, Christie had begun his career as an apprentice machinist, entering the automobile industry where he spent most of his life as a design engineer building pilot models, automobiles, and front-drive steamers for the New York City Fire Department. During the golden age of automobile racing, his peers considered him a daring and reckless driver. At first, the chief of the Tank Corps was not interested in developing new tank types, such as the Christies. He was more interested in completing the present production of the American M1917 Renault and Mark VIII. Then he changed his mind, submitting a paper entitled "Tanks: Functions in Relation to Design," which followed the outline suggested by Carlson and the Ordnance Department. The paper became the Tank Corps' statement of policy.[19]

With new medium tanks, Rockenbach believed the Tank Corps could execute its mission better than with the numerous ones on hand, both light and heavy, used during the war and currently being manufactured in the United States. He emphasized, however, that the tanks were to assist the infantry, and all other arms were auxiliaries. This action removed consideration of a medium tank for the cavalry, an idea Carlson had presented.

Meanwhile, General Williams wrote General Rockenbach suggesting that a Tank Board be created along the lines of the Westervelt Board so it could recommend a permanent tank development program. The Tank Corps chief failed to respond, and no board was created. He did approve, however, specifications for a medium tank program, especially involving the Christie.[20] The support for the Christie was influenced by General Summerall's comments and criticism of Rockenbach's lecture "The Role of Tanks in Modern Warfare." In a letter to the chief of the Tank Corps, Summerall complained of the useless wearing out of tanks owing to the long distances required to arrive at the point of employment.[21] In July 1919, the Ordnance Department investigated Christie's "wheel and track design." As a result, the department decided to proceed along the lines indicated by General Rockenbach and the Subcommittee on Tanks, which was created to oversee new tank development and report to the Ordnance Committee.

Thus, by 1919, the Ordnance Department, especially Williams, Westervelt, and Carlson, began to set forth an important direction for a future tank policy and mechanization of the army. Their efforts, however, were soon to be contained.

On November 26, 1919, the Army General Staff lost interest in the medium tank program. Colonel Robert E. Wyllie of the General Staff's equipment branch sent a memorandum to Chief of Staff Peyton March, calling attention to the new medium tank program. Wyllie accused the Ordnance Department and General Rockenbach of not consulting other army branches. "The Tank Program is being formulated by the Tank Corps independently of any other service, including the General Staff," he wrote March. Furthermore, he noted, tanks were intended solely for preparing the way for the infantry and not to be used for independent action. The memorandum suggested that General Rockenbach not be permitted to dictate the entire tank program. Instead, the General Staff should decide on the program after a thorough study of the tactical role of the Tank Corps. The memorandum recommended that the War Plans Division, General Staff, make a study of tanks and prescribe general principles to be followed by the Tank Corps and the Ordnance Department.[22]

In January 1920, Rockenbach presented his case at a conference of senior army commanders. He commented that during the last year of the war few AEF commanders had considered the role of tanks. Rockenbach further pointed out wide differences of opinion over the use of tanks. "It runs from enthusiastic support of the two divisions with whom we had an opportunity to train and to fight, to damning by those who did not get us (tanks)," he reported, adding, "The weapon of the Tank Corps is the tank, just as the weapon of the cavalry is the horse." He emphasized the importance of keeping the corps independent. Rockenbach believed that to plunge the Tank Corps into some other arm, such as the infantry, would not only hamper the foot soldiers but also "emasculate the Tank Corps." Along the lines of new tank development, he outlined the importance of continuing the Ordnance Department's medium tank programs. In the discussion that followed, the head of the War Department's War Plans Division, Major General William G. Haan, protested Rockenbach's comment regarding ignorance about tanks. Those officers who "have not seen tanks, have thought about them, read about them, and their uses." He added that consideration was also given to the type of organization that would be most desirable.[23]

General Rockenbach continued to assert his position. In a series of articles in the *Infantry Journal*, he emphasized that tanks were not a substitute for artillery, though tanks should have airplanes detailed to assist in attacking antitank guns. For the first time he suggested tanks be used with cavalry. Such a force, he reasoned, combined greater mobility and firepower, especially when strong opposition overcame dismounted troopers. Tanks in cooperation, he stated, could remove the resistance. He also suggested tanks could pull a cart or wagon loaded with long and short forage, rations, and demolition materials. It was evident Rockenbach looked for ways to replace the horse with a machine.[24]

The conflict over the type of postwar organization went back to January 1919, when Secretary of War Newton D. Baker and Chief of Staff Peyton C. March presented to Congress the War Department's bill to reorganize the army. The proposed bill called for a standing army of more than five hundred thousand men, a program of universal military training giving the army its own reserves, and the consolidation of bureau management in the General Staff. The congressional debate that followed aroused the traditional American fear of a large standing army. Comments were made that there was too much General Staff domination, and that the danger of "militarism" was rampant. Many military affairs committee members were concerned that the Baker-March Bill would "Prussianize" the army.[25] The first AEF veteran elected to Congress, Kentucky congressman King Swope, can best explain the prevailing attitude. Later he would sum up the congressional attitude at the time, "Everybody had a bellyful of the damn army."[26] America was entering a new era—the organizational shift to corporatism in American business. Liberal roots, such as capitalistic expansion, democracy, and free trade, were looked at as the wave of future America. The U.S. military was profoundly affected by this "new age" tired of war. Within this setting, the military was unable to clearly define its image as the new mass culture reflected traditional isolationism. All these events led to institutional self-preservation for the army ground forces, including the cavalry that was striving for identity. The skeletonized army would not see its numbers grow to five hundred thousand until the United States went to war again decades later.

In the meantime, the continuation of the Tank Corps as an independent combatant branch was challenged by the chairman of the House Committee on Military Affairs, Republican representative John C. McKenzie, who at one time expressed doubts as to the need for a Tank Corps in peacetime. He suggested that it might be advantageous to combine the Tank Corps with the field artillery. When General March appeared before the House Committee on Military Affairs, he disagreed, stating that the functions of the Tank Corps were not similar to those of field artillery, adding that the equipment and training of tankers was different from those of infantry and cavalry.[27]

During the House Army Reorganization Hearings in September, Herbert Parson, formerly a lieutenant colonel in the AEF, read a letter he had written to Republican senator James W. Wadsworth Jr., chairman of the Senate Military Affairs Committee, who also held a lieutenant's commission in a National Guard cavalry troop. He emphasized that the infantry "must be kept in mind in legislating particularly with regard to making any provision; it does the fighting, gains the ground and holds it and suffers the greatest losses." He further stated that the cavalry, artillery, and air service were infantry auxiliary troops.[28] The AEF chief of engineers, Major General William M. Black, predicted this arrangement would become problematic because of the anticipated autonomy of the branch system and the relationship between auxiliary arms and the dominant fighting arm, the infantry.[29] His prediction would come true but not before the debate over compartmentalization or branch autonomy resolved itself.

General Rockenbach, who was unfamiliar with the Baker-March Bill, appeared before the Senate hearings. He nevertheless supported a proposal for a separate status for the Tank Corps. He also reiterated that the function of tanks was making a path for the attacking infantry.[30] Major General James W. McAndrew, commandant of the General Staff School and wartime chief of staff to General Pershing, approached the organizational position of tanks by agreeing with Rockenbach that they were adjuncts of the infantry. He added, however, that tanks were at the time only in their infancy and could be better developed in a separate corps rather than under the infantry.[31]

The *Army and Navy Journal* reported at the end of September that confusion in military legislation would be placed more in perspective with the appearance of General Pershing before both congressional committees. His presence was critical. No one officer in the army has a wider knowledge or more complete grasp of military lessons learned, the article explained. The journal editorialized that the postwar congressional attitude against a standing army of five hundred thousand proposed by the War Department created confusion over army legislation. It was having an effect on army morale and its future effectiveness.[32]

The debate continued. In October, General Summerall spoke before the Senate military hearings and continued the argument for a separate Tank Corps. He supported a branch chief for the infantry, cavalry, and field artillery, along with special training for each arm. Optimistically, Summerall proposed large-scale maneuvers to bring the combatant arms together for coordination and cooperation.[33]

Finally, the heated debates were resolved with the appearance of popular and venerated General Pershing before a joint session of the Committee on Military Affairs. He testified, "The Tank Corps should not be a large organization, only of sufficient numbers to carry on investigations and conduct training with the infantry and place it under the infantry as an adjunct of that arm." Under questioning from committee members, the general reasoned that placing the Tank Corps under control of the infantry was wise because during the war it was closely allied with that arm. More so, he noted, tank development would be more effectively encouraged by the infantry "perhaps to a greater extent than if left to itself." He emphasized that development be continued because "there is a great future for tanks." To establish more organizational power, he advocated a peacetime combatant arms chief called "inspector" for the cavalry, field artillery, and infantry.[34] Pershing's testimony doomed an independent organization of the Tank Corps and encouraged the establishment of line branch autonomy that had an impact during the postwar period on how the army would fight its next war.

In June 1920, the National Defense Act was signed into law. Congress was determined to reduce military spending and eventually approved an army of two hundred thousand, which never materialized. It would be 1939 before Congress finally authorized funding for increasing the army. The 1920 act restored the autonomy of the bureaus and extended compartmentalization of the army into combat arms and services, each having specific duties, such as training, branch schools, equipment, and tactical functions. Branch offices for the line troops were established on the same administrative level as the service bureaus. Thus, the Offices of the Chiefs of Infantry, Field Artillery, and Cavalry were added to the existing Office of the Chief of Coast Artillery. Major General Willard A. Holbrook was appointed the first chief of cavalry, as was Major General Charles S. Farnsworth chief of infantry and Major General William J. Snow chief of field artillery. Thus, the new branch chiefs were responsible for developing their own tactical doctrine, organization, and management of personnel, training, and equipment. The chiefs also established their own equipment boards and service schools. During the postwar period, the line branch organization developed into a deeply fragmented, self-serving, elitist culture within the army.

This compartmentalization of combat functions, as we shall see, was a major obstruction later in establishing a combined arms mechanized force at an operational level of warfare, meaning how to fight the next campaign. Subsequently, it hampered the ability of the peacetime army to move away from an attritional warfare doctrine consisting of the infantry-artillery team. In dealing with its peacetime dwindling force, the army failed to recognize that even a reduced force with mechanized mobility could defeat a larger enemy force that relied on World War I experience. Relying on two levels of war, such as war strategy and battle tactics, the army failed to understand the art of using combined arms maneuver-oriented tactical techniques in a campaign. Though there were many fiscal constraints along with public apathy affecting the army after the war, there existed a historical motivation to rely strictly on a World War I doctrine that had a French influence. Forgotten were the ideas on the impact that mechanized warfare would have on a future battlefield, especially those expressed by Generals Parker and Fuller in the summer of 1917 and forgotten as soon as the war ended.

Furthermore, during the postwar years the branch chiefs and their successors had to defend their autonomy because of a continuous reduction in funds, personnel, training, and retaining obsolete surplus equipment. Congress and the presidents during the 1920s looked at the navy as the first line of defense, while the country in general was no longer concerned about war or national defense.[35] After all, the war to end all wars was over, and Congress and the public looked with disdain at anything suggesting militarism. In 1921, an unknown American soldier was buried in Arlington National Cemetery. Idealistic hopes in America had been realized that the war was in fact the last world war.

Meanwhile, General Pershing's recommendations before the joint session of the Committee on Military Affairs had carried, and the Tank Corps ceased to exist as an independent arm. By law, the National Defense Act placed all tank units under the control of the chief of infantry. Pershing's influential endorsement of the Superior Board's 1919 report on organization and tactics indicated he personally felt that in the future the army would probably operate on the American continent. He expected mobility to be more of a necessity in America than in western Europe, where static warfare hampered mobile operations. Infantry and cavalry were expected to develop greater mobility. The infantry was to train for open warfare, fostering the offensive spirit with rifle firepower and successful bayonet charges with supporting tanks and artillery. The cavalry, in turn, was to maintain the cardinal principles of mobility and firepower by developing good horse mounts and thorough training. It was anticipated that cavalry would find useful employment in other fields and under different conditions than in France, especially in the southern border areas with Mexico. Many traditional cavalrymen believed tanks had no effect whatsoever on their future. The feeling was that tanks were valuable weapons for the infantry only in trench or siege warfare.

In contrast, a small minority thought otherwise. Along with Parker and Carlson, the war produced a number of ambitious officers who sought to change postwar doctrine based on the advent of new technology. Colonel George S. Patton Jr. of the AEF Tank Corps and Captain Dwight D. Eisenhower, who trained tank troops at Camp Colt, Pennsylvania, and Camp Meade, Maryland, both wrote in the *Infantry Journal* in 1920 in defense of tanks, forecasting their value in future wars.[36] Both supported an independent Tank Corps and attempted to address doubts among many officers who questioned its value. Patton's war experience and his cavalry background emphasized light tanks for mobility and maneuver. Later, Eisenhower wrote that he and Patton were given the order to desist. According to Eisenhower, he was called before General Farnsworth, the first chief of infantry, and ordered to conform to infantry doctrine or face a court-martial.[37] The young officer conformed. In a manuscript apparently prepared while at Fort Meade, Eisenhower argued that tanks could never take over the mission of the infantry. Their utility was assisting the advancing infantry.[38]

With the resolution over the Tank Corps resolved, Patton returned to his first passion, which only the cavalry could offer, a love for the horse, polo, and saber. Visions of employing cavalry en masse providing shock power, maneuver, and exploitation still gripped the romantics. Historian and editor of *The Patton Papers,* Martin Blumenson, stated that Patton—in the interest of economy—understood that if tanks went under a traditional arm, he preferred the cavalry. According to Blumenson, "Patton intuitively understood that tanks operating with the cavalry would stress mobility, while tanks tied to the infantry would emphasize firepower." Patton also realized that during peacetime Congress probably would restrict funds necessary to construct new tanks and support a modern tank policy. Tanks in peacetime, he noted, were like coast artillery, "a lot of machinery which never works." Until World War II, Patton became associated with conservative cavalrymen; however, he judiciously engaged in a balancing act between the horse versus motorization and mechanization.[39] Another historian argued that it was not career preservation but an unsteady and equivocating debate between the machine and the horse. Patton was right in the middle of the debate. His instincts, nevertheless, reflected his concerns over the realism of mechanization. For the next two decades, he opposed any scheme and machine not sufficiently tested. He realized that postwar budget restraints would affect the army.[40]

The tank debate continued. Bradford Chynoweth, a young maverick infantry officer who graduated from West Point and liked to call himself an "army brat," was assigned to the engineers in the States during the war. After the war he spent a year in civilian life, then returned to the army and was appointed a major of infantry. Young Chynoweth called to mind having a few fuming incidents with General Rockenbach, whom he called "Old Rocky." He viewed "Old Rocky" as a traditional army martinet. After completing a quick course at the Tank School, Chynoweth served as Rockenbach's tank intelligence officer in charge of publicity. According to Chynoweth, Patton was the only officer who could handle Rockenbach. Patton explained his method. When he came up with an idea, Patton wrote it up in a letter and left it on the general's desk. Then he waited a week, returned to the general's office, and heatedly argued against his proposal. It worked. "Old Rocky" would rise from his chair, arms whirling, and support Patton's original idea. Chynoweth claimed Patton would then gracefully concede.[41]

Meanwhile, Chynoweth entered the tank controversy with two articles he wrote in 1921. "Chen," as his friends called him, argued in the *Cavalry Journal* for cavalry tanks. Tanks, he declared, offered a balance of mobility and firepower. Chen's postwar studies convinced him that tanks were more of a cavalry evolution than infantry. Tanks, he maintained, were a new combat arm that changed the role of the mounted trooper. It was apparent to him that mechanization doomed the horse soldier. Chynoweth suggested in the *Infantry Journal* that infantrymen be trained for maneuver depth and area tactics rather than the linear approach that characterized methodical and attrition warfare of World War I. Mass tanks were more suitable for affecting a breakthrough and exploiting success. Traditionally exploitation was allocated to the horse cavalry. He argued, however, that tanks were a new arm that changed the role of cavalry. Furthermore, tanks were not a substitute for artillery. Nevertheless, he wrote, they forced cooperation with other combat arms and then went on to add that infantry needed to be motorized. In addition, he argued for a new combat arm for tanks. He accused the army of being too conservative because it was endeavoring to adjust newly acquired techniques to a framework of prewar principles.[42] Chynoweth, who was considered "a great brain," continued arguing against an infantry tank doctrine that placed tanks in a supporting role. Years later this position led him to exile from the tank service and deployment to the Philippines, where he was imprisoned by the Japanese after the fall of the islands.[43]

The *Cavalry Journal*, meanwhile, continued arguing for its branch's reason to be. An "Editorial Comment" in October 1921, noted, "The cavalry is destined to a splendid future of glorious achievement. We must not permit a transitory predominance of gasoline and technical novelties to obscure that truth."[44] The cavalry still planned to practice mounted combat, charging with sabers and pistols. Nevertheless, it was realized with the advent of machine guns that the mounted charge was an unrealistic tactic in modern warfare. After the war, many horsemen still liked the charge if just for its glamour effect. However, realistic tactical training then consisted of a mobile mounted force riding to action and then taking cover to fight dismounted. The lead horses were then taken to the rear and placed under cover. Generally, cavalry was used for stealth reconnaissance referred to as "sneak and peek." When necessary, the cavalry engaged in probing attacks to determine the enemy's situation and strength. After contact the horsemen disengaged from the main infantry force and supporting artillery and deployed to a security mission, protecting flanks or operating in the enemy's rear, disrupting his logistics. This doctrine was a throwback to how the U.S. Cavalry, both Union and Confederate, generally fought during the Civil War. Embedded in its thinking were the exploits of noted cavalrymen, such as Nathan Bedford Forrest, John Hunt Morgan, James Ewell Brown (Jeb) Stuart,

Joseph Wheeler, John Buford, Benjamin H. Grierson, Philip Henry Sheridan, and James Harrison Wilson, to mention a few. The Civil War tradition, however, collided with new opportunities for warfighting due to the introduction of rapid-firing weapons and new mechanical mounts developed during World War I. The failure to absorb new opportunities regarding advances in technology was not obvious to the cavalry after the war. It still looked to the past.

At the Cavalry School during the 1920s, two influential books emerged, giving traditionalists a reason for optimism about their future and fuel for mounted innovators. The traditionalists, with sabers in hand, accepted R. M. P. Preston's book *The Desert Mounted Corps: An Account of the Cavalry Operation in Palestine and Syria 1917–1918*, as their bible. It argued that the horse soldier was more valuable in modern warfare than in the past.[45] Field Marshal Viscount Allenby's successful campaigns against the Turks were held up as a perfect demonstration of how cavalry should be used. Laid out were examples of mobility, aggressiveness, daring, quickness of perception, firepower in dismounted action, the support of mounted machine gun units, and the endurance of horse and man in a sandy and almost arid terrain. Allenby's campaigns were mentioned as modern examples of mobility and firepower as opposed to a methodical infantry linear approach.

The other book was George T. Denison's award-winning *A History of Cavalry from the Earliest Times with Lessons for the Future*, which dealt with morale and command. Denison wrote that cavalry leaders needed to possess strong inventive genius. By this, he meant cavalry leaders needed to be self-reliant enough to adopt reforms when necessary.[46] This was taken by a few cavalrymen to mean that the future role of horse soldiers would change with advances in technology, especially with the arrival of a new mount, the tank.

In defense of the future mission of the horse cavalry, a litany of opinions on its use during the 1914–1918 war was debated during the 1920s. It became a defense against the perceived diminution of the use of cavalry. Prominent in this effort was the Cavalry School, which issued a pamphlet, "Cavalry Operations during the World War," which listed a number of examples of the proper use of cavalry as well as faulty employment and tactics, especially by the British, French, Italians, and Germans. A few operations were selected demonstrating cavalry actions during the war: the western front in July–August 1914, the campaigns in Palestine and Mesopotamia, and the Italian campaign of 1918.[47]

On the western front during the early stages of the war, German cavalry, like the French cavalry, favored mounted action with lances rather than dismounted combat. A German cavalry corps was responsible for the successful screening of the infantry as it moved through Belgium. The cavalry's pursuit of retreating French and British armies, however, lacked aggressiveness. The

French cavalry failed in its primary mission of reconnaissance because it never discovered the strength of German infantry. Unlike French and German cavalry, the argument was that British cavalry was trained to fight mounted as well as dismounted. This dragoon ploy saved the British army from an early defeat. It successfully covered the advance and the withdrawal from the Mons position. Field Marshal Allenby's successful campaign against the Turks was favored to dispel the perceived demise of horse solders. His use of cavalry in September 1918 was brought up as a historical example of how the arm should be employed in pursuit against a weak and demoralized enemy. The Italian campaign of 1918 was also used, explaining how the basic requirement of cavalry mobility and firepower successfully prevented the Austrian rear guard from organizing a retreat.

In testimony before the U.S. House Committee on Appropriations, a cavalry officer explained the military value of cavalry by quoting opinions from Marshal Ferdinand Foch of France; Marshals Douglas Haig, Allenby, and Lord John French from England; Marshal Erich Ludendorf from Germany; and Generals Pershing and Hunter Liggett from the AEF. All these prominent individuals emphasized the future importance of horse cavalry.[48]

The point was strongly made that cavalry of all nations, except England, entered the war trained and primarily equipped for mounted action. This type of warfighting was criticized as being impotent. Historically, American cavalrymen were trained to fight mounted and dismounted when necessary. Since the war, it was realized that there was a need for greater firepower without sacrificing mobility. The political mind-set, however, accepted the notion that the last conflict was in fact the war to end all wars. All the cavalry could do was argue for their "reason to be."

In his 1921 annual report to the War Department, General Holbrook said that the cavalry had lost none of its importance regarding the defense of the United States: "For open warfare, in campaigns where armies do not have their flanks resting on perfect obstacles, there is more use for cavalry than ever before in modern war." He continued, "The broad principles underlying the conduct of open warfare have not been changed by the experience of the late war, and it is for open warfare that our army is being trained."[49]

The money problems that beleaguered the military after the war had the effect of diminishing the cavalry as well as the infantry to skeletal organizations. The 1st Cavalry Division was the only active division added to the regular army in 1921. The 2nd Cavalry Division was added but listed as inactive. The divisions contained two brigades, with two regiments each. Each regiment had two squadrons with three troops each designated as A through F. In addition, other combat elements were added, a 75mm horse artillery battalion, a mounted engineer battalion, an ambulance company, and division trains along with special troops such as headquarters, signal, ordnance, and veterinary.

In fall 1923, the 1st Cavalry Division held its first postwar maneuvers in Texas. The purpose was to gain data and experience on organization and equipment. It was suggested that machine guns be organized as a troop in a squadron rather than as a platoon in a troop. This reorganization consolidated firepower for a rapid mounted approach or mounted attack against a hostile force. At division headquarters command and control radio sets were carried in light trucks. The emphasis, however, was on increasing firepower and horse mobility with some motorization of equipment, meaning the employment of vehicles for transportation of supplies and horses. A young cavalry officer, Major Adna R. Chaffee Jr., wrote up the maneuvers in the *Cavalry Journal.* Soon to be a dominant figure in American military history, Chaffee romantically concluded, "The division, coming into its home station with its officers and men sun-tanned and hard, sitting lightly and easily in their saddles, with its horses in good condition and its organization well knit, had every reason to be proud of its efforts to round out and complete its training years."[50] His article sounded like all the cavalrymen experienced the proverbial good time.

After the 1923 cavalry maneuvers, Patton wrote in the *Cavalry Journal* that current armored cars and tanks could not keep up with the cavalry's horses.

T1 light armored car on a Pontiac chassis built in 1928. Note the two .30-caliber machine guns. From the Hunnicutt Collection, Patton Museum.

For the cavalry's future, he suggested that armored cars be equipped with .30-caliber machine guns in a 360-degree traverse.[51] This was an echo of Parker and Boyd's argument in February 1916 to add armored cars to the cavalry. By the mid-1920s new armored car development began again, and as during the pre–World War I period was based on truck chassis and commercial vehicles called "cross-country cars."

In 1923, the cavalry began evaluating a few stripped-down commercial cross-country cars as administrative vehicles for command, liaison, and transportation for headquarters troops. The armored car on the cavalry insignia was established in 1924, and only part of a squadron from the 1st Cavalry Division was activated. In 1928, the Cavalry Board tested the first scout car, the T1, manufactured by Pontiac. This vehicle, designed for highway and road reconnaissance, was a simple bucket-seat commercial car. Unfortunately, the vehicle had too many defects.[52] Nevertheless, the results were promising, and a new, improved light scout vehicle was designed for possible reconnaissance.

Regarding one of the essential cavalry missions of reconnaissance, an award-winning study to deal with the issue appeared in the October 1923 edition of *Cavalry Journal*, the same year the 1st Cavalry Division engaged in maneuvers. Captain Edward M. Fickett, who would become a prominent cavalry group commander in World War II, argued for greater cooperation between the air service and cavalry in three areas of reconnaissance other than command and artillery spotting. The first was classified as distant or strategical, which determined the concentration of the hostile force, its strength, general composition, route movement of main columns in time and space to battle, reserves, supplies, and communication. In addition, distant reconnaissance supplemented information already on hand regarding topographic, geographic, economic, and political features. Next, the article defined close or tactical reconnaissance. It was executed when the opposing forces were within striking distance. Reconnoitered were the terrain, enemy's location, strength, distribution, composition, and its direction and progress of movement. In addition, tactical reconnaissance included location of the enemy's flanks, local reserves, lines of communication, supplies, and troop morale. This information was necessary, the author maintained, so commanders could prepare the tactical plan and allocate their assets for combat. The last was battle reconnaissance, where all troops engaged in combat. At this stage, the author wrote, flank reconnaissance was of special importance for both air and ground elements. As the three areas of reconnaissance merged, the author argued, it was imperative that the air service and cavalry understand each other's missions and work together for command information. Concluding, the cavalry could possess eyes in three dimensions where formerly it had two. Coordination could be a problem when both the air service and the cavalry failed to support each other.[53] The significance of this article was that it gradually channeled in on air-ground reconnaissance at the expense of tactical ground support for an attacking force. The cavalry gradually accepted this doctrine until World War II, when the German blitzkrieg demonstrated the importance of tactical air support for attacking panzers.

War-weariness brought about a new national community that viewed war as an anachronism. Also facing the army's training and readiness was the prosperity of the Roaring Twenties, bootlegging, and the female flapper that generated a new gender morality. The country was undergoing a second industrial revolution with the rise of the modern corporation and a culture of management elites. Welfare capitalism, the auto age, and a new mass culture of advertising, radio broadcasting, and movies were all a part of the new morality, which consumed most of the United States. This new American age of prosperity relegated the regular army to an insignificant role in defense of the country. The postwar 1922 Washington Conference even affected the U.S. Navy, the country's first line of defense, by scaling down battleships and tonnage of the major powers. Americans and their politicians had other things on their minds, least of all a strong military. Thus, the regular army was tasked to defend the continent with an inadequate force and leftover war equipment for years.

Meanwhile, the War Department in 1924 issued the first postwar *Field Service Regulations: Operations (FSR 1923)*, which was virtually a translation of the French doctrinal manual that underlined the methodical attritional style of warfare in future land operations. One historian noted that *FSR 1923* was influenced by two competing ideas: the experiences of World War I, and the desire to avoid another European world war because of a national revulsion by the government and people over the colossal casualties of World War I.[54] The doctrine, however, did reflect Pershing's "open battle" dominated by infantry linear maneuver and rifle firepower. The *1923 FSR* was flawed because it was not an engine of change based on what was written and taught. It was based on the postwar boards whose members testified what they and their units had done during the war. The AEF board reports were not conceptually played back against the *1923 FSR* to determine if army doctrine was right regarding equipment, organization, logistic support, training, and education, as a result of postwar legislation that created the cavalry, field artillery, and infantry branch systems that remained supreme until World War II.

Regarding the mission of the combatant arms—infantry, cavalry, artillery, engineers, air service, and signal corps—*FSR 1923* emphasized combined employment of all arms. *FSR 1923*, however, specified, "The special mission of other arms are derived from their powers to contribute to the execution of the infantry mission." Furthermore, it stated, the "infantry is the essential arm of close combat, executing fire and movement as a means of action." Like other combat arms, cavalry was expected to contribute to the execution of the infantry mission.[55]

Because of its high degree of horse mobility and relatively modest firepower, the cavalry was expected to execute the mission of reconnaissance, counter-reconnaissance, and security in the service of large units, the *FSR 1923* stated. Armed with rifles, machine guns, pistols, and sabers, the cavalry was expected to execute "rapidity of movement that enables it to strike a sudden blow at weak points in hostile dispositions." Furthermore, the regulation stated, "Cavalry may be employed to attack hostile forces in the process of concentration, to operate against exposed enemy flanks or rear, and to exploit by pursuit the success obtained by other arms." The regulation noted that the efficiency of cavalry depended upon the condition of the horses, especially when engaged in arduous duty.[56] To most cavalrymen this was somewhat amusing. After all, the horse was the cavalryman's noblest companion.

Covered in *FSR 1923* was the role of the air service in reconnaissance missions, a role assigned to the cavalry before the war. Because of having a higher degree of mobility than horse cavalry, observation aviation was expected to provide enemy information to higher command, to the infantry, artillery, and air service. No mention was made of cooperation between the cavalry and the air service.

Two years later, Lieutenant Colonel Daniel Van Voorhis, an accomplished professional cavalryman representing the Office of the Chief of Cavalry, requested an air observation squadron and one photo section for the cavalry division. This was an inkling that he had special organizational skills and an idea of what the future battlefield meant to the cavalry branch. He became an innovator. The past war had demonstrated that distant reconnaissance absorbed a large force and put a considerable strain on the horses and men, which was out of proportion to the results obtained. An air service was a suitable substitute, which Van Voorhis understood.

Van Voorhis entered the army as a private and served in the Spanish-American War. Through determination and dedication, he moved up through the ranks, serving in France during World War I as a cavalry officer. He soon became a dominant figure in moving the cavalry toward mechanized warfare by implementing a change in cavalry doctrine, moving it away from the horse mentality to mechanization. He became one of the early agents for change and adjustments to a new means of warfighting with new mounts.

In 1925, Van Voorhis had a strong sense that the air service could improve the reconnaissance mission of the cavalry. Secretary of War John W. Weeks, however, did not deem the idea of an air service reconnaissance squadron for a cavalry division advisable. The argument was that a squadron would require a significant number of ground installations, causing serious housing problems and distribution of troops. This situation, the argument went, would tie down a cavalry division at the expense of mobility.[57] By 1926, the Army Air Corps had a number of observation squadrons. Arrangements were made with the 16th Observation Squadron and the cavalry to work together at Fort Riley on some tactical issues, but no provision was made to organically attach an observation squadron to division cavalry. Later Van Voorhis would try again.

Meanwhile, Frank Parker returned to the United States, graduating from Command and General Staff College at Fort Leavenworth, Kansas, in 1922 and subsequently becoming an instructor for a short period. In 1923, he graduated from the Army War College and then became an instructor, lecturing on tactical maneuvers with a combined arms force. Because of World War I, he argued, mechanical development in warfare would enter its greatest phase, meaning large units of combat arms, including aviation, would all maneuver together in massive concentrations. The increasingly mechanized warfare, he explained, would result in economy and diffusion of fighting personnel.[58]

In October 1925, Parker, now a brigadier general in command of a brigade, was summoned before the president's Aeronautical Board of Inquiry, generally known as the Morrow Board. It was named after its chairman, Dwight W. Morrow, a partner in the J. P. Morgan investment firm and classmate of President Calvin Coolidge at Amherst College. The board was supposedly convened to establish a national civil aviation policy and determine the best means for utilizing aircraft for national defense. Aviation gained notoriety because it was a time when considerable congressional interest was directed to the dismal record of American aircraft production during the war, the role and status of the Army Air Service, and an emerging commercial airline industry. In addition, aviation drew considerable civilian interest in various aerial contests and the cross-country exhibitions of barnstormers. Even though the Air Service was the more glamorous and technological of the services, it too would have to deal with scarce funds for experimentation and testing of new air equipment. More so, the board was created to anticipate Colonel William "Billy" Mitchell's defense at his upcoming court-martial.[59]

The issue began in September with a written statement to the press after the crash of the dirigible *Shenandoah* during a severe storm over Ohio and the loss of three aircraft on a first flight from Los Angeles to Hawaii. Mitchell commented on what had been on his mind for years, stating, "These accidents are the direct result of the incompetent criminal negligence, and almost treasonable administration of our national defense by the Navy and War Departments." The uproar was brought about by Mitchell's persistent impatience with military leaders who had not shared his views for a separate branch for the Army Air Service and the role of strategic airpower as the decisive force in warfare. Clearly, such comments were a court-martial offense. The fiery Mitchell's solution was to create a Department of National Defense, where each service, army, navy, and air service, had equal status. He symbolized the belief that airpower was the new military factor, thus subordinating the army and navy. By order of President Coolidge, he was to appear before a court-martial in Washington. Mitchell was charged and found guilty for bringing discredit upon the War Department, insubordination, and conduct to the prejudice of good order and military discipline. As a result, he was "sentenced to be suspended from rank, command, and duty with forfeiture of all pay and allowances for five years." Disheartened, he resigned from the army on September 1, 1926.[60]

At the time of the Mitchell affair, Parker was commanding the 2nd Infantry Brigade, 1st Division. He told the Morrow Board of the importance of the Army Air Service as part of the combat team. Parker disliked hearing, he said, the infantry claiming that it was the most important branch of the service or that the infantry and artillery together were the most important branches of the army. Parker likened the combat branches to a baseball team where every player was equal. The infantry, artillery, cavalry, and aviation should all be organized and administered alike, he argued. Parker did not agree that aviation should have the same status as the other military services.[61] What he was trying to convey was that a lack of teamwork would surely degrade the army's missions. This position was in response to Mitchell's proposal for an independent air force beyond army control. Where Parker looked at attack aviation as part of the combined arms team, Mitchell was more interested in strategic airpower and a separate branch. This split of doctrinal interests had a lasting effect on formulating a mechanized combined arms doctrine with tactical air support until well into World War II.

A few years later witnessed a strong effort by the infantry to retain its operational control over tanks, as mechanization became more of an issue. The infantry, the dominant fighting arm as defined in the *FSR 1923*, reasoned that only it could function under all conditions of terrain and weather. This argument, evolving from tradition and mission, again was emphasized in reaction to the revolutionary thoughts of J. F. C. Fuller and his ideas on an all-tank force for offensive operations. The controversy was set in motion in his article "Tactics and Mechanization," which appeared in the May 1927 edition of the *Infantry Journal*. That month the journal devoted the entire issue to mechanization.

As weapons change, Fuller wrote, so must tactics or the art of warfighting change, including organization and training. Yet development of mobility through offensive power remains constant, he argued. He criticized the attrition warfare brought about by the infantry-artillery approach on the western front during the last war, stating, "Tactics ceased to be an art, they had generated into a mathematical calculation of bullets and shells." Regarding the infantry, Fuller noted that it lost its ability to attack and now had engaged in a slow pursuit or had become more of a follow-up force to massive artillery barrages. To him the most important tactical lesson of the last war was the introduction of petrol power in the form of tanks that enabled the primacy of the tactical function of the offensive to be reintroduced. The question was how could mobility be regained. For Fuller it was mechanization, replacing and supplementing muscle power with mechanical means that offered greater firepower. He wrote, "First and foremost we must get the present form of war out of our heads," meaning forget thinking in tactical terms of cavalry, infantry, and artillery, because these combat arms are only "chloroformed" by their separate organizations. With the advent of mechanization, he held, the army must think in functions. Here Fuller meant that battles fought by traditional infantry, cavalry, and artillery were no longer suitable on the modern battlefield. He suggested the military mind become more receptive and flexible and start thinking more in terms of tactical functions. He listed five: to discover, to hold, to hit, to protect, and to smash. Continuing, he noted that the armies, which can develop more mobility, have the decisive advantage.[62]

Weapons that Fuller listed as most important were airplanes, tanks, armored cars, tractor-drawn and self-propelled artillery, antiaircraft guns, machine guns, and light automatic weapons. The modern tank, he held, possessed the highest freedom of movement because it could hold and hit more economically than the infantry. For smashing, Fuller believed tanks were more suitable than cavalry. Concluding, he recommended thinking in names and weapons and abandoning the idea of traditional combat arms. "History is a great laxative, it frees our minds from shibboleths, and happily it has a habit of repeating itself."[63]

Fuller's article provoked considerable discussion among American army officers. Colonel C. H. Miller, the commandant of the Tank School, referred to Fuller as one of the enthusiasts who regarded cavalry, and to some extent infantry, as obsolete. He concluded that cavalry and infantry as organized in the U.S. Army still had a place in the combat team. Colonel Frank Cocheu, assistant commandant of the Infantry School, reacted by defending the use of rifle firepower and the bayonet attack and positioning tanks only as auxiliaries. An instructor at the Army War College, Lieutenant Colonel W. B. Burtt, did not like Fuller's brushing aside old views of cavalry, infantry, and artillery, stating, "We cannot subscribe to the view of a tank or mechanical army." The army of the future, Burtt maintained, would depend upon mobility and enlightened personnel who represented field artillery, the "Queen of Battle." Lieutenant Colonel P. S. Bond, a faculty member at the Infantry School, disagreed with Fuller's idea of using mechanical power to replace manpower. A member of the War Plans Section, Office of the Chief of Infantry, concluded that the cost of mechanization was not acceptable to Congress in peacetime. He maintained that tanks were still a hobby for Fuller, who seemed to have abandoned the bayonet. Furthermore, he upheld the chief of infantry's direction for developing light tanks and to some degree armored cars for reconnoitering, security, and pursuit, a traditional mission of cavalry. He noted, improvements in tanks will result in corresponding advances in antitank weapons.[64]

Even General Rockenbach, who at the time was commanding general, District of Washington, strongly supported the view that all other combat arms were auxiliaries of the infantry, which he emphasized could function under all battle conditions and on all terrain. Rockenbach referred to Fuller as a romantic prophet. However, the former Tank Corps commander noted, mechanical warfare would change tactics, organization, and training for all combat arms, including infantry auxiliaries, such as cavalry, artillery, airplanes, and tanks. In his opinion, it was essential that the army have two types of tanks. He proposed a light, fast 6-ton tank armed with a 37mm cannon or machine gun capable of accompanying the infantry and engaging in reconnaissance. Apparently, he delegated the cavalry mission of close-in reconnaissance to infantry elements. The second proposal was for a fast, durable medium tank under 20 tons with cross-country capabilities and more armaments with all-around fire. This type, he suggested, was essential as an accompanying tank for the cavalry and for reserve of armies, corps, and divisions.[65] Rockenbach's suggestion for a medium tank for the cavalry surely raised the concerns of the chief of infantry, whose idea was for lightness in tank development and maintaining complete control of tank policy.

The only dissenter was Major Merrill Spaulding, also a faculty member at the Infantry School. He suggested, without being specific, that the military was too complacent. As a result, he warned, the military mind would have a tendency to return to the security of old ideas. He supported Fuller's views, stating that his opinions should not be ignored. With the advent of tanks and mechanization, Spaulding believed a new type warfare now emerged beyond the role of tanks during the last war. Fuller's advice, Spaulding suggested, should be considered or the army would continue to loiter sleepily in the past.[66]

Lecturing at the War College after the Fuller debate over mechanization and presenting the current infantry point of view was the chief of infantry, Major General Robert H. Allen, whose primary interest was rifle marksmanship. He did, however, attempt to set the tone for an infantry tank policy when he directed efforts to develop light tanks, arguing that lightness enhanced the infantry's maneuverability. This move toward lightness, he argued, could restore the balance of power to the infantry. The problem, Allen maintained, was Rockenbach, who after the war attempted to dominate tank policy, calling for an independent Tank Corps armed with light, medium, and heavy tanks. The chief of infantry was critical of Rockenbach's policy, noting that it proved to be a costly mistake. He was referring to early Christie convertible tanks, M1919 and M1921, and the ordnance-designed medium M1921 and M1922. Allen argued that a light tank program should have had priority as an infantry-accompanying weapon, whereas Rockenbach had argued that tanks were no longer special weapons for the infantry. Of interest, he suggested tanks were just as important for the cavalry. Recall that Patton's war experience also suggested lightness to enhance mobility. Allen viewed light tanks as suitable for breaking the stabilization that existed during the last war and ideal for the army's tradition of warfighting, open warfare and rifle firepower. During the discussion after the lecture, the point was emphasized that tanks belong to the infantry. It was conceded, however, that tanks could operate with the cavalry.[67] Allen's attitude was quite different from the recommendations proposed by Fuller. It continued placing the development of tanks under infantry control and set a policy of lightness that dominated infantry and later cavalry doctrine and tactics until World War II.

While the infantry and its chief debated Fuller and light tanks, the 1st Cavalry Division conducted its first maneuvers since 1923 in Marfa, Texas, during September 1927. After the war, the division's main mission was to patrol the Mexican border. Because of a lack of funds, the maneuvers were only the second held since the end of the war. The purpose was to experiment with motorization of cavalry units in the border area, such as moving personnel, animals, and equipment by truck for varied distances. This was the cavalry's first experience with *portée,* or motorized transportation. A motorized field artillery battery also participated in the maneuvers, marching overland from Fort Sill, Oklahoma. Some cavalry officers, however, questioned the value of motorized artillery organic to a cavalry division.[68] Based on the 1927 maneuvers, the chief of cavalry, Major General Herbert B. Crosby, introduced a new, smaller cavalry organization.

General Crosby's proposed cavalry organization was designed to reduce overhead by improving firepower with increases in machine gun troops and mobility by motorization. Undoubtedly, he was concerned about the future role of his branch. He believed some form of mechanization would create a greater demand for cavalry. As the army grappled with the issue of motorization and mechanization, a climate developed that began to challenge many of the cavalry's traditional missions. The features of Crosby's new skeletonized peacetime division of about five thousand men called again for one organic air observation squadron for long-distance reconnaissance, a photo section, an armored car troop, an increase in the firepower of the machine gun troop, and the possible addition of an attached light tank company operated by cavalry personnel. Crosby did clarify that the tanks were infantry weapons as the "cross-country commercial armed cars were cavalry weapons." Cross-country cars were substituted for motorcycles. He consented to furthering the augmentation of the scout car, which at first led to the development of commercially built vehicles, then to more powerful and heavily armed armored cars. The M1917 tanks, however, demonstrated that they were too slow for cavalry operations. In addition, partial motorization of the division trains, quartermaster, and motor repair and transport elements were planned. Also under consideration was organization of a cavalry corps for strategic reserve capable of operational mobility with increased firepower. The primary element was the introduction of commercial light and medium armored cars that were used for the first time by the 1st Armored Car Troop in July 1928.[69]

Little did Crosby realize that his reorganization would begin the process of eliminating the horse by gradually increasing the firepower of the mounted attack. It set a trend for the cavalry, which now had more warfighting equipment to fight mounted and dismounted, thus lowering the utility of the mounted attack with pistols. The introduction of the commercial cross-country armored car would have an important impact on the future of reconnaissance as the dominant mission for mechanized cavalry. Certainly, Crosby made many of his peers take notice that mechanization was the cavalry's future.

While Crosby was struggling with a new organization, Major H. H. Arnold, a recipient of the Mackay Trophy before the war for successfully completing a reconnaissance flight, wrote an attention-grabbing article in the January 1928 issue of the *Cavalry Journal*. Arnold, a future proponent of strategic airpower and a separate air arm, argued for greater cooperation between the air service and cavalry. Like Fickett's award-winning study in the *Cavalry Journal* in 1923 and Van Voorhis's request for air reconnaissance, Arnold saw at the time the value of aircraft for the cavalry and praised the Air Corps for its willingness to work with the troopers at Fort Riley. Aircraft, he wrote, could be the cavalry's best friend; however, it also could be a dangerous enemy. Arnold was referring to low-flying attack aircraft successfully engaging cavalry riding their horses. He argued that cavalrymen should be closely associated with observation and attack planes, especially the latter. This reasoning was prompted by the recent developments in air tactics. Arnold saw the importance of airpower operating against hostile cavalry, machine gun nests, and heavy points of resistance. Bombardment aviation was a weapon of destruction and thought to be of little value to cavalry operations. Pursuit planes were expected to clear the air of hostile planes. He concluded by encouraging a strong cavalry–Air Corps team that understood each other's missions.[70] Unfortunately, the country's sleepy acceptance of isolationism and the ongoing efforts of the Air Corps to gain autonomy and overall control of military aircraft—along with the cavalry's dedication to and ongoing love for polo, ritual, and ceremony—precluded the army from realizing the importance of developing a theory and practice of tactical airpower.

More controversy occurred as the army continued its turbulent debate over the role of mechanization. The issue was over who controlled the army's future. It began when the British Mechanized Force, established in 1927, impressed Secretary of War Dwight D. Davis. As a result, he made an important decision. He directed the chief of staff, General Summerall, to initiate a preliminary study on the employment of a mechanized force on the future battlefield to determine how the United States could effectively prepare for such a national contingency. Thus began the army's first experimental test force in the twentieth century to bring into being the embryo for a new doctrine. This event challenged the infantry's self-proclaimed right to be known as the dominant warfighting arm.

Subsequently, Summerall assigned the challenging task to Parker, who became assistant chief of staff, G-3 (Operations and Training). Parker in turn directed his G-3 section to study the application of mechanization. By the end of 1927, Parker had completed a preliminary study that proposed placing the mechanized force under the chief of cavalry. This idea, however, was stricken because Crosby was more concerned with his upcoming maneuvers in 1928. In addition, Brigadier General Campbell King, the G-1 (Personnel), objected, arguing that line branch chiefs should cooperate. He believed tanks, artillery, and infantry should form the backbone of a proposed mechanized force.[71]

On March 20, 1928, Summerall was presented with another Parker study, "A Mechanized Force," which outlined its employment in the future battle. The essence of this study defined the advantages of mechanization as increased mobility, protection, firepower, and offensive power that decreased casualties. The tactical needs for a self-contained mechanized force were spelled out as being solely offensive. These units would form the spearhead of major attacks whenever the terrain permitted. Also, the mechanized force would be useful as strategic advance guards of large forces for deep penetration. Moreover, their mobility could enable the force to rapidly seize and temporarily hold distant positions, along with the capability to provide flank guard. Of importance, the study mentioned that a mechanized force be considered a corps, thus providing army units with sufficient size and penetrating power to fight beyond the enemy's corps artillery and deny the use of his reserves.

General Parker proposed the tank as the key maneuver element, stating, "The tank has pointed the way . . . [and forms] the backbone of mechanized forces." The study also called for self-propelled artillery, mechanized infantry, air support, and armored cars to protect flanks as the attack progressed ahead of the main force. Parker's study had little to say about the mission of the cavalry other than that the force nearly paralleled the employment of traditional cavalry but the meaning had changed. Parker, however, stated that the proposed mechanized force was neither cavalry, infantry, artillery, nor any other branch; it was a separate entity. Its tactics, Parker claimed, might be considered those of armored cavalry. He believed the "cavalry's viewpoint and characteristic of mind" that implied bold and aggressive leadership was essential for the proper development of the force. Parker agreed that infantry did require accompanying tanks; however, he noted that tanks should not be viewed solely as infantry weapons.[72]

The most innovative conclusion from Parker's G-3 study called for a mechanized force of combined arms organized at corps level with tanks constituting the main maneuver element in deep offensive operations. This was a significant suggestion, since it moved the idea to an operational level of warfare, a series of proposed tactical events with a combined arms mechanized force. It was almost a replay of Parker's August 1917 ideas: calling for a tank-driven, self-contained mechanized force with self-propelled artillery and motorized infantry supported by attack aviation. These ideas were reflected again in his 1928 G-3 study. It was a radical plan, calling for a balanced combined arms grouping of tanks, self-propelled field artillery, mechanized infantry, engineers, air support, and a service detachment. The study viewed the proposed mechanized force as demonstrating more speed and mobility when compared to the pulverizing strategy of attrition warfare and infantry tactics of open warfare, the doctrine of maneuver and rifle firepower. By suggesting a force structured upon a combined arms organization, Parker's study was the first rational attempt challenging the autonomy of the postwar combat branch chiefs. As before, Parker attempted to move the army to an operational level of mechanized warfare and away from the strategy of attrition warfare.

General Parker's study created a heated debate between Summerall and his staff, the chiefs of infantry and field artillery, and the chief of the War Plans Division. The issue was how to organize and equip a mechanized force. To resolve the issue, Parker tasked Major Adna R. Chaffee Jr., an imaginative officer assigned to the G-3 Training Section in June 1927, to draft a memorandum to Summerall. The memorandum, "A Mechanized Unit," was dated April 2, 1928. Of importance was Parker's perception of the future warfighting potential of mechanization and his ability to harness the talents of Chaffee toward that goal. This action was a major contribution to the secretary of war and the chief of staff's directive on mechanization.

Key to Parker's goal was Chaffee, a loyal cavalryman who initially wanted to revitalize the horse cavalry. By 1928, Chaffee had realized the importance of mechanization and deep operations in a future war. He sincerely admired James Harrison Wilson, a noted Civil War cavalry officer. Wilson's deep raid through Alabama and Georgia with the largest mounted force assembled during the war struck at the economic and industrial base of what remained of the Confederacy in the western theater at the end of the war. Wilson also had neutralized one of the Confederacy's top cavalry raiders, Nathan Bedford Forrest. Most of Wilson's troopers were armed with repeating Spencer carbines. Usually his method of fighting was to move rapidly to the scene of battle and then dismount and fight as infantry. Wilson was a strong proponent of operational mobility and total war. In his eulogy of Wilson in the July 1925 *Cavalry Journal,* Chaffee concluded, "He knew the great value of firepower, and how to combine it with rapid movement, and with mounted assault when practical." Continuing, "Let's hope that the powers that be will assemble the biggest mounted force they can every year and order us all there to learn the things that Wilson knew so well and that he had rapidly learned in the hard school of war."[73] One of Wilson's close friends and a cavalry division commander during the celebrated raid, Emory Upton, became a pivotal figure in American military theory and during the war he commanded in three line branches—artillery, cavalry, and infantry. Upton said of Wilson, "I would commute the rest of my life for just six months of such service."[74]

Historian John T. Broom persuasively argued that Wilson's type of warfighting influenced Chaffee, who in the late 1920s began developing his own theory of mechanized warfare and deep operations. It was Wilson's operations in Alabama and Georgia, Broom suggested, that formed the basis of Chaffee's vision for mechanized cavalry. Broom maintained that Chaffee was a loyal cavalryman and a product of the nineteenth-century army's experience in the Indian wars and the Civil War, an experience he absorbed through his father. He then applied new weapons of the twentieth century to mechanize the cavalry. The American Expeditionary Force introduced again the application of direct power through the application of attrition warfare inherited from Grant's Overland Campaign in Virginia. The use of accompanying tanks suited the infantry in the application of attrition warfare. Chaffee, however, tried to reconcile the issue of attrition by exploiting and destroying the enemy's command, control, and logistical network similar to Sherman's and Wilson's army-sized raids through Georgia, Alabama, and the Carolinas as opposed to Grant's methodical attrition campaigns in Virginia in 1864 and 1865.[75]

Like Robert Grow, an emerging and inspired forward-looking cavalry officer who also studied history, Chaffee was familiar with the award-winning book by George T. Denison, *A History of Cavalry* (1877). Denison challenged the prejudice among professional cavalry officers for not using firearms instead of holding on to the traditional lance. According to Grow, one sentence that influenced him, which he and Chaffee talked about later at Fort Knox, was, "A cavalry general should be possessed of a strong inventive genius, and be self-reliant to strike out a new line and adopt reforms where he sees them necessary."[76]

Meanwhile, in an attempt to head off a branch debate, a paper, "A Mechanized Unit," was prepared, addressing the problems and concerns regarding Parker's "A Mechanized Force." In general, it met approval from the secretary of war, Summerall, the assistant chiefs of staff: the G-1, General King; the G-2, Colonel Stanley H. Ford; the G-4, Brigadier General Ewing E. Booth; and the chief of the War Plans Division, Brigadier General George S. Simonds. General King, who had objected to placing the development of mechanization under the chief of cavalry, suggested this concept be first tested through an exhaustive series of war games at the Army War College. The chief of artillery, in turn, was concerned about whether his arm ought to use self-propelled or tractor-drawn mounts. Generally, General Crosby accepted Parker's paper, as did the other branch chiefs—except for the chief of infantry, who vehemently opposed setting up another branch with tanks as the focus. General Allen recommended that tanks remain with his branch, that the development of armored cars remain with the chief of cavalry, and that the development of self-propelled artillery remain with the chief of artillery. Allen reasoned that the tank was specifically introduced to bring back the infantry's tactical mobility by providing fire and movement at a critical stage in the attack. He also objected to Parker's using the phrase "cavalry viewpoint and characteristic of mind," which suggested bold and aggressive leadership. These traits were by no means limited to cavalry, he argued.[77]

Parker responded that if the infantry continued to control tanks and tie them to the speed of foot soldiers, it could obstruct the creation of a new mobile warfighting doctrine driven by a mechanized force. He believed that history had proved that nations survive only if they keep ahead of their country's enemies in quality as well as quantity in their fighting men and their warfighting equipment.[78]

Secretary of War Davis approved in principle Parker's "A Mechanized Force." Subsequently, the War Department, based on Parker's recommendation, made the next move. It resolved the issue of mechanization by appointing on May 10 a board of officers "to make recommendations for the development of a mechanized force within the army and to study questions of defense against such forces."[79] In addition, an organization was to be established within the existing strength of the army without appealing to Congress for new legislation and funding. The board was composed of eleven members representing the General

Staff, and all combat and service arms. The board was expected to complete its findings by October 1. One of the key members was Major Chaffee.

While the board was dealing with all issues regarding mechanization, an Experimental Mechanized Force (EMF) referred to as the "Gasoline Brigade" was assembled early in July at Fort Meade, Maryland, under the command of infantry officer and former commandant of the Tank School, Colonel Oliver Eskridge. Its mission was to develop and test tactical and strategic capabilities for spearheading a deep operation, disorganizing the enemy's command system and reserves. In addition, the EMF was tasked to study and test organization and equipment, and to determine the limitations of motor vehicles as compared to those that were animal drawn. The main element of the force consisted of a light and a medium tank battalion equipped with re-engined M1917s, Mark VIIIs, and a new light tank, the Cunninghams, which were issued to the 4th Light Tank Company of the EMF. The M1917 carriers were World War I Liberty trucks referred to as "mechanical bastards of the first order." Added was a motorized infantry battalion mounted in World War I Liberty trucks, a field artillery battalion porteed in Liberty trucks, an engineer company, and an antiaircraft artillery detachment. In addition, the chief of cavalry reluctantly tasked to the force an armored car troop using commercial "cross-country cars." Unfortunately, the force lacked tactical balance because it was plagued by obsolete and insufficient equipment, making it difficult to determine proper combat methods for road travel. The force was more motorized than mechanized, making the tactical field test lessons unclear, especially during mock battles and sluggish marches over Maryland and Pennsylvania roads. In spite of numerous handicaps, the motley force did struggle to complete its short mission, providing helpful information on offensive operations and technical matters.[80]

In the final report, Colonel Eskridge admitted the critiques following the technical and tactical field exercises brought out a great number of concerns. In fact, he noted, the experiment was pitiful. World War I tanks were too slow, as was the motorized force that lacked road reconnaissance for convoys. These conditions made the road-bound force more vulnerable. The EMF lacked highly specialized command and staff training. This deficiency resulted in a lack of command and control for a fast-moving force. As a result, a better system of command and control was suggested. What Eskridge failed to mention was the problem of quickly bringing forward groups of infantry under suitable armor protection to consolidate positions taken by attacking tanks. The exercise also demonstrated the need for aircraft for security and reconnaissance. Finally, it was noted that the exercise opened a new field of study on how to supply new and faster vehicles, fuels, spares, and rations. The experience with the EMF, however, had not resulted in the adoption of a new doctrine outside the control of the combat arms chiefs.[81]

The Cunningham tanks, however, were enthusiastically received by most of the officers, including Colonel Eskridge. On the initiative of the chief of ordnance, an Ordnance Advisory Committee of the Society of Automotive Engineers had earlier designed a dual-purpose vehicle, a two-man tank and a cargo carrier with the engine in front to take advantage of a better center of gravity and placement of armaments. The chief of infantry, obsessed with lightness, was for a light tank and therefore approved its design as an accompanying tank. There was neither cavalry input nor interest in the development of the Cunninghams for the mounted force. On June 9, 1928, the War Department Mechanization Committee tentatively approved the tank. On the committee was a General Staff officer representing the cavalry, Major Chaffee, who apparently reserved comments on the infantry tank.[82] Perhaps he saw no use for it in the cavalry.

In September, Parker, who believed the EMF provided considerable information and experience, recommended to Summerall that the force be dismantled as planned. It was evident command, communication, and control needed to adjust to the speed of the vehicles, their deep penetration, and wide areas of operations. Accordingly, Parker proposed a permanent, self-contained, and highly mobile offensive force capable of independent action.[83]

On October 1, the Mechanized Board, as part of its mission, visited the Experimental Mechanized Force. Summarizing its report, the board stated that tanks should be the backbone of the attack, while the infantry should be carried forward and protected by mechanical means. In addition, the infantry organization should be built around more firepower, with additional machine guns and semiautomatic weapons. Regarding artillery support, the board suggested self-propelled mounts. The board emphasized that effective cooperation with the Air Corps was essential for ground attacks and reconnaissance. The cavalry, it was agreed, had sufficient mobility to advance through light resistance and hold ground if necessary. It was expected that a large number of automotive horse transports would be required for employment to secure an adequate holding position. In addition, it was noted that the cavalry did not have the necessary march speed to be incorporated in a mechanized force. What became disturbing to the chief of infantry was section III, "General Discussion of the Organizational Needs of a Mechanized Force," which suggested that the cavalry was more suited to lead a mechanized force because of its tactical mobility.[84] A second experimental force was assembled at Fort Meade in 1929, which led the War Department to recommend instead that a permanent mechanized force be organized.

Because of the board's report and Parker's recommendations, Summerall moved to approve the establishment of a permanent mechanized force. Congress, however, was in no mood to provide funds for a mechanized force in the

1930 budget. Consequently, a study was immediately undertaken, resulting in a recommendation to include provisions for a mechanized force in fiscal year 1931. The secretary of war and Summerall were notified that at least for the next two years, all experiments and training intended to keep abreast of developments in mechanization had to be conducted with troops, materials, and facilities currently available.[85]

How did the chief of cavalry deal with the issue of motorization and mechanization for his branch? General Crosby found armored cars useful for reconnaissance and communication purposes, though he questioned their off-road capabilities. Armored cars, he held, depended more on speed for protection than on the thickness of their armor to deflect hostile fire. Crosby believed there was a growing appreciation for cooperation between the mounted force and the use of airpower for reconnaissance to extend the view of the battlefield. On the issue of tanks, he admitted that fast light tanks might on occasion be useful for a cavalry division when dealing with rough terrain. He acknowledged that tanks were infantry weapons just as armored cars were cavalry weapons. Regarding mechanization, the chief of cavalry believed no machine now or in the future could replace the horse cavalry. Nor could any machine ever replace the foot infantry. More specifically, he argued that tanks, armored cars, airplanes, and all means of warfighting could not take the place of man and horse. The following year the 1st Cavalry Division Board at Fort Bliss, Texas, determined that the medium armored car was more promising than the light or even a heavy armored car. Because of a lack of funds to acquire additional armored cars, efforts were made to experiment in overland tactics with the existing medium armored cars. It was then suggested that the armored car troop continue experimenting with this vehicle.[86]

General Crosby had reason for concern about the future status of his branch over the issue of mechanization, which he believed would create a greater demand for cavalry operations. The chief of the automotive section in the Office of the Chief of Ordnance, Major Levin H. Campbell Jr., wrote, "The horse does not possess either the power or the rate of speed desirable in a modern army." The trend of thought, he believed, was more toward the use of motor-propelled vehicles. Though the army might still use horses, mechanization and motorization were steadily circumscribing their usefulness, he held. Campbell based his argument on the fact that the United States possessed the largest automotive industry in the world and therefore the army must take advantage of this favorable position. During the 1920s America made approximately 85 percent of the world's passenger cars; by 1929, it was producing almost 5 million cars a year. Campbell concluded by suggesting a new organization capable of working out suitable tactics for deployment of motor-driven vehicles of great mobility.[87]

Captain John K. Christmas, also from the automotive section, wrote that the basic elements of a mechanized force were artillery, airplanes, petrol, and tanks. He received a degree in mechanical engineering from Lafayette College and pursued postgraduate work at the Massachusetts Institute of Technology and at the Ordnance School in ordnance engineering. He would become one of the Ordnance Department tank experts. Christmas recalled that during the last war, tanks were merely regarded as infantry auxiliaries. This view, he wrote, now had changed, making the tank the basic element of the mechanized force. He believed the army moving to a mechanized force provided the mobility and tactics of the cavalry without the actual designation of cavalry.[88]

Lecturing at the War College in September 1928, Parker again stressed that war was becoming increasingly mechanical. He strongly supported the establishment of a permanent mechanized force because the tank—as the main maneuver element with supporting airpower and artillery—could provide means for restoring mobility to the battlefield. Considering the army's current inventory of tanks, he emphasized that their role in mechanization was impractical because of age, size, and unwieldiness. Parker proposed the gradual phasing out of the army's World War I equipment because it was no longer suitable for the modern battlefield.[89] In December, he wrote the commandant at the War College that little had been done to advance mechanization since the war. Because tactics depend on armament, he wrote, it behooved the college to introduce situations that define the role of mechanization in a future war.[90]

At the time Parker and Chaffee were crystallizing a combined arms concept, similar thinking was reflected in tactical lectures given by Captain Walter McAdams at the Fort Meade Tank School before the school was moved to Fort Benning. One of the attendees at the last course of 1928–1929, West Point graduate Sidney R. Hinds, recalled the enthusiastic McAdams being one of a few infantrymen who "kept alive the search for a better idea." Sid Hinds claimed McAdams and a few other progressive instructors at the school were already reflecting the combined arms thinking of Fuller and B. H. Liddell Hart, the latter a military theorist who proposed the mobile tactic of the "expanding torrent" as opposed to linear thinking. Later, historian Martin Blumenson maintained that after the world war, Patton wrote in essence on the concept of the expanding torrent and the importance of tank penetration tactics that called for widening the hole and moving to secure flanks in depth if necessary. In addition, Liddell Hart advocated the strategy of the "indirect approach," which called for deep penetration by fast-moving armored forces against the lines of least resistance as opposed to a direct frontal attack. Like Fuller, Liddell Hart argued for a form of lightning war consisting of tanks, infantry, and airpower. The lectures on "a better idea" by McAdams, according to Hinds, influenced a few cavalry officers who attended the school, such as Charles Unger and C. C. Benson—all of whom later played a role in mechanization of the cavalry.[91]

While the Tank School was in the process of moving to Fort Benning, where tanks now came directly under the Infantry School and its linear thinking, Chaffee delivered a famous lecture at the Army War College in September 1929 entitled, "The Status of Mechanized Combat Organization and the Desired Trend in the Future." It was an elaboration on "A Mechanized Force," in which Chaffee had made a major contribution per instructions from Parker. Chaffee stated that the principal conclusion reached from the Experimental Mechanized Force was that little could be accomplished tactically unless the army had equipment that was modern. More exactly, the argument was for the army to devise a plan and stick to it until modern equipment was acquired. The G-3-driven studies and Chaffee's lecture raised an important, persistent question: Which came first, doctrine or equipment?[92]

Chaffee emerged as another outspoken advocate of expanding the role of tanks. He also questioned the infantry's tactical doctrine of open warfare and the reliance on rifle firepower, arguing that infantry tactical movements were exceedingly slow and costly. Therefore, he noted, new methods must be devised ensuring mobility without the disastrous losses experienced during the last war. One of the problems, he maintained, was that tanks were tied to the slow pace of the infantrymen. He held that the infantry, as a vital component of the mechanized force, needed to organize and have a marching speed equal to that of tank elements. For greater balance and supporting power, he made a case for self-propelled and tractor-drawn artillery. Once a balance was established, the mechanized force could be organized as a self-contained, highly mobile, mechanized unit with great striking power but limited holding power. The mechanized unit, he told the attendees at the War College, should develop the capability for rapidly seizing and then temporarily holding distant key positions. His idea of such a unit was a corps with a self-contained mechanized force, capable of penetration or envelopment in a deep operation, thus facilitating a decided advance for the corps as a whole. Here Chaffee, like Parker, was reaching out to a new art of war, an operational level to fight and win a campaign with a series of tactical events. One area that Chaffee failed to consider in his lecture was the role of tactical air-ground support for the mechanized force.

Major Chaffee, however, was careful in treading on the autonomy of the line branch chiefs. He had not argued for a separate branch for a mechanized or armored force. Rather, he called it a grouping of troops from all the branches in combat vehicles under a new tactical idea and a new system of tactics. Yet he did add that the idea must be as elastic as cavalry, a branch instilled with the spirit

(Above) The Christie M1928 convertible chassis with the vehicle operating on tracks during cavalry and infantry tests. From the Hunnicutt Collection, Patton Museum. *(Below)* Christie M1928 convertible chassis with the cavalry for a brief test, May 1929. Note: tracks were removed and machine gun mounted above the chassis. From the George S. Patton Papers, courtesy of the Library of Congress.

of mobility, rapidity of action, and simplicity of control. What type of soldier did Chaffee have in mind to execute the new idea? His answer was a progressive leader with a creative mind and one not afraid of radical changes. This had not set well with the chief of infantry.

Benefiting from the influence and support of Parker, Chaffee became one of a number of visionary thinkers. To restore mobility to warfare, Parker and Chaffee argued for introducing a new doctrine that would bring the army into the future. This doctrine involved rapid and deep attacks by a balanced combined arms team built upon the backbone of the main fighting element, the tank, supported by mechanized infantry, self-propelled artillery, and tactical airpower. As noted, the new doctrine provided an opportunity for moving to an operational level of warfare, a theory of larger unit operations with a mechanized force capable of independent maneuver. This was a problem. The line branch chiefs were not willing to move in this direction because it challenged the security of their compartmentalized organization.

In the fall of 1928, a major event occurred that would cause considerable friction over the nature of mechanized warfare, distressing the elitist's culture of the line branch chiefs for years. What followed in the 1930s was an example of the folly of pursuing cutting-edge tank technology at the expense of down-to-earth practicality, which was hindered by the compartmentalized organization of the branch system. The new Christie high-speed convertible chassis, the turretless M1928, was unveiled. At first, it was called an armored car, not a tank.[93] With the introduction of the M1928, the whole issue of a tank policy became obscured due to bickering between the Ordnance Department, the infantry, and the cavalry. It set in motion a major debate over who would direct a future tank policy. The ensuing argument over who would acquire the Christies launched a long struggle over how to tactically use the vehicles by either the cavalry or the infantry and who would determine the technical merits of the vehicle. Unfortunately, this organizational debate brought out in great intensity the self-serving antagonism of the two line branch chiefs and the technical service. The debate was ongoing until World War II. For the cavalry, it brought into question who should develop and tactically control new warfighting equipment, such as tracked vehicles, and the role of horses in modern warfare. This intense internecine conflict stymied development of a centralized combined arms doctrine for the cavalry, which was still tied to the importance of the horse in modern warfare.

J. Walter Christie's unofficial demonstration at Fort Myer in October 1928 impressed Summerall, who circumvented the Ordnance Department and verbally ordered the Infantry Tank Board to test the chassis. In February 1929, the War Department approved a request by Major General Clarence C. Williams, the chief of ordnance, in cooperation with the chief of cavalry, to procure the M1928 as an armored car equipped with .50- and .30-caliber air-cooled machine guns. At the time, the Ordnance Department had $15,000 budgeted under the upcoming 1930 ordnance appropriations for two armored cars for the cavalry.[94]

The U.S. Cavalry was aware that foreign armies during and since the last war had devoted considerable attention to the development of armored cars for use with their horse cavalry. Prior to 1928, the U.S. Army had expended no funds for armored car development. At most, several types of improvised armored cars were produced using commercial chassis. By 1928, General Crosby became serious about developing a cavalry armored car able to extend reconnaissance and establish contact with a hostile force far beyond the capabilities of horse-mounted troops. Though the feeling was that armored cars could never replace the horse, they were expected to play an important reconnaissance role in the cavalry's future.[95]

After the dramatic demonstration of the M1928, Crosby opted for the Christie chassis, providing necessary cavalry requirements for armor, armaments, ammunition, accessories, and suitable crew space and overhead protection. These requirements were made before the cavalry had the opportunity to test the chassis. However, Crosby did request the Christie chassis be equipped as recommended and then subjected to shop tests under the supervision of representatives from the Ordnance Department and cavalry. In June 1929, the Infantry Tank Board loaned the M1928 to the cavalry representatives at Fort Meade for a few weeks.[96] Tests of the Christie chassis occurred at a time when the cavalry was anxious to evaluate armored cars and was considering the incorporation of an infantry tank platoon for duty with the 1st Cavalry Division.

Competition over the Christie vehicles now began in earnest between the cavalry and infantry. Though the vehicles proved less than satisfactory, the Infantry Tank Board recommended to the new chief of infantry, Major General Stephen O. Fuqua, the adoption of the Christies as standard and included them in the current production program for manufacture of five or six tanks. Fuqua was a die-hard, traditional infantryman who relied on the open warfare doctrine generated by Pershing during and after World War I. His concern was how to adapt Christie tanks to open warfare in support of the infantry attack. Sid Hinds, an infantry tanker who witnessed the tests, echoed the good-show emotional attitude Fuqua possessed: "Good god, the man [Christie's driver along with his companion] doesn't see the logs! Merciful Heavens, they will all be thrown out and killed." Hinds claimed the Christie vehicle "oozed over the logs with hardly a ripple except for the beautiful snake-like action of the wheels and tracks with the occupants in the open body nonchalantly undisturbed and not

Major General Stephen O. Fuqua, chief of infantry, March 28, 1928, to May 5, 1933, and military liaison to Republican Spain during the civil war, 1936–1939. Courtesy of *ARMOR* magazine.

even hanging on to a handle."[97] The noted late tank historian Colonel Robert J. Icks recollected Fuqua telling him he wanted the fast Christies to serve strictly as infantry-accompanying vehicles.[98]

The cavalry expressed at the same time an interest in acquiring one chassis for additional testing as an armored car for reconnaissance. Per instructions from General Fuqua, Colonel H. L. Cooper of the Tank Board sent a letter of endorsement to the chief of cavalry, General Crosby, for comment. Crosby sent the endorsement to the Cavalry Board and requested Major Patton to reply. Board members, especially Patton and Major C. C. Benson, saw a potential for the Christie vehicle for cavalry operations. As a result, the vehicle was seen as logically compatible with cavalry doctrine because it provided greater mobility, firepower, and shock action. Patton found the vehicle promising; however, he notified Cooper that he found a few limitations and felt that the inventor overstressed the speed of his vehicle with his usual "histrionics." Nevertheless, Patton supported purchasing the chassis as an armored car. His comments to the Cavalry Board were that there existed some uncertainty regarding the possible purchase. This was due to "the fact that in the Christie car we are buying a principle not a vehicle." Furthermore, he noted, "It was unwise at the present time to say more of the Christie car than that it is a combined wheel and track vehicle capable of high speed across country and on the road." His concern was that the chassis was a commercial product and, therefore, expensive and slow to manufacture.[99]

Though Patton was impressed with the Christie and mechanization, he believed at the time that "Oil and Iron do not win battles—Victory is to men not machines." His argument was that machines could not operate in certain terrain and in bad weather. Patton was for motorization and mechanization as long as they did not compromise the horse. According to his biographer, Martin Blumenson, this attitude eventually led to using machines with horses, which occurred during the 1934 cavalry maneuvers. Blumenson also suggested that Patton's interest in the Christie vehicle prompted him to privately subsidize J. Walter Christie. However, Patton's son claimed his father never gave money to Christie, adding that he met the tank designer when he was a young boy at their home on Cleveland Avenue in Washington, D.C., "once when Christie was sober and the other time he was drunk."[100]

While the infantry and cavalry were testing and debating the merits of the chassis and suggesting modifications, negotiations began with Christie to purchase the vehicle for the cavalry. The cavalry was the first to formally request procurement of the chassis. The negotiations lasted for months. At first, the tank designer demanded $82,500, which included his development costs. This amount was more money than the Ordnance Department had in its budget for cavalry appropriations for an armored car. Finally, in October, Christie offered to sell his "National Defense Machine," as he called it, for $26,500 providing additional vehicles were purchased.

Shortly the chief of infantry requested, as did General Williams, that $250,000 appropriated for Cunningham light tanks be used instead to purchase Christies. The infantry was not impressed with the ordnance-designed Cunninghams because of the suspension system and the vehicle's terrible vibration. This vibration presented a serious problem for the gunner, who found it difficult to keep his eye on the sight. Consequently, many infantrymen from the Tank School had "gunner's black eye." One sergeant described the Cunningham as "the finest sports model tank in existence and can run away from any of 'em; the only trouble is that the driver can't see where he is going and the gunner can't stick on his seat to shoot." In addition, the Cunningham did not shield the driver from the front-mounted motor's smoke and mist. One reason the infantry rather than the cavalry had an interest in the Cunninghams was because of General Allen's obsession with light tanks to accompany foot soldiers across roads and bridges, as they existed in the United States. When General Fuqua became chief, he wanted at least five Christie-type tanks because that number constituted the infantry's smallest tactical unit, the platoon. He was also interested in tactical tests to enhance command, control, and communication between foot soldiers and tanks, especially with the fast Christies. By the end of the year, General Williams was asking the Subcommittee of the House

Committee on Appropriations for funds to procure at least five Christies for the infantry and one for the cavalry.[101]

In the meantime, a cavalry officer, Major C. C. Benson, assigned to the Historical Section, General Staff, had enthusiastically endorsed mechanization and the new Christie convertible chassis through his writings. He commented in the *Cavalry Journal* that mechanization and cavalry were a natural combination and bemoaned the preoccupation of his branch's love of the horse.[102] In the *Infantry Journal*, Benson argued that tanks would change warfare, and their presence would necessitate drastic changes in current infantry and cavalry tactics.[103] He referred to the Christie chassis as the "Model 1940" because he believed it was ten years ahead of its time. It had "the agility of a wildcat," he wrote. Benson held that the Christie could provide a valuable service with the cavalry for exploitation, pursuit, and raids. His overenthusiastic article centered on the vehicle's strategic and tactical value due to the wheel-track concept. The vehicle, as in Christie's earlier models, professed to use its wheels for strategic deployment and its tracks for tactical employment, especially in rough terrain. The most innovative feature was its suspension system, which employed four large rubber road wheels on both sides of the chassis, each mounted on pivoted arms, which bore a long, adjustable coil spring. This considerably increased the speed and provided a more stable gun platform than in any previous tank developed, according to Benson.[104]

General Williams, however, was not in agreement with Benson because the M1928 had not undergone the required ordnance service tests. Benson's zealous interest in the M1928 and mechanization also had ramifications in Congress. Williams told the Subcommittee of the House Committee on Appropriations in December 1929 that Benson's articles lacked information on mechanical performance tests expected of a new vehicle. Furthermore, the chief of ordnance noted that the War Department would not be satisfied with testing just one vehicle, which Christie had subjected to numerous flashy demonstrations in and around Washington. Though the M1928 made an impression on the infantry and cavalry, Williams told the committee members about the potential mechanical problems that developed after being tested for only a short period. He recommended a more thorough ordnance proving ground test.[105]

Throughout the cavalry and infantry tests during 1929, the M1928 went through many modifications at Christie's plant in Rahway, New Jersey. The original weight went from 5 tons to 8 tons mainly due to correcting numerous mechanical failures. The lightly armored vehicle still had neither a turret nor a suitable driving compartment for combat operations. Christie was adamant that it remain this way. He strongly disagreed with putting a turret on the vehicle or any subsequent vehicle. He viewed a turret as not desirable for his high-speed machine. In addition, his asking price was too high and surpassed the War Department's annual budget for the cavalry. Subsequently, Christie claimed he scrapped the M1928 beginning in September 1929 and started a new design recommended by the Tank Board. Finally, because of congressional hearings in December 1929 over funding and General Williams's comments, the decision was made not to purchase the M1928 model.[106] The following month, the secretary of war, Patrick J. Hurley, closed the debate, withdrawing the procurement of the M1928-type chassis for the cavalry because tests by the cavalry and infantry were inadequate. What irritated many who dealt with Christie was that he refused to sell his chassis on what they considered reasonable terms, especially when he asked for additional funds to compensate for years of developmental expenses.

While the debate over the Christie chassis was going on, now Major General Parker found the M1928 concept a solution for fast and deep mechanized operations. Since his experiences in World War I and discussions with Fuller in 1917 and as G-3, Parker had harbored the idea of moving a mechanized maneuver-oriented dogma to a new level of warfare. Parker, however, again worked himself into a controversy when he attempted to get the army thinking about an operational level of war. As Summerall had ordered, the assertive G-3 endeavored to supplement the *FSR 1923: Operations*. This regulation reflected the primacy of the infantry's tactical doctrine of open warfare and relegated all other combat arms to auxiliaries. *FSR 1923*, according to Parker, was vague in that it made only minor distinctions between small and large units. To fill the void, an instructional manual for commanders of large units, corps and higher, was prepared. It reflected the 1924 manual, which was a translation of French postwar doctrine on larger unit operations. The new manual was immediately challenged when commandants of the War College and the Command and General Staff School disagreed. They argued it was cumbersome and had again been plagiarized from the French manual for larger units that prescribed a rigidly controlled operation called the methodical battle.[107] Summerall supported his G-3, and after a review it was released in May 1930 as the *Manual for Commanders of Large Units (Provisional)*, vol. 1, *Operations*. This provisional manual was the first effort to articulate a doctrine for operations of larger units in the field.

Shortly before the provisional manual was released, Parker was ordered to Chicago, where he took command of the VI Corps and, in 1932, the Second Army. This move ended his activism for a mechanized maneuver–oriented force. He had alienated many in the infantry community, especially the successive chiefs of infantry, who viewed him as a threat to their autonomy. It is important to note that Parker launched Chaffee, who along with other proponents for new mounts continued the struggle for a mechanized cavalry into the 1930s. Frank Parker saw beyond the tactical level to an operational level of warfare.

He understood the importance of mechanization at a level of warfare where campaigns and major operations are planned. More so, Parker conceptualized beyond a broader dimension of time and space to a series of maneuver-oriented operational events rather than at the tactical level of battle. He set the tone for today's armored cavalry. The chiefs of infantry and cavalry had not accepted this because it would depend on cross-coordination of their line branches. It was typical of branch chief intransigence. They marched to their own drums. They saw the little picture because they were more concerned with retaining a tactical orientation for their branches that focused on traditional doctrine emphasizing Pershing's idea of infantry-dominated open warfare and rifle firepower. In addition, reduced budget appropriations and an indifferent public during the interwar period cast a cloud over future military innovations in equipment and doctrine. This also caused the branch chiefs to continue defending their compartmentalized organization as established by the 1920 National Defense Act and *1923 FSR*. By 1930, it was evident that the branch chiefs were too embedded in compartmentalization to support the ideas of a few forward-looking officers. Branch autonomy and a Western Hemisphere defense mentality thwarted any effort to move doctrine to an operational level to deal with a major war. It was impossible to provide a new, dynamic way of mechanized warfighting in deep operations, as suggested by Parker. The war experience and devastating casualties had produced a strong antimilitarism attitude. The United States was geared to idealistic normalcy and isolationism because of the expectation that the war of 1914–1918 was the war to end all wars.

3

The Struggle for an Innovative Doctrine and a Combat Car

By the end of the decade, the legitimacy of mechanization for the army as a whole was questioned. During Calvin Coolidge's presidency, the country experienced general prosperity, which meant reducing federal spending, lowering taxes, and balancing the budget. In the year 1928, the Kellogg-Briand Pact was signed by twenty-three members of the League of Nations renouncing aggressive war; however, it made no provision for sanctions. In March 1929, Herbert Hoover became president and assumed the responsibility—as commander in chief—for national defense. His propensity for an orthodox economic theory calling for a balanced budget drove his policy on national military strategy, which focused on defense of the Western Hemisphere. The feelings of the president and Congress were that the two large oceans, the Atlantic and Pacific, were the country's major defensive barriers against foreign powers. Furthermore, the national attitude was that distance was the country's greatest defensive weapon. The Monroe Doctrine of 1823, which called for the end of European colonization of the Western Hemisphere, sustained this belief. As a result, the president believed the country's principal military needs for a national military strategy were a navy and an army air corps. Hoover intended the army ground forces to remain skeletonized, relying on quick expansion in time of war. He insisted on two changes regarding the army. First was the conversion of the cavalry to a mechanized force. He believed the cavalry was too conservative and obstinate when it came to change. The president criticized the cavalry for believing "battles could not be won without the horse; defeated enemies could not be pursued; and the breeds of horses would deteriorate without the cavalry." His second proposed change affected the Air Corps, which he maintained had the same conservative attitude as the cavalry. He proposed increasing the active Air Corps by 40 percent. These decisions made funds more competitive for the other branches. Finally, the president's national strategy included efforts at world disarmament, especially offensive weapons, such as tanks and large mobile field guns.[1]

The lack of congressional funding for the military during the 1920s was determined by a Republican administration that believed a limited government and a small army best served the public good. This attitude seriously affected the army, including the cavalry, which saw its traditional mission seriously compromised and its troop strength cut by more than half. By the time the army entered the 1930s, the country was not even prepared for a defensive war in the Western Hemisphere. Until the shadows of another world confrontation arose in the late 1930s, the army struggled with years of stagnation and deterioration because every year Congress significantly appropriated fewer funds than requested by the War Department and General Staff. By imposing budget restraints, Congress produced a stifling defensive orthodoxy in the military that was further intensified by the line branch chiefs, who had to defend their compartmentalized institutions and at the same time compete for funding.

Thus by the end of the 1920s, the course of developing mechanization had been deflected by the president's national military strategy, budget restrictions, the Depression, and the mind-set of the traditional line branches. In his annual report to President Hoover, General Summerall discussed the importance of a mechanized force for the army. "Any nation," he held, "which fails to provide for the utilization of mechanization to the utmost practicable degree must suffer the consequences of neglect in future war. Furthermore, failure to prepare to meet an attack by a mechanized force may result in defeat by troops organized and equipped as of today."[2] His words were prognostic but met with political indifference.

As the army approached the next decade, it was faced with the problems of dealing with obsolete equipment, peace requirements, and political and financial dynamics of a depression. This climate created a built-in attitude determined to continue to do as in the past. The lessons of World War I became a guide for future warfare. This approach produced rigidity and an attempt to reaffirm traditional infantry open warfare and maneuver that had not risen from World War I experience. Instead, the lessons referred to a pulverizing strategy of attrition warfare at the expense of mobility, maneuver, and speed. This perpetuated a traditional doctrinal way of thinking, and as a result, muted attempts at any form of institutional change. In 1930, the idea of a mechanized force became a goal because it offered a new way of warfare based on flexibility, maneuver, mobility and speed, and surprise. This new doctrine of mechanized maneuver-oriented warfare at an operational level raised fears among some that it would compromise the compartmentalization of the branch system as established in 1920.

General Summerall, the army chief of staff from November 1926 to November 1930, deserves credit for influencing and galvanizing a staff, especially Parker, for examining the ever-increasing application of scientific means for warfighting during a period of budgetary restrictions. Not only did Summerall have to deal with the onset of the Depression and a lack of funds for the army, but he also had to follow President Hoover's national defense policy. In addition, he had to contend with the growing autonomy of the combat arms chiefs, who were victimized by continued manpower downsizing and an obsession with victory in World War I. This attitude led to a longing for the symmetrical doctrine of infantry open warfare and firepower that was sanctioned by General John J. Pershing. However, as the past war demonstrated, attrition warfare was still the cornerstone of traditional American doctrine. Summerall understood the need for a new approach to this doctrine by encouraging the development of a mechanized force. He saw firsthand the power of defense and attrition in the Great War. He criticized the huge expenditure of artillery that had not surprised the German defenders. More so, he remembered the ghastly infantry losses during the short period the AEF was in the war.

Later, as a major general, Adna R. Chaffee Jr. commented that Summerall had early on recognized the value of tanks as independent and mobile weapons and the importance of their development.[3] In January 1943, Summerall recalled that when he was chief of staff he was convinced that the tank was an important element of "the artillery-infantry-machine gun team."[4] During his tour as chief of staff, he had recognized the improvements of automobiles and in aviation, and increases in the speed of tanks. These developments, he held, could restore mobility to warfare and return to the offense a more effective means to overcome the defenses characteristic of the western front during World War I. As the main maneuver element in a mechanized force, tanks now provided the capability to move at a faster rate than foot infantry. Rather than becoming infantry auxiliaries, "tanks will become a weapon exercising offensive power in [their] own right." He viewed the infantry as providing great holding power over the ground gained by tanks. As for the artillery, he proposed a self-propelled system having the same mobility as tanks. Summerall did not neglect aviation; he saw the importance of attack aviation in support of the mechanized force. The highly mobile capability of airpower, he held, should be harnessed to engage in tactical and strategic combinations with other arms in carrying out the army's mission. One of his goals as chief of staff was to develop a self-contained force with engineers, signal corps, antiaircraft weapons, chemical weapons, and transports to deploy the force for a particular mission. One of the most interesting events during Summerall's tour was the attempt by the infantry and the Ordnance Department to develop a light, common tank chassis that could also be utilized as cargo carriers, as well as artillery and chemical tracked motor carriages, a commonality Summerall found ideal for a mechanized force as it potentially kept pace with fast-moving tanks.[5]

General Summerall believed the army's future doctrine depended on all combat elements exercising rapidity of movement. Proceeding in this direction, he wrote, had caused extremes of opinion and much diversity of views regarding details of equipment, organization, and tactics. It caused increased tensions with the combat arms branch chiefs, especially the infantry. As a result, "We are in a transitional period, the precise outcome of which no one can foresee." He believed that for national defense no major foreign power was prepared to abandon the mass army. Therefore, it was imperative that attention be given for a mechanized force.[6]

Before the mechanized force was established, Colonel James K. Parsons, commandant of the Tank School, who was also the officer in charge of field development for a mechanized force, completed a proposal in April 1930 calling for the creation of six tank divisions to be included with the mobilization of six field armies. Years later an officer in the War Department stated, "Parsons' proposed table of organization was astoundingly similar to our Armored Division T/O [table of organization] of 1940, both in composition and strength. Had I discovered it sooner, it would have saved me untold hours of labor." Parsons's effort was the first ambitious proposal for a self-contained tank division during the interwar period. His recommendations, however, created some serious concerns in the office of the chief of cavalry because Parsons indicated no need for the horse branch. Parsons's tank divisions challenged a number of the cavalry's traditional missions, such as covering the advance and flank of the main force, exploiting a breakthrough and attacking the enemy's flanks and rear, seizing strategical positions and filling gaps in the line. Parsons, like Parker and Chaffee and a few ordnance officers, foresaw the importance of tank forces in the next war. He rationalized that with the introduction of the Christie chassis as an armored fighting vehicle, it was now possible to expand maneuvers providing it mounted a 47mm gun rather than the standard 37mm gun. The introduction of tank divisions, he maintained, was now appropriate because speed, radius of action, and mechanically dependable tanks had improved considerably since World War I. This suggestion could provide the army with greater mobility in deep operations. He proposed tactical aviation to support the tank divisions. Parsons, however, did not include foot infantry in his proposed division because he felt they needlessly restricted the radius of deep operations for a tank division force. Regarding artillery, he believed it needed no armor, since it was usually deployed in the rear.[7]

Like Campbell and Christmas from the Ordnance Department, Parsons argued that the United States had the necessary industrial production capacity to build armored fighting vehicles. Much to the dismay of air enthusiasts, he concluded that the cost to field tank divisions would be small as compared with the interwar development of the Army Air Corps. The cost of creating a tank division, however, was considered too ambitious and expensive, and Congress was in no mood to fund it.

The Office of the Chief of Cavalry, like the other autonomous office branches, questioned Parsons' recommendations. That office was now under the command of Major General Guy V. Henry Jr. who replaced General Crosby in March 1930 as the new chief of cavalry. Crosby was appointed commissioner of the District of Washington, D.C., and a few years later he was caught up in the ill-fated march of the Bonus Army on the capitol. Meanwhile, Henry tasked Major George S. Patton Jr. to respond to Parsons. Patton, who joined the Office of the Chief of Cavalry in 1928, debunked Parsons's recommendations, undoubtedly to defend traditional cavalry. Patton stated that Parsons's "opinion is as dangerous as it is impossible." Furthermore, he rebutted, deep operations with a tank division could end in a disaster, noting, "A stalled tank is junk." And like battleships, "tanks become obsolete." Patton also disagreed that tank divisions could replace traditional combat arms, especially foot infantry. Patton instead emphasized the importance of rifles, machine guns, and artillery firepower as the dominant fighting means, and the importance for each combat arm to develop self-propelled or tractor-drawn weapons. Patton held that the .50-caliber machine gun would make a suitable antitank weapon and that it was cheaper to produce in quantity. He alleged that large costs invested for the .50-caliber lightweight machine guns would probably do much more toward winning the next war than expenditures for tanks, especially given the improvements in armor-piercing ammunition. Consequently, tanks would be more vulnerable than in the past. Additional negative complaints he included dealt with terrain issues and huge logistical requirements for tank divisions. Patton argued that an attempt to use tank divisions at high speeds over great distances in an independent operation would result in a disaster. He believed transportation of tanks overseas would be too difficult. As a consequence, it seriously reduced the ship's man-carrying capacity.[8]

Patton insinuated in his criticism of Parsons's proposal to Henry that deep independent operations with a tank division could not take over the traditional role of cavalry. Nevertheless, serious misgivings in the Office of the Chief of Cavalry over the possible creation of another combat arm were also quite evident in other branch offices. "The organization of one or more tank divisions at the present time and based on present data is unwise," Patton believed. His only concession was suggesting the creation of a small combined arms mechanized force, which could give the best promise of implementation. Resorting to history, Patton maintained an independent tank force must be utilized in a manner analogous to that employed by the cuirassier of Napoleon's armies, meaning to be deployed "as an offensive reserve for the delivery of a rapid and powerful blow over a limited front at a carefully selected time."[9]

Aside from challenging the cavalry's mission and congressional funding, Parsons's proposal was interesting in that his idea moved the employment of a tank division for each field army to an operational level by providing a broader dimension of time and space beyond tactics. His recommendations in 1930 reflected Parker's ideas as early as World War I in potentially moving mechanized warfare to an operational level.

In his final report to Secretary of War Patrick J. Hurley, General Summerall wrote, "A new element foreseen as a development in armies of the future is the mechanized force." Consequently, as a last act as chief of staff, he ordered the establishment of such an organization. Yet he was realistic enough to reject the tank division Parsons proposed. The army had "neither a standard tank nor satisfactory means of commanding and controlling fast-moving mechanized units." Nevertheless, Parsons's division was studied for years at the Command and General Staff School.[10] His idea of tank divisions predates the implementation of German panzer divisions. There was some speculation, though unconfirmed in the Army Ground Force study "Prelude to Armor," in *The History of the Armored Force Command and Center* (1946) that the Germans might have studied Parsons's organization for a tank division. Unlike the congressional budget restrictions of the U.S. Army, the Wehrmacht and the Soviet Red Army by the mid-1930s had unlimited means to start translating their ideas into tank divisions.

Meanwhile, a permanent mechanized force was forming up. General Summerall's choice for commanding the force was Colonel Daniel Van Voorhis, from the 12th Cavalry Regiment, who had command experience mainly along the Mexican border. He had the reputation of being a strict disciplinarian and a spit-and-polish cavalryman. Soon the task-oriented cavalryman became the driving force that welded dissimilar elements and obsolete equipment into a coherent formation. Within a year, he would move the embryonic force from theory to the beginning of a new fighting force and mechanized doctrine. Van Voorhis had little experience with motor vehicles, yet he understood the potential mobility of these new mounts. One key attribute was his ability to handpick

Major General Sereno Brett, Mechanized Force executive officer and an infantry tanker, circa 1931. Courtesy of Elwyn B. Robinson Department of Special Collections, Chester Fritz Library, University of North Dakota, George S. Patton Papers.

Robert W. Grow as a cavalry captain, circa 1920s. Courtesy of *ARMOR* magazine.

the right people to execute his mission. As executive officer, he selected an infantryman, Major Sereno E. Brett, a World War I tanker and a prewar carnival motorcycle rider who briefly lectured on his battle experience with the Tank Corps at the Infantry School after the war. Brett, as one officer noted, "knew no fear while defying Sir Isaac Newton's Laws of Gravity." Brett had the ability to fire the imagination of a number of young infantrymen.[11] Unfortunately, he would fade over the years because of a drinking problem.

As the S-3, Operations, Van Voorhis picked Major Robert W. Grow, a stone-faced, businesslike cavalryman with a budding determination to move his arm into the future with new mounts. Bob Grow had been appointed as a second lieutenant of Field Artillery in the Minnesota National Guard in November 1916. The following year he graduated from the University of Minnesota and was commissioned in the cavalry. In 1922 Captain Grow was senior cavalry instructor at the ROTC Detachment at the University of Illinois. During Grow's days at the Cavalry School in the early 1920s, he began to believe the horse that served the cavalry for centuries could no longer survive the devastating fire unfurled during World War I. The horse, he held, no longer served as a platform for modern weapons. He began to think about "better" horses yet retained the cavalry tradition of mounted warfare. For the next ten years, Grow was in the center of cavalry transition, challenging the traditional doctrine of his branch. He would play a dominant role in the development of mechanization in the cavalry and then transfer to the Armored Force.[12] His diaries of these events became a historical treasure regarding the critical period of cavalry and armor history. Grow, like Patton, had a strong sense of history and the lessons it taught.

On October 1, 1930, the Mechanized Force began assembling at Fort Eustis, Virginia, as a combined arms force equipped with old and experimental material that was readily available. World War I Liberty trucks were acquired to tow artillery. Also acquired were M1917 American Renault light tanks and armored cars. These three elements constituted the striking power of the Mechanized Force. Other troops were to seize and temporarily hold distant key positions, engage in turning and enveloping movements, undertake flank or rearguard

The T2E1 medium armored car was built in 1928 and was based on a LaSalle chassis. It was used by A Troop, 2nd Armored Squadron, and deployed with the Mechanized Force early in 1931. This troop formed the mounted nucleus of the Detachment for the Mechanized Cavalry Regiment in November 1931. Courtesy of the Patton Museum.

Civil War veteran drives a 75mm self-propelled howitzer alongside a Cunningham light tank. Note the extended hood, which housed the engine department. Courtesy of the Patton Museum.

missions, and engage in breakthrough and exploitation. The new force, despite its obsolete equipment, was organized as a rapidly deployable force. War Department intentions were to establish a mechanized force capable of executing expanded strategic and tactical mobile missions beyond the support of division infantry. According to Grow, the force had all the makings of the cavalry's mission. He was, however, concerned about acceptance because the Mechanized Force did not seem to receive support from the "horsemen" in the War Department and in "the field."[13] Meanwhile, in November 1930, General Douglas MacArthur succeeded Summerall as army chief of staff.

By early 1931, the Mechanized Force had finalized its organization.[14] The most important element was the armored car troop (A Troop, 2nd Armored Car Squadron), which was to provide ground reconnaissance far beyond what had been previously accomplished by horse cavalry and ground troops. More so, with the addition of mounted machine guns the troop had more fighting power and road mobility. Its wheeled cars, however, were lightly armored. Nevertheless, A Troop, 2nd Armored Car Squadron was the forerunner of all future cavalry armored reconnaissance elements. With a greater emphasis on extending reconnaissance with armored cars, Van Voorhis set the stage for a policy that had an impact on tactical thinking during the 1930s and through World War II.

The element on which offensive power of the Mechanized Force was built was a tank company (A Company, 1st Tank Regiment) equipped with obsolete M1917s and new Cunningham light tanks. They were the striking power of the force. For march speed, the tanks were transported on wheeled carriers. A chemical detachment, using the Cunningham tank chassis, was to provide smoke for the tank attack. Unlike armored cars, the tanks of that time were obsolete, and the few new Cunninghams proved useless for a combined arms mechanized force.

Since a mechanized force was vulnerable to air attacks, an antiaircraft de-

Company H, 34th Infantry-Machine Gun Company, Mechanized Force. Courtesy of the Patton Museum.

World War I Liberty truck and 75mm field gun, the type commanded by Captain Harry S. Truman in World War I. Courtesy of the Patton Museum.

tachment was provided, mounting dual .30-caliber Browning water-cooled machine guns on 2-ton trucks. In addition, one commercial White truck chassis carried two multi .50-caliber Brownings.

The machine gun company (H Company, 34th Infantry, Motorized) with its .30-caliber Brownings was carried on six-wheeled cross-country ¾-ton Chevrolet trucks. The crews and machine guns were capable of detrucking for holding action, which was the infantry's primary mission.

An artillery battery (A Battery, 6th Field Artillery, Portee) provided supporting fire, using self-propelled and portee artillery, the latter with French 75mm guns carried on World War I Liberty trucks. The self-propelled vehicles were armed with American 75mm guns and one vehicle with a 75mm pack howitzer. For an ammunition carrier, the field artillery used a T1E1 chassis. The purpose of this mix was to evaluate self-propelled and portee artillery for a Mechanized Force.

The engineer company (C Company, 13th Engineers) was responsible for special bridging. The ordnance company maintained, repaired, and salvaged broken vehicles incurring special technical and mechanical problems in the field. A quartermaster repair unit provided upkeep of the forces' motor vehicles.

(Above) Mechanized Force takes a break. Note the T7 armored cars (1930–1931). Courtesy of the Patton Museum. *(Below)* Civil War veteran and young lady pose alongside T7 armored car. Courtesy of the Patton Museum.

As these units gradually assembled, they were expected to provide higher commanders with a hard-hitting, rapidly deployable force capable of sustained independent action at a tactical level. Van Voorhis, however, was confronted with the problem of performing the force's mission with an improvised organization using mainly obsolete equipment. Later he would comment on those early days of mechanization. The mission was more fitted to the organization and its obsolete equipment rather than first defining the mission and then building up a force with new equipment. In other words, "We got the cart before the horse and made the mission conform to the equipment at hand." Continuing, Van Voorhis expanded on the problems facing the army as it approached the next decade. He maintained that some army elitists favored a modest organization that became a tremendous, slow-moving, powerful force, rolling majestically forward against strong enemy resistance. This linear or horizontal approach, he stated, was challenged by "the ardent enthusiasts" who visualized a fast-moving mechanized force, hard-hitting and capable of rolling up hostile flanks. His approach was vertical by virtue of the mechanized force's mobility that would

Officers' wives pose on top of a T7 armored car to publicize the Mechanized Force. Courtesy of the Patton Museum.

operate deep into enemy territory. He questioned the traditional horizontal approach. The reason was that the characteristic that made possible the execution of the horizontal linear mission prevented the accomplishment of a vertical penetration with a mechanized force.[15]

Meanwhile, Chief of Cavalry Guy Henry attempted, as Van Voorhis had in 1925, to have an observation squadron organically assigned to a cavalry division. He was advised that the Air Corps idea at the time was not to assign any squadron to cavalry and infantry divisions except for attached duty. The Air Corps was determined to continue to maintain strict control over all air elements.[16]

During the next several months, the Mechanized Force held a series of marches and field exercises. In addition, special public, political demonstrations and ceremonies were held explaining the Mechanized Force's mission. Amid much publicity, the new Christie tank reached Fort Eustis for field tests with the force in April. Those who witnessed the new fast tank were impressed. Earlier General King, Summerall's G-1, had visualized the Christie tank as the main ar-

Armored "steeds" equipped with radio communication halt near Newport News, Virginia, during Mechanized Force maneuvers. Courtesy of the Patton Museum.

mament for the mechanized force. King called the Christie tank a "combat car," so as not to violate the 1920 National Defense Act, which placed tanks under the control of the chief of infantry.[17] The experience with the Christie tank, however, indicated the scope and variety of problems between the user and provider. During the first long march, the new ordnance-designed Cunningham tanks supposedly performed well. It was noted, however, that tanks required close and automatic gun support. Uniform signals and the lack of accurate maps were also a problem. The engineers required cross-country vehicles, and an antiaircraft detachment was required to disperse and cover the march column. One of the major problems was supplying a variety of spare parts for some new but mostly old equipment.[18]

The future of Cunningham tanks was now questionable. It was an infantry tank that had considerable deadeye space forward of the crew and lacked the speed of the Christie, with its new suspension system. It was not suitable for mechanized cavalry. The demise of the Cunninghams can ironically be explained by an interest shown by the Amtorg Trading Corporation, the Soviet Union's purchasing agent in New York City. Late in 1929, a representative from the Red Army operating through Amtorg, I. A. Khalepskii, visited the United States for the purpose of acquiring fifty or more Cunningham tanks. Khalepskii was in line to become the Red Army's director of mechanization and motorization. At first, he was not successful, because at that time, the Soviet Union was not recognized by the United States. In March 1930, Khalepskii returned to the United States in another attempt to purchase Cunninghams. No sooner had he arrived than he lost interest. He felt they were too slow and cumbersome, thus not suitable for deep operations, an operational level of war the Red Army was developing. Like General Fuqua, Khalepskii found the Cunninghams "poorly set up," because the forward position of the engine increased the temperature in the crew compartment to unacceptable levels. The engine's frontal position also impaired the driver's field of vision and the gunner's ability to lay the tank cannon on target. What caught Khalepskii's attention was the article written by a U.S. cavalryman, Major C. C. Benson, who enthusiastically endorsed the new and fast Christie convertible tank in *Army Ordnance* and *Infantry Journal*. Khalepskii was impressed with the "Model 1940." Subsequently a contract was negotiated and signed in April 1930 between Amtorg and J. Walter Christie for two tank chassis. Khalepskii witnessed the signing, which was executed in New York City.[19]

While the Ordnance Department, the cavalry, and the infantry were struggling with negotiations, the high-handed Christie in late December 1930 surreptitiously shipped two tank chassis as farm tractors to the Soviet Union before the United States had the opportunity to purchase the Christies for the cavalry and infantry.

(Left) Red Army's I. A. Khalepskii and one of the two Christie chassis *(below)* purchased by the Armtorg Trading Company in April 1930, arriving in the Soviet Union in January 1931. Seen here is the Christie undergoing tests in the Soviet Union in 1931. The vehicle's suspension system and armor shape were the nucleus for the evolution of the famous World War II T-34 battle tank. Courtesy of the Red Army archives.

The two chassis for the Red Army evolved into the BT (fast tanks) series. With the Christie design, the Red Army, with Khalepskii's expertise, leapt to the forefront of world tank technology. The BT series was the main maneuver element for deep operations, which contributed to a new field of Soviet military study. This meant successive mechanized operations of a deep battle of annihilation above a corps level called operational art. With the Christie design and subsequent improvements, the Red Army—unlike the U.S. Army, which was upset over Christie's duplicity—was able to complete the equilibrium between doctrine, tank technology, and warfighting at an operational level. Through product evolution begun by Khalepskii, the Christie suspension and sloped armor eventually saw its final realization in World War II with the introduction of the famous T-34 tank. Unfortunately, Khalepskii had not lived to see the results of his efforts, the final development of the Christie system. He was executed in 1938 along with thousands of his comrades during Stalin's purge of the Soviet military. His peers considered him the "father of Red Army armor."

At the time Amtorg was finalizing the Christie contract, General Williams retired as chief of ordnance and was replaced by Major General Samuel Hof, who opposed the purchase of Christie tanks because he felt it was his department's function to procure and test one model. For years after the Christie debate the Ordnance Department and General Hof were criticized by armor historians for stifling the purchase of the Christie system and providing the Red Army the opportunity to develop it into one of World War II's best medium battle tanks. Nevertheless, Hof's position had merits that were overlooked by the combat branches. Furthermore, the Ordnance Department never had the opportunity to test the M1928. The chief of ordnance also questioned the technical capabilities of the combined wheel and track system after a report by one of his engineers, Captain John K. Christmas, an ordnance tank expert from the Automotive Section. It was argued that too much stress was placed on the speed of the Christie by the cavalry and infantry without considering the ruggedness and fighting ability of a vehicle that did not have a turret to support a gun mount. A turret, armaments, and crew space, according to Hof, would produce a different vehicle where the question of weight exceeded the dimension for a light tank.[20] Military characteristic and tank doctrine driven by the infantry hinged on lightness. Added weight would move the vehicle to a medium class.

Only after they met all military requirements, successfully completed technical service tests, and had all their defects corrected would General Hof suggest purchasing additional Christie tanks. He was quite irritated that ordnance was not given the opportunity earlier to test the M1928, as were the cavalry and infantry. On June 6, 1930, Christie was advised his bids for five infantry tanks and one cavalry armored car were rejected. Instead Hof requested Christie provide the cost for one test model. He responded by demanding $135,000 for one tank and future manufacturing rights. Hof could not agree. On June 18, 1930, the obstinate tank designer attended a conference over the tank issue with Captain Christmas and Major W. P. Boatwright, also from ordnance. He again reiterated his firm position. Christie claimed earlier several officials from the War Department, mainly Summerall and Williams, had promised him an order for six tanks. He warned the officers that he would bring political pressure if nothing came from the negotiations; then he made an extravagant statement that "he had spies, or agents, who kept him posted on developments in this situation, in all branches of the military service." Furthermore, he warned the officers that if the Ordnance Department attempted to build a combination wheel-track tank he would make every effort to stop it.[21]

At this point General Summerall stepped in and ordered the parties involved to cooperate on the tank impasse. In no uncertain terms, General Hof explained that the infantry had neither the right nor the ability to comment on the technical features of tank design. He also reminded the attendees that after the last war the Ordnance Department purchased from Christie his manufacturing rights, "present and future." Hof was referring to a contract dated June 15, 1920, whereby the United States was entitled to the free use and ownership of the original patents and pending patents covering all improvements made by Christie. Christie was paid $100,000 for the patents and royalties. Also, between December 22, 1916, and September 23, 1921, Christie received from the government $739,240 for his vehicles and modifications.[22]

The heated conference, which saw accusations flying across the room, aggravated Summerall. The arguments now were mainly centered on the issue of speed and weight, especially when a turret was added. Now quite irritated, Summerall ordered the infantry and Ordnance Department to work out a contract, otherwise the remaining funds would revert to the Treasury. Finally, it was reluctantly decided to work out an arrangement with Christie. A contract was signed on June 28 for one tank at a cost of $55,000. An additional $6,000 was reserved for tests and $1,000 for modification of the Liberty engine, which ordnance furnished Christie at no cost. It was agreed that the tank would be delivered by September 1 with a turret. A War Department Board was appointed in October following the suggestion of General Hof. Representatives from the Ordnance Department, cavalry, and infantry were to submit and make recommendations regarding the acceptance tests.[23] The chairman of the board was Brigadier General William H. Tschappat, General Hof's assistant. Christie's challenger, Captain Christmas, was appointed ordnance inspector. Representing the cavalry were Colonel Van Voorhis, Lieutenant Colonel Chaffee, and Major Patton. Van Voorhis anticipated competing with the infantry over the

Christies in the event that additional ones passed the tests and were then ordered by the War Department.

Christie, however, stalled on his promised delivery date. He advised Captain Christmas early in December that 90 percent of the tank was completed, estimating a January delivery. Finally, on January 19, 1931, the tank arrived at Aberdeen Proving Ground (APG), Maryland, four and a half months late, but not before Christie again subjected the vehicle to flashy demonstrations in and around Washington as he had with the M1928. This delay, which was noticed with serious trepidation by Ordnance Department personnel, was due to the priority Christie gave in delivering two tank chassis to the Soviet Union. Upon their arrival at APG, Christmas noted discrepancies from the contract. Complicating matters, Christie advised General Hof that the Liberty engine supplied by ordnance in July was no longer serviceable and needed to be replaced or thoroughly overhauled. Christie then decided to return his tank to the factory for repairs and alterations. General Hof was now even more disgusted with the undependable Christie.

The debate over the merits of the Christie now moved into the public arena. On February 11, 1931, the *Chicago Tribune* strongly criticized General Hof and the Ordnance Department for failing to purchase a number of Christie tanks for tactical exercises. The article indicated that General Fuqua and Colonel Harry L. Cooper, commandant of the Tank School, were in favor of purchasing the speedy tank over the Ordnance Department's Cunningham light tanks. The article concluded by recalling wartime experiences when the Ordnance Department was busy redesigning the French 75mm field gun rather than producing American-built guns, sarcastically noting, "It is easier for a camel to pass through the eye of a needle than for a good weapon to get into the United States Army."[24]

General Hof responded in a memorandum to the volatile Major General George Van Moseley, General MacArthur's Deputy Chief of Staff and close friend. He informed Moseley that the military requirements had yet to be met. He added, "A fast tank can easily be made, if desired. The only real problem is to make a tank, which is reliable and durable at such speeds." Furthermore, he argued, "What value is an automobile that can go 60 mph if it can only be relied on for a short run?"[25] Hof also advised Moseley that the acceptance tests were again suspended due to another mechanical breakdown. Shortly after Hof's comments on the *Tribune* article, the Army Appropriation Act for the fiscal year was passed, making $250,000 available until June 30, 1932, for the purchase of tanks.

The Christie tank acceptance tests had not gone well at APG due to numerous mechanical failures and the vehicle's inability to pass all required tests

J. Walter Christie standing in his M1931 vehicle. The officer is Captain L. W. Tharp. January 10, 1931. Courtesy of Chester Fitz Library, University of North Dakota, George S. Patton Papers.

and specifications set forth in the contract. Meanwhile, General Hof provided Christie with another Liberty engine, but it did not resolve the mechanical failures. Finally, the vehicle passed the majority of acceptance tests, except a 140-mile run with tracks over rough terrain. On March 6, the chief of ordnance orally offered Christie $54,000 for the tank "as is."

Congress now entered into the controversy. General MacArthur received a stern letter from Henry E. Barbour, the chairman of the Subcommittee on Appropriations for the War Department, requesting information on the status of the Christie tank. MacArthur was advised that the War Department Appropriation Bill recently passed by Congress provided that $250,000 should remain available until June 30. He indicated that the money was to be used for purchase of Christie tanks (even though the bill did not so state). Barbour and the majority members of the committee were of the opinion after interviewing army officers at the December 1930 hearings that at least four or five tanks were necessary for service tests.[26]

Undergoing extensive questioning during the hearings were General Hof and Captain Christmas. Hof was asked whether in the area of research and development there was anything for the Mechanized Force. "No, sir; not as such," was Hof's reply. Cavalry officers had no input at the hearings, since this was a tank-infantry issue as prescribed by the 1920 National Defense Act. The ordnance argument was presented in detail, meaning a mechanical test was necessary before first ordering additional tanks for tactical tests. The chief of ordnance also told the congressmen that the whole issue was further muddled over weight classification. After the war, the definition of a light tank was a two-man vehicle capable of being transported by a tank carrier, such as the M1917 and the Cunningham light tank series. Realizing this, General Hof stated that the weight of the Christie tank with added armaments and a turret made it comparable to a medium tank, but it "has the armament and manpower of a light tank." In one of the heated exchanges, committee member Ross A. Collins reminded Hof that it is right and proper to respect the wishes of his commander in chief of the army regarding tank development. However, "since Congress is called upon to provide for national defense, you should take into consideration likewise the enactments of the Congress."[27]

The officers who influenced the committee's decision were from the infantry, General Fuqua, Colonel Cooper, Major Brett, and Captain L. W. Tharp, a test officer from the Tank Board. The four officers expressed their belief in the superiority of the Christie tank and requested it. Brett made an interesting statement that because of the convertible principle and speed of the Christie "we are not worried particularly about aviation." The attitude was that aviation bombs do not affect a fast-moving tank unless from a direct hit. Cooper advised the committee members that the infantry had no money appropriated for tanks in 1929, but the cavalry did have some funds for purchasing an armored car of a similar model as the M1928. Cooper also interjected a personal plea, stating, "Mr. Christie is a poor man. He spent his entire fortune in the development of this vehicle." When asked about the effectiveness of tanks, Cooper said they were a strategic reserve for General Headquarters. Because of certain terrain, tanks were normally held in GHQ reserve and sent to any place an attack was made, he stated. Cooper noted that because of the convertible principle, the Christie concept had a strategic and tactical advantage over other tanks.[28]

Barbour looked at MacArthur and stated that the wishes of Congress "will be carried out."[29] MacArthur responded, describing the number of mechanical failures occurring during the acceptance tests, "necessitating a great deal of repair and replacement." Barbour was advised that purchase negotiations were under way, or would be as soon as Christie's company agreed to the changes dictated under the terms of the contract. The contract for a fighting vehicle listed combat specifications that were expected by the cavalry and infantry. Concluding, MacArthur mentioned to Barbour that it would be "unwise to greatly expedite procurement of additional tanks, if by doing so, we fail to incorporate the changes which may be indicated by a proper test."[30]

During this critical period, J. Walter's adopted son, J. Edward Christie, claimed his father supplied Prohibition whiskey and women while entertaining some members of Congress at the Willard Hotel in Washington, D.C. Naturally, the goal, according to J. Edward, was to get congressional funding for his tanks. Like his father, J. Edward despised the Ordnance Department and described in some strong, profane terms the likes of Christmas and especially General Hof.[31]

Because of political pressure, Hof formally offered Christie $54,000 on March 10, 1931, for the vehicle, including a complete set of drawings. Again, this was not to Christie's liking. The next day he wrote Secretary of War Hurley, requesting an early end to the tests and return of his vehicle so it could act as a pilot model for further production. His goal was to completely dismantle the vehicle, correct any mechanical weaknesses discovered, and alter the blueprints requested by the Ordnance Department.[32] The problem, Captain Christmas maintained, was that historically Christie never furnished a complete set of drawings. He always submitted his vehicle without blueprints, since they were custom made.[33] Christie reminded Hurley that General Summerall assured him that his company would receive an order for four or five vehicles. The adamant designer complained that General Hof blocked the order. Christie's goals were driven by his expectation of receiving orders for additional custom-made vehicles. Christie did not have the production facilities to mass-produce.

Agitated about not hearing from Christie and his attorney, Raymond Tiffany, Hof requested a conference. On March 13, it was held in the office of the assistant secretary of war. In attendance were Generals MacArthur and Hof, Captain Christmas, Christie, and his attorney. Also there were MacArthur's deputy chief of staff and chief of the Manufacturing Service, General Tschappat, the president of the War Department Board. Tschappat reviewed results of the tests to date. As before, the meeting became heated, with accusations flying. Christie virtually demanded an order right away, without competing bids. General Hof expressed his intention to purchase additional vehicles at the appropriate time. However, this depended upon bids submitted in compliance with the comptroller general's guidelines. Much to the irritation of Christie and his attorney, no agreement was reached. MacArthur intervened and suggested that a special War Department Board be reconvened, submitting a final report on the acceptance tests for disposition by Secretary Hurley.[34]

Three days later General Tschappat submitted the War Department Board's report. Van Voorhis, Chaffee, Patton, ordnance officers, and Colonel Cooper and Major Brett from the infantry found that as a result of the acceptance tests, "the basic features embodied in the pilot model of this tank are sound and that their development will lead to the securing of a tank which will meet the needs of the army."[35]

Still Hof had not received a reply from Christie's attorney regarding yet another conference. Finally, on March 18, the meeting occurred. In attendance were Christie and his attorney, Generals Hof and Tschappat, and Major Boatwright. After asking for a reply to Hof's March 10 offer, Tiffany stubbornly responded. He requested an end to the tests and that the vehicle be returned immediately to the factory. Then and only then, Tiffany stated, would drawings be made available. Christie's attorney vowed the company would break the contract, refuse to deliver the tank, and bring a lawsuit unless Christie had the tank back to dismantle it. As the conference heated up, the chief of ordnance was accused of breaking faith and assurances given earlier by General Summerall. Hof responded that Summerall spoke without authority. He advised Christie and his attorney that they should have dealt with the proper authorities—meaning the Ordnance Department—rather than engaging in lobbying and circumventing the chain of command. Frustrated, Hof offered $53,000, but without drawings. Finally, a few days later Christie and his attorney agreed to the desired mechanical tests provided they were completed in two weeks. On March 27, tests were completed and the tank was released for dismantling. On April 1, Christie advised ordnance the dismantling was in progress.[36]

The U.S. government never acquired the tank that went through the acceptance tests during the winter of 1931. After it was released, the maverick designer in anger supposedly placed the vehicle in the War Department courtyard in the parking space of Assistant Secretary of War Frederick Payne. Months later Christie claimed he retook possession of the vehicle, which had been moved to an Ordnance disposal yard.[37] In a few years, he would sell the vehicle to England.[38]

While the debate was going on over the acceptance tests, the chief of infantry became quite concerned over Van Voorhis and the Mechanized Force usurping his branch's prerogative as the dominant fighting arm. As being solely offensive, General Fuqua stated, the Mechanized Force "shows entirely too much enthusiasm." Fuqua emphasized that there was no such animal as armored cavalry. He argued that if horses were eliminated there would be no cavalry. Fuqua informed the Adjutant General's Office, which served as the army's administrative service, that turning the function of testing weapons and equipment over to the Mechanized Force retarded development. The chief of infantry based his argument on the long-standing policy that this was the right of the chiefs of the combatant branches. He strongly noted that fire and movement were the responsibilities of the infantry, which required proper synchronizing at the critical stage of the attack. Furthermore, he argued, tanks were developed to

Morris Motors in England purchased a Christie M1931 in October 1936, which was not accepted by the Ordnance Department. British officers were impressed after viewing the 1936 Red Army maneuvers, where hundreds of fast Christie-type tanks were demonstrated. Courtesy of the Royal Armoured Tank Museum.

bring tactical mobility back to the infantry. Fuqua wanted the Christies for a fast breakthrough of the enemy's line or for fast movements around the enemy's flanks. He objected to the attitude that only cavalry had the viewpoint and characteristics of mind to develop a Mechanized Force. This attitude, he said, lacked conviction. Concluding, "Every branch should be deeply interested in the development of a Mechanized Force in order that its battlefield mission may be correctly gauged, and its effect upon the combatant arms as now organized."[39] Fuqua based his conclusion on the 1920 National Defense Act and the *1923 FSR*, which referred to infantry as the predominant combat arm and indicated that all others arms and services were auxiliaries.

The following month Fuqua sent a memorandum to General Moseley, who was tasked to review and coordinate all major War Department projects. Fuqua advised Moseley he was trying to channel infantry thought into the same doctrine of open warfare adopted by General Pershing during the Great War. He reiterated his argument that de-horsing cavalry by creating a new organization, such as the Mechanized Force, would cause irretrievable loss to the army's traditional mounted arm. Regarding the Christie tank, he maintained that it should be procured for the infantry because the Infantry Tank Board had taken the initiative in its early development.[40] All of Fuqua's arguments reflected his determined desire to maintain the preeminence of the infantry as the predominant combatant arm.

About the time Moseley received the memorandum from Fuqua, he visited Fort Eustis to view the Mechanized Force and the Christie tank, which was undergoing acceptance tests. During one of the exhibition runs, which were covered by a moving camera mounted on a truck, a rubber tank tire was thrown off a road wheel, cutting short the demonstration. Moseley informed ordnance officers and members of the Society of Automobile Engineers who were watching the exhibition that "future mechanization lay along the lines of an auxiliary to the established functions of the line rather than a separate entity for battle maneuver."[41] This confirmed the rumors that the force would be disbanded. Major Grow wrote, "Hopes for a strong independent mechanized brigade of all arms were dashed." The feeling was that Moseley was the driving force behind MacArthur's efforts to separately modernize each combat arms branch and service. The chief of infantry, who actively argued for infantry control over tanks, no doubt influenced Moseley.[42]

In spite of the mechanical problems, Van Voorhis and Chaffee strongly supported the Christie tank for its mobility and the stable gun platform that the suspension system provided. Van Voorhis, however, was unable to attend most of the War Department Board's meetings due to his preoccupation with the Mechanized Force. Nevertheless, he was kept informed on the acceptance tests. The commander of the Mechanized Force had the opportunity to witness a demonstration of the Christie at Fort Eustis in April 1931 and was impressed. He wanted the tank as a combat car and for its wheel-track principle. This categorization as a combat car was an innovative way for the cavalry to get around the tank issue because the 1920 National Defense Act had placed all tanks under the control of the chief of infantry.

After reviewing for months the basic War Department policy for modernizing the army, General MacArthur, on May 18, 1931, released to the press his intent to mechanize and motorize the army. He was moved by current dire economic conditions that precluded congressional funding for the War Department. In his opinion, an independent mechanized force was too expensive to equip and organize.[43] In addition, ground forces were losing personnel and funds to the more glamorous Air Corps. MacArthur's solution was to decentralize mechanization. The new directive was more practical, according to MacArthur, than a centralized mechanized program. The objective was to ensure that each of the ground forces adopted mechanization, which would allow each combat branch to develop its doctrine independently. Unfortunately, it served to boost branch autonomy at the expense of a mechanized combined arms doctrine.

In addition, the new policy was an attempt to move army organization from the perspective of equipment types to a position of missions assigned to each combat branch. Since the infantry's mission was to close with the enemy, it remained the decisive arm of decision to be supported by auxiliary and other arms. Tanks remained assault, breakthrough weapons for the infantry, and when employed with the cavalry were designated as "combat cars" for pursuit and exploitation. Regarding cavalry reorganization, two types of regiments were planned, one horsed and one mechanized. Modern firearms, the release stated, had eliminated the horse as a weapon. Only in special situations the horse—supported by motor transportation—might furnish the best mobility, it stated. The mission of cavalry—as in the past—was long-distance reconnaissance, tactical reconnaissance, seizing points of strategic and tactical importance, pursuit or delay of the enemy, exploitation either alone or in conjunction with other combat arms, and forming part of a strategic or tactical reserve. The last mission assigned reflected a combined arms operation where mechanized cavalry was composed of motorized infantry, tanks, and mobile artillery all capable of moving at a uniform speed. To develop an organization and equipment to meet the assigned missions of the traditional combat arms, the Mechanized Force was disbanded and reorganized as a reinforced cavalry regiment. The new cavalry force absorbed all the appropriate equipment.[44] General Henry recalled that during his first visit with the new chief of staff, "MacArthur pointed out of his office window to the parked passenger cars and said, 'Henry, there is your cavalry of the future.'"[45]

In spite of the War Department announcement in June, the Mechanized Force engaged in one of its most extensive maneuvers undertaken. The scenario was to deploy on a long strategic march designed to play out command and control at higher levels. The Mechanized Force, designated as the Red Army, covered a distance of ninety miles from Fort Eustis to Camp Lee via Richmond to engage the opposing Blue Army. An airplane from Langley Field, attached to the Red Army, coordinated with the armored car troop and reported on the disposition of the enemy.[46] Eventually all elements of the force fused together and deployed in deep operations as a combined arms unit, taking all objectives and dispersing the enemy many times in heavy and continuous rain. This rudimentary maneuver again had all the trappings of an operational level of war. Even so, the concept of pursuing the art of war at a new level was short-lived when the force was disbanded.

One area the Mechanized Force failed to deal with was greater use of aviation, which was not part of the force's table of organization and equipment. Major (later general) George C. Kenney, who was a strong advocate for Air Corps independence, was at the time one of the most influential teachers of attack aviation. He was known for developing tactics in actual flying practice. Kenney met with Grow after the June maneuvers, commenting that the Mechanized Force was not a good target while on the move. After flying over the Mechanized Force, he suggested a seventy-five- to one-hundred-yard vehicle distance on road marches. Grow did add that while at Fort Eustis the force had excellent relations with the Air Corps.[47] However, Air Corps zealots continued to argue for complete control of air assets and branch autonomy, making coordination with the ground troops difficult.

Although the Mechanized Force would disband, training programs nevertheless continued. Meanwhile, General Henry passed on to Moseley a tentative table of organization and equipment for a mechanized cavalry regiment. The proposed table was prepared by Grow, who was assisted by Brett. It was based on "Report of Mechanized Force 1931" during the period of November 1, 1930, to June 30, 1931. The conclusion was that a mechanized force could be a powerful mobile instrument if properly equipped and organized. Since the force had been organized, experience had demonstrated that it was not suitably structured or equipped, and lacked sufficient strength. Despite this situation, the current force had all the potential elements essential for accomplishing its mission of extended maneuvers involving fast tanks supported by other mechanized and motorized arms.[48]

The "Report of Mechanized Force 1931" contained a number of observations. The experience of A Troop, 2nd Armored Car Squadron, indicated its value in distant road and bridge reconnaissance in advance of the force in spite of the fact that its motor vehicles were framed in boilerplates. This type of reconnaissance was considered critical because the force was terrain sensitive. The troop was also given missions of close reconnaissance, security, covering withdrawal, delaying action, and seizing key positions. Experience also indicated that the armored car troop needed to carry a sufficient crew for dismounted reconnaissance. Since tanks had become the most important tactical element of the force, considerable criticism was directed at tank carriers. It was apparent they decreased mobility by their presence, causing considerable anxiety. A detachment was required for their protection. The Christie wheel-track system was expected to solve this problem. The artillery was tasked to provide close tactical support for tanks. Because the battery rapidly displaced from one position to another, the report noted, self-propelled mounts containing armor protection against small arms fire were recommended. The machine gun company fulfilled two tactical functions. First was march and bivouac security. Second, it operated as a holding force, securing captured ground. The final report, however, noted that the infantry was not organized to perform both of the assigned missions.[49]

In Weimar Germany, the Reichsheer had more than a curious interest in the Mechanized Force. In the summer of 1930, a *Fuhrerstabsofficiere* (general staff officer with troops) in the Truppenamt (Troop Office, operations and training), Captain Adolf von Schell, arrived at the Infantry School at Fort Benning, Georgia. He had served in the Kaiserheer during the maneuver battles that characterized the war against Russia during the early stages of World War I and was a former member of the Free Corps Movement.[50] Beginning in 1931, the Reichsheer was in the process of formulating a new tactical doctrine based on speed and maneuver. The manual reflecting this doctrine, *Die Truppenfuhrung,* was being prepared by Major General Ludwig Beck, who spoke fluent English, was affable to America's industrial might, and understood the country's war-fighting potential. Von Schell was a colleague of Major Heinz Guderian at the Reichswehr Defense Ministry, where the future practitioner of armored warfare was working on problems of troop motorization.[51]

Captain von Schell successfully completed the infantry advanced course in June 1931, receiving a superior rating in military history. As a student, he contributed to the school by sharing his tactical experiences learned during the Great War. In addition, he was skilled in discussing modern concepts regarding a mobile warfare doctrine, which was being considered for *Die Truppenfuhrung.* In 1933 the Infantry School published his influential book, *Battle Leadership,* which reflected his informal lectures delivered at the advanced and company officers' classes and to officers from the 29th Infantry. The lectures echoed his experiences in the last war, including ideas on the borderline between open and trench warfare and his current observations of the U.S. Army. He stated that the U.S. Army, because of the country's geographic location, did not need to worry about a sudden invasion. As a result, "the United States does not need a regular army in proportion to its size to insure its national safety," because "the role of the theater occupies the foreground of attention." He gave considerable credit to the U.S. Army's school system for training regular officers when it came time to deal with an expanding reserve army. Von Schell, nevertheless, did criticize leadership by noting that the situation in the American army did not allow for practice with the tools of warfare, such as playing out doctrine with new equipment in maneuvers.[52]

The most important aspect of von Schell's lectures was his explanation of "out of the tactical mission" called *Auftragstaktik.* The term is not easily defined in English; however, von Schell called it "mission tactics." What von Schell meant was in mobile and fast-moving battles where situations change, sometimes dramatically, effective command and control was inevitably compromised. How the mission would be carried out was the senior commander's responsibility by issuing written orders, but not in the minutest detail. Rather, containing a clear statement of intent that is passed down to subordinate commanders. "Orders should be short, but as clear as possible." He added, "This is done because the commander on the ground is the only one who can correctly judge existing conditions and take the proper action if a change occurs to the situation. There is also a strong psychological reason for these 'mission tactics.'" Trust and loyalty are the key because the commander or even the subcommander and noncommissioned officer "will accomplish more for what he does and act in accordance with his own psychological individuality," because limits caused by minutest orders are removed. Thus, German military tactics as were being developed in the small Reichswehr encouraged initiative, ingenuity, innovativeness, and audacity. "Every soldier should know that war is kaleidoscopic, replete with constantly changing, unexpected, confusing situations. Its problems cannot be solved by mathematical formulae or sets of rules," he stated, adding, "We must teach our men in peace that battles differ greatly from maneuvers and that there will often be critical periods when everything seems to be going wrong." During peacetime, von Schell maintained, training and maneuvers need to be programmed to allow subordinate commanders more freedom to make tactical decisions. He added, "One can do in war only what one has learned in peace."[53]

His legacy while at the Infantry School emphasized that leadership training should encourage soldiers to think for themselves and allow them to develop tactical events on their own initiative. Von Schell thanked then Lieutenant Colonel George Catlett Marshall, head of the Academic Division, who later became chief of staff of the American army in World War II and responsible for molding and mobilizing the country's manpower for war. Von Schell also thanked fellow infantry officers at the Infantry School for "their encouragement, advice and assistance in the preparation of the book," which was based on his wartime experiences and tactical concepts. One of the officers who encouraged von Schell to publish was Major Truman Smith, who continued their relationship in Germany when Smith was posted as the American military attaché in Berlin from 1935 to 1939. Smith became an astute observer on the growth of the Luftwaffe and panzer units.

Even Patton found von Schell's book ideal to his way of thinking because the German had an understanding of leadership and motivation. Patton wrote in his copy (dated July 28, 1934) that a commander must lead personally by example. At the time, however, according to one historian, Patton did not understand von Schell's revelations about the German concept of *Auftragstaktik*.[54] This is understandable because of the difficulty the U.S. military had interpreting the ongoing concept of developing *Auftragstaktik*.

Earlier in May, von Schell had requested a two-week attachment to the Tank School at Fort Meade and the 34th Infantry (Motorized), which had a machine gun company detached to the Mechanized Force. No doubt, von Schell was interested in what was going on at Fort Eustis with the Mechanized Force and its emphasis on independent operations with a mobile force that went beyond traditional infantry and cavalry formations. Though approved, his attachment was later canceled because of the impending change of station of the 34th Infantry and the breakup of the Mechanized Force. When the War Department began rethinking mechanization in the spring of 1931, von Schell's request was withdrawn. Later he again visited the United States and assertively inquired about army mechanization and activities at Fort Knox. Von Schell returned to Germany, where he eventually became an instructor in tactics at the Kriegsakademie.

Major Patton, in the meantime, continued his balancing act between the horse and the machine in a *Cavalry Journal* article. In 1930, he claimed less than 6.5 percent of the roads in the continental United States were improved. Because of this condition, he questioned armored car off-road mobility, finding it tactically useless except for supply. He took issue with reconnaissance aircraft operating with the cavalry because airmen were limited to daylight hours and by weather conditions. He accepted the use of mechanization, both ground and air, as a means for getting horse cavalry over the difficult task of executing a breakthrough so troopers could charge through the breach, moving over all kinds of terrain and through all weather conditions, day or night. Armored cars at this stage of exploitation, he held, were useless because they could not operate at night or hold captured ground. Furthermore, he argued, tanks and combat cars would quickly outrun their supplies, not to mention the exhausting effects the initial attack had on crews. Patton took issue with the infantry when he stated that horse cavalry is more than an auxiliary arm. It was "capable of separate and wholly independent action." He added that war is killing, and the appropriate cavalry battle tactics for this can best be described as "Grab the enemy by the nose and kick him in the pants."[55] George C. Scott delivered this comment in the movie *Patton* (1970). The comment became a motivating speech, with the word "ass" substituted for "pants," when Scott, as Patton, was addressing American troops during World War II.

Examining the exercises and composition of the Mechanized Force and the roles played by the combat arms and service troops, it is evident that a new tactical doctrine was born at an operational level of warfare. A number of officers, such as Van Voorhis, Chaffee, Brett, and Grow, could smell the future battlefield. To them the Mechanized Force possessed a shock effect and mobility greater than traditional cavalry and infantry. The "Report of Mechanized Force 1931" demonstrated that such a force had a greater view of the battlefield in depth and time. It emphasized that once a new doctrine was developed, modern equipment was required to perform missions established by doctrine. Equipment—mostly left over from World War I—used during the exercises was obsolete or always breaking down. As a result, outdated equipment and weapons handicapped the possibility of a new meaning for a mechanized force. Unfortunately, some lessons were not learned. The chief of infantry still adhered to the doctrine that the sole function of all arms was to support the advancement of the foot soldier. This attitude was based on the slow pace of World War I tanks, and a mind-set derived from the 1920 National Defense Act and the *1923 FSR*.

Finally on June 4, 1931, General Tschappat, president of the special War Department Board that evaluated the Christie tank acceptance tests, sent a memorandum to General Hof that recommended acccepting Christie's bid for seven tanks. Christie was the only serious bidder. Undoubtedly, congressional influence played a part, since Congressmen Barbour and Collins indicated at the December 1930 budget hearings that for the next government fiscal year they expected $250,000 to be used to purchase Christie tanks. The contract was signed by Christie, representing his company, on June 22 for the manufacture of seven convertible tanks, equipped with mild steel plates. The infantry was to receive three Christie T3s armed with a 37mm tank cannon to serve with the 67th Armor. The cavalry was to receive four T1 combat cars armed with .50-caliber machine guns. The order of distribution was altered in favor of the cavalry due to the impending mechanization of the cavalry at Fort Knox.

The long controversy over J. Walter Christie and his convertible principle and helical suspension system caused the using services, cavalry and infantry, to wallow for years over developing a tank policy. The debates made it almost impossible to match standardized equipment with a predetermined doctrine.

Not satisfied with the seven M1931s produced for the U.S. Army, Christie started building a turretless 5-ton M1932 with a simulated bow-mounted cannon. He called it a "flying tank" because of its enhanced speed and lightness. Christie made an ambitious statement that the lightly armored vehicle could be carried beneath a heavy bomber. However, the M1932 was never tested with an airplane as Christie had boasted. He attempted to sell his new vehicle to the U.S. Army and the Red Army. Because Khalepskii found Christie's behavior atrocious, the Red Army decided to cut all future ties with the designer.[56] Neither the U.S. Cavalry nor Infantry were interested. After receiving the seven M1931 tanks from Christie, the army also cut all official ties with the civilian designer.

The cavalry attitude, meanwhile, was instinctively hostile toward any machine that replaced the horse, while the field artillery failed to understand the importance of developing self-propelled equipment to meet the needs of a mechanized force engaging in deep operations. In response to media criticism, a strong defender of the horse cavalry, Brigadier General Hamilton S. Hawkins, wrote in the September–October 1931 issue of the *Cavalry Journal* that a mechanized force could never replace horse cavalry in any terrain. Such a force, he wrote, required horse cavalry and infantry for support. Though he agreed on the combination of the two, he argued that a mechanized force could not alone take over large operations in support of the infantry in battle as could the cavalry, adding that cavalry was more useful now than ever before in history. Furthermore, he wrote, "The horse can still remain on the battlefield and will remain as long as men engage in open warfare." He upheld the Civil War cavalry tradition of mounted and dismounted action.[57] His attitude typified the beliefs of many conservative cavalrymen posted in the southern border states and at the Cavalry School at Fort Riley, Kansas.

None of the traditional combat arms wanted another branch separately established to impart maneuver-oriented operational mobility, such as employing firepower, shock action, and sustained combat, and to deal a decisive blow in battle while fighting mounted. If this policy was perpetuated, it could lead to a new doctrine based on combined arms with tanks providing the main maneuver element, the branch chiefs believed. Nevertheless, the Mechanized Force had demonstrated that all the elements were there for establishing an operational level of warfare.

Finally, the Mechanized Force was scratched. It was too expensive to maintain. Congress concurred but spared the Air Corps from funding reductions, leading to further strained relations with the line branches. The new chief of staff, General Douglas MacArthur, formally announced, "Every part of the army will adopt mechanization and motorization as far as is practical and desirable." He was not as adamant as Summerall in support of tanks as the main maneuver element in a separate mechanized force. As chief of staff, MacArthur did support the mechanization of cavalry, believing the horse would eventually be eliminated. He believed cavalry would move from horse-mounted troopers to mechanized cavalrymen, sometimes referred to as "cushion pounders." MacArthur gave credit to Christie tanks for awakening the cavalry to possibilities of supplanting the horse. Over the next years, he alleged mechanization would take time, because development of technology was slow and laborious. This being the case, he stated, the intricate technical nature of equipping the army was time-consuming and costly. He anticipated it would take twelve months to reach quantity tank production during a major mobilization.[58] Thus MacArthur's policy statement, which reconciled and sustained the autonomy of the branch bureau system, dominated the army's mechanized development program until the establishment of the Armored Force in 1940.

Certainly, Van Voorhis expressed the most perceptible lesson from the Mechanized Force. It was a remarkable declaration of mobile warfare and organizational skills. It was an expression of combined arms coordination and the potential role of technology in restoring mobility to the battlefield. His "Methods for the Tactical Employment and for the Training of a Mechanized Cavalry Regiment" formed the focal point for moving the cavalry toward mechanization. The mechanization of one cavalry regiment, he noted, was the first step in applying machines to cavalry missions. Bordering on branch heresy, he visualized replacing the horse with machines. The mission was to provide a mobile combat element in tactical or strategic cooperation with other arms and fighting primarily mounted or, if necessary, dismounted. He continued, "On favorable terrain, mechanized cavalry should extend the sphere of action to much greater distances and increase the speed of performance of its missions" without altering the accepted mission of the mounted arm. Van Voorhis believed the employment of a mechanized cavalry regiment would greatly extend offensive power, such as firepower and shock action. He focused on the characteristic of mechanized mobility of mass maneuvering and economy of force rather than on traditional horse mobility. Furthermore, he noted, with great speed, firepower, and armor protection, the mechanized cavalry had the potential to carry the attack through to an early decision.[59]

Colonel Van Voorhis's proposed organization consisted of a combat car squadron that provided the main offensive power of exploitation, pursuit, and shock action. His objectives were to move the cavalry away from the basic tenet of reconnaissance, security, and dismounted infantry. Van Voorhis wanted to make the mechanized force more of a decisive force on the future battlefield by swaying the course of combat in time and space. For a covering squadron he planned for an armored car troop for distance reconnaissance and a scout troop for close security on the march and during combat to provide detailed reconnaissance. A machine gun troop was to provide firepower in combat and to act primarily as a holding force. Its offensive mission was more limited because of an emphasis on defensive combat power. A headquarters troop with a signal platoon was to provide the personnel and equipment for command, communications, and supply.[60]

Thus, Van Voorhis, more than any cavalrymen, was the first marketeer to implement change leading to today's army of combined arms cavalry regiments and their mounted warriors.

In spite of all Van Voorhis's efforts to bring the cavalry into modernity, branch journals still presented articles that argued for traditional horse cavalry. As the cavalry entered the 1930s, it still tenaciously looked for reasons to be for the horse. Patton and Benson were suggesting a marriage between the horse and the machine.[61] Some officers went so far as to condemn the machine, which would supplant the horse and destroy the effect and tradition of cavalry.[62] As one cavalry officer indicated, the machine "injects a feeling of uncertainty and uneasiness as to the future of our Cavalry."[63] The traditional conservativeness embedded in the cavalry between wars can be summed up in a comment by Major John B. Smith: "The inspiring sight of horsed Cavalry on the march is preserved for the eye of the man to behold."[64]

4

Cavalrymen Looking for New Mounts and a Tactical Doctrine

As expected, Colonel Van Voorhis was unhappy over the breakup of the Mechanized Force and the loss of tanks. Years later he commented that he had explained in person to General Douglas MacArthur and his deputy, George Moseley, "that to assign the mechanization mission of the army to one particular branch would be a great mistake; that mechanization was a problem which concerned all branches of the service and that they should not be deprived of the opportunity to develop mechanization as applied to their respective branches in a coordinated all-out mechanized effort; that I could not conceive of branches developing mechanization within their own respective spheres."[1]

Van Voorhis strongly believed cavalry influence was the strength of the Mechanized Force. Nevertheless, he understood it was not cavalry but a composite of combined arms and services. It was the decision of the War Department that a new combat branch such as a mechanized force was not in the best interest of the army, especially when the country was experiencing its worst economic depression in history. To resolve this issue, a new cavalry unit was established, Detachment for the Mechanized Cavalry Regiment. Because of this move, the combat arms' compartmentalization was still intact and still driven by seperate doctrines and missions.

On November 2, 1931, the new unit of about four hundred officers and men left Fort Eustis and began a four-day march to Camp Knox, thirty-five miles southwest of Louisville, Kentucky. Union soldiers early in the Civil War had occupied this area, as had the notorious Confederate cavalryman John Hunt Morgan and his 2nd Kentucky Cavalry. Early in the war, Brigadier General William Tecumseh Sherman, who was to rise to the pantheon of outstanding Union generals during the war, took command of five thousand troops at Muldraugh Hill before an invading Confederate force converged on the area. Near the small river town of West Point, Fort Duffield was built in 1861 at the confluence of the Salt and Ohio rivers to interdict a Confederate move toward Louisville, a major Union supply base. Fort Duffield was Kentucky's oldest and largest fortification. During World War I a field artillery training center was established at the camp of thirty-three-thousand-odd acres and named after Major General Henry Knox, the chief of artillery for the Continental Army during the War for Independence. Later Knox became the first secretary of war in President George Washington's cabinet. Close by, along Otter Creek, is one of Kentucky's oldest and most scenic inns, built between 1780 to 1821 and first used as a woolen mill, then a gristmill, and finally a flour and grain mill. An old record book lists a monetary payment to a stonemason named Tom Lincoln, the future president's father. In 1927, it became a country inn noted for serving cured Kentucky hams. During the 1930s, the inn, known then as Sulfur Wells Hotel, became a favorite place for many cavalrymen from Fort Knox and foreign visitors to meet and informally discuss ideas on mechanized tactics. In 1947 the name was changed to Doe Run Inn.

Colonel Van Voorhis and his post executive, Lieutenant Colonel Adna R. Chaffee Jr., liked Camp Knox's varied terrain and large acreage. They found the camp both scenic and primitive. Major General Guy V. Henry Jr., the chief of cavalry, had some reservations. He considered it far removed from the Cavalry School at Fort Riley where doctrine was taught. Van Voorhis nevertheless found it more suitable than Fort Eustis for organizing and training the expected mechanized cavalry regiments. Van Voorhis planned that the new home not be dominated by any of the line branch chiefs. From the beginning, the mission of the mechanized cavalry at Camp Knox was to provide cavalry with a significant role in warfighting, fighting with new mounts in deep mobile-oriented operations. This mission fell to Van Voorhis, whose duties were to develop a tactical and training doctrine for all units expected to be attached to their new home in Kentucky. The move provided him with a new breed of young cavalrymen who began to develop a more aggressive way of warfighting than the horse cavalry and infantry. Accordingly, Camp Knox now began the great experiment of combining cavalry tactics with modern armor warfighting. Eventually the military camp grew to about 109,054 acres and became the breeding ground for ideas on mobile combined arms operations well into the twenty-first century. Little did Van Voorhis realize at the time that this new station would become the home of the U.S. Army Armor Center and School of Mounted Warfare, and the Patton Museum of Cavalry and Armor.

Colonel Van Voorhis and Camp Knox were now the recipients of the 1931 directive that mechanized cavalry would be organized to fulfill the normal cavalry role substituting vehicles for horses. The cavalry missions Van Voorhis was to undertake at the new camp were as follows:[2]

Reconnaissance
Distant reconnaissance
Battle reconnaissance
Counterreconnaissance

Security
Covering concentration
Covering movement of other forces
Covering flanks during battle

Offensive combat
Pursuit
Exploitation of a breakthrough

Defensive combat
Delaying actions

Special operations
Providing liaison between larger units
Securing advance positions for less mobile forces
Advance, flank, and rear guard
Raids
Minor warfare
Attack or defense of a river
Convoying other troops

Undoubtedly, these new missions were too ambitious due to a lack of manpower, inheriting obsolete equipment, and fiscal restrictions. It was a case of establishing a doctrine before developing suitable equipment. Even with these problems, it was the cavalry's beginning in determining the application of new tactics and techniques of mechanization for a new mounted arm. It was the beginning of innovative opportunities, such as integrating doctrine and technology. In the meantime Major Robert W. Grow, Van Voorhis's S-3, began to draw up a table of organization and equipment for the anticipated 7th Cavalry Brigade (Mech). However, he was far from optimistic, expecting little to happen under current economic conditions. Cavalry regiments that were to be converted to mechanized were slow in coming. Years of debates and congressional budget restrictions occurred before the 7th Cavalry Brigade finally formed up with two regiments.

The newly arrived detachment for the Mechanized Cavalry Regiment was composed of headquarters and an armored car troop (A Troop, 2nd Armored Car Squadron, commanded by Captain Charles H. Unger), which was in the cavalry tradition of having a reconnaissance element with each major unit or command. The detachment also consisted of C Company, 13th Engineers, 19th Ordnance Company, and the 28th Motor Repair Section from the Quartermaster Corps.

Bob Grow later recalled that pre–World War II cavalry training directives emphasized the "sneak and peek" methods of stealth reconnaissance. What began at Camp Knox, however, was an innovative attitude by Van Voorhis, Chaffee, and Grow, who believed that the cavalry's new undertaking was not stealth reconnaissance but to fight, and if necessary carry out various types of combat missions.[3] This developing attitude by a new breed of marketeers at Fort Knox collided not only with traditional doctrine and the horsemen at Fort Riley but also with the established combat arms branch system, especially the infantry.

What the Detachment at Camp Knox found waiting for it were leftover World War I dilapidated troop barracks, a few shops, a warehouse, and a rundown clubhouse referred to as the Central Mess. The cavalrymen and support troops had their hands full that wet autumn setting up the camp for the anticipated arrival of the 1st and 4th Cavalry Regiments.

By the end of November word had been received that only the 1st Cavalry was coming to Camp Knox, but not until May or June 1932. Meanwhile, crossed sabers in yellow were adopted for vehicle markings. By the end of the year, low morale began stressing the troopers. Eleven armored cars of various designs constituted the armored strength of the detachment, along with Franklin personnel carriers. World War I modernized Liberty trucks and Harley-Davidson motorcycles were used for traffic control and communication. With no exercises or military training scheduled for the upcoming winter, restlessness and boredom soon took their toll. Buildings for a permanent garrison also were slow in coming.

On January 1, 1932, Congress designated Camp Knox as a permanent garrison, and the name was changed to Fort Knox. A new post commander, old horseman Brigadier General Julian R. Lindsey, arrived on February 6 so Van Voorhis could devote more time to mechanizing the cavalry. Lindsey was "quite a guy," one young officer noticed. He rode his horse all over the base, and in the winter he wore a pair of boots with electric heaters plugged into a battery behind his saddle. Lindsey's senior aide, a future distinguished general officer, was Lieutenant Isaac Davis White, who also took an active part in the devel-

General Julian R. Lindsey, Fort Knox commander, and Colonel Barney at old Tiptop at Fort Knox after the departure of the 1st Cavalry (Mech) for the 1934 Fort Riley maneuvers. Courtesy of the Patton Museum.

opment of mechanized cavalry. Shortly Lindsey began dealing with improving living conditions rather than getting involved in mechanization and establishing a cavalry brigade at Fort Knox. Nevertheless, he was vocal in certain areas, emphasizing the importance of artillery. He met with Van Voorhis, Chaffee, and Grow about adding an artillery unit to the mechanized cavalry. Much to the disappointment of Van Voorhis and Chaffee, Lindsey preferred motorized rather than self-propelled artillery for force balance.[4] This became an ongoing equipment issue of motorized versus self-propelled artillery until the army was drawn into another war years later.

At the time, Van Voorhis received a memorandum from the Cavalry School regarding the future of mechanized cavalry. It outlined the expectation that future mechanized units were to assist the horse cavalry. This memorandum upset Van Voorhis, who could not fathom the people at Fort Riley writing regulations for the proposed redeployment of cavalry regiments to Fort Knox. He believed the elites at Fort Riley knew nothing about mechanization and the importance of employing mechanized regiments as a whole rather than combining them with horse elements. This compromise diluted the principle that change was ever present and provided new opportunities, such as using mechanized regiments as an effective force without horses. The attitude at Fort Riley and of the chief of cavalry, Grow recorded, was not to carry out full substitution of "iron horses for flesh and blood horses in an existing cavalry regiment."[5]

There was now a sense of importance that increased morale at Fort Knox. It was announced that the 1st Cavalry Regiment was expected to arrive from Fort D. A. Russell, Marfa, Texas. The army's oldest mounted unit's history goes back to 1833 as a regiment of Dragoons in the regular army. Early in the Civil War, the unit was redesignated the 1st Cavalry. Now the officers were reluctant to substitute vehicles for their horses; however, the soldiers were pleased that they no longer had to look after the animals.[6] They were to travel to Fort Knox in a truck convoy.

Due to pressure from Van Voorhis and the establishment of the mechanized cavalry at Fort Knox, four of the seven Christies ordered were received and designated as combat cars CC T1. The last one arrived in the summer of 1932. All four were to be assigned to the combat car platoon, which was in the process of being organized. The CC T1s were organized for tactical training as the main striking element for the cavalry. The first combat car platoon commander was Lieutenant H. H. D. Heiberg, who arrived at Fort Knox in January 1932 and remained until 1936. Heiberg recalled his first close contact with Chaffee at a Founders Day dinner at the Brown Hotel in Louisville. Since Heiberg was the youngest West Point graduate present at the affair, it was customary for him to make a speech at the dinner. "Mr. Heiberg," Chaffee said, "will you give us the definition of leather?" Heiberg froze and then tried to "hem and haw" his way through. "Sit down, Mr. Heiberg," said Chaffee, drawing laughter from the attendees. Finally, the embarrassed Heiberg remembered that the cadet definition started with "The Pig" and ran through various procedures of tanning that concluded with "This sir, is leather!" Nevertheless, Heiberg said, they all thought highly of Chaffee.[7]

Like Grow and White, Heiberg was one of several young cavalry officers who became closely involved with development of the mechanized cavalry during its formative years. With the acceptance of the seven Christies, the army was now committed for years to make an effort to develop the combat feasibility of the convertible design. This occupation with the Christie convertible design between the two branches had a debilitating effect on a common tank policy during the 1930s.

Heiberg, like Van Voorhis, Chaffee, and Grow, was at first impressed with the Christies, mainly because the vehicle provided a more stable gun platform than any tank in existence. This was attributable to the vehicle's large road-wheel articulation, allowing it to negotiate rough terrain. Because the 1920 National

Defense Act placed tanks under the control of the chief of infantry, the Christies were armed not with turret tank cannons but with .50-caliber machine guns. The gunner stood on his feet, which were spread when firing. The driver sat in the pointed bow. The Christies were designed to carry a crew of two, a driver and a gunner. According to Heiberg, there was an attempt to add a third member to the crew, the commander, who stood behind the gunner. This was a most uncomfortable position and almost impossible to accommodate, because he had to stand on toolboxes or ammunition racks to reach the turret's vision slots in the cupola. There was also the problem of the double sidewall required for enclosing the independent suspension system. This made the size of the fighting compartment small compared to the overall dimensions of the vehicle. There were no intercoms; shoulder taps and foot signals were used for internal communication. Before long, Heiberg and Grow saw problems with the Christie power train's multiple steel clutches, which were not properly aligned, causing numerous gear failures. The vehicles were powered by a twelve-cylinder Liberty engine, which imparted the drive through a master clutch. The transmission was a selective sliding gear type with two steering clutches and a reduction gear linked to each road wheel at the rear of the vehicle. Problems with the CC T1s began to mount throughout the year due to extensive usage and the cost of keeping them in operation.

Finally, Van Voorhis was notified by the War Department that a new organization called Detachment, 1st Cavalry (Mech), was to go into effect on July 1, 1932. A few months earlier Van Voorhis had talked to Grow about the 1st Cavalry being the beginning for organizing the 7th Cavalry Brigade (Mech), a size necessary for the cavalrymen to engage in major maneuver-oriented operations. Grow considered the detachment a turning point in this direction. This was essential for troopers at Fort Knox if they were to become an effective combat force with new mounts. It would be months, however, before the 1st Cavalry moved to Fort Knox and not until years later was the 7th Cavalry Brigade finally was formed by an additional cavalry regiment.

For most of 1932, activities centered on small-unit organization, command post exercises, and development of equipment, especially the CC T1s. In September, J. Walter Christie visited Fort Knox after a stay at Fort Benning, Georgia, to discuss his new M1932, a 5-ton, two-man turretless so-called "Flying Tank" that used pneumatic tires. The vehicle differed considerably from the proposed tank envisioned by the Ordnance Department and using arms because it was not considered a suitable fighting vehicle.

Before visiting Forts Benning and Knox, Christie, in a meeting with ordnance officers, made a significant statement that "he had spies, or agents, who kept him posted on the development" of tanks. Boldly, he warned the officers that if they did not agree with his type of convertible vehicle, he surely would not hesitate to utilize political pressure to get his way. He told a surprised Van Voorhis and Grow that he was the only man who knew how to build tanks. Adding that he had political friends, and if he did not get the contract for tanks in fiscal year 1933, he could make trouble for the company that did. Van Voorhis felt Christie was more wrapped up in the M1932 and had lost interest in the cavalry's CC T1s and infantry's T3s. Van Voorhis and Grow were not too impressed with Christie the man or with his new and smaller M1932, which seriously compromised the crews' fighting space and armament.[8] As before, Christie, in his usual boisterous and independent way, emphasized speed over military requirements and fighting capabilities.

Christie eventually lost the battle with the cavalry and the Ordnance Department over his new M1932. After four T1 combat cars and the three infantry T3 tanks were acquired by 1932, the army ended its relationship with the eccentric Christie and decided to take on internal development of the Christie system. Undoubtedly, Christie was a man with ideas but without common sense. He continued to display a lack of understanding regarding military requirements for armored fighting vehicles. Throughout the 1930s, it was always his way or no way at all, and forever and a day he showed contempt for the Ordnance Department. Another consistent and serious problem over the years was that Christie never had the facilities or capabilities to mass-produce armored fighting vehicles. All he offered to the Ordnance Department, cavalry, and infantry were vehicles constructed by piecework with a handful of employees in a small garage.

Many in the army who came in contact with Christie during the interwar period praised his vehicles, especially Van Voorhis, Chaffee, Benson, and Patton from the cavalry, the chief of infantry, General Fuqua, and tankers from the Tank School. However, respected contemporaries of Christie, such as a future chief of ordnance, General Levin H. Campbell Jr., and Generals John Christmas and Joseph Colby; and highly regarded tank experts and noted historians Colonels Robert J. Icks and George Jarrett, viewed the civilian designer with trepidation because of his obnoxious personality and wild machinations. Bob Icks summed up the relationship when he appropriately stated, "J. Walter Christie's personality antagonized those whom he came in contact with. The vehicle itself was very fragile [and] overshadowed his good ideas." Icks recalled that Christie "goddamned everyone and everything." Christie was so adamant about his capabilities that he refused to consider users' requirements and required military service tests.[9] General Campbell recalled at the time that Christie was a good inventor but disconnected when it came to users' needs, adding, "a fine man when sober." Regarding the Ordnance Department tank expert John Christmas, whom Christie criticized in no uncertain terms,

Campbell stated he "was a man of highest integrity, and did all he could to help Christie develop his tanks."[10]

The problem with the M1932 was that it was too small to be an effective fighting vehicle for the cavalry or infantry. Furthermore, its armament capabilities were never tested. The vehicle just had not met cavalry and infantry requirements. Christie believed he was an idea man, and it was up to the Ordnance Department or the government to mass-produce his vehicles, providing he was paid for development costs. Brigadier General Sidney R. Hinds as a young infantry officer witnessed the M1928 demonstration. He had another take on the Christie situation that reflected the attitude of a number of progressive infantry officers. Hinds believed the Christie tank was the "best in the inter-war years." The problem, he claimed, was the Ordnance Department, which was more interested in developing its own designs than those of a civilian. The reason, Hinds believed, was that ordnance had "their own designs and had to keep their arsenals busy."[11] The argument over the value of the Christie system for the army is still debated today.

After numerous exercises with the CC T1 during 1932, it was determined that the fighting compartment was not suitable for cavalry operations because observation and control were a problem, along with a complicated multiple radio net. Throughout the year, numerous breakdowns continued. Grow complained that the CC T1s were not built as cavalry fighting vehicles. He called the CC T1 "a mobile cradle for an engine." General Henry agreed. He accepted that Christie was a smart engineer, but the CC T1 was not adaptable for cavalry missions and was far too costly for a proposed mechanized cavalry brigade to procure and maintain. Instead, Henry decided to present the Ordnance Department with characteristics for the next pilot combat car, the wheeled-track convertible CC T4, which called for a lightweight vehicle not to exceed 8.5 tons and with enough space for a crew of four, two more than the CC T1.

Young Heiberg recalled gathering around the piano in the evenings at the central mess, reinforced with Kentucky moonshine, singing to the tune of the traditional artillery caisson song the "Lament of the Cavalry Tanker."[12]

> First on wheels then on tracks
> As we break our bloody backs,
> Keep those Christies a-rolling along.
> In and out, mostly out,
> While you hear the Colonel shout,
> Keep those Christies a-rolling along.
> For its Hi-Hi-Hee—
> In the Horse-Tank Cavalree—

The officers' mess where Lieutenant H. D. D. Heiberg and fellow officers sang the "Lament of the Cavalry Tanker." Note the piano to the left of the stone fireplace and the condition of the chairs and tables. Courtesy of the Patton Museum.

Heave on your clutches hard and strong.
Where e're you go
You will always know
That those Christies are rolling along,
Lord keep them rolling,
Keep those Christies a-rolling along.

A strong interest was also shown in armored cars. Van Voorhis's guidelines for fighting vehicles were for offensive weapons that strike hard with accurate firepower in all directions. The first standardized armored cars were the M1s, built by James Cunningham, Son and Company, which were beginning to arrive at Fort Knox. By October 1932, ten had been delivered. No tactical tests, however, were conducted. Instead, they were regularly employed in training exercises. Obviously, Van Voorhis realized wheeled vehicles had less cross-country ability

M1 armored car was the first to be standardized by the cavalry after tests by the Ordnance Department in 1931. It was built by James Cunningham, Son and Company. From the Hunnicutt Collection, courtesy of the Patton Museum.

than the combat cars. Nevertheless, he believed the first essential mission for armored cars was observation and the ability to provide heavy firepower when necessary. Second in importance was the ability to lay down fire in all directions, promptly. Third in importance was the volume of fire. Finally, maximum armor protection was required. In other words, Van Voorhis's primary uses of armored cars were as fighting reconnaissance vehicles providing the terrain was suitable. Once these fighting conditions were met, he maintained, every effort must be made to provide survivability for the crews from hostile fire.[13]

Fort Knox was not the only place where ideas on mechanization were emerging. At the Army War College, where students received professional education for pivotal positions at higher commands, Colonel Selwyn D. Smith, cavalry, presented a memorandum to the assistant commandant for a proposed new cavalry division. Most likely, he noted, theaters of cavalry operations would be on the American continent, along the borders and seacoasts and large centers of population. Most areas considered had roads of limited quality. He questioned the usefulness of motorization in certain parts of the borders with Mexico and in Central America. Nevertheless, he believed, the cavalry should adopt mechanization and motorization to increase its mobility, security, and striking power. He suggested radios be developed to aid the cavalry on the march and in combat. Air observation was invaluable to cavalry, as was attack aviation, he argued. This memorandum was significant not only for the importance of mechanizing the cavalry but for command and control. In addition, tactical aviation was suggested to round out the combat capacities of mechanized cavalry.[14]

At Fort Clark, Texas, another avid reader and young maverick cavalry officer from the 1st Cavalry Brigade who excelled in polo told some ranking officers at a briefing that the horse was "through" and had no place on the modern battlefield. Captain Truman E. Boudinot's statement was considered sacrilegious, especially coming from a cavalry officer, and his peers did not hesitate to tell him so. It was also evident to Boudinot that the cavalry should mechanize. He began to criticize the lethargic attitude of horse-bound superiors regarding mobile warfare. Boudinot was also interested in the Christie design because he felt the vehicle had speed and agility, which he compared to the dash of a horse on the polo field. His developing philosophy was "decision followed by rapid action."[15]

In spite of limited tactical exercises at Fort Knox during 1932, concerns about future equipment requirements for a mechanized force and the army as a whole were questionable to say the least. The initiative for world disarmament was part of President Hoover's foreign policy to reduce land, naval, and air forces. Representing the president, Ambassador Hugh Gibson, the acting chairman of the American delegation at the 1932 Geneva Disarmament Conference, proposed to the International Land Commission "the total abolition of tanks and all heavy mobile land artillery over 155mm in caliber." This process of peace and reduction of armaments was designed to place emphasis on the increasing power of the defense through decreases in the power of the attack. General MacArthur concurred; he was ready to give up tanks as they were considered offensive weapons of war. With this possibility, Secretary of War Patrick J. Hurley considered substituting light armored cars for tanks for use by the mechanized cavalry. This political position at the time created uncertainty regarding a mechanization policy for national defense. A year later concerns still beset the cavalry at Fort Knox that combat cars could be ruled out by the Geneva Disarmament Conference because they were considered offensive weapons. The mechanized cavalry was built around combat cars that were considered the key maneuver element. The debate was somewhat muddled by limiting the weight of combat cars to 8.5 tons and the belief that a mechanized force would always require the close support of the foot and horse soldier for resisting a hostile landing and control of the frontiers.[16] Charles L. Scott, who later became a regimental commander at Fort Knox, claimed the disarmament conferences and depression had a negative impact on the creation of an armored force.[17] Eventually, the uncertainty was resolved when efforts at disarmament came to naught, mainly because of international apathy and bickering, along with President Hoover's inability to continue his foreign policy after losing the election.

Concerned about congressional funding, the cavalry during 1932 made an effort to switch its interest to cross-country cars. That spring the Bonus Army of World War I veterans began its descent on Washington, D.C., to demand the cash bonuses now that had been promised by Congress in 1924 but not payable until 1945. Led by General MacArthur, army elements, including a squadron from the 3rd Cavalry, forcefully evicted the veterans. The spectacle of cavalry moving against the Bonus Army with drawn sabers, infantrymen with bayonets, and the use of tear gas grenades fueled the liberal pacifists in the country who distrusted anything military. One young cavalry officer later wrote that the American communists were circulating handbills at Fort Myer urging the troopers to "refuse duty if you are ordered out against the bonus marchers and the unemployed." More than anything these propaganda bills created anger among the cavalrymen.[18]

Hoover's administration was doomed, as was further congressional funding for the military. The army's role in breaking up the bonus march did nothing to enhance its reputation. The army now reached its lowest point, therefore putting at risk the safety and security of the country. It could no longer function as a fighting force on a modern battlefield. The American fear of a standing army blended with an attitude that embraced isolationism.

In the meantime, the dauntless and visionary cavalryman Major General Frank Parker, who, as noted earlier, was one of the driving forces for a separate mechanized force, had been ordered to Chicago, where he took command of the VI Corps and in 1932 the Second Army. Though he was out of the mechanization debate, he continued to show a strong interest in the employment of tanks in mechanized warfare. In another one of his studies, he issued a warning that the current lack of knowledge on tank tactics and their limitations by officers could lead to impossible situations. Because of the army's lack of interest in modern tank development and budget problems, it was reluctantly accepted that the slow and obsolete heavy Mark VIIIs and the 6-ton M1917s were still essential auxiliary arms for the infantry. This was the early 1930s, and the army was still struggling over a future tank policy. Parker maintained that a future tank, such as the new, fast convertible Christie, was important for rapid deployment to the battlefield and more suitable for deep pursuit.[19]

Back in Texas the political bickering was finally resolved. The move to Fort Knox had been in a state of uncertainty for months because of political squabbling by congressional representatives from Texas and Kentucky over the transfer during an election year. Leading the objection was a rural Texan and vice president–elect, John Nance Garner.

Once the political issue was resolved, the 1st Cavalry began preparing for its long march to Fort Knox. On December 14, the Black Hawks bid an emotional farewell to their horses. In the morning the troopers left their stables, the buglers sounded adjutant's call, and the squadrons moved at a trot to positions for the final mounted formation with sabers drawn. The commander of the regiment, Colonel William A. Austin, delivered the farewell remarks, "We bid, at this hour, farewell to our faithful, willing and noble champions of march, maneuver, campaign and garrison—our beloved horses. A new era opens before us. The Regiment carries on . . . keep steadfast in the path of right, hold your head high, heart and hand clean, fear naught, be true to yourself and be worthy to wear the insignia of the Black Hawk Regiment." After this last mounted ceremony the stirring notes of "Boots and Saddles" were sounded by massed buglers, taps were played, and, finally, the band played "Auld Lang Syne."[20] No longer would cavalrymen saddle up and adjust reins and curb bits.

Officers of the newly formed 1st Cavalry (Mech) pose for a group picture at Little Rock, Arkansas. Colonel Daniel Van Voorhis is at the center, and Major Robert W. Grow is on his left. Courtesy of the Patton Museum.

Under the command of Van Voorhis the winter march to Marfa, Texas, began on December 17. A truck convoy left Fort Knox and headed for Fort D. A. Russell to bring the 1st Cavalry to its new home. The long and tedious motorized movement was tasked as a training exercise. While men of the 1st Cavalry were turning in their horses and saddles, the convoy from Fort Knox arrived on December 30 and, before Congress reconvened to reconsider the move, started the long move back to Fort Knox on January 2, 1933. Thus began a new assignment for the 1st Cavalry. The army's oldest cavalry regiment entered a new era, mechanization.

Quoting from a Marfa press item, Grow reflected on the local bitterness over the loss of the 1st Cavalry. "Three hundred of the 'dandy first' cavalry regiment left here today for their new post at Fort Knox, Kentucky. As they departed, leaving behind 250 common law wives and some mixed Mexican-American babies, the citizenry professed to regard the virtual abandonment of the post with concern."[21] On January 16, the convoy arrived at Fort Knox. At the time, the motor march under severe weather conditions was remarkable, because it was the longest made by an American army in less then thirty days. Finally, the first mechanized cavalry organization came into existence. Efforts were now made to mold the new organization into a balanced, self-contained unit capable of delivering a quick, hard, stunning blow.

General Julian R. Lindsey and staff greet the arrival of the 1st Cavalry (Mech) at Fort Knox. Courtesy of the Patton Museum.

Combined arms balance, as we shall see, now became an ongoing problem, especially because of budget restraints. In addition, there were problems associated with developing reconnaissance and tactical air support, suitable mobile field artillery, and the role of mounted infantry. All these issues became intractable due to the compartmentalization of the offices of the branch chiefs and army chiefs of staff on how to handle a separate branch.

As soon as the 1st Cavalry was in place, a Mechanized Cavalry Board was set up under the label of a technical committee. Grow was one of the key members of this committee, which became increasingly important. Its function was to deal with experimental equipment used by the 1st Cavalry, such as armored cars, scout cars, combat cars, semitracked vehicles, and motor carriages necessary to exploit mobility. At the time the board was established, the Tank Section of the Infantry School at Fort Benning (formerly the Tank School at Ford Meade) was still fixate on tanks as strictly infantry support weapons. Eventually, the technical committee became the Cavalry Board and then the Armored Board.

Another event—along with the political considerations discussed at Geneva—occurred in April 1933 that again created uncertainty regarding a mechanization policy. It had an impact on the development of doctrine and equipment, especially the CC T4. The secretary of war, George H. Dern, ordered a limitation on the weight of combat cars and tanks to 7.5 tons. The order took into consideration bridge-holding capabilities and culverts located in the United States.[22] This action was spurred by the government's desire to contain costs. Not only was the army being subjected to budget restraints, but it also was facing limitations on the type of vehicles it could develop. Lightness was an issue in the 1920s with the chief of infantry, General Allen, and now lightness was again emphasized. This restriction plagued the army's tank policy until the eve of World War II. It again clouded the issue of which came first, doctrine or equipment.

The ground force structure during the 1930s was muddled with ambiguity driven by budget restraints, the branch system, and a lack of political decisions. In addition, public support for the ground forces in defense of the country was dampened by pacifist elements that still believed America's geographic defensive position was sustained by its two ocean moats that isolated the country. Ominous war clouds were certainly apparent in 1933 with the rise of fascist regimes in Europe and Japan's aggression in Asia.

German curiosity about U.S. Army efforts at mechanization and implementation of a mobile warfare doctrine had not ended with von Schell's attempted visit to Fort Meade in 1931. Van Voorhis, Chaffee, and Grow recalled other German officers' visits to Fort Knox in 1933. In the spring a German general staff officer, who gave a favorable impression of Hitler, was given a ride in the CC T1 and an armored car. Grow noted that the German (referred to as Major Phillips, an ordnance tank expert) expressed ideas on mechanization that agreed with developments at Fort Knox, where the 1st Cavalry was in the early process of forging a self-contained mobile force capable of deep operations and fighting mounted.[23]

Grow, however, commented at the time, "It was pleasing to note that the German's ideas agree with ours," but not with General Henry. This was an interesting statement but not all that precise. Captain R. W. Strong, who was chief of the cavalry equipment branch from 1930 to 1933, claimed Henry was probably the finest and ablest horse soldier the cavalry had produced. He was also encouraging studies and tests deemed helpful to the development of armor for cavalry missions. Strong recalled that cavalrymen in the Office of the Chief of Cavalry were paying special attention to events in Germany and Reichswehr ideas on mechanization in the early 1930s.[24]

In the summer of 1933, the same German officer returned for a few days with a general staff officer, Major Hans von Greiffenberg, who had just left the Command and General Staff School at Fort Leavenworth, Kansas, after a year of observation. His purpose at Fort Leavenworth was to study the school's educational material for proposed historical lectures at the Truppenamt, the clandestine German General Staff School. He was mainly interested in methods and organization, the doctrine of employment of large units, and American military literature, especially on the Civil War. The Annual Report of the Command and General Staff School noted that von Greiffenberg provided information on the projected tactical war strength of proposed German divisions.[25]

Before their arrival, Van Voorhis discussed with Chaffee and Grow what they should tell their guests. It was decided to keep the discussions open except those on the .50-caliber machine guns. They were secured from the German visitors. After rides in the M1 armored car and viewing a radio display for command and control, the visiting officers and their hosts retired for dinner at the historic Sulfur Wells Inn. There an interesting discussion began comparing mechanized developments in both countries. Van Voorhis recalled that the Germans were not particularly interested "in our equipment," but more interested "in our views on the proper tactical and strategic employment of mechanized forces." It was Grow's mission to talk with the German visitors about semitracked vehicles and their coordination with cavalry, which he later referred to as the early development of armored infantry. Within this context of combined arms, they also discussed with von Greiffenberg and Phillips the importance of coordinating field artillery with armored cars or tanks. Talks then centered on efforts by the 1st Cavalry to coordinate cavalry, field artillery, and engineers with combat cars. Grow believed at the time that the German military establishment placed more emphasis on tanks carrying the offensive load. At the time of the German visit, most activities were centered on command post operations rather than on maneuvers. Nevertheless, ideas did began to blossom. Grow noted, however, that the German army was more advanced in the development of vehicular equipment. He agreed then and later with Van Voorhis that ideas emerging from Fort Knox—though primitive—were ahead of the Germans regarding the employment of balanced, self-contained fighting units. Of interest, Grow wrote in his diary that the German military was preparing in earnest for a major war in Europe. Van Voorhis and Grow's comments, however, were overstated. The Germans at the time had already developed the idea of a blitzkrieg war at an operational level, as demonstrated during the 1930 and 1932 maneuvers.[26]

What von Greiffenberg and Phillips saw when they visited the 1st Cavalry (Mech) was a combat organization that included the armored car troop for long-distance reconnaissance, the scout troop for close-in reconnaissance and security, the assaulting or striking squadron of combat cars, and their holding unit, the machine gun troop. Communications were carried out largely by a simple code system over voice radio supplemented by motorcycles, automobiles, and hand signals. It became obvious that given the right equipment the mechanized cavalry could perhaps effectively train to fight mounted. Dismounted action was only considered when it provided a tactical advantage.[27]

Years later von Greiffenberg was part of the operations team that planned Operation Barbarossa, the invasion of the Soviet Union. He then became chief of staff for Field Marshal Fedor von Bock's Army Group Center in the drive to Moscow in 1941. After the war he collaborated with other German officers and a historian in writing, from the German perspective, *The German Campaigns in the Balkans (Spring 1941),* which was published by the Department of the Army in November 1953. Von Greiffenberg died in June 1951.

During 1933, the 1st Cavalry (Mech) had undergone a period of vast activity. Organization, training, maintenance, and small unit tactics were all part of the adjustment to mechanization. This was the first army effort to determine if the machine could replace the horse, which was outlined in the Cavalry School's *1932–1933 Mechanized Cavalry* manual that reflected the attitude of the chief of cavalry, Major General Guy V. Henry. Meanwhile, through the efforts of General Lindsey, barracks were renovated and built. Personnel shortages and management became a challenge when the Civilian Conservation Corps sent companies to Fort Knox for training and to establish a district headquarters to administer and supply the various camps in western Kentucky. Equipment also was slow in coming. A chemical platoon arrived with its 4.2-inch mortars mounted on Ford truck chassis, providing smoke screens for the assaulting combat cars. The machine gun troop was in the process of receiving cross-country carriers called semitracked vehicles, or half-tracks. The armored car troop had already received its first M1 armored cars. The combat car platoon had to do with its four combat cars T1. By now Grow was convinced the Christies should not be driven in the convertible mode. Earlier during the Christie acceptance tests, the chief of ordnance, General Samuel Hof, objected to the convertible principle because it was too much of an engineering compromise. Years later the Ordnance Department concluded that the Christie convertible system was not feasible and deleted the wheel-track convertible principle. Meanwhile, the Ordnance Department began developing a new combat car, the T4, based on Christie's convertible principle and suspension system.

In late 1933, the early cavalry theorist on mechanization General Parker was posted military commander, Philippine Department. There he became involved in the strategic importance of another dimension of mechanization—airpower as a means of defense and promoting commercial interests. Though Parker believed American control of the archipelago was important for economic stability of the region, he nevertheless found the islands indefensible.[28] He promptly became aware of Japan's aggressive intentions in the region, reporting to the War Department a marked increase in the arrival of reserve Japanese military officers familiarizing themselves with the archipelago. He also reported to the War Department that local newspapers were concerned that Japan's aggression in China was a forerunner of a more ambitious plan for guardianship over the Far East; and as a result, Japan could strike the Philippines sooner than expected.[29] These concerns caused Parker to expand his interest in mechanization by focusing more on the possible use of airpower in the defense of the Philippines.[30] These reports by Parker should have alerted the War Department that mechanization in another dimension was becoming more important as Japanese aggression began to threaten the Pacific bastion. During this period, the Japanese were interested in developing radio-controlled tanks and replacing their foreign-purchased Vickers medium and light Renault tanks with their own designs, emphasizing lightweight tankettes. In addition, the Japanese were interested in developing amphibious tanks, having acquired plans from Christie in the mid-twenties for his amphibian, which also was unofficially tested by the U.S. Marine Corps off of Culebra Island in February 1923.[31]

By January 1934, the 1st Cavalry was on its way to solidifying doctrine, tactics, and an organization for mechanization. Doctrine visualized cavalry elements operating in small groups dispersed over a broad frontage with armored cars providing reconnaissance and the combat cars forging a spearhead supported by mobile supporting troops and engineers. Anticipation and activities picked up as the troopers at Fort Knox began training for the important spring

Combat Car T1 overturned during a command exercise before the trip to Fort Riley, Kansas. The officer with the look of dismay and hands on hips is H. H. D. Heiberg. Courtesy of the Heiberg Collection, Patton Museum.

maneuvers at Fort Riley, Kansas, home of the Cavalry School and the center of the army's foremost horse supporters and elites. All officers were expected to jump or play polo. It was anticipated that the faculty and students of the Command and General Staff School would observe the maneuvers designed to resolve the issue of marrying the horse with the machine.

The early years at Fort Knox prepared cavalrymen to experiment with a new mission to fight mounted rather than being relegated to the fringe of the battlefield engaging in reconnaissance, security, and defensive actions. The mechanized cavalry was undergoing a fundamental change because the horse was not an effective means for dealing with modern firepower. The organization of the 1st Cavalry (Mech) for the maneuvers consisted of an armored car troop for long-distance reconnaissance, a scout troop for close-in reconnaissance, the combat car squadron for the assault, and a holding element, the machine gun troop. Clearly, the organization was established to fight with new mounts. These cavalrymen planned to demonstrate this new approach to warfighting at the

Combat Car T1 being uprighted. Combat exercise before trip to Fort Riley, Kansas. Courtesy of the Heiberg Collection, Patton Museum.

upcoming maneuvers. Unfortunately, too many cavalry leaders at Fort Riley lacked the vision to realize that new mounts were required, because modern warfighting was driving the horse from the battlefield. The revolution was under way. As the saying went, "There are no foxholes for horses."

Most of the officers and troopers at Fort Knox considered themselves cavalrymen. Even though combat cars were new weapons for the cavalry—as tanks were originally for the infantry—there was a distinction. The combat car permitted cavalry to carry out its mission to fight mounted, and if necessary dismounted when it offered a tactical advantage. The combat car was the essential striking element. Even so, it was not the sole mounted element. A variety of new mounts were required and tactically tested, such as wheeled armored cars and scout cars, all supported by towed field artillery. The armored and scout cars were deployed for reconnaissance to facilitate the successful employment of the mechanized cavalry's striking force, the combat car squadrons. These organizational and equipment requirements for fighting mounted meant more to the mechanized cavalry than the infantry, whose primary mission was fighting dismounted. From the establishment of the Mechanized Force in 1930 until the arrival of the 1st Cavalry (Mech) at Fort Knox, Van Voorhis and Chaffee aimed to develop a mechanized arm capable of fighting with new mounts and independent of the main force. This required training and the ability to think mounted.[32] New means of waging mobile warfighting required rapid mental processing in executing command and control in an expanding time-and-space battlefield. During the early 1930s cavalrymen at Fort Knox realized this and worked under the most trying situations to formulate a new doctrine. Cavalrymen began more and more to demonstrate that change was ever present and provided new opportunities. This attitude was shortly to be tested at Fort Riley.

Before the maneuvers, the *Cavalry Journal* published in the September–October 1933 issue lectures by Major George S. Patton Jr. In these lectures, which had been delivered to regular and reserve officers at Fort Myer, Virginia, Patton explained that mechanized and horse cavalry could function together. He did see some utility in the independent use of mechanized cavalry; however, its deployment was the exception. Continuing in the traditional role of strategic reconnaissance cavalry, armored cars served in an intermediate position between the horse and the airplane. Those nations that start a war with machines, Patton wrote, will see their mechanized forces cease to exist in a matter of months. Yet fighting will go on, he continued, because "God takes care of horse replacements."[33] General Henry, a highly respected equestrian, felt comfortable with Patton's lectures. As chief of cavalry, he was considered preserving the so-called golden mean between overenthusiasm for new mounts and taking into consideration the limits of mechanization. The *Cavalry Journal* wrote that he endeavored to make the best of the situation. The chief of cavalry appeared before a congressional committee early in 1934, stating that both mechanized and horse cavalry were required in proper proportions and used in cooperation and coordination with each other.[34] Major General Malin Craig, a former chief of cavalry and later chief of staff, told Henry on one occasion, "You are making a great mistake in these changes you are introducing into the cavalry. It should remain armed and equipped as it was in the days of our fathers."[35]

Prior to the important Fort Riley maneuvers, the Ordnance Department had ready for experimentation a new combat car for the 1st Cavalry, a 9.5-ton convertible CC T4. Like Christie's CC T1, the CC T4 employed the convertible wheel-and-track and helical spring suspension system. Before the Fort Riley maneuvers, the Technical Committee briefly tested the T4 at Fort Knox, recommending that it, with modifications, be declared standard. It was a decision Chaffee strongly supported based on his earlier experience observing the Christie tank acceptance tests and comparing those with the combat car's operational mobility and speed. Earlier in September 1933, the secretary of war had imposed a 7.5-ton limit on tracked fighting vehicles; as a result, the Ordnance Department continued developing a lighter combat car and tank. The T4 weight was in excess of 8 tons.

While events were unfolding at Fort Knox and Fort Riley, the director of the Cavalry G-2 Division told the attendees at the War College war plans course in April that mechanized and horse cavalry would operate together. He noted, however, that only in exceptional cases could mechanized cavalry operate independently. Furthermore, he added, with the expansion of airpower and long-range artillery, mechanized and motorized cavalry would be compelled to operate at greater depths and wider dispersion. This being the case, he held, more cavalry with greater mobility would be required to accomplish the mission. Concluding, he told the attendees, "Don't let the performance of one machine, over selected peace-time terrain, give you false ideas as to the ability of masses of machines on battlefields." Besides, "remember the mud, the roads, and the traffic." What was being confirmed at the War Plans Division was that mechanization had a limited place within the cavalry's organization. It was emphasized that horse cavalry with the addition of greater firepower was not obsolete and was more suitable to operate in all types of terrain and weather. Chaffee's mechanized cavalry soon challenged these attitudes.[36]

Combat Car T1 makes a pit stop during the Fort Riley trip in 1934. Courtesy of Jon Hornbostel.

In spring 1934, the Iron Horse 1st Cavalry was commanded by Colonel Adna R. Chaffee Jr. Van Voorhis had left in February and had been reassigned as chief of staff, Hawaiian Department. Grow remained as the regimental executive. The Iron Horse cavalry, composed of approximately six hundred men, demonstrated its operational mobility by traveling overland from Fort Knox to Fort Riley, the column covered about 7.5 miles of road space and contained 189 vehicles, including 103 that were used for supply and maintenance. Six combat cars, including the Christies, made the trip. However, the four Christie CC 1s in Heiberg's platoon broke down and had to be loaded on a railway flat car and sent on to Fort Riley for repairs. Other vehicles of various types supplemented the six combat cars. The cavalrymen were greeted by Fort Riley's commanding general, Colonel Bruce Palmer from the Cavalry School, the chief of cavalry, Major General Leon Kromer, and Lieutenant Colonels Charles L. Scott, George S. Patton Jr., and Jonathan "Skinny" Wainwright, along with other high-ranking cavalry officers.[37]

The maneuvers were designed to determine how far cavalry, both mechanized and horsed, had progressed with mechanization and motorization, and in new weapons development. Because there was a lack of combat cars, trucks with yellow painted stripes were used to make up the deficiency, according to Chaffee's biographer. Of importance was resolving the type of missions the mechanized cavalry could execute with horsemen. It raised the imperative issue: Could the cavalry still be the "eyes and ears" of the army, or should it strike out in a new direction because of partial mechanization?

A tactical scenario was organized with the armored car troop engaging in distant reconnaissance with M1 armored cars and motorcycles. The scout troop was organized for close-in reconnaissance and security. The assault squadron was composed of the four CC T1s and other vehicles of various types. The holding unit was the machine gun troop mounted on T1E1 semitracked vehicles. Communication was primarily by voice radio, supplemented by motorcycles and automobile messengers.[38]

In the first series of maneuvers, the 1st Cavalry engaged a horse regiment; in the second series, the mechanized and horse regiments operated together against a hostile opponent. On May 8, the T4 and T5 arrived and went into a reserve position, resulting in General Henry establishing a board to evaluate the two new combat cars. The next day, A Battery, 1st Field Artillery, and the T4 and T5 operated for the first time with the 1st Cavalry. Heiberg remembered the Christies' stability was impressive when the .50-caliber machine guns were firing as the vehicles moved cross-country. On May 14, the cavalrymen from Fort Knox engaged in a wide envelopment maneuver that demonstrated the remarkable strategic mobility of the mechanized cavalry in deep operations, farther than any ground troops. This operation clearly demonstrated the extended sphere of action of 1st Cavalry. All these activities produced a divergence of opinion on the proper deployment of mechanized cavalry. At issue was to what extent the machine actually replaced the horse or extended the sphere of horse cavalry without altering the traditional mission of that arm.

The 1st Cavalry (Mech) report on the maneuvers concluded that during the exercises, mechanized cavalry in general "carried out all normal cavalry missions, such as reconnaissance, counterreconnaissance, delaying actions, seizing and holding positions, cooperation on the flanks of battle, and defensive and offensive combat." Most important, however, the maneuvers established that mechanized units greatly extended the sphere of cavalry action. Regarding mobility, the maneuvers brought out two important principles: speed on favorable roads was a requirement for armored cars, while tracked combat cars needed speed for shock action, especially over rough terrain. On the negative side, the mechanized unit proved least effective in counterreconnaissance and defensive actions. There were also problems with the combat cars because of equipment substitutions. The T11 armored car was recommended over the M1. Full-track carriers were recommended to replace the T1E1 for the infantry. Motorcycles, either with or without sidecars, proved ideal for communication. For artillery to be effective, the report noted, it needed to be moved forward. Those used in the maneuvers were drawn by commercial trucks and proved to lack mobility because they could not maintain march speed with the regiment.[39]

The report noted that attached aviation was valuable to the mechanized cavalry for distant air and ground reconnaissance. The report also recommended more peacetime training between these two branches because coordination was still a problem. As the maneuvers indicated at Fort Riley, tactical air support was still wanting. This was a situation the mechanized force had trouble dealing with throughout the 1930s. The reason was that the status between tactical and strategic airpower became more problematic for the line branches as the Air Corps was moving to the big bomber and strategic airpower. The chief of infantry, Major General Edward Croft, reflected this attitude during a discussion at the War College. He questioned the "mania for the machine that relegates the human being to a place of lesser importance." He told the chief of the army's strategic planning agency in the War Plans Division and attendees that the mission of the Air Corps was destruction of the enemy's air force and strategic bombing of its industrial centers. Regarding logistical support for an enemy force, Croft did not expect the Air Corps to attack "little field trains way up front." His solution for the army and the one adopted by the cavalry was to train men with machine guns, such as the .50-calibers, to "keep the planes pretty well up." To quote the chief of infantry, "I don't believe airplanes

Major General Leon B. Kromer, chief of cavalry. March 26, 1934, to March 25, 1938. Courtesy of *ARMOR* magazine.

are going down to hunt a sparrow when they have something else to hunt. I believe that is the teaching of the Air Corps Tactical School."[40]

The report on the maneuvers by the Cavalry School's Academic Division outlined tactical considerations, which mechanized cavalry might employ for independent action. These included delaying action, seizing and temporarily holding critical terrain, general reconnaissance, limited defensive action, offensive combat, flanking operations, and pursuit. Outlining unfavorable characteristics for independent action, the Academic Division listed the problems of night operations and a lack of dismounted firepower. The same tactical problems occurred when the mechanized cavalry operated with the horse cavalry. The report also acknowledged that mechanized cavalry operating with horse cavalry could provide an assaulting unit of great force over suitable terrain and therefore increase the zone of action. In one of its recommendations, the report suggested the Fort Knox troopers in the near future participate in more extended maneuvers and exercises with the Cavalry School Brigade horse units.[41]

The recommendations from the Academic Division clearly indicated that a serious attempt was being made to marry the machine with the horse. Major General Leon B. Kromer, the new chief of cavalry, commenting on the Riley maneuvers before the students at the Army War College, spoke favorably and optimistically of future combinations of horsed and mechanized cavalry. Regarding future wars, he stated, "As advancing tanks absorb hostile fire and enable the following infantrymen to cover the ground to the enemy with lessened casualties, so will the combat cars (of mechanized cavalry) under like conditions assist the horsed cavalry in closing with the enemy."[42]

The Infantry Board, which also witnessed the maneuvers, concluded that the cavalry had not appreciated the cross-country ability of track-laying vehicles, nor did it take full advantage of that ability during the maneuvers. Furthermore, the board reported, "The power of the mechanized cavalry regiment for independent action is overstated by many." The infantry observers nevertheless were impressed with the interest and energy displayed by all cavalry troopers participating in the maneuvers. The board was also impressed with the high morale and endurance of the mechanized cavalry and the horse cavalry's remarkable marching ability and discipline. In most of the maneuvers excessive speed was emphasized rather than finding and fixing the hostile force. The board concurred with the chief of cavalry that "the principle role of mechanized cavalry will probably be to supplement the action of horsed cavalry in performing recognized cavalry missions." Furthermore, the board noted, independent missions should only occasionally be assigned to the mechanized force. Regarding equipment, the board was impressed with the CC T4 and the CC T5, suggesting they be modified as infantry weapons. What the infantry had in mind was developing mobile tactics with a light tank that adhered to the restricted weight limit.[43]

What the various reports did not stress, according to Chaffee and Grow, was one important fact: that for the first time, a fully self-contained mechanized unit with largely substitute equipment was given the job of carrying out the cavalry role in direct confrontation with the horse cavalry. It was realized by a number of horsemen that mechanized cavalry could carry out the cavalry role and fight independently. No longer could it be said that the horse was indispensable or that mechanization when fully implemented could not carry out the cavalry mission. The Fort Riley maneuvers marked a turning point. "It was obvious to most of us that we had the right picture: a fully contained combat unit. It needed changes and equipment but all the elements were there," Grow later recalled.[44]

Chaffee commented, "Those fellows at Riley ought to understand that the definition of cavalry now includes troops of any kind equipped for highly mobile combat and not just mounted on horses. The motto of the [Cavalry] School says, 'Through Mobility We Conquer.' It does not say, 'Through Mobility on Horses Alone We Conquer.'"[45]

During and following the maneuvers, the Henry Board service tests demonstrated that the CC T4 outperformed the ordnance CC T5. The T5 displayed a double "Mae West" turret and a new rigid suspension system. The strange-looking

(Above) Convertible Combat Car T4 prototype, Fort Riley maneuvers, spring 1934. Courtesy of Jon Hornbostel. *(Below)* Nonconvertible Combat Car T5 prototype, 1934. Note the "Mae West" turret configuration and the vertical volute suspension system for the road wheels. This suspension system would be used in virtually all U.S. Army combat cars and tanks before and during World War II. Fort Riley maneuvers, spring 1934. Courtesy of the Patton Museum.

The World War I light tank M1917 used to demonstrate antitank weapons at Fort Riley, probably in the spring of 1934. The M1917 was obsolete, fragile, and could be immobilized by a crowbar. Though it was the most numerous tank in the U.S. Army, it was disliked by the cavalry. Courtesy of the Patton Museum.

double turret—named after a busty and bawdy entertainer of that era—was designed to make use of two machine guns for scouting missions. The T5 was a radical departure from the Christie design. The vehicle was fully tracked and nonconvertible, employing a volute spring or bogey suspension system with a divided power train. During the tests the Christie-type suspension system provided a more stable gun platform with better ditch-crossing capabilities, while the ordnance-designed vehicle was more maneuverable but so choppy in cross-country performances that accurate marching fire was impossible.[46] Understandably, observers at Fort Riley had not favorably viewed the Mae West profile.[47] The arrangement was necessary due to the divided power train where the engine was mounted in the rear and the transmission in the front of the combat car. A long drive shaft connected the two units. This was a problem because the tunnel that enclosed the drive shaft bisected the crew, causing an obstruction, hence the two side-by-side mounted turrets. General Henry was not impressed with the T5; instead he approved the CC T4, sending Heiberg to Rock Island Arsenal as the top consultant on its development.[48]

At year's end, however, Henry and Chaffee were overruled. The decision was made to acquire a modified T5 (minus the Mae West) for the cavalry. Generally, combat car proponents at user levels favored the T4. At the staff level, however, the War Department favored the weight and lower cost of the T5, thus taking advantage of the opportunity to produce a less expensive vehicle manufactured at the government's Rock Island Arsenal. With the CC T5, ordnance was trying to develop standard components mainly to get more for a buck, to simplify maintenance, and have components available in the event of mobilization. In addition, the T5 avoided the engineering dilemma imposed by the wheel-track convertible design. Heiberg recalled that the decision to adopt the T5 was made in the War Department "by officers [who had] probably never ridden in a tank, much less fired from one."[49] Heiberg's comments reflected what cavalrymen at Fort Knox had been thinking about since their arrival. They were looking for equipment that gave them agility and reliability. It became an issue of user requirements versus the technocrats in the Ordnance Department.

A War Department directive imposing a weight limit drove the decision for less than 8 tons. This solidified the army's preference for lightness and placed cavalry emphasis on speed and mobility over firepower and armor protection. The combat car was standardized for production as the M1. By fiscal year 1937, 89 M1s plus the T5 had been manufactured at Rock Island Arsenal for the cavalry. The Ordnance Department had its own designers and in addition had to keep its arsenals busy. Shortly the T2A1 was standardized as the light tank for the infantry. These vehicles shared basic components, such as the ordnance-designed volute suspension system, which remained characteristic of all U.S. tanks until late in World War II.[50]

A few years after the Fort Riley maneuvers, a German tank expert, Heinz Bach, commented on the cavalry's T5 Mae West combat car. He wrote in the *Militar Wochenblatt,* and repeated in the *New York Times,* that in spite of its high speed, the combat car was a "perfect example of bad construction. It rolls from side to side . . . its stability is limited. The two turrets interfere with each other. The Americans had repeated all the mistakes European tank and armored car builders had made" Furthermore, he criticized American tank armor as being too light to resist modern weapons.[51] Heinz Guderian, who was one of a number of emerging German army practitioners of armored warfare, noted that the Christie tank developed by the Red Army since the early 1930s was also too light. He did comment that it was a well-designed and well-tested machine with great speed.[52] The War Department, however, defended its mechanized

Combat Car M1 in woods, circa 1936. Courtesy of the Patton Museum.

equipment, especially the T5, claiming it compared favorably with that of any nation.[53]

Also, in 1934 the Indiana Motortruck Company, a subsidiary of White Motor Company, Cleveland, Ohio, developed the four-wheeled open-top scout car. A truck company rather than a touring car company now made the vehicles, which eventually became standardized as the M1 Scout Car. Seventy-six were produced and began to arrive at Fort Knox as the major reconnaissance vehicle for the mechanized cavalry. In 1935, a new modified design was introduced, the M2, that eventually resulted in the evolution of the M3 Scout Car, which was mass-produced by the White Motor Company. The scout cars were armed with .30- and .50-caliber machine guns.[54]

Meanwhile, the 1st Cavalry returned to Fort Knox after the maneuvers and underwent personnel turnovers. At the end of June 1934, Chaffee departed Fort Knox for duty in the War Department budget section, where he became instrumental in selling the mechanized force to the War Department and Congress. Bob Grow was reassigned as an instructor at the Command and Staff School, where he lectured on developments at Fort Knox and mechanization of the cavalry. His educational experience at Command and General Staff School at Fort Leavenworth and later at the War College in Washington, D.C., was reflective of the intellectual assets the army was building for future military requirements. Even though the army was small, Bob Grow recalled, peer union and education were important in developing leadership, especially required for higher commands. "When you play bridge, golf, softball, hand ball and ride horses with people, as well as see them in class, you get to know them pretty well. We all learned from each other and this had a profound effect when we went to war," he recalled.[55]

During this intellectual formative period the future army chief of staff, J. Lawton Collins, strongly believed, like Grow, that the school system—branch schools, the Command and General Staff School, the War College, and the Army Industrial College—saved the army. "If it weren't for the army school system I don't know what in the world we would have done." He did comment that the basic branch schools "concentrated on the techniques and tactics of their particular arm."[56] This arrangement of intellectual compartmentalization of the combat arms eventually caused serious organizational problems when the army rapidly expanded on the eve of the country's entry into another world war.

General Lindsey was on the verge of retirement, so he went on terminal leave. His legacy was that he attempted to improve base facilities, but he is remembered more for entertaining and playing golf, as well as designing the Lindsey Golf Course. Though the general had not condoned openly the use of liquor on the base—during prohibition years—he did find ways to imbibe when entertaining. According to Lieutenant White, his boss turned the other way and did not engage in witch hunts during Prohibition because liquor was readily available from moonshiners who were easily accessible around the hills of Kentucky.[57] With the general's departure, White was assigned to A Troop as an armored car platoon leader. General Henry now commanded the yet-to-be-formed-up 7th Cavalry Brigade (Mech). According to Grow, General Henry, who had a superb reputation as an equitation instructor, at first had been a cautious supporter of mechanization but now became enthusiastic. His perceptions had changed. He understood the new opportunities offered the cavalry branch, yet he still enjoyed riding and participating in horse shows. As the *Cavalry Journal* noted, he "made the best of the old and the new."[58] After the maneuvers, additional towed artillery and quartermaster elements were added to the expanding cavalry at Fort Knox. Air observation units were attached as needed.

Colonel Bruce Palmer, a promising horse cavalryman who entered the army as a private and eventually became a brigadier general, took command of the 1st Cavalry. Palmer was another key figure in developing cavalry mechanization.

A T5 half-track towing a 75mm howitzer M1, which was its prime mover, 68th Field Artillery (Mech-Towed). Courtesy of the Hunnicutt Collection, Patton Museum.

He was a close friend of the army's future chief of staff during World War II, General George C. Marshall, who had been a classmate at Fort Leavenworth's Cavalry and Infantry School and General Staff School. Palmer, according to his son the late General Bruce Palmer Jr., was an adored Irish romanticist who decided to challenge the status quo. Though he loved horses, he accepted their limitations.[59] He had an important impact on developing vehicles for cavalry tactics while at Fort Knox. Also joining the 1st Cavalry in 1934 was a future general officer, Willis D. Crittenberger, a graduate of West Point and the Cavalry School. Like many of the cavalry officers at Fort Knox during the 1930s, Crittenberger became a strong and influential supporter of mechanization in the army.

Months after the Riley maneuvers General Kromer admitted that there was no common understanding of cavalry combat. He said, "In no era of history has science placed at the disposal of military men such possibilities for increasing mobility." When discussing the fundamental tactical principles of cavalry, the chief accepted the augmentation of horse cavalry through mechanization but emphasized that the horse soldier, by taking advantage of terrain and making efficient use of .50-caliber antitank weapons, could successfully engage mechanized units.[60] What he said had some significance, since the Cavalry School was limited when it came to employing mechanized equipment, and the teaching regarding mechanization was not at all inclusive.

In the fall of 1934, the 1st Battalion, 68th Field Artillery (Mech-Towed), was placed on the active list at Fort Knox as a mechanized unit, adding to the all-arms force multiplier. The field artillery provided the 1st Cavalry with two towed firing batteries equipped with 75mm guns, passenger sedan cars, motorcycles with sidecars, and 1.5-ton trucks. By the end of the year, the cavalry had plans for mechanizing another regiment, the 13th, and combining it with the 1st, finally forming up the long-anticipated 7th Cavalry Brigade (Mech).

The next five years at Fort Knox were periods of continuous refocusing on convincing the chiefs of cavalry and the army staff of the importance of mechanization and the role of the mechanized cavalry brigade as a self-contained all-arms fighting force. This required army planners to reconsider the traditional doctrine of attritional open warfare driven by infantry rifle firepower and maneuver. Unfortunately, intellectual conservatism caused hesitation in spite of dynamic advances in technology. The army elites failed to appreciate the import of changes that might be expected on the modern mechanized battlefield. One could persuasively argue that the Great Depression had an impact because of budget restraints. Certainly, this hampered top army leaders when it came to equipment, quarters, and personnel funding. Even so, it should not have precluded the combat branch chiefs and the army staff from thinking about how the next war was to be fought and with what, especially in light of the emerging war clouds in Europe and the Pacific. Only when World War II was under way was the antiquated *1923 Field Service Regulation: Operations* revisited. Meanwhile, the struggle became more intense over the combat value of the horse over the machine.

General Malin Craig, a former chief of cavalry who had succeeded General MacArthur as chief of staff in October 1935, was now responsible for managing the mechanization of the army. During MacArthur's tour as chief of staff, tactical doctrine as to the use of tanks had undergone inadequate changes since World War I. Brigadier General Bradford G. Chynoweth—then a major—recalled MacArthur at the time being referred to as the "Champion of Modernization." After World War I Chynoweth was one of the nonconformists who had argued for cavalry tanks. At first, he placed the chief of staff on a pedestal, but later he developed mixed feelings. "Doug had stomped on every suggestion for change for four long years," Chynoweth remembered.[61] There were no new trends in organization during his tour except at Fort Knox, where he supported mechanization of the cavalry. As chief of staff, MacArthur was more interested in the "extraordinary advances" in airpower, especially those regarding a heavy bomber. They are "marvels of speed and endurance," he believed. He also believed airplanes had "an unquestioned ability to perform many of the reconnaissance missions that before the world war fell exclusively upon the cavalry." Furthermore, MacArthur strongly supported the establishment of a General Headquarter for the Air Corps so all elements of that service could occupy an independent status under a commander subject only to the orders of the army's General Staff.[62]

In addition, during MacArthur's tour two distinct opinions emerged from the Infantry School at Fort Benning regarding the tactical employment of tanks. The difference was over the role of accompanying or leading tanks. The conservatives or hard-liners, who were in the majority, were firmly against the independent use of tanks, reasoning that leading tanks impaired the support of other combat arms. The progressives or dissenters argued for leading tanks because their mobility and speed made them capable of independent action. This was why the progressives favored the Christie system. Chynoweth liked to call himself a progressive dissenter, favoring deep penetration for exploitation.[63] Patton once told Chen, "You are a Hell of a good soldier, with a Hell of a big mouth"[64]—a strange comment coming from Patton.

When Craig began his tour in 1935, he maintained a similar parochial attitude toward a workable tank or combat car policy. As a result, the mechanized cavalry spent the next years refocusing and defending its role and mission. Craig believed that mechanization of cavalry was correct. Nevertheless, he stated, "Motorization and mechanization have not driven the horse and mule from the battlefield. Their proven worth is still too valuable." His comments were in response to an article in the *Cavalry Journal* in 1936 that discussed the mechanization of cavalry in England and the Soviet Union. The author recognized the importance of mechanization; however, he outlined its limitations, such as trees, ravines, streams, darkness, and antitank weapons. As a result, mounted troopers always retained their traditional mission, he concluded.[65]

The attitude that the horse was not outmoded found its way to Fort Knox. Following the branch line, Colonel Palmer, the 1st Cavalry commanding officer, commented that new equipment and motorized units added to the versatility of horses by increasing maneuverability in all weather conditions and terrain, day or night. Careful not to disturb the status quo of senior cavalry officers, Palmer diplomatically stated that the horse still had a place on the battlefield, a comment reflecting the attitude of General Craig and the chief of cavalry.[66] While serving in the Office of Chief of Cavalry, Grow, like Palmer, diplomatically contained and tempered his inner beliefs about the new meaning of cavalry so as not to appear a nonconformist to traditional doctrine or unloyal to his chief. Yet Grow strongly believed in change, as did many other cavalrymen, especially those who were posted to Fort Knox. He believed the new meaning was defined as one cavalry arm with new mounts capable of fighting mounted and dismounted when necessary. It was a time when modernity and its proponents came in somber conflict with parochial military elites.

In spite of the obstinate attitude of army elites regarding mechanization, the years 1935 and 1936 were periods of extensive training and changes at Fort Knox. During this period tensions became more evident as the world moved closer to war. The year 1936 saw Hitler and Mussolini proclaiming their ambitions for a fascist empire. The year before, Hitler had denounced the disarmament clause of the Versailles treaty, and in March 1936, German troops reoccupied the Rhineland. Meanwhile, Mussolini also ignored the collective security of the League of Nations and invaded Ethiopia without a declaration of war in October 1935. By the end of 1936, a civil war in Spain had broken out in full intensity, involving Germany, Italy, and the Soviet Union. Hitler in turn signed the German-Japanese Anti-Comintern Pact, which became an extension of the Berlin-Rome alignment directed toward the France and Soviet Union alliance.

At Fort Knox these foreign developments were closely watched. Tactical exercises, meanwhile, continued through 1935, including placing strong emphasis on night operations and firing the .50-caliber machine guns at moving targets. In the spring, General Kromer, along with Major General Upton Birnie Jr., chief of field artillery, visited Fort Knox. Birnie was interested in the field artillery's organization as it participated in the Fifth Army maneuvers with 1st Cavalry. The maneuvers indicated that the 68th Field Artillery was capable of integrating with the mechanized force. The following year, training and exercises were conducted involving night marches and antiaircraft firing at towed targets with .50-caliber machine guns. In June the 1st Cavalry with the support of the 68th Field Artillery concluded a tactical march of 350 miles to Nashville. Troop A, with its armored cars, was given a difficult reconnaissance mission on the Ohio River, covering a 350-mile front from Paducah, Kentucky, to Cincinnati, Ohio.[67] In addition, aviation was considered in planning missions for the mechanized force. All these activities were in anticipation of the Knox-Michigan Second Army maneuvers scheduled for August 1936.

In July 1936, the 1st Cavalry issued an organization mission statement for a mechanized brigade. Prepared by Crittenberger, it reflected the unit's principal characteristics of mobility, firepower, and shock action similar to those of horse cavalry but with different dimensions in time and place. The mission statement noted that because of speed, greater firepower, and protective armor, the mechanized force had superior capabilities. Conversely, the statement indicated the force had limitations when it came to cross-country mobility, supply, and maintenance. The mission of the proposed brigade was to deploy armored cars as reconnaissance elements capable of extending fifty to one hundred miles in advance of each regiment. The combat car elements furnished the striking power, while the machine gun troop provided the necessary ground fire support. Their mission was to fire on hostile antitank weapons. In addition, the machine gun troop was displaced forward with its semitracked vehicles on the heels of combat cars to organize a holding pattern, cover the assembly area, and then redeploy for an attack. The machine gun troop was also given the mission of repelling hostile mechanized forces and covering a withdrawal if necessary. Mortar elements were to provide smoke on the hostile antitank area, thus neutralizing hostile gunners. Meanwhile, mechanized and motorized artillery supported the attack with suppressing fire. Command and control was by means of radio and motorcycle messengers.[68] This organization was significant because it demonstrated how the mechanized elements at Fort Knox were moving toward a tactical maneuver–oriented combined arms team. The mission statement of the 1st Cavalry was soon put to the test in August. However, one question remained, namely, the role of aviation.

Because of funding problems, the 1936 maneuvers were split between the V Corps at Fort Knox and VI Corps in Michigan. Grow—now assigned to the Supply and Budget Section in the Office of Chief of Cavalry—attended the Second Army maneuvers, as did General Kromer and Colonels Chaffee and Charles L. Scott. The mechanized cavalrymen were determined to demonstrate how far they had progressed since the Fort Riley maneuvers in 1934. For the first time the mechanized force was opposed by other arms of the army. Both opposing forces tactically employed their units in larger numbers. Numerous types of equipment and mounts were used, such as horses, trucks, combat cars, armored cars, scout cars, machine guns, towed artillery, mortars, motorcycles, and radios.

The first phase of the maneuvers was held in V Corps. It paired the mechanized cavalry (Red Force) against a hostile force (Blue Force), which included a horse brigade moving along the Ohio River toward Elizabethtown. Colonel Palmer, commanding the mechanized force, sent out his reconnaissance armored car troop and observation aircraft to locate the Blue Force. The planes employed were initially from the 113th Observation Squadron, Indiana National Guard, and later replaced by a flight from the 12th Observation Squadron from Brooks Field, Texas. Once located, Palmer deployed a force of motorized infantry and artillery to block the advancing enemy. He then enveloped the hostile force with the 1st Cavalry and supporting artillery. In another scenario, Palmer initiated a difficult night march and attacked the withdrawing Blue Force early in the morning before it could deploy in a defensive position. Palmer's action against the Blue Force demonstrated the elasticity of the mechanized cavalry's organization and its weapons. Attack aviation was to be employed with the mechanized force during the Fort Knox maneuvers according to prearranged plans with representatives from GHQ Air Force. Established in March 1935, GHQ Air Force became the combat arm of the Army Air Corps. With this new tactical organization, however, the agreement to cooperate with the mechanized force was difficult to execute. In addition, the distance between the airdrome and the mechanized force compromised the time element and frustrated the cavalrymen at Fort Knox.[69] The main reason, however, was a lack of interest and positive attitude on the part of GHQ Air Force.

At the time, the air force paid considerably more attention to developing a unified, independent, strategic striking doctrine than to providing tactical ground support. To understand the problem, it is necessary to appreciate the development of air doctrine. By the mid-1930s, ground attack aviation emphasis gave way to high-speed, long-range heavy bombers. The commander of GHQ Air Force, Major General Frank M. Andrews, a classmate of Chaffee and potential protégé of the future army chief of staff, General George C. Marshall, best explained the application of airpower during this period.

After graduating from West Point, Andrews served as a cavalry officer for eleven years. He was prepared to go with the 2nd Cavalry to France in World War I, but because of an interest in flying, he transferred to the Aviation Section of the Signal Corps. He, like Billy Mitchell earlier, became one of a number of airmen who argued for independent airpower and advocated the heavy bomber. Drawing on European developments, he argued that bombardment aviation in a future war must be emphasized because it is the principal force employed in modern warfare, especially against civilian populations and their economic centers. Unlike ground forces, there are no flanks, no rears, and different logistics for airpower. Besides, he argued, there are no air frontiers, adding, "It is hardly necessary to discuss here GHQ Air Force functions in direct support of ground forces." Besides, ground forces took too long to mobilize. His solution was to rely on bombardment aviation to neutralize enemy air bases and airplanes on the ground. Only a well-equipped air force with modern bombers was capable of mobilizing a force for immediate independent operations upon a declaration of war, he held.[70]

The march to Michigan and the Allegan County portions of the maneuvers, however, had not set well with a number of cavalry observers, who claimed no scenario was set up to demonstrate the power of a mechanized force. Rather, the force was restricted to road movement and used only for harassing and delaying tactics. According to Grow, no opportunity was offered to deploy Palmer's mechanized force as a mechanized brigade as during the Fort Knox phase. Time limits were imposed that were inconsistent with actual war conditions, thus further restricting mobility and maneuverability.[71]

Grow's comment had not taken into consideration the whole picture. There were disagreements. A quick learner, Colonel George C. Marshall commanded the enemy force, a brigade. During one part of the maneuver, I. D. White's armored car troop was assigned to Marshall's brigade. Marshall took notice. Unlike Grow, White claimed the army learned a lot from the split maneuvers.[72] In fact, the VI Corps commander and former chief of the War Plans Division, Major General Charles E. Kilbourne, reported to the War Department that the mechanized force, commanded by Colonel Palmer and his executive Lieutenant Colonel Crittenberger, demonstrated its mobile power by striking deep into the enemy's territory, disrupting communications and supplies. Regarding defensive actions, Kilbourne noted that the mechanized force could spoil an enemy's offensive by threatening its vital points, resulting in distracting or the redeployment of the hostile force. Some consideration, however limited, was paid to growing airpower. A three-plane formation from Langley Field simulated an air attack on the Allegan Airport by dropping a flare to simulate bombing.[73] Kilbourne's interesting report, which outlined the basic components for a lightning war, was issued three years before the Germans launched their blitzkrieg.

Colonel Marshall called the maneuvers an outstanding success. They were impressive and realistic, he noted. He praised Palmer's mechanized force, noting it "has been great, and has learned a lot, operating for the first time on an extended front in strange terrain."[74] Palmer said that the participation of his mechanized force marked an epoch in the history of the army. What he meant was that for the first time troopers from Fort Knox were able to effectively maneuver against all other tactically employed arms. The mechanized cavalry's mobility was proved. Palmer commented how his reconnaissance troop rapidly maneuvered around the flanks and rear and between the Blue Forces. He emphasized that his reconnaissance troop was seldom observed or attacked, adding that the troopers were astute in locating the enemy and promptly reporting their movements. In general, Palmer deployed a flexible organization, meaning rapidly attaching other units and mixing them with his regiment to form a combat team to deal with various phases of the maneuvers. Palmer, however, did recognize that there was a need for riflemen to protect flanks and rear areas from horse units and provide close-in protection for supporting artillery. He suggested the inclusion of a rifle troop in the mechanized cavalry organization for dismounted action. This recommendation was the forerunner that created armored infantry battalions, which became part of the combined arms team used by armored divisions in World War II. Finally, he suggested that aircraft be detached permanently to Fort Knox for air-ground observation and communication with the mechanized force.[75]

Because of the maneuvers, it was also recommended to eliminate heavy armored cars and semitracked vehicles from the mechanized cavalry's table of organization and equipment and replace them with four-wheeled scout cars fitted with some armor and machine guns for reconnaissance missions. White recalled that the M1 armored cars "were very heavy . . . and if you get them off the road in mud or steep terrain they were just impossible."[76] The move to a light scout car was to utilize the automobile industry's ability to mass-produce an effective fighting vehicle from standard light and cheap commercial vehicles.[77]

After the maneuvers, Scott took command of the "It Shall Be Done" regiment, the 13th Cavalry. Scott had been commissioned in the cavalry and served in various horse units until he became an instructor at the Cavalry School. The regiment was to be transferred from Fort Riley in September 1936 and mechanized. Before the move, the 7th Cavalry (Mech) was composed of Headquarters and Headquarters Troop, the 1st Cavalry (Mech), 68th Field Artillery (Mech-Towed), and supporting service units. Scott was a strong proponent of the M1 scout car over the M1 armored car, which he felt did not have the required flexibility for reconnaissance missions. Modified scout cars were the last open-top wheeled armored vehicles used by the cavalry. Eventually, Bantams or jeeps and the semitracked vehicle called a half-track, which provided better off-road mobility, replaced the scout cars.[78]

General Kromer, who was generally pleased with the maneuvers, endorsed the expansion of the mechanized force at Fort Knox. It was the beginning of the efforts to create a self-contained mechanized cavalry division capable of deep operations. The efforts at Fort Knox demonstrated that the mechanized cavalry possessed a doctrinal identity that soon doomed the horseman. As to be expected, it came in conflict with traditional army doctrine and the bureaucratic autonomy of the branch organization.

After returning from the Philippines in 1936, Frank Parker briefly resurrected his interest in combined arms mechanized warfare. He provided the assistant chief of staff, G-3, a dossier based on his last twenty years of military experience. In doing so, he did heed history's admonitions. Parker again supported the argument for a manual for commanders of large combined arms units and services. Regarding maneuver warfare, he emphasized the importance of rapid and deep exploitation that seeks to turn or surround the enemy's entire rear. Parker argued that until a system of large unit instruction was established, the small peacetime army would continue to function as a baseball team that conducted only individual practices rather than practicing as a whole team.[79] Not until the army's grand maneuvers of 1941 did small units have an opportunity for the first time to experience large-unit teamwork with combined arms. As one historian noted, "It was a turning point in the history of the U.S. military."[80] Parker was indeed a visionary who constantly looked for new opportunities to develop a maneuver-oriented mechanized force for the army at an operational level.

In spite of an urgency to redefine cavalry doctrine beyond traditional reconnaissance, pursuit, and exploitation, there were also equipment issues that altered a maneuver-oriented mechanized doctrine. The impediment was due to the emphasis on the dominant role of infantry, as directed by the 1920 National Defense Act and the *1923 Field Service Regulation: Operations*. In addition, there were severe limitations of congressional financial resourses for research and development of modern equipment required for mechanized warfare.[81] When Van Voorhis took command of the Mechanized Force in 1930, vehicles for the cavalry were driven by three military characteristics, mobility, fighting power, and minimal protection, all influencing the traditional policy of lightness. Once at Fort Knox, the tactical applications of cavalry fighting vehicles were based on scout cars for reconnaissance, combat cars for fighting mounted, towed artillery for support, and infantry personnel carriers and mortar mounts for fighting dismounted.[82] The emphasis, however, was still on fighting mounted in deep maneuver-oriented operations. Unfortunately, much-needed equipment for this type of operation was hard to come by. The process of mechanization was moving at a snaillike pace. Not only was it a time when the country was dealing with a depression, it was a time when Congress started passing neutrality acts that prohibited the export of arms and munitions to belligerents outside the United States.

Meanwhile, the Wehrmacht's interest in military developments in the United States continued. Writing in the *Militär Wissenschaftliche Rundschau* in January 1936, Colonel Guderian—now considered by the U.S. Army attaché in Berlin as one of the foremost experts on motorization and tanks—noted that the United States occupied first position among all countries in the world regarding the technical production of its automobiles. Its army, however, has not yet participated in this economic development, he wrote. Guderian criticized the U.S. Army for not paying special attention to Christie tanks, which were given their greatest realization in the Red Army as fast tanks for deep operations. He was also critical of the autonomy of the U.S. Army branch system that gave control of tanks to infantry and to a cavalry reference of tanks as combat cars. Concluding, he rightly noted that a consolidated authority for tank development was lacking.[83] This was an interesting observation from a foreign officer, who apparently had a greater understanding of the obstacles caused by branch compartmentalization than did the U.S. Army. To him this organizational condition stymied efforts to coordinate an all-arms tank policy at an operational level.

In November 1936, Guderian published another article in the *Militär Wissenschaftliche Rundschau,* which reflected his strong interest in mechanized warfare as articulated in the United States, England, France, and the Soviet

Union. The article represented the official doctrine regarding the employment of tanks in mechanized warfare. Guderian emphasized that the striking power of armored troops must rely on firepower, speed, and armor protection. Though the tank was the main maneuver weapon, it must also rely on the cooperation of other combat arms, he argued. Guderian quoted the famous British tank proponent J. F. C. Fuller, who stated that tanks tied to infantry decreased the value of that weapon, a problem he found in the American and French armies. The mission of motorized infantry supported by both towed and the new self-propelled mount was to augment the effect of a mass tank attack. Regarding airpower, Guderian saw the importance of providing support for ground attacks. Concluding, he stressed opposition to infantry-accompanying tanks. He emphasized the significance of tank mobility, mass, and surprise and the importance of auxiliary combat arms as organic to tank forces. What Guderian was saying was that armor needed to be deployed en masse to ensure the effects of combined arms combat power at the decisive place and time.[84] Guderian's comments found their way to the G-2, Military Intelligence Division, and the War College. His ideas about a maneuver-oriented operational level of warfare at a campaign level were not digested. The closest to these ideas had been and continued to be formulated at Fort Knox.

Nevertheless, attempts were also made in the U.S. Army to deal with the issue of mechanized divisions similar to those being developed in Germany. In 1936, the Command and General Staff School at Fort Leavenworth published a tentative table of organization for a mechanized division. The driving force behind these instructional tables necessary for studying organization and tactics of a mechanized division was made up of the assistant commandant of the school, Colonel J. A. McAndrew (infantry) and Lieutenant Colonel Allen F. Kingman (infantry).

Bob Grow remembered when he was an instructor at Leavenworth in 1934–1935 that McAndrew and Kingman were sympathetic to combined arms and mechanization.[85] While at the War College, Kingman discussed tactical characteristics and missions of a mechanized force with tank units. He argued that a mechanized force possessed a high degree of independent strategic and tactical mobility with great striking power. This type of mission, he argued, can be more effectively executed by it than by the other combat arms. He suggested the mechanized force could restore movement in higher commands, such as at the army level and the groups of armies. Kingman talked about wide envelopment and turning movements, deep exploitation, and distant pursuit. Though articulated, he had not realized he was placing the mechanized force at a potential operational level of warfare—the campaign to achieve strategic objectives. In contrast, he believed tank units at a tactical level had a different mission, supporting the assaulting riflemen in the ensuing battle. He concluded, "A mechanized force must never be tied to assaulting rifle units."[86]

In 1936, Sereno E. Brett replaced Kingman and continued spreading new ideas regarding mechanized warfare. The following year, when Brett joined the faculty at Command and General Staff School, an instructional text was published entitled *Tactical Employment of the Mechanized Division (Tentative).* It described the organization and tactical employment of a mechanized division. The mechanized force was described as an all-arms, self-contained force capable of deep, independent operations, leading to pursuit and exploitation of combat successes. This doctrine was similar to what the mechanized cavalry had been working out for years at Fort Knox. Adding to force mobility, the text described the importance of aviation for command and control, reconnaissance, and tactical ground support.[87] Unfortunately, the mechanized division text was only for instruction because it was doubtful the chiefs of cavalry and infantry would agree to another combat branch at their own expense. Army elites still failed to realize what was happening; new mechanized tactics were a breakthrough that lead to a new operational or strategic means of warfighting.

Another text, *The Tactical Employment of Cavalry*, was published at the school, stressing principles and methods applicable to both horse and mechanized cavalry. Although the text emphasized combining horse and mechanized cavalry, it noted that mechanized cavalry was a powerful striking force of great strategical and tactical mobility over favorable terrain. Regarding a mechanized cavalry doctrine, the text stated that the sphere of action was much greater and offered increased speed in the performance of missions provided it did not alter the fundamental mission of traditional cavalry, which depended on the horse. This attitude sustained traditional doctrine, which was stealth "reconnaissance; interference with or obstruction of the enemy's movements, supply, or control; deception of the enemy; and participation in the decisive phase of the battle to assist in gaining the decision."[88]

What made *The Tactical Employment of Cavalry* interesting was that it challenged the army's traditional strategic doctrine of open warfare driven by the infantry-artillery team. The essential characteristics of cavalry mobility and firepower were defined in the cavalry text. It was in disagreement with the long-held idea of General Pershing's open warfare and maneuver with rifle firepower, which was expected to solve the problems of stabilized and attrition-type warfare. However, this was not the case. The pulverizing strategy of attrition warfare was still the historical norm for the army, with the infantry as the dominant combat arm. Because the cavalry was trained for a high degree of mobility, it was expected to execute open warfare with the correct employment of its mounted horse units along with mechanized cavalry, all in support of the infantry. It made clear, however, that mechanized cavalry was vulnerable to antitank weapons and unsuitable terrain. In addition to these criticisms of mechanized units, the text noted, "unlike horses, mechanized units cease to function as soon as the fuel is exhausted." Furthermore, the development of fighting vehicles was constantly in a state of limited progress and, unlike horses, was not definitely fixed. The text noted that, as in the past, mounted horse units on occasion were expected to fight dismounted, similar to infantry. Though reluctantly accepting mechanization, many conservative cavalrymen still clung to the past, trying to rationalize a future for their adored horses.

The cavalry text also defined the role of armored and scout cars as both having the primary mission of distant reconnaissance beyond the sphere of action or important combat actions that could not readily be performed by horse or aviation units. Combat cars with their striking power were considered to add greatly to the power of horse elements through augmentation. Unlike the more farsighted but doomed *Tactical Employment of the Mechanized Division*, *The Tactical Employment of Cavalry* paid little attention to the role of aviation other than battle reconnaissance and communication. The horse soldiers significantly influenced the text by underlining the importance of combining horse and mechanized cavalry elements, suggesting the latter be held in corps reserve to be employed in cooperation with the infantry in exploitation of a breakthrough or encirclement of the enemy's rear.[89] This came in conflict with the Leavenworth text *Tactical Employment of Infantry (Tentative)*, which defined the mission of tanks as twofold, supporting elements and leading elements. The former was to operate closely with foot soldiers, whereas the latter was to engage in a breakthrough and disrupt the enemy's defense system in the rear area, similar to the cavalry mission of exploitation.[90] Clearly, this was a reflection of infantry-dominated leadership and doctrine that continued throughout the 1930s.

5

The "Great Cavalry Debate" over New Opportunities

By the end of 1936, the 1st Cavalry (Mech) and the 13th Cavalry (Mech) had merged into the 7th Cavalry Brigade under the command of now Brigadier General Van Voorhis, who had returned to Fort Knox after his tour in Hawaii. Colonel Charles L. Scott from the Office of Chief of Cavalry in charge of material and equipment was appointed commander of the 13th Cavalry. Meanwhile, Major General Leon B. Kromer began thinking about a table of organization for a mechanized cavalry division in anticipation of expansion, provided it was supplied with sufficient equipment and supporting assets.[1]

When Van Voorhis returned, Congress, the president, and the American people still firmly believed the country could avoid future wars in spite of the imminent threat of another major world conflict. The consensus was that the nation's goal of hemisphere defense could be achieved with a minimum of military strength and by avoiding entangling commitments to former wartime allies. It was expected that these goals were obtainable through diplomatic efforts that promoted peace and arms limitations. Finally, reality set in and efforts were made to strengthen the army. General Malin Craig's Protective Mobilization Plan (PMP) emphasized the manpower factor over research and modern equipment, yet still maintaining serviceable equipment to deal with future contingencies.[2] However, the army chief of staff did authorize under the PMP 292 M2 combat cars for the cavalry and the modernization of the M1 combat cars on hand. His efforts, though, were generally conservative due to financial restraints; as a result, he had serious problems keeping the army prepared, at least in comparison with the military buildups in Germany, Italy, Japan, and the Soviet Union.

In 1937, the Cavalry School issued a book entitled *Cavalry Combat*, copyrighted by the U.S. Cavalry Association. Kromer wrote the foreword, arguing that there was a need for a compilation of historical examples illustrating various types of cavalry combat. General Kromer realized that mobility was the antithesis of static warfare. With the creation of open flanks, he wrote, today's cavalry could perform its time-honored mission by increasing the magnitude of mounted combat. A reviewer for the *Infantry Journal* called the book "a monument to the horse and its gallant rider." He praised numerous historical examples describing the exploits of horse cavalry, reluctantly adding, "last and least," a chapter on mechanization. This chapter occupied only 9 out of a total of 512 pages. The reviewer, quoting from the book, stated, "The use of armor must result in a compromise between the opposing factors of protection and mobility; the attempt to provide absolute protection by adding armor (weight) can result only in decreasing the mobility of the vehicle to such a degree that it becomes valueless to the cavalry." In the book's conclusion, motorization and mechanization were accepted. Horse cavalry, however, was still believed to be more sustainable on the modern battlefield and not dependent upon fuel refinement and mechanical perfection. Thus, by the end of 1937, there was a concerted effort by some to refocus on the reason for the existence of the horse in spite of continued advances in mechanization and motorization.[3]

Because of a lack of funds, the 7th Cavalry Brigade's scheduled maneuvers and exercises at Fort Riley in 1937 were canceled. Instead, a small detachment from the Cavalry School visited Fort Knox in May and witnessed a number of demonstrations and night marches. Most prominent was the extent of operations developed by the mechanized cavalry brigade. The reconnaissance troop demonstrated its capabilities in gathering intelligence. The mortar platoon placed smoke hundreds of yards in front of the attacking combat cars, while the machine gun troop demonstrated its ability to move forward and hold vital terrain captured by the attacking force. One of the most interesting operations was when the mechanized brigade made an all-night march south to Horse Cave on historic Highway 31 East, once a buffalo trail. The following morning, with its advance guard out, the brigade simulated an attack on the Fort Knox military reservation. The direction of attack, supported by smoke from the mortar platoon and the 1st Battalion, 68th Field Artillery, was along Wilson Road. The exercise demonstrated the brigade's ability to engage in deep operations at night, marching 125 miles in darkness and then launching a morning attack. Van Voorhis was pleased with Colonels Bruce Palmer and Charles Scott, who handled their troops and columns with great ability.[4]

What was the reaction of the visitors from Fort Riley?

They concluded that cavalry had two types of similar missions with two different means available to accomplish them, the horse and the machine. The Fort Riley visitors deduced that a soldier's education was not complete until he realized he was both a horseman and a mechanized cavalryman. The visitors also noted that horse cavalry had not changed in a number of years and that in the current combat environment it could still operate under all conditions. Mechanized cavalry—even though it could travel about five times faster than horse cavalry—was still subject to terrain obstacles and gas, oil, and water requirements. The visitors, reflecting the Cavalry School's attitude, suggested that both elements could operate together, accomplishing a single mission under ideal conditions where speed, action, and surprise were important.[5]

After the visitors from the Cavalry School left Fort Knox, Germany's Lieutenant Colonel Adolf von Schell returned to the United States. At the time, he was chief of staff of the Inspector of the Panzer Corps and Army Motorization. Previously he had been a tactical instructor at the Kriegsakademie. In January 1936, Truman Smith, the U.S. Army attaché in Berlin wrote then Colonel George C. Marshall, senior instructor of the Illinois National Guard, that his friend von Schell had become one of the most respected instructors at the Kriegsakademie. The chief of general staff of the army, General Ludwig Beck, considered von Schell "one of the half-dozen finest tactical instructors of the German Army." Smith added that von Schell "remains a devoted friend of America." However, Smith confidentially informed Marshall that the Nazi government "will be backed by one of the if not the most powerful armies in the world. I look for eventual trouble," noting that Germany probably would expand to the east. In March 1935, Hitler had announced the expansion of the army to thirty-six divisions and three hundred thousand men. Conscription was reintroduced, and in May a new armed forces law established a separate air force. In March 1936, Hitler took a major gamble when he sent German troops across the Rhine River and occupied the demilitarized Rhineland, thus securing again an invasion route into Belgium and France. Interestingly, Smith told Marshall that England and France would probably give Hitler a free hand. Meanwhile, Marshall had congratulated von Schell on his promotion. Smith added that this compliment pleased von Schell and no doubt will "help my position immensely." Apparently, the amiable relationship between the American and German military ended when Smith later reported that General Beck was retired in 1938 for political reasons, meaning that there was a lack of understanding between Beck and Hitler over foreign affairs. Beck let it be known that Hitler's bellicose foreign policy of conquest, especially the annexation of Austria into the Nazi Reich and claims to the Sudetenland areas of Czechoslovakia, would eventually end in a total disaster for Germany.[6]

Arriving in June 1937, von Schell's mission from the Reichskriegsministerium (War Department) was to report on mass-production methods and standardization used by the American automobile industry. The Wehrmacht at the time had far too many makes of motor vehicles, and von Schell realized that changes had to be made to meet military requirements as the army rapidly expanded. Von Schell's other mission, however, was more important. He had one month to visit twelve military bases to assess the degree of U.S. Army mechanization and tactics, especially at Fort Knox and Fort Benning and other bases that had tanks or combat cars.[7]

The visits were a result of courtesies extended by the German government to Truman Smith. When Smith was an instructor in tactics at Fort Benning, he befriended von Schell, who was in the advanced course during 1930–1931. This relationship brought to Smith's attention at the time how revolutionary Reichswehr ideas were and the changes being made in German military doctrine. Smith believed at the time that the "German Army would outstrip our own unless we kept abreast of their new weapons and tactics."[8]

The mutual arrangement also provided an atmosphere for selected U.S. Army personnel to attend the Kriegsakademie. Smith arranged visits to Wehrmacht military establishments to inspect their armored fighting vehicles and military developments. While in Germany, Smith and his staff placed heavy emphasis on reports concerning the new panzer divisions. They even reported that the Wehrmacht had practically abolished the cavalry as a result of the creation of panzer divisions. The attaché reported that special favors were given to American officers because the anti-Nazi General Ludwig Beck was convinced of the need to keep the United States out of another European war. Smith noted that many in the German officer corps were aware of the American intervention in World War I that brought about their country's defeat. They were especially cognizant of America's industrial might.

Meanwhile, von Schell's specific goals during his visits from June 21 to July 23 were to examine the infantry's new light tank and the cavalry's combat car, observe their maneuvers, take short rides on roads and cross-country, and take external photos. The only restriction was that technical details concerning design and characteristics were to be avoided if possible. Twelve military bases were on his list to visit.[9] Only three visits, however, provided interesting exchanges on tactics and mechanization occurring in both countries. Fort Riley was not suitable to visit at the time because the Cavalry School was not in session and little activity was going on at the base. He also had to skip a number of bases due to time. The underlying purpose for von Schell's visits was to compare the substance of the German *1933–1934 Field Service Regulation, Die Truppenfuhrung,* which emphasized mission-directed tactics or *Auftragstaktik* of the all-arms team and local command and control initiative, to determine if similar developments were occurring in the U.S. Army. Before beginning his tour, he and the German military and air attaché in Washington paid a courtesy visit to

Kromer; the chief of infantry, Major General George A. Lynch; and the chief of ordnance, Major General William H. Tschappat.

At Fort Meade the observant von Schell visited the 66th Infantry (Light Tanks), observed a combat demonstration, rode in a M2 light tank, and inspected tank parks and repair shops. Summaries of his impressions were submitted to the War Department Military Intelligence Division (MID), then under the staff command of the assistant chief of staff, G-2, Colonel E. R. Warner McCabe. Regarding tactical doctrine, von Schell criticized the U.S. Army for attaching tanks to infantry and suggested instead they be given an independent mission so a breakthrough could be made that was broad and deep enough for successful exploitation. Though he commented little on technical details, von Schell found that the light infantry M2s had a noisy gearshift and thin armor. Nevertheless, he thought the M2 was a smooth-riding tank and was impressed with its speed and reserve power.[10]

Von Schell's remarks on European tanks and doctrine also were illuminating. The 66th Infantry's commander, Colonel S. B. Buckner Jr., stated that they merited serious consideration in connection with the army's future tank doctrine. Almost predicting the military success of the German invasion of France in May 1940, von Schell commented that French doctrine contemplated scattering tanks over wide fronts. Accordingly, he forecasted they could lose most of their tanks in the first battle. He gave credit to Red Army tanks used in the Spanish civil war but criticized the poor performance of Spanish tankers who had not used tanks en masse, preferring instead to employ a few at a time. Nevertheless, he inferred the Red Army had a sound tank doctrine because it believed in the principle of deploying tanks en masse at an operational level. He added that the army's leadership was rather weak because Stalin's regime had begun purging top military leaders. Von Schell had little respect for the Italians, claiming they were not fighters and knew little about tank deployment. Regarding the British, he thought they had dropped behind in tank development and tactics. Interestingly, though, von Schell found British military leaders inclined to be more visionary than realistic regarding tactical doctrine.[11]

Next on his schedule was the Infantry School at Fort Benning, Georgia, where he had the opportunity to again examine the M2 light tank. He respected the tank's speed but criticized its high silhouette. The "Mae West" turrets created too many blind spots, necessitating that the crew stand erect. He also found the tank's armor and armament insufficient for modern combat. This, he noted, would make American tanks too vulnerable. Von Schell was puzzled that infantry and cavalry should employ the same vehicle for different tactical uses. Again, he was critical of the U.S. Army's idea for attaching tank companies, battalions, or even regiments to accompany the infantry. This, he believed, squandered a mobile asset. Regarding German panzers, he alleged that they would be employed en masse at decisive points, preferably as panzer divisions or corps. What von Schell explained to his American hosts was the German doctrine of the point of main effort, known as the *Schwerpunkt.* He also commented that periscopes were indispensable and that all tanks should be equipped with radios for communication and command and control.[12]

Von Schell found the development of local command and control initiative wanting in young officers who attended the Infantry School. "They have a long road to travel," he noted. Reflecting back to what he explained in *Battle Leadership*, he told his hosts as they walked through the Academic Building that young officers should be thoroughly familiar with tactics and techniques of smaller units, the company and the battalion. He even stated that young officers should know the mission of the battalion and regiment and "be competent to function as regimental staff officers." The problem, he felt, started in the Infantry School's large class sizes instead of manageable classes of twelve to fifteen students. This arrangement "is not desirable." More could be accomplished by a "preceptorial system of instruction." This system, he felt, produced more thoroughly instructed students and permitted the commandant to exercise all the control necessary to assure such uniformity of doctrine as desired. In the German army, he stated, officers were trained to act rather than to deliberate, adding, "The quest for perfection spoils the attainable good."[13] These suggestions were akin to the German tactical philosophy of *Auftragstaktik*, or mission-directed tactics that encouraged the coordination of the commander's intent with the lowest subcommander's ability to demonstrate the initiative and agility as the combat environment developed.

Next on his list was the largest tank or combat car formation in the U.S. Army, the 7th Cavalry Brigade at Fort Knox, which von Schell visited for a few days at the end of June as a guest of Colonels Palmer and Scott. Van Voorhis, who was in Washington at the time of the visit, viewed von Schell as a "prominent German General Staff Officer" and an expert on the subject of shock divisions that were composed of all combat arms.

Von Schell arrived in Louisville by train, dressed in civilian clothes, and was met by Lieutenant Colonel Willis D. Crittenberger, the brigade's operations and training officer, and Captain Gustave W. West, Van Voorhis's aide. They drove in an army car to the base, where Colonel Bruce Palmer, who was acting in Van Voorhis's place, welcomed von Schell, although not before the visitor had changed into his dress uniform, which displayed the coveted red leg stripes of a staff officer and decorations, including an Iron Cross and a medal indicating four battle wounds received during the last war. His highly polished black boots were fitted with silver spurs. Noticeable was the short sword dangling from his

Colonel Bruce Palmer greeting Colonel Adolf von Schell at Fort Knox, Kentucky, June 29, 1937. Photograph copyright © *The Louisville Courier-Journal.*

side that displayed the Nazi swastika.[14] His erect, steel-like demeanor and comments were not lost on observant cavalrymen, especially Crittenberger, who reported the visit to Van Voorhis when he returned from Washington.

During his visit, von Schell witnessed extensive tactical demonstrations and maneuvers of the brigade's personnel and their use of mechanized equipment. He examined the CC M1 and observed a tactical exercise in which the 1st Cavalry, one squadron of the 13th Cavalry, and the 68th Field Artillery participated. Afterward, von Shell made a number of comments to Palmer, Scott, Crittenberger, and other officers. He found the .50-caliber machine gun used on the combat cars as an antitank weapon impractical, arguing that in the next conflict the cavalry could not avoid heavier tanks in an infantry fight. In this combat environment, he argued, the .50-calibers were useless. Furthermore, von Schell added that self-propelled howitzers must support tanks rather than 75mm guns, which he felt should be replaced if the U.S. Army was to field a new mechanized division.[15]

He questioned the U.S. Army's autonomous branch system that was dominated by the infantry's linear idea of open warfare and rifle firepower. This compartmentalization degraded the all-arms concept because artillery, cavalry, and infantry commanders knew only subordinate commanders in their branches. The cross-coordination of combat arms was important to von Schell because leadership, knowledge, and trust within the command structure allowed all to understand how others would react to various combat situations. The cavalry officers advised von Schell that the World War I experience of pounding attrition was not planned for operations in the North American theater. He hesitated, considering this defensive hemisphere attitude impractical because in the future the U.S. military might find itself again in a European war. If this were the case, he argued, the army should plan for an organization to meet the combat environment that potentially existed on the continent, which by then might contain a preponderance of tanks and mechanized forces.[16]

Regarding German mechanization and motorization programs, he stated that their development was placed under one head. This, he claimed, eliminated a duplication of effort, equipment, and expense. For example, the Panzer Corps of three armored divisions drove German mechanization. Each division had a mechanized brigade capable of employing hundreds of tanks and a motorized *Schutzen* brigade for holding. A *Schutzen* brigade was composed of motorized and motorcycle infantry units, 105mm howitzer batteries, motor reconnaissance, engineers, and signal and antitank elements. The three panzer divisions were organized to perform the infantry role as well as the cavalry role, all to be deployed en masse.[17] This assessment of the German war machine was a reiteration of Guderian's earlier concepts on a new mobile warfighting doctrine that found its way to the U.S. War College. As the German military establishment was moving to a new concept of warfare at an operational level, U.S. Army elites vacillated between the infantry and cavalry over who was to control tanks at a tactical level. This situation seemed ludicrous to the Germans.

General Van Voorhis's comment later regarding von Schell's visit was "that there is indicated a necessity for a fixed organization as a heavy striking force composed of all elements" in the U.S. Army. Most important, he stated, equipment must be designed to meet the needs of this type of doctrine. What he stated was important: that doctrine needed to be first clarified before equipment was procured. Equipment was to be the means to carry out the defined doctrine. When asked how German armored divisions with their ponderous masses of mechanized and motorized equipment and personnel would operate around hostile flanks and rear, von Shell replied, "There will be no flanks in a major European war except those flanks created by the breakthrough and subsequent rolling up of hostile lines, in which case the German Armored Division would be able to operate very effectually." Van Voorhis found this statement of great interest. However, he said, "It may not be wise to base our expansion of mechanization upon our observations of European armies." Nevertheless, he continued, "The power and versatility of the combined efforts of ground troops and combat vehicles must be recognized."[18]

In regard to his assessment of developments at Fort Knox, von Schell also mentioned to Palmer, Scott, Crittenberger, and other officers, "You are searching and experimenting along the same lines as my army in your efforts to overcome hostile antitank weapons. It is a combination of speed, armor and all other means we can devise, including smoke and mass attack to deal with this serious problem." The German staff officer stated that European tanks in the near future were likely to be heavier and carry more armor plate, whereas light tanks, like the combat cars, probably would be relegated to a reconnaissance role. He felt, however, that major European powers were not ready for a war in the near future. The French were too provincial, the Italians too tempestuous, and the Spanish too decadent, he claimed. However, he was concerned about England because of its ability to control key points in Europe, such as Gibraltar and the entrance to the Mediterranean Sea, the Suez Canal, and outposts of the northeastern Atlantic.[19] Von Schell was very perceptive.

The reports compiled by Crittenberger on von Schell's visits were of great interest, especially to the cavalry at Fort Knox. Crittenberger also advised the pro-military senator from Massachusetts, Republican Henry Cabot Lodge Jr., of the visit and von Schell's comments. Lodge was a strong proponent of military preparedness and served on the Senate Military Affairs Committee. Crittenberger also wrote Lodge he was pleased that his recent orders to the Command

and General Staff School finally had been revoked. "I am delighted that I am to be permitted to remain here, at least another year, continuing my association with mechanization," he stated. Crittenberger wanted to stay with Van Voorhis as his operations officer so he could "follow through several projects in the development of mechanization which need strong support at the present time." His goal, as operations officer, was to make extensive efforts for the 7th Cavalry Brigade to receive appropriate equipment, which was slow in coming. He advised the senator, "Not until it is completely equipped will we be able to determine its true value as a combat element."[20] Crittenberger and Lodge became longtime friends.

Overall, most of the officers were impressed by von Schell's comments. That he was genuinely pro-American was the usual comment. Besides criticizing the U.S. Army's autonomous branch system, he felt its boards were self-serving and took too much time, thus limiting cross-coordination between the combat arms. The German army, he added, had no Cavalry, Infantry, or Field Artillery Boards. He let it be known that American tanks and combat cars were too light and too vulnerable, with inadequate armor. They made better reconnaissance vehicles, he believed. He explained to his hosts the principle of mass, meaning concentrating panzers at the decisive point rather than deploying them as infantry-accompanying tanks, which he called "frittering away." Tanks, he believed, were GHQ and army units; the brigade of two tank regiments was the smallest tactical unit, he indicated. All this information became available from the military forts he visited and was passed up to McCabe at G-2 and circulated through the army staff system.

Bob Grow commented that attaché reports dealing with foreign mechanization were also extensively studied at the Office of the Chief of Cavalry, especially reports from Truman Smith in Berlin. Based on this information, he believed at the time that U.S. mechanized cavalry was ahead, in some respects, of the Germans and way ahead of the French in the doctrine of employment.[21] Obviously, the cavalry leadership at Fort Knox had the opportunity to grasp von Schell's ideas of moving from a tactical to an operational level that the Wehrmacht was perfecting with its panzers and tactical air support. However, the move to this form of operational art was not possible then in the United States because of funding, the attitudes of the chiefs of staff, the heavy-bomber proponents, and a nation bent on neutrality and isolationism, but more so, due to the branch chiefs and their dogma and compartmentalized organization. Though the cavalrymen at Fort Knox were voices of change and building a consensus, their concepts never became an engine of change as happened in the Wehrmacht.

Scott later claimed that Van Voorhis and Chaffee had already independently thought out this type of organization and the importance of mass deployment of armored units. In a few years, these ideas were the ones implemented by the

Germans, though expanded, during the blitzkrieg, where panzers played the dominant role. All this was made known to the U.S. Army combat arms and the army staff when von Schell visited the United States in the summer of 1937.[22] Van Voorhis and Grow, however, implied that it was the Germans that learned from the tactical changes going on at Fort Knox.

Von Schell's trip to San Francisco was canceled due to a lack of time. Marshall was disappointed that von Schell did not visit, writing, "I would have welcomed the opportunity to talk over the things in general, and military matters, particularly." Von Schell also expressed his regret to his American friend. He had to return to Germany before the Wehrmacht's fall maneuvers, which were expected to be the largest held in Germany since World War I and designed in part to reassess the effectiveness of tactical air supporting tanks in independent operations.[23] Von Schell also wrote Marshall he always remembered his days at Fort Benning with great pleasure and fond remembrances. He kindly informed Marshall that their relationship "brought much to me that was highly important to my profession."[24] By that time, Brigadier General Marshall had taken over command of the 5th Infantry Brigade, 3rd Division, at Vancouver Barracks, Washington.

Certainly, von Schell took serious notice of American military developments during his visits in the summer of 1937, as did Major Hans von Grieffenberg when he and a staff officer visited Fort Knox in 1933. Their visits and experiences were surely reported to the German army high command. After von Grieffenberg returned to Germany in 1933, the Wehrmacht published its *1933–1934 Truppenfuhrung* and continued revising its philosophy of *Auftragstaktik* when von Schell returned to Germany in 1937. Most important, however, von Schell provided numerous examples and ideas for change to improve tactics when he was in the United States. The chiefs of cavalry, field artillery, and infantry were too self-serving to implement the changes; they could not perceive the problem was that their compartmentalization hindered cross-coordination of ideas.

Apparently, the issue of tactical air support for the ground forces was not discussed with the cavalry officers. However, in August 1936, Truman Smith reported on the development of the Junkers 87 "Stuka" dive-bomber. In 1937, it entered production. German interest in dive-bombing had begun as early as 1934, and a few years later Ernst Udet, chief of the Luftwaffe development branch, showed a marked interest in the U.S. Navy's implementation of close-support dive-bombing. Development of the ground attack mission was also improved by German experience in Spain. These events led to the successful tactics that integrated the Luftwaffe with mobile ground forces by providing close air support.[25] The marriage of tactical aviation with the mechanized force at Fort Knox had not progressed as it did in Germany due to the emphasis on developing a heavy strategic bomber. The compartmentalization of the ground forces and the airpower proponents stifled any attempt to follow the German direction in perfecting a lightning war doctrine. Instead, the 12th Observation Squadron was transferred to Fort Knox from Brooks Field, San Antonio, Texas, in the summer of 1937 to augment distant reconnaissance and cooperation between air and ground units. Major John C. Kennedy, whose mission was to develop such cooperation, commanded the squadron. Van Voorhis commented that the transfer finally filled a long-felt need in the development of mechanized cavalry reconnaissance. This resolved a request he made in 1925 that had been denied by the secretary of war.[26]

The problem of integrating tactical aviation with ground elements arose late in World War I and had not been resolved as expected because of neglect, technical problems, and the controversy over mission and air tactics. The *1923 Field Service Regulations* attempted to clarify this situation. It specified that one of the missions of aviation units was to attack hostile ground forces and their supporting units, including supply columns. No direction, however, was given regarding a tactical effort against enemy tanks or in support of an infantry assault with breakthrough and accompanying tanks. This was partly a result of the growing influence of the controversial Billy Mitchell, who had gone public with his criticism of army and navy leadership allegedly for not understanding the role of strategic airpower. He questioned the future application of ground attack aircraft and argued that airpower should focus more on deep strategic operations, such as attacking the enemy's supply concentrations and manufacturing areas.[27]

Even as the air force was emphasizing strategic bombing during the 1930s, a U.S. Army officer attending the Kriegsakademie during this period of amicable exchanges touched on developments in mechanization at an operational level. After returning to Washington, he reported on Germany's development of panzer forces for deep operations with a combined air-ground mechanized force. The army's chief of staff, General Malin Craig, greeted his observations with apathy.[28] As a result, when the United States entered the war, cooperation between armored units and aviation essential for the successful execution of a blitzkrieg was wanting. Neither the Army Air Corps nor the Army Staff had a clear objective regarding tactical air support for a mechanized force.

Meanwhile, von Schell, whom Guderian considered an energetic and indefatigable man with many stimulating ideas,[29] was, at the request of Adolf Hitler, appointed by Field Marshal Hermann Göring to head the German automotive industry while retaining the positions of inspector of the German Panzer Corps and inspector of army motorization. Truman Smith found these appointments of great importance, from both a military and a commercial viewpoint. The appointments were indicative of Germany's further emphasis on industrial and military mobilization. The attaché reported, "Colonel von Schell's views as to tank organization and tactics appear to have recently made great headway and to be strongly supported by both Commander in Chief of the Army and the Chief of Staff." Smith also reported to McCabe, the G-2, on Germany's rush to construct factories for producing tanks, heavy artillery, motor vehicles, artificial gasoline, rubber, and fibers.[30]

It seemed that only at Fort Knox was the potential for improvements in tactics and materials to be taken seriously. However, after von Schell left, the summer tactical exercises with the National Guard began and demonstrated otherwise. These exercises were the fourth since the Fort Riley maneuvers in 1934 when the mechanized cavalry tactically operated with the horse cavalry. The purpose of the exercises was to permit the 7th Cavalry (Mech) to engage horse units from the 54th Cavalry Brigade (107th Cavalry, Ohio National Guard, and 123rd Cavalry, Kentucky National Guard). The National Guard units were reinforced with six T7 scout cars with radio communication from the 13th Cavalry and artillery from the 68th Field Artillery. Bruce Palmer commanded the Red Force, and Brigadier General Newell C. Bolton commanded Blue Force, the 54th Brigade, which was provided with a large number of simulated antitank guns. Colonel Scott from the 13th Cavalry (Mech) acted as an adviser to General Bolton. Both brigades were provided flight reconnaissance by the 12th Observation Squadron. The mission was for both brigades to seize tactical points such as a bridge over the Salt River and delay a hostile advance toward the crossing. The observant and analytical Crittenberger, Van Voorhis's operations and training officer, described the cavalry maneuvers at Fort Knox in the September–October 1937 issue of the *Cavalry Journal*. Surely it made an impression on the Fort Riley contingent, which now firmly believed there was a place for mechanized vehicles in supporting horse units.

Bolton outlined certain instructions for defeating Palmer's mechanized force by coordinating the use of his antitank guns that were mutually supported and prepared to move rapidly against any threat. Most important, Bolton made sure his riflemen, gun crews, and horses were concealed in places inaccessible to Palmer's combat vehicles. When Palmer's Reds made a breakthrough, Bolton expected his horsemen to fall upon the hostile rear or flank with antitank guns. His goal was to demonstrate the mobility of his horse units, the effect of antitank weapons and appropriate communications to "hold together" widely scattered mounted elements that were concealed in the woods and hills.

When the tactical operation began, Palmer's command group had not seen the hostile horse cavalry for three hours, but it had encountered antitank guns. Bolton immediately used his few scout cars to extend the horizon of his horse brigade and held his machine gun troops from both regiments centralized so he could dispatch them to threatened points. He then secured critical terrain that flanked Palmer's force. Bolton demonstrated that his horse cavalry proved of great value in certain terrain that hampered mechanized vehicles. Palmer was unable to seize the bridge over the Salt River.

Palmer commented, "The cavalry combat car is primarily purposed to destroy or hamper the movements of hostile personnel, and secondarily to resist or overcome opposing mechanization." Next he noted, "Reconnaissance mechanization is purposed for reconnaissance, not for combat, when combat can be avoided, nor for command." Palmer's position was not in line with some innovators who were willing to fight for information if necessary.

What was Van Voorhis's take on the tactical exercises? He believed they prepared the National Guard for reinforcing the Regular Army on mobilization day. Van Voorhis emphasized the need for better mobile communication for mechanized forces due to their rapid movements. Contrary to Palmer, he added, "These maneuvers again demonstrated the usefulness of armored reconnaissance vehicles to gain contact with a highly mobile enemy and maintain this contact," meaning fight for information if necessary and not use scout cars for command and messenger service. He gave credit to Bolton for his detailed instructions to his units but neglected to give credit to Palmer for handling the Red Force.

By mid-1937, the president of the Cavalry Board and commandant of the Cavalry School at Fort Riley, General Henry, finally admitted that for some time a slight fissure had developed between mechanized and horse cavalry. One issue of concern had to do with Fort Knox's Mechanized Cavalry Board, which had a separate existence from the Cavalry Board at Fort Riley. Henry argued that the chief of cavalry was tasked to formulate and develop the tactical doctrine for his branch according to a War Department regulation. The attitude at Fort Knox was that a separate board detached from the horsemen at Fort Riley was better at solving many problems facing the mechanized force. As a result, Van Voorhis was reluctant in communicating with the horsemen who dominated the Cavalry Board. The functional issues over boards, Henry argued, further accentuated the rift and the will to move the mechanized force to a separate branch status away from the cavalry. He believed this action could destroy the combat arms branch system. In addition, Henry was concerned that the infantry was copying ideas from the mechanized cavalry regarding certain tactical missions, such as pursuit and exploitation. If the rift continued, he believed, the infantry might try to take over the mechanized cavalry.[31]

Under the direction of General Kromer, Grow, who was assigned to head the Supply (mainly equipment) and Budget Branch in the Office of the Chief

Major Truman Boudinot next to an M2 light tank and wearing a polo helmet from 1st Squadron, 13th Cavalry (Mech), circa 1941. Courtesy of Lt. Col. Burton Boudinot, USA (Ret.)

of Cavalry, prepared a solution. Grow was a supporter of one cavalry. Subsequently, the Mechanized Cavalry Board was made a subdivision of the Cavalry Board, whose function was to consider subjects that pertained to both horse and mechanized cavalry. This solution was designed to establish better communication with the chief of cavalry regarding the development of techniques and tactical doctrine common in mechanized and horse cavalry. In addition, the solution was to charge the chief of cavalry with direct supervision over the 7th Cavalry Brigade rather than having it under the command of the tactical unit, the V Corps. That situation had been a sensitive issue with the chiefs of cavalry since 1934.[32]

While at Fort Knox, the recently promoted Major Truman Boudinot, who had graduated from the Command and General Staff School, took command of the 2nd Squadron, 13th Cavalry in the summer of 1937. Since his sacrilegious remarks in 1932, Boudinot had begun to challenge the tendency to deploy armored cars in traditional horse formations. To him there was one axiom: "Vehicularly mounted troops had to be supported by coordinated infantry and artillery; combat arms has to be integrated." At this time, Boudinot discovered, as did many cavalrymen, that transferring from horse to armored cars was not easy.[33] Boudinot was impressed with Van Voorhis's leadership as a combined arms integrator. His son, Burton S. Boudinot, remembered Van Voorhis as a tall, snow-headed man with deep character lines in his face and piercing eyes. Young Boudinot was impressed with Van Voorhis's highly polished cavalry riding boots and his Sam Browne belt that crossed his jacket. When on duty, Van Voorhis always carried a riding crop, noted Boudinot. He befriended the brigade commander because they had a mutual interest in bird dogs.[34]

Unfortunately, one day young Boudinot and a friend went into Van Voorhis's doghouse looking for some straw. It was dark, so a match was lit. Shortly, Boudinot's mother came running out of the house. Van Voorhis's doghouse was on fire. The firemen came and contained the fire, but the doghouse was totally consumed. The prized dogs, however, were safe. Boudinot was now in deep trouble. He felt "his life was over and my father would be sent to prison." Fortunately, Van Voorhis got to Boudinot's father and diplomatically explained the misfortune of his son. The punishment was harsh; however, the pressure on the young boy slowly let up. Van Voorhis's military bearing, and at times his kind nature, left a favorable impression on young Boudinot, who later became editor in chief of *ARMOR* magazine.[35]

As a brigade, the mechanized elements at Fort Knox now constituted a larger organization with potential to execute greater operational mobility and firepower. This was important because it provided the prospect of moving a self-contained mechanized force from a tactical level to an operational level of war, at which major operations are executed to accomplish strategic objectives. What this entails is that commanders from corps and higher use tactical events to strike at the center of gravity as defined by strategy. Van Voorhis visualized a large force capable of executing missions based on speed, firepower of the .50- and .30-caliber mounted machine guns, shock action, and deep operations with supporting elements. The transition to an operational level necessary to fight a campaign, however, was questionable because of congressional funding, the confusion over mobilization, and the autonomous culture of the combat arms branch system.

In Europe, events were developing that threatened army mechanization and a workable tank policy in America, further reinforcing the parochial attitude of the branch chiefs and the chief of staff. The Spanish civil war (1936–1939)

General Daniel Van Voorhis inspects 68th Field Artillery (Mech-Towed), 7th Cavalry Brigade, fall 1937. From the Hunnicutt Collection, Patton Museum.

provided numerous reports from attachés on the employment of tanks. One of the officers influencing General Craig and the army's military school systems was the former chief of infantry, Major General Stephen O. Fuqua, who had earlier objected to the creation of a separate mechanized force with the tank as the main maneuver element. Fuqua was now the U.S. military attaché in Spain. It was his opinion, and that of his peers in England and France, that tanks had not proved themselves in separate offensive operations because antitank guns effectively challenged them. The officers concluded that the only value of the tanks was supporting the attacking infantry.[36]

One of the horse cavalry's strongest and most influential spokesmen at the time was retired Brigadier General Hamilton S. Hawkins. He wrote in the *Cavalry Journal* that the war in Spain proved army critics wrong when they argued tanks and airpower supplanted the traditional combat arms. Furthermore, he held, many military men were as ignorant as the American people were about the deployment of horse cavalry. Besides, he said, mechanization could never replace horse cavalry. Hawkins believed tanks were not suitable for the modern battlefield because they were vulnerable when halted, and if they went too fast, they lost their infantry support. As well, he wrote, tanks were expensive and thus could not be manufactured in sufficient numbers for independent operations, adding that mechanized forces had not yet proven themselves and were vulnerable to air attacks. Hawkins's attitude can be summed up by his statement "When the tests of war have been made and the facts are clear, we shall certainly regret having to replace cavalry with mechanization, if we are so foolish as to carry this project too far."[37]

As events were unfolding in Spain, Van Voorhis paid a visit to the War College in October 1937 to present a lecture on mechanization. He commented on von Schell's views, stating that the ponderous German panzer divisions were designed to create a breakthrough rather than roll up the flanks. Thus, German plans began emphasizing the mass use of heavier tanks as the strong striking force. This questioned, if not challenged, the U.S. Army's tank policy of lightness. In America, Van Voorhis acknowledged, there existed a "wide divergence of opinion within our Army as to a suitable mission for American mechanization." He raised a question regarding the mission of tanks as an organic part of a division: Should tanks support the infantry or the infantry support tanks? Until this issue was settled, the issue of a mechanized division could not be resolved. Reflecting on early American efforts at mechanization Van Voorhis stated, "The mission was fitted to the organization and its equipment, instead of first defining the mission and then building up a force with characteristics which made it capable of executing this mission." When the Mechanized Force was reorganized as cavalry in 1931, he noted, "Clarity of purpose in organization and equipment was established." He further stated that the strength of mechanized units lay in the power of maneuver and firepower, therefore requiring prompt and quick decisions, "and to develop this faculty one of the greatest problems which confronts us is making the mobility of the mind equal the mobility of the machine." He also brought up the question of infantry holding power, which he believed was more of the mission of GHQ for attachment at a time when necessary. This was his idea of an effective use of force for deploying mechanized elements. It was evident that the emphasis was being placed more on new mounts at the expense of infantry holding power. Van Voorhis again lamented the lack of equipment necessary to perform the mission of mechanized cavalry.[38]

The same month Van Voorhis's lecture was presented at the War College, students discussed and reported on the issue of mechanization. Their conclusion was that mechanization had not revolutionized the conduct of warfare. Tanks and armored cars only extended the range of action. Furthermore, it was their opinion that in the gun-armor race antitank guns were ahead. Undoubtedly, this observation came from the numerous reports from Spain. Mechanized units, the students stated, were particularly dependent on a constant supply of oil and gasoline, road conditions, terrain, and antitank defenses. At the operational level, it was assumed conditions favoring the employment of a mechanized force began in the opening phases or the closing phases of a campaign. On trends for a mechanized force, students suggested it become a GHQ weapon used for deployment when necessary as an economy of force with the infantry. More so, it was confirmed that mechanization was not just confined to the cavalry but representative of all combat arms. The issue of an infantry battalion holding force was applicable to the impeding continuous mobility of the mechanized force. The organic infantry battalion, it was noted, served as a "guinea pig" for the study of problems associated with a combined arms mechanized force.[39]

By end of 1937, the role of the infantry in a mechanized force was still debatable due to the opposition of crossover cooperation between line branches. It was, however, conceded that the army was divided over how the line branches should deal with modern trends in warfighting.

Meanwhile, the stubborn J. Walter Christie did not go away. A British mission led by Giffard Le Q. Martel, the assistant director of mechanization, attended the Red Army September 1936 maneuvers and had the opportunity to see hundreds of Christie-type BT fast tanks executing deep penetration. The visiting mission found the BTs developed by I. A. Khalepskii's directorate most impressive. Martel, however, was more interested in the BT's suspension performance and its agility than in its armor. This enthusiasm was reported to the War Office. By the end of the year, Lord Nuffield of Morris Motors had acquired a M1931 from Christie as a prototype tank for the British military, but not before paying off a lien on the vehicle. This was the same vehicle that impressed George Patton, Daniel Van Voorhis, and Adna Chaffee, as it was going through Ordnance Department acceptance tests in January 1931, but never purchased. Once acquired by Morris Motors, the suspension system was used in a series of cruiser tanks intended to deploy as fast and lightly armored cavalry tanks after heavier infantry tanks established a breakthrough. For years, the British were to use the Christie suspension system in their cruiser-designed tanks, many of which saw action in North Africa early in World War II.[40] One even made it to Fort Knox early in the war for evaluation by the Armored Board.

Before the year 1937 ended, the tall but emaciated seventy-year-old inventor emerged again and attempted to promote his new 6-ton machine to the U.S. government. On a cold afternoon in December, with much fanfare, Christie demonstrated the new machine at Westfield Airfield in New Jersey in the presence of a hefty crowd, including reporters and a few members from the Ordnance Department. He told the press his machine could be used only for

(Above) The British cruiser tank A13, 1938, reflecting the Christie suspension system. From the Ickes Collection, Patton Museum. *(Below)* British Crusader tank, Fort Knox test by the Armored Board, September 1942. Note the Christie suspension system. This tank fought in North Africa. Courtesy of the Patton Museum.

peaceful purposes and indicated it was available to the U.S. government. An ordnance officer noted that there was no security surrounding the demonstration. As before, Christie emphasized speed over military requirements. The vehicle was attractive; however, it was too light. This raised, as before, the question of what added weight might do to the superstructure when armor, armaments, turret, and crew space were added. Upon inspection, it was determined the vehicle could not be converted to a wheeled mode. Another ordnance officer complained Christie was so distracted that it was impossible to talk to him. He was more interested in talking to the press and pontificating. The inventor, in his usual way, bragged that he was working with Morris Motors, Limited, in England to develop a new tank.[41] Eventually, the vehicle ended up in England, but it was rejected. Dejected and angry, the aging J. Walter returned to the United States with his chassis. One of England's foremost armor historians noted that Christie's greatest failing was that he had not understood that there must eventually be a finite stage of development in any tank program.[42] He constantly tried to re-create.

Meanwhile, the cavalry was still interested in a smooth-riding cross-country wheel-track convertible vehicle capable of sustaining high-speed mobility, similar to the Christie system. It was evident the cavalry was not giving up on the convertible principle. In April 1937 a design for a new convertible combat car was approved for manufacture as a pilot model, the "car, combat convertible, T7." Thus began a two-year struggle to solve the technological compromise of the convertible system. Though the hull arrangement was similar to the CC M1, the vehicle employed pneumatic tires and a Christie-type independent suspension system. The T7 was tested at Aberdeen Proving Ground and at Fort Knox. In 1939, before the Plattsburg maneuvers, Chaffee decided the vehicle was not

Christie again tried to sell Morris Motors his latest chassis, the M1938. It was not accepted. He tried then to sell it to the U.S. government but again it was rejected as not meeting military requirements. Courtesy of the Royal Armoured Tank Museum.

The T7, the last convertible vehicle tested by the Ordnance Department and cavalry. Courtesy of the Patton Museum.

suitable as a combat car due to problems in converting from track to the wheel mode.[43] Bob Grow, who in 1933 was convinced the cavalry needed an all-track vehicle, identified this problem long before Chaffee made his recommendations. At that time, Grow questioned the engineering compromise of the convertible principle.[44] In October 1939, the Cavalry Board recommended the development of the T7 be canceled, thus ending a long interest in the Christie convertible principle for strategic deployment and tactical employment.

While the cavalry was looking at a new combat car, the attitude at the War College began to reflect the lessons learned from the Spanish civil war. By January 1938, attaché observations were circulating throughout the military school systems. Generally, it was noted, tanks in Spain were not deployed in large masses, and their tactical employment was faulty. One reason was that tank crews were unskilled, resulting in unwarranted mechanical troubles. In addition, there existed a lack of skill in the aggressive deployment of tanks. The reports of tanks not supporting the infantry, or the infantry not supporting tanks, were one of the observations coming out of Spain from military attachés.[45] Most influential were the dispatches from the U.S. military attaché in the Republican-controlled parts of Spain, Stephen Fuqua. Tanks were no longer decisive weapons due to the effectiveness of the defense and antitank guns. Most of the tanks used were lightweight. It was explained that an attempt to use Red Army tanks in an independent attack by the Republicans west of Madrid in April 1937 resulted in a number captured after falling into tank traps. Moreover, others broke down due to mechanical failures. The conclusion was that independent attacks without supporting infantry were doomed to failure. This assessment sustained the accepted U.S. tactical doctrine driven by the *1923 Field Service Regulation* that infantry was the dominant fighting arm and that the mission of tanks was to support and operate in close cooperation with the infantry. In addition, Fuqua questioned Giulio Douhet's theory of massive bombings and shattering civilian will to resist as a solution to the stalemate that existed during World War I. In December 1936 and January 1937, the Nationalists extensively bombed Madrid. The results, according to Fuqua, were few or nonexistent. He claimed the bombing increased morale. People shook futile fists at Franco's bombers and called them "Swines."[46] Thus, the lessons brought back from Spain were that infantry needed supporting tanks with heavy armor, supported by artillery and tactical airpower. The U.S. Army, nevertheless, still adhered to the time-honored tank policy of lightness and branch autonomy. Meanwhile, the air force moved ahead, planning and preparing for strategic bombing.

The Command and General Staff School, per instruction from the War Department Military Intelligence Division, circulated a detailed description of the current development of German panzer divisions that was to be seen and used by military instructors. This was not a surprise to the cavalry officers at Fort Knox, who had spent time with von Schell in June 1937. The panzer division was considered the most important element to be used en masse supported by the *Schutzen* brigade. The focus for the instructors to digest explained that panzers when operating as a division were expected to be rapidly deployed into the interior of a hostile country at the outbreak of a war. Then the panzers were expected to be the main striking force, spearheading an important envelopment movement aimed at strategic points. Interestingly, the report for the instructors mentioned that panzer divisions were "to operate independently in a country of great expansion, such as Poland and Russia." Special note was made giving credit to von Schell's efforts for improving the organization and tactics that made mass employment of panzers the heart of the offensive, which was reported by Smith to have the support of General Ludwig Beck and his chief of staff. Credit was also given to Guderian, who was called a pupil of J. F. C. Fuller.[47]

General Van Voorhis especially responded to the attaché reports on panzer divisions circulated by the War Department, stating, "They are considered as informative only and not to be taken as a guide for us in our development of mechanized cavalry here at Fort Knox." He strongly noted that the tactical employment of mechanized cavalry "was peculiarly American," and not influenced by events in Europe. The difference he meant was that at Fort Knox mechanization "was designed to utilize the great power and advantage of maneuver, which

Colonel Willis D. Crittenberger at the microphone, General Daniel Van Voorhis pointing, Senator Harry S. Truman, and Marvin McIntyre with hands folded, during visit to Fort Knox, January 22, 1938. Courtesy of the Truman Library.

is inherently American." While the Germans were thinking mass employment with their panzers, the mechanized cavalry was experimenting with "mobility, firepower and shock in a wider and speedier employment with the necessary facilities for striking a hard blow when the occasion demands." Van Voorhis frequently stated that mechanization should be approached with an open mind and that "we should not freeze on anything."[48]

On January 22, 1938, a World War I artillery captain with the American Expeditionary Force and future president, the senator from Missouri Harry S. Truman, visited Fort Knox along with other government officials. President Franklin D. Roosevelt prompted the visit and tasked his important appointment secretary, Marvin H. McIntyre, with leading the officials on a political circuit in Louisville and a military tour of Fort Knox. The goal of the visit was to inspect the activities and progress of the 7th Cavalry Brigade. In addition, the other visiting New Deal democrats were Senators Joseph F. Guffy of Pennsylvania, Matthew M. Neely of West Virginia, and Lewis B. Schwellenbach of Washington. The visitors met General Van Voorhis at Tip Top, the entry to the base off Route 31 West, and began an inspection of the area arranged by Willis Crittenberger. Included were the facilities of the 1st and 13th Cavalry, 68th Field Artillery, and the 12th Air Observation Squadron. Afterward, the entourage witnessed a review of the 7th Cavalry Brigade, where 250 vehicles participated. After the review, the government officials were given a cross-country ride near Wilson Road in M1 combat cars from the 13th Cavalry. The visit included a tour of the bullion depository. Afterward, a final meeting with Van Voorhis was held in his quarters.[49]

Certainly, Van Voorhis discussed with President Roosevelt's representatives the importance of expanding the 7th Cavalry Brigade into a complete division. The question of funding was brought up, as was the importance of bolstering the army's fighting ability. Van Voorhis noted that army tanks or combat cars were inferior to those of most European countries. He explained that mechanized units were meagerly composed of reconnaissance troops with wheeled armored cars; four troops of light combat cars; a machine gun troop with more than one hundred machine guns, and for support a number of mortar mounts for firing smoke projectiles and towed artillery for firing high explosives. Surely, Van Voorhis made the government officials aware of the importance of Fort Knox's thirty-three thousand acres as the experimental laboratory and center for army mechanization. At the time, the army was still skeletonized; Van Voorhis noted that only 125 officers, 2,900 enlisted men, and their families made up the garrison.[50]

President Roosevelt's representatives then returned to Louisville to attend Senate majority leader Albin W. Barkley's testimonial dinner at the Brown Hotel. A few years back, when Barkley had visited Fort Knox, Bob Grow recalled, he showed no interest in mechanization and what was happening at the garrison.[51] The evening event ushered in the important congressional campaign for 1938 and a critical year for army mobilization. The event was planned to forestall Senate aspirations of the democratic anti–New Deal governor of Kentucky, Albert Benjamin (Happy) Chandler.

It would be interesting to know what comments regarding preparedness, equipment, and the looming international situation Van Voorhis exchanged with McIntyre, Truman, and the other government officials. More so, what information did McIntyre relay to the president about developments and changes at Fort Knox? By the end of year, President Roosevelt had laid out his most important plans for army expansion to deal with the growing military powers of Germany, Italy, and Japan. However, he still placed emphasis on hemisphere defense, stressing greater support for airplane production for the Air Corps rather than opting for a well-balanced army as proposed by the Army General Staff and outlined in General Craig's Protective Mobilization Plan. As Europe was moving closer to war, plans for expanding the mechanized cavalry to a division were considered essential but not critical at the time due to the priorities that depended on congressional funding.[52] For the time being, the army had to make do with obsolete equipment, much of which was still left over from World War I.

Meanwhile, two months after his visit to Fort Knox, Senator Truman addressed the George Washington Post of the American Legion in Washington, D.C., emphasizing the importance of military preparedness. "We must not close our eyes to the possibility of another war, because conditions in Europe have developed to a point likely to cause an explosion at any time," he told the veterans and guests. Almost foretelling the future, Truman stated, "In the coming struggle between democracy and dictatorship, democracy must be prepared to defend its principles and its wealth." He concluded, "I hope we shall never have to fight again, and the best way to keep from it is to be adequately prepared for all contingencies."[53]

In January 1938, Chaffee, still in the War Department, received a memorandum from Colonel McCabe, the G-2. He was made aware of tank operations and casualties that occurred during most of the military operations in Spain to date. The memorandum pointed out that the limited combat success of tanks was due to infrequent terrain reconnaissance, failure of infantry and artillery support, poor discipline and training of personnel, mechanical failures, questionable use against villages and strong obstacles, and tanks deployed "in too small numbers and the superiority of antitank weapons 20mm and larger."[54] Chaffee wrote Van Voorhis over the current tank policy of lightness, noting that all tanks deployed during the Spanish civil war were in fact light and thus vulnerable to antitank fire. The mechanization program in the U.S. Army's budget still reflected lightness, and little effort was being made to develop medium tanks, "which are our most serious deficiency in mechanization at the present time."[55] To deal with the issue of antitank guns, the mechanized force began to rely more on mortar mounts to concentrate smoke on hostile positions. Meanwhile, General Craig's austere equipment program adopted the design of the German 37mm antitank gun, which soon became obsolete.

During General Kromer's tenure as chief, Grow drafted a proposed organization of a mechanized cavalry division based on the Leavenworth text. Van Voorhis agreed with the proposal. Because of financial restrictions and problems of control, he was firm about a two-regiment mechanized cavalry division. Kromer instead preferred a three-regiment division. The proposed division was approved by the War Department for planning purposes; it was only a paper organization. The issue was resolved when Brigadier General Walter Kruger, chief of the War Plans Division, opposed a mechanized cavalry division because of its anticipated size. He reminded Kromer that the mission of cavalry was recon-

Major General John K. Herr, chief of cavalry, March 26, 1938, to March 9, 1942. Courtesy of *ARMOR* magazine.

naissance and security. Now Van Voorhis had real doubts about complete cavalry mechanization. He began thinking about a separate mechanized force.[56]

In March 1938, General Kromer retired at a critical time. He had finally accepted mechanization of the cavalry arm and, according to Grow, who was closely associated with Kromer, "could have made cavalry the mechanized arm had he been supported by the army's general staff and senior officers in his own branch."[57] A number of cavalrymen believed Bruce Palmer from the mechanized cavalry would be the next chief. Palmer himself felt the same way. This, however, was not the case. Major General John K. Herr, a tall, lean, and rugged horseman with considerable energy and enthusiasm, now took over as the Chief of Cavalry. For years Palmer resented General Malin Craig's selection. Herr became the chief at a critical time when the cavalry was still experiencing a rift over a new mount and mechanization.

No sooner had Kromer retired than the War Department introduced policies governing mechanization and the tactical employment of mechanized units. These policies, which were in part influenced by the events in Spain, had

General Malin Craig, chief of staff, October 2, 1935, to August 31, 1939, and former chief of cavalry, July 24, 1924, to March 20, 1926. Courtesy of *ARMOR* magazine.

not reflected the ideas of Van Voorhis or Chaffee. The April 1938 policy stated, "The greatest value of mechanized cavalry lies in its ability to conduct distant reconnaissance and to create an initial combat success . . . which will form a basis for further action or exploitation by the higher commander." The War Department acknowledged that the combat arms would fight in their traditional roles. This policy reflected the long-established role of cavalry as a reconnaissance force. This traditional vision created even more friction among cavalrymen over whether to replace the horse with a new mount and provide a greater role for mechanized cavalry. The other faction was just as adamant about replacing the horse. Their spokesman was Johnny Herr.

As in World War I, tanks were still required to provide close support for the attacking infantry and not move far beyond the effective fire support from division artillery. The role of armor, the policy ordained, was solely that of infantry-accompanying tanks. This tank policy darkened the fate of the infantry progressives, such as Sereno Brett, Chen Chynoweth, and Sid Hinds, and their desire for leading tanks over supporting tanks. After an infantry breakthrough with accompanying tanks, the cavalry was expected to pursue, envelop, and exploit—naturally with mounted horsemen. Reconnaissance and security missions, as before, were left to the cavalry.[58]

As noted, this significant policy was strongly influenced by attaché reports from Spain, especially from Fuqua. His reports from Republican Spain were circulated at the Command and General Staff School and the Army War College for analysis and war planning. By now Fuqua's reports, plus articles by other military attachés and newspaper reporters, noted the eclipse of the tank because of the emergence and dominance of the flat trajectory antitank gun.[59] General Craig summarized his feelings before a congressional subcommittee hearing on military affairs. Regarding military operations in Spain, he observed that tanks were not successful because of antitank weapons, insufficient armor, mechanical defects, poor tactical deployment, and inadequate support from artillery and tactical aviation. Craig believed future military operations "must be carried out by the traditional arms; that well-trained infantry and artillery form the bulk of armies. Air and mechanized troops are valuable auxiliaries."[60]

General Van Voorhis took a different view of the April policy statement. He always believed that the continued application of MacArthur's 1931 directive could only result in scattering the efforts of mechanization over too wide an area. The role of mechanized cavalry now should emphasize combat rather than any secondary missions such as reconnaissance. Greater emphasis, he believed, was in the value of mechanized mobility that increased the fighting power, radius of action, and "adding a strategic mobility far beyond that of horse cavalry alone." In a positive way, Van Voorhis believed that the April policy would "serve to continue the practice of fitting the organization and equipment to the mission rather than fitting the mission to the equipment." The policy, he believed, "will serve to clarify the distinction between the employment of mechanized cavalry and infantry tanks." Van Voorhis felt that this was important because "no longer will either type of mechanization be required to perform the role of both."[61]

While the army was digesting the War Department's policies, George Patton—now executive officer, Academic Division of the Cavalry School and 9th Cavalry at Fort Riley—wrote Crittenberger at Fort Knox about seeking a light 37mm gun and mount for the cavalry and the advantage of having a more effective light machine gun. This, he argued, could give horse and mechanized cavalry greater fighting ability rather than "the position of a scouting force." He looked at added firepower for the cavalry only if the cavalry "is to fight Europeans or Asiatics." Then he suggested the cavalry dispense with all its pack weapons, because we can "do our work very handsomely with the rifle, the pistol and saber . . . if the sole purpose of our existence is to fight Mexicans." Patton believed that the success of cavalry "mechanization was due to taking what was available, rather than copying infantry and seeking for perfect blueprints before we turned the wheel." Responding to a request by Crittenberger for names of "horsey cavalrymen" for duty at Fort Knox, Patton highly recommended Major

Lucian K. Truscott Jr. Patton considered him an excellent horseman, a splendid troop commander, a good instructor, and, what was expected of a cavalryman, a superior polo player.[62]

A few months later Crittenberger contacted Grow in the office of the chief of cavalry regarding the critical equipment situation. Grow agreed with Crittenberger regarding the equipment problem; however, the cavalry, he responded, should not strive for the best, but in "war the cavalry has to take what it can get." He summed up his response to Crittenberger regarding all cavalry reconnaissance armored fighting vehicles: "We should never compromise with the maximum available efficiency. Instead of lauding the 'light cheap commercial vehicle,' we should keep our feet on the ground, set up the requirements, and draw up the characteristics to meet them then design the vehicle. This is the standard. Nothing else should be considered except in emergencies."[63] Grow was placing doctrine before equipment.

Meanwhile, back at the War Department Budget and Legislative Planning Branch, Chaffee "conducted a one-man campaign for mechanization."[64] He determinedly worked to acquire meager funds for the mechanized force, to be

General Daniel Van Voorhis, pictured second from the right, with his staff and regimental commander preparing for a command exercise. Colonel Charles L. Scott is pictured second from the left, and Colonel Bruce Palmer is pictured first from the left. From the Van Voorhis Collection, Patton Museum

used for equipment, housing, and supporting the addition of another cavalry regiment at Fort Knox. In this important capacity Chaffee had close associations with the chief of staff, the secretary of war, and many congressmen. According to his biographer, "The mechanized cavalry could not fail to profit from such resourceful behind-the-scenes manipulation," as Chaffee engaged in many weary and perplexing hours championing his cause.[65]

Back at Fort Knox, the mechanized force engaged in its usual spring maneuvers. This time more than five hundred vehicles of various sorts were employed to test new combat equipment and tactics, and to demonstrate how far the force had progressed since its conception. In May 1938, the 7th Cavalry Brigade's three thousand soldiers were ready for a seven-hundred-mile march with hundreds of combat cars, armored cars, scout cars, command cars, towed field artillery, and supporting maintenance and kitchen trucks required to sustain the movement.[66] The brigade was to strike deep into the south at a mythical invader who landed at Charleston, South Carolina. As the brigade formed up, Van Voorhis was the first commander to control an attacking force by radio from a two-seater Douglas O-46A observation plane. His control of the mechanized columns from the air predated Patton's experiences in incor-

General Daniel Van Voorhis standing under the wing of his Douglas O-46 from the 12th Observation Squadron. He used it to command his split cavalry elements during the 1938 maneuver. From the Van Voorhis Collection, Patton Museum.

porating planes in all his armored exercises before the United States went to war.[67]

Van Voorhis planned, prepared, and executed the operation by splitting his command into two striking forces. There was, however, no effort made to combine horse units with the mechanized force. It was the beginning of the long march into the south and then back to Fort Knox. The west column under the command of Colonel Palmer from the 1st Cavalry, supported by a battery from the 68th Field Artillery, moved through Elizabethtown and Glasgow, Kentucky, and on to Cookeville, Tennessee. The east column was under the command of Colonel Scott from the 13th Cavalry. His column was also supported by a battery from the 68th Field Artillery along with the addition of trains, quartermaster, ordnance, and medical elements. Scott moved his force through the foothills of the eastern Kentucky mountains, passing by Hodgenville, Columbia, Somerset, and Monticello. Both columns threw out their respective reconnaissance elements, ground and air, to the front and flanks. When the two columns finally bivouacked, they were nearly a hundred miles apart. Meanwhile, Van Voorhis was advised of a simulated enemy in the Chattanooga-Knoxville area. Changing the tempo, he took advantage of the opportunity, shifting his two columns to converge at Crossville, Tennessee, on the second day in preparation to en-

The Douglas O-46, used to drop messages in tomato cans. From the Van Voorhis Collection, Patton Museum.

gage the enemy, which was now reported marching on Chattanooga. Reacting rapidly, Van Voorhis now deployed his consolidated force to Chattanooga and on to Fort Oglethorpe, Georgia, successfully terminating the simulated tactical operation.[68]

For the next part of the maneuver, the mechanized brigade took on the role of an invading force by heading back to Fort Knox and engaging the 10th Infantry Brigade supported by an attached tank platoon. After crossing the mountains, the force arrived at Nashville, Tennessee, where Van Voorhis was informed of a simulated enemy concentration in the Dayton-Indianapolis–Terre Haute area. He immediately dispatched a plane from the 12th Observation Squadron to reconnoiter the hostile concentration. By radio and messages dropped in one-gallon tomato cans with ribbon streamers attached and weighted with sand, the plane reported hostile forces moving toward the Ohio River. Subsequently, Van Voorhis dispatched the 12th Observation Squadron under the command of Major Kennedy to reconnoiter the river from Cincinnati to Cairo, Illinois, while ground reconnaissance proceeded to reconnoiter a front of about 200 miles along the Ohio River from Carrollton to Henderson, Kentucky. Following the reconnaissance, the mechanized force advanced in two columns northward under the cover of darkness over Routes 31 West and 31 East subsequent to being advised that a hostile force was moving to Fort Knox. Using maximum security, Van Voorhis completed a 150-mile night march. Concentrating, the mechanized force then attacked an antitank accordion defense set up by the 10th Infantry Brigade. The 12th Observation Squadron conducted battle reconnaissance and provided for artillery adjustment as the 13th Cavalry, supported by a battery from the 68th Field Artillery, delivered the main blow at the Fort Knox military reservation. Forest Hill and Wilson Canyon were the typical maneuver areas, as was Muldraugh Target Range. Scott's force rapidly passed through the defenses, linking up with the 1st Cavalry behind hostile lines. Consolidating his force, Van Voorhis ordered his two regiments to reverse and assault the hostile force from the rear following a simulated artillery preparation. The 10th Infantry Brigade was disposed of and the maneuvers called off.[69]

What was the significance of the 1938 spring maneuvers as compared with previous maneuvers?

General Van Voorhis effectively executed command and control over his two striking forces by monitoring at times the progress of his planned deep operations from the air. He seized the initiative, taking opportunities to change the tempo of operations and varying his method of attack. The operation again provided the framework that had all the tactical elements of moving to operational mobility necessary to achieve a new level of warfare with a self-contained force. It was an inventive plan of depth, boldly executed and synchronized with the agile use of mechanized equipment that produced decisive results. Van Voorhis's two-column deployment and then consolidation of his mechanized force for the attack became the tactical policy of flexibility established by the combat commands of World War II armored divisions.

Apparently, Van Voorhis's effective command and control over the two maneuvering columns impressed the observant Patton. He wrote that when contact with the enemy was evident, "the commander or his executive must be in the air in order to maintain that instantaneous personal touch with the enemy situation that is demanded for the successful employment of rapidly moving mechanized units." Patton concluded that Van Voorhis set the example. Did Patton learn from Van Voorhis's control of his mechanized columns from the air during the May 1938 maneuvers?

Kennedy's squadron superbly handled air-ground communications, especially the tomato-can dropping of messages over the moving columns. Van Voorhis also made sure that reconnaissance by air and ground was paramount before rushing into a mounted assault. Planning by his officers was also evident by the smoothness with which the brigade went into bivouac and the effort at keeping all the vehicles rolling. Most important, Van Voorhis and his two regimental commanders, Palmer and Scott, after a rapid night march covering many miles to the area of deployment, made sure that the cavalrymen did not relax or lose their momentum. Van Voorhis anticipated redeployment as detailed intelligence arrived on the enemy's intentions. The night march was one of the recommendations that came out of the Fort Riley maneuvers in 1934 and became part of improved tactical training at Fort Knox. Van Voorhis took the initiative by striking the hostile force as soon as possible before it could organize for an assault or set up defenses.[70]

Summarizing the 7th Cavalry Brigade's maneuvers in 1938, it was the strongest expression to date of deploying in depth a self-contained mechanized maneuver-oriented force. The method of attack was worked out by Van Voorhis and his commanders, especially by his operations officer, Willis Crittenberger. Scout cars usually made the first contact and then moved to guard the flanks. The 12th Observation Squadron radioed fire control data to the 68th Field Artillery and information on the enemy to brigade headquarters. Brigade headquarters, in their noticeable "buggy whip" scout cars, through various methods of communication maintained constant contact with maneuvering units. The 68th Field Artillery, initially deployed in the rear, pounded the hostile force with high explosives from .75-caliber howitzers and modernized 75mm guns, while the mortar troops mounted in scout cars or trucks laid down a smoke screen from the protection of a hillcrest. Combat cars formed the spearhead of the attack after the artillery neutralized antitank defenses. The combat cars then moved to exploit. Meanwhile, the half-track personnel carriers dropped off machine gun troops to hold ground gained and then redeployed to aid in the attack. While the attack was progressing, maintenance and repair units with their shops on wheels maintained contact with the attacking brigade. Noticeably lacking was tactical air support.

An article in *Shell Progress* commenting on this type of an attack by the 7th Cavalry Brigade noted, "There wasn't a single feed bag or bale of hay on their mighty picket lines. Instead Shell tank trucks moved down the lines of combat cars feeding them with 92 octane aviation fuel."[71]

In September, Van Voorhis's obervations were presented again at the War College. Commenting on the analysis of tank casualties in Spain, he noted that recent intelligence reports related to actual war conditions rather than to theory. Most important, he said, was that the use of mechanization in Spain had not been entirely satisfactory due to the vulnerability of tanks to flat trajectory antitank guns and poorly trained tank crews. He reiterated what had already been reported by the military attachés—problems of mechanical failures, errors in tactical employment en masse, failure to reconnoiter in advance of the main force, exposure before and after an attack, lack of surprise, and failure to neutralize hostile antitank fire.[72]

Colonel James P. Pritchard, who served as a junior aide to Van Voorhis at Fort Knox from 1936 to 1938, claimed he knew his boss's tactical intentions because he saw him almost every day. As a result, Pritchard was able to understand and comment on Van Voorhis's tactical intent. Van Voorhis, Pritchard said, always wanted to do more tactical planning because, "If we're going to have a war, and we're just about in it, we're going to expand, armor is going to grow, mechanized cavalry is going to grow and we need to build up Fort Knox because this is the home of it all. This is where it's going to be developed." Regarding his striking force of combat cars, Van Voorhis believed they should never be split up over a wide front but deployed decisively and with "enough depth so that you can exploit your success."[73] No doubt Van Voorhis more than Chaffee was one of the most forward-looking cavalryman of the time.

While the 7th Cavalry Brigade was executing deep maneuver–oriented operations and attempting a breakthrough to an operational level, the Germans were accelerating their industrial and military mobilization and finalizing a new concept for mobile warfighting. U.S. Army elites were too fixated on tradition and thus failed to understand that moving to an operational level of war between strategy and tactics was a new form of military art. A student at the Army War College in October summed up this attitude: "I hold here a pamphlet, *Tactical Employment of the Mechanized Division*, used as a text at Leavenworth during the past few years. The April directive consigns the booklet to the school

Brigadier General Daniel Van Voorhis, commander of the 7th Cavalry Brigade, Major General Hugh A. Drum, commander of the First Army, and Colonel Bruce Palmer, commanding officer, 1st Cavalry (Mech), examine an 81mm mortar and its carrier. Courtesy of the Patton Museum.

archives. There will be no Panzer Division in our Army."[74] This was in reference to the April 1938 War Department policy governing mechanization and tactical employment of mechanized forces. The policy avowed that recent operations in Spain demonstrated that "combatant arms will fight in their traditional roles." It further emphasized that the mechanized cavalry was expected to adhere to its traditional mission of exploiting the infantry's success.[75]

By the end of 1938, it was apparent that the fate of cavalry as a separate combat arm was in jeopardy. This was due primarily to the emerging antitank mentality in the army and the conservative attitude of General Herr, who had a personal and fervent emotional attachment to horse cavalry. His traditional attitude was uncompromising.[76] One prominent historian wrote that Herr "was interested in the Cavalry's nineteenth century traditions [more] than in modern realities."[77]

Herr's defense of his branch's fate and his love for horses over mechanization were expressed in an October 1938 memorandum to Chief of Staff Malin Craig. "It [mechanized cavalry] has not yet reached a position in which it can be relied on to displace horse cavalry. For a considerable period of time, it is bound

Now commanding general of V Corps, a rare smiling Daniel Van Voorhis is shown visiting Fort Knox and checking out a combat car. During this period, it was usual for officers to travel in civilian clothes. Note the combat car helmet, circa 1938. From the Van Voorhis Collection, Patton Museum.

to play an important but minor role while the horse cavalry plays the major role so far as our country is concerned. . . . I feel that the psychology of the public as well as that of important key men in our legislative branches and men in the army itself has mistakenly become unfavorable to the horse. . . . We must not be misled to our own detriment to assume that the untried machine can displace the proved and tried horse."[78]

That same month, in a tense and pessimistic meeting with Craig's G-3, Major General Robert McC. Beck, and his staff members and members from the Office of the Chief of Cavalry, including Grow, Herr indicated he was unwilling to give up a single horse or man from the horse cavalry. More so, Herr stated he "had never yet committed himself as to the type of mechanized division which he considered best." The G-3 called for a two-regiment division, as did Van Voorhis. Herr wanted this division to be experimental.[79]

In addition, a question came up at the conference about the location of the Cavalry School. Both Van Voorhis and Chaffee had opposed a mechanized school at Riley and supported its location at Fort Knox. Their feelings were that a mechanized school at Riley would not have the support of the elite horsemen.

The two-school development, according to Grow, had the potential of causing a greater rift in establishing a separate combat arm. Grow, however, agreed with Herr that there should be one cavalry school.[80]

Brigadier General Walter Krueger, chief of the War Plans Division, had other ideas regarding a cavalry division. Krueger became the principal author of *1939 Field Service Regulations (Tentative)*, a conservative manual that had the support of Herr and chief of infantry, George A. Lynch. The manual differed little from the *1923 Field Service Regulations*, of which Lynch was one of the architects. Krueger opposed the proposed mechanized cavalry division because of its anticipated size. He reminded Herr about what Grow called the "sneak and peek" mission of cavalry, which was reconnaissance and security. Now the dismayed Van Voorhis continued to have serious doubts about complete mechanization of the cavalry.[81]

By now the conflict over the proper mission of cavalry had intensified. Was it to be deployed in an auxiliary or in a battle role? The battle role found strong support from the mechanized force at Fort Knox, while the army generally supported the auxiliary role, placing cavalry as an adjunct to the dominant combat arm, the infantry. However, there was an effort at a compromise. "It is equally true that Cavalry, as a whole, is more effective when horse and mechanized units are employed in cooperation than when either is employed alone," wrote Grow in the *Cavalry Journal*. He maintained that this teamwork between horse and machine would be the most important issue confronting senior cavalry commanders, concluding, "We do not want a mechanized force of tanks in the European sense. We do want a mechanized cavalry that is cavalry in every sense of the word."[82]

Writing later in the *Cavalry Journal*, he stated that combat cars were the backbone of mechanized cavalry but their independent deployment was never contemplated. Combat cars formed the nucleus of a balanced homegeneous attack, he wrote, and cavalry must be capable of fighting unsupported by infantry. It must fight mounted, dismounted, or by combined action. The problem, he argued, was that cavalry had failed to obtain the "ideal horse." There existed various types of equipment for special purposes, such as combat cars, armored cars, scout cars, half-tracks, and mortar mounts. In addition, he believed that for years to come proper teamwork between the horsemen and mechanized cavalry constituted the most important issue confronting senior cavalry commanders. What was the solution? Only strong cavalry leadership and perception of the future role of mechanization could save the new mounted arm from extinction, he believed. Grow thought the hidebound tradition of the horseman versus the mechanized cavalryman's "scatter-brained enthusiasm without his feet on the ground" was the greatest problem. He pleaded, "Let us be cavalrymen not horsemen or mechanized cavalrymen."[83]

Herr now had the opportunity to change the course of cavalry history by moving it into the future with a mechanized cavalry division and new mounts. Yet in spite of major developments abroad, the horse mentality was strong on tradition.

Future general James H. Polk recalled days at the Cavalry School during 1937–1939 when Guy Henry was the commandant. They were "gorgeous years," training with both horses and mechanized vehicles. Polk, a graduate of West Point, claimed the philosophy of the army in the 1930s was "work hard; play hard." Polk, who became a mechanized cavalry group commander in World War II, viewed the problems with horse cavalry differently than did the "cushion pounders" at Fort Knox. When he and his peers were junior lieutenants with the 8th Cavalry Regiment, they thought their squadron commander, Major Ernest N. Harmon, was too zealous. Certainly because Polk and three of his officer friends got sloshed on tequila one weekend and kicked a resident officer's sign over the head of another squadron commander.[84] Undoubtedly, Harmon's words over the incident were not censured and surely were spoken as a true cavalryman. Polk never forgot the incident and eventually considered Harmon one of the most prominent armor commanders of the war.

Polk believed the training at Fort Riley was sound and basic. He loved everything about horses, especially the training and polo. Polk was recognized as one of the best horseman at the fort. In the morning horses were exercised, and in the afternoon junior officers attended classes and engaged in equestrian riding, shooting, and saber practice, terrain rides, shoeing horses, and packing mules, all filling out a cavalryman's day, he recalled. Polk praised some of his instructors such as Patton, who taught tactics, and Terry de la Mesa Allen, a future infantry division commander in World War II, who taught night tactics. Instruction was mainly in theory and was challenging. Polk loved it. He believed serving in the horse cavalry provided troopers with a better sense of terrain than when mounted on an auto or tank.[85]

Meanwhile, Chaffee, recently promoted to brigadier general, was back at Fort Knox to take command of the 7th Cavalry Brigade when Van Voorhis took command of V Corps late in 1938. Brigadier General Marshall, now assistant chief of staff in the War Plans Division and soon to be deputy chief of staff in the War Department, congratulated Chaffee on his promotion. "There is no one in the Army I would rather have seen promoted, for the good of the service, than your self," he wrote. Marshall let Chaffee know that from "all sides I hear expression of satisfaction at your advancement." Major General Stanley D. Embrick, deputy chief of staff, told Marshall that he had Chaffee "down for chief of staff." Marshall added, "It would be a splendid thing for the Army if you could take that course."[86]

Senator Harry S. Truman, a member of the Appropriations Committee, again visits Fort Knox, Kentucky, in November 1938. General Adna R. Chaffee Jr. is showing a 75mm howitzer to the senator. On Truman's right is Lieutenant Bob Lee, Chaffee's aide and personal pilot. From the Chaffee Papers, Patton Museum.

Chaffee, who had just returned from the Fort Riley horse-mechanized maneuvers, thanked Marshall for his encouraging letter. After his experience there, he also wrote Marshall that "any consideration of brigading or divisioning horse and mechanized elements is faulty organization. You can just as well put a yoke around the tortoise and the hare and expect them to race as a team."[87]

Chaffee inherited another problematic issue that dealt with the Cavalry Board. The board recommended replacing towed artillery with self-propelled guns. It was believed self-propelled artillery was necessary to neutralize antitank weapons, while providing general supporting fire for combat cars and targets of opportunity. Chaffee approved, and Herr was advised of the need.[88] The chief of field artillery, Major General Robert M. Danford, however, disagreed. The year before, his predecessor, Major General Upton Birnie Jr., had argued that despite tremendous improvements in mechanization and transportation, "horse-drawn is a little better than motor-drawn" artillery. Danford nevertheless supported towed artillery, believing that it could deliver far more supporting fire. However, he regarded the mechanized cavalry's appeal for self-propelled artillery as no more than a request for a vehicle with all the essential characteristics and limitations of a tank. The solution, he argued, was a combat car armed with a cannon and sufficient armor to withstand shelling from antimechanized weapons.[89] Danford's attitude can be summed up as, "The horse can be used by us, and he still remains superior to the motor. . . . to discard him during peace in favor of the motor . . . is an unsound policy."[90]

Nonetheless, with support from the chief of ordnance, a 75mm pack howitzer was mounted on a combat car M1 and classified as the T3, 75mm howitzer motor carriage. The field artillery, however, considered the T3 unsuitable because of limited crew space. As a result, no additional ones were built for the mechanized force.[91] Another aspect of the problem rested with General Craig's Protective Mobilization Plan, which limited research and directed the army to retain its serviceable equipment. Not until Major General Jacob L. Devers replaced General Chaffee as the chief of the Armored Force in August 1941 was the issue of self-propelled field artillery finally resolved.

The Christie M1938 (T12) tested October 1938 at Hempstead, Long Island. Later Christie offered Chaffee a performance run at the end of August 1939. Courtesy of the Patton Museum.

While the Ordnance Department and cavalry were still experiencing problems with the convertible combat car T7, the annoying Christie once again attempted to sell one of his vehicles to the army. A conference was held in the Ordnance Department in March 1939. This came about because of a test of his reshaped M1937 that was rejected earlier by the British and observed by a U.S. Army ordnance officer in December 1937. The vehicle (also referred to as the Christie Tank, Model T12 and M1938) underwent a performance test in October 1938 at Hempstead, Long Island, in Christie's effort to settle a contract dispute. Two ordnance officers witnessed the test. The usual fanfare preceded the test run, which shortly ended after the right side of the vehicle broke when it was pushed to excessive speeds. Consequently, the tank required extensive welding if it was to run again. The ordnance officers reported this latest vehicle, as before, had not met military requirements for an armored car or light tank because its low silhouette compromised the size of the fighting compartment (it had no turret), the vehicle armor was too light, and it lacked armament. Like cavalry and infantry observers in 1929, the officers were impressed with the suspension system. It was noted, however, that the wheel-track principle had been dropped. Skeptical, the attendees were interested to hear Christie's next ploy to interest the government in his reshaped tracked vehicle. One of the attendees was his old nemesis, Major Christmas, the department's technical tank expert. The ordnance officers told Christie that he had not presented anything new, other than exaggerated claims and publicity for speed at the expense of military characteristics required by the cavalry and infantry. Like many times before, Christie became angry and stated that the vehicle was satisfactory in every respect. He now presented his most preposterous threat.[92]

If the Ordnance Department did not facilitate production of his tank now, Christie threatened to bypass that service and go directly to high officials in the War Department, as well as visiting President Roosevelt to gain approval for undertaking construction of the vehicle. Christie then stated that if there was no response, he was prepared to use an organization of civilians who could build a large number of his tanks for training purposes. He then jumped from his chair and, displaying his usual fiery temper, shouted to the officers as he left the room that he was off to see the president.[93] Fortunately, his threats did not materialize. Later the army, in cooperation with General Motors Cooporation, briefly showed an interest in the Christie suspension system when Tank Destroyer Command was created as an answer to the German blitzkrieg.[94]

In a later interview with the *Elizabeth (N.J.) Daily Journal,* Christie stated his lifetime philosophy: "All I want the army authorities to do is say: Give him [Christie] the money and let's see what sort of machine he can turn out. I won't let them change a thing on it. If they try to, I'll walk out. I won't let them tinker with it." This interview summed up his long association with the army.[95]

Unfortunately, Christie's uncooperative efforts and demands that the army accept first his convertible principle and then speed and greater lightness confused any efforts to develop and solidify an effective tank or combat car policy during the interwar period. His demands became a historical struggle with the Ordnance Department and the cavalry and infantry. The heated debates with Christie over technology and tank guidelines also demonstrated the fallacy of combat arms autonomy, which led to the inability of the army as a whole to solidify a uniform tank policy during most of the 1930s .

By mid-1939 Europe and Asia were moving toward another world war. Patterns of aggression increased with marked intensity. In March, the internecine struggle ended in Spain with the surrender of Madrid and Valencia to Franco's fascist forces. Germany annexed the rest of Czechoslovakia and looked toward Poland. Mussolini, meanwhile, conquered Albania on the Adriatic Sea. Japan, after years of fighting an inconclusive war in China, changed to a strategy of economic attrition.

Taking notice of domestic and international developments, a respected, amiable friend of General John J. Pershing and an experienced army commander during World War I, retired Lieutenant General Robert Lee Bullard commented that the country was not prepared for war. He surely understood the quagmire of trench and static warfare. "We have entered each of our wars prepared for the last one. We modernized in equipment, training and methods as in the last war. Then we were content to remain almost at a standstill for fifteen years. Our arms became obsolete rapidly." He continued, "War technique has advanced swiftly. Machine power and fire power determine strength in battle." He could not understand how a country such as the United States could produce five hundred thousand motor vehicles a year yet not mechanize a single division. To Bullard, this was an absurdity, especially while European and Asian nations were rapidly mechanizing divisions for a new method of warfighting.[96]

Even as European and Asian fascist nations were engaged in aggression, the U.S. Army prepared to mobilize its largest number of troops since World War I. Attempting to anticipate preparations for war, plans were made for First Army maneuvers in the Lake Champlain region of New York involving fifty thousand soldiers from the regular army, army reserves, and the National Guard. Chaffee's brigade from Second Army was invited to participate.

Even as preparations were being made for the Plattsburg maneuvers, the chief of staff, General Craig, following earlier recommendations of General Krueger, disapproved a project expanding the mechanized cavalry into a division. In spite of this setback, Herr was able to acquire additional vehicular equipment for Chaffee. For the mechanized cavalry at Fort Knox, Herr consented to it receiving sixty combat cars. The chief, however, showed more enthusiasm for motorcycles. He told Chaffee that there was a vast field of possibilities for both solo and tricycle motorcycles as substitutes for the rifle troop and reconnaissance scout cars.[97] Certainly, this was not what Chaffee wanted to hear.

The 7th Cavalry Brigade again had the opportunity to test and expand operational and tactical mobility during the upcoming maneuvers in August. The force was organized around the two iron horses: the combat car for the assault, and the scout car or half-track for reconnaissance, security, dismounted action, and transportation of supporting troops and their weapons.[98] The month before the Germans launched their blitzkrieg against Poland in September 1939, the 7th Cavalry Brigade had engaged in a wide enveloping movement, completing a successful deep maneuver–oriented operation, cutting across the hostiles' lines of communication. The brigade's deployment confirmed what Van Voorhis had already demonstrated the year before: that a fast-moving mechanized force could decisively influence a battle. The maneuvers proved that rapidity of action and agility with a mechanized force were an antidote to the traditional strategy of attrition warfare, which was the traditional American way of warfare that determined the direct approach, requiring superior manpower, massive firepower, and logistics to position, close, and destroy the enemy.

Unfortunately, the issue of lightness that drove American tank and combat car policy during the interwar period proved too questionable for modern warfighting, as events in Spain and Poland demonstrated. It was expected, nevertheless, to use up the large number of obsolete and outdated infantry light tanks and cavalry combat cars until the production of medium tanks got underway. Nevertheless, events in Europe and the astonishing German lightning war technique soon gave Chaffee the rationale he needed, but not before the "great debate" was resolved.

Reporting on the maneuvers, Chaffee concluded that now there was no doubt that mechanized cavalry was capable of long, rapid, strategic moves under cover of darkness and without lights. He saw the full value of what he called a combat team employing all elements, such as ground and air reconnaissance, and combat cars and artillery. He argued that the mechanized cavalry, with its great firepower, rapidity of action, and striking power, had a decidedly adverse effect on the morale of opposing ground troops. Chaffee noted that because of the expansion requirements of the Air Corps, no combat aviation of any kind was made available for the maneuvers. He also took a shot at the horse cavalry, noting that its deployment in an antitank cordon defense was just like the infantry. This, he felt, created a defense mentality at the expense of mobility. Furthermore, he wrote, "Infantry tank units do not possess the auxiliary means of reconnaissance and support to successfully oppose a strong force of mechanized cavalry."[99]

An M3 scout car with Major General John K. Herr on the left and Brigadier General Adna R. Chaffee Jr. on the right. Notice the body language. From the Hunnicutt Collection, Patton Museum.

In his critique of the maneuvers Lieutenant General Hugh A. Drum, who commanded the First Army, commented that Chaffee's brigade was a powerful asset and that many lessons had been learned. Furthermore, he noted, the 7th Cavalry Brigade was "a psychological weapon as well as a tactical one." As the battle progressed, "my troops first called it 'the mosquito' then 'the hornet' then 'the devil' and at the end a name I dare not mention." What was apparent, though, was that the brigade did not have suitable supporting infantry, and its tanks were too light and lacked firepower and armor protection. In addition, the brigade was weak in reconnaissance elements. After the maneuvers, it was obvious that the 7th Cavalry Brigade under Chaffee's leadership was a potent striking force. It was, however, too small and light for a mechanized cavalry division and had not yet reached the mission status of an armored division.[100]

The chief of cavalry was neither pleased nor in agreement with Generals Drum and Chaffee about the First Army maneuvers. Herr maintained the lesson that it was futile to hold maneuvers when not all the combat elements were included, meaning horse cavalry. He advised Brigadier General Frank M. Andrews—now the assistant chief of staff, G-3 and the first airman to head a War Department general staff division, and former commander GHQ air force—

that Chaffee's brigade was sent to the maneuvers at great expense and were entirely unsuited for off-road operations, which had been attempted with great difficulty. Herr was annoyed because the one horse cavalry squadron available at Plattsburg was deployed only on picket line duty to make horses accessible to umpires and directors of the maneuvers. No doubt, the cavalry chief was looking from the eastern portion of the lake to the rolling country that gradually sloped away and then west upward to the Adirondack Mountains. The terrain around the Lake Champlain area was more suitable for horse cavalry, he argued. Concluding his memorandum to Andrews, Herr complained that his horse squadron was deprived of valuable training and was "frittered away doing nothing." He then chastised commanders for squandering away cavalry in World War I and during the postwar period.[101] Soon Andrews and Herr were on a collision course regarding how the army mobilized and the role of the combat arms chiefs.

Before the Wehrmacht's invasion of Poland, Herr had been adamant about retaining his horses in the face of a rapidly increasing move toward mechanization. Two weeks after the fatal charge of the Pomorsha Cavalry Brigade armed with lances against German panzers in Poland, the disconnected Herr stated, "It seems obvious that the machine cannot eliminate the horse."[102] The following year the prominent military writer Brigadier General Henry J. Reilly, who wrote in 1931 that the horse could remain on the battlefield as long as men engaged in open warfare, defended Herr's position on the utility of horse cavalry in a modern war. He speculated that if Poland's horse cavalry had possessed modern armament and united under one big cavalry command with adequate mechanized forces and supporting aviation, the German blitzkrieg could have been defeated.[103]

General Herr, however, did accept the scout car as part of his horse regiments for general reconnaissance, command, and communications, and a light tank company for a horse-mechanized division. After the successful blitzkrieg in Poland, the cavalry chief communicated with General Craig's successor, General George C. Marshall, attempting again to expand Chaffee's mechanized brigade. He told General Marshall that the mobility and exploitation of the breakthroughs made by German mechanized troops after the assault by infantry and less mobile troops were a "typical cavalry mission." Therefore, his argument went, the 7th Cavalry Brigade required expansion immediately to a division modeled after the German panzer division. He persuasively presented this argument at a War College lecture at the end of September 1939. Such a mechanized division, Herr argued, should be charged to his combat arm along with additional mechanized divisions under cavalry development and leadership. However, the most unyielding statement he made to General

A platoon from the 13th Cavalry Regiment (Mech) demonstrated its combat cars to the people visiting the 1939 New York World's Fair. After a strenuous first day at the fair, the cavalrymen woke up to the German blitzkrieg. It was September 1 and the war in Europe had begun. Courtesy of the Patton Museum.

Marshall was that mechanization was designed to expand the scope of existing combat arms and not to replace them. The chief of cavalry suggested that any increase in personnel for the proposed mechanized divisions should not come from any existing units but "should be in addition thereto." He advised

Marshall that Chaffee was aware of the situation and that he agreed with his recommendations.[104]

While Herr was in a titter over the First Army maneuvers, the weary soldiers went to their home bases or just home, except the 7th Cavalry Brigade. Toward the end of the maneuvers, the whole brigade assembled in anticipation of participating in the New York World's Fair at the request of Mayor Fiorello La Guardia. The War Department approved, and Chaffee now had an opportunity to exhibit his proud brigade of six hundred vehicles to the nation. On the way, the brigade with all its combat cars, armored cars, scout cars, mortar mounts, and half-tracks were reviewed and inspected at West Point. The superintendent told Chaffee his brigade was "most impressive. Even more impressive was the outstanding evidence of the efficiency of its officers and men of their military bearing, exemplary conduct, and high morale."[105] Chaffee's aide, H. H. D. Heiberg, remembered, "We tied up cross town traffic for four hours." However, he noted, a platoon demonstration was held every afternoon and viewed with interest by hundreds of people.[106]

While these demonstrations were in progress, Christie contacted Chaffee and invited him to view his M1938 that was tested in October 1938 at Hempstead, Long Island. The vehicle was supposedly parked at the Hempstead Welding Company undergoing repairs. Christie planned to take Chaffee to the welding company and arrange for a performance run. However, the vehicle was gone, and Christie had no idea of its whereabouts. According to a Federal Bureau of Investigation report, the vehicle had been sold at public auction to pay off a mechanic's lien. Christie had earlier asked the FBI to get involved because he believed foreign spies were implicated in the theft of his armored vehicles. The agents advised him that the FBI had no jurisdiction over this issue and that he should take up his request with the War Department.[107] This was Chaffee's last contact with Christie.

On a more humorous note during this period of intensive debates, Heiberg recalled one occasion when he turned out to go on a tour of Fort Knox with Chaffee after they returned from a trip to Fort Riley. Heiberg was wearing a yellow scarf, which was quite popular at the former base. Chaffee looked at his aide and quietly said it was a beautiful scarf; do not wear it again, though, he said, because "as soon as I let you wear that, every damn artilleryman will turn out looking like his throat has been cut."[108] Red was the artillery's color as yellow was the cavalry's color.

Since army expansion was slow and budget appropriations were still questionable, it was unlikely the War Department would approve an additional mechanized cavalry unit. Herr believed mechanization should be an integral part of the line branches and argued throughout his tour against a separate force. His concept of executing open warfare or a war of movement was focused on defense of the homeland, where wide spaces existed, rather than in Europe, where he visualized static defenses that could repeat the expenditures of large numbers of warfighting materials as happened in World War I. The chief of cavalry was relying on the availability of more than 12 million horses and 4.5 million mules, along with the largest motor industry in the world. This was suitable for his cavalry if deployed on the American continent. He argued that cavalry should be used in large masses so as to perform major combat roles at a corps level and "used as Army or GHQ Cavalry." Regarding mechanization, he stated, "It has not yet the capacity for sustained mobility that is inherent in horse cavalry. Horse cavalry, on the other hand, has great strategic mobility when transferred by train or by motor." Once deployed, he claimed, cavalry "has greater tactical mobility in all conditions of terrain and weather than any other ground arm." Like the unyielding pro-horse Hawkins in his earlier *Cavalry Journal* articles, Herr concluded by chastising the American people for their vast ignorance of the worth of American cavalry. "We as cavalrymen," he held, should make an effort to demonstrate to the American people the history and worth of cavalry. This was his "abiding faith" in the future development of modern cavalry.[109]

General Herr's inflexibility, the problems of creating a mechanized cavalry division with self-propelled artillery, and the failure to coordinate tactical air-ground support other than reconnaissance were disappointments to Van Voorhis and then to Chaffee. Furthermore, Willis Crittenberger, who was now on duty with the chief of cavalry in Washington, wrote Van Voorhis at V Corps that the whole project for expanding the mechanized cavalry brigade into a division had been disapproved by the War Department. Instead, consideration was given to another brigade. Crittenberger speculated that "some outside influence in the higher echelons of the General Staff" was responsible for the decision. Not only was funding a problem, but also the attitudes of Herr and Lynch. Both opposed the conversion of existing units for mechanized expansion, Crittenberger advised Van Voorhis. Since his arrival in the Office of the Chief of Cavalry in 1938, Crittenberger observed a great interest in mechanization among his fellow officers. Continuing, "Were it not for the bugaboo of personnel, I do not doubt that several moves would have been made in the mechanized field long since." Crittenberger—now quite frustrated—told Van Voorhis that it appeared mechanization was moving for a showdown, and that he hoped "common sense would then prevail."[110]

At the time Van Voorhis was commanding V Corps, he caught the attention of General Marshall, who had accepted a post with the general staff in the War Department. Marshall was advised that of all the corps in the continental United States, Van Voorhis's corps was considered the most efficient in producing results. Marshall asked Van Voorhis "to give us a brief statement of just how he goes about this particular phase of the business" in handling reserve corps affairs.[111]

The inability of Van Voorhis and Chaffee to integrate close tactical air ground support, other than long-range reconnaissance, was a problem, as was integrating self-propelled artillery and armored infantry. These issues were not resolved until World War II. Tactical air support became more of an issue with the establishment of GHQ Air Force in 1935 under the command of Brigadier General Frank M. Andrews, who was a strong and vocal proponent of the heavy bomber, Boeing's YB17. During this period, a lack of coordination and execution of a tactical doctrine gave way to a doctrine of long-range strategic bombing. Part of the problem was that the GHQ Air Corps was organized as a separate force, virtually gaining an equal status with the line branches. This resulted in airmen solidifying their own specific branch doctrine, which placed emphasis on strategic bombing. Furthermore, a technical revolution in heavy long-distance aircraft and equipment favored a strategic bombing doctrine that gained prominence in the 1930s. Air Corps leaders viewed the commercial aircraft industry as an economic convenience that favored strategic needs for long-range bombing doctrine. A result of these developments, along with congressional funding problems and shortage of personnel, caused GHQ Air Force to make trade-offs in equipment, leading to the development of a light bomber for tactical bombing at higher levels beyond the ground-support coordination line where they were least exposed to enemy fire.[112]

In addition, the Spanish civil war had created little interest at the Air Corps Tactical School at Maxwell Field, Montgomery, Alabama, where emphasis was on precision strategic bombing of the enemy's warmaking industrial centers. Also, there was a concern that low-level attacks against enemy ground forces were too risky because of improvements in antiaircraft weapons.[113] Even Major George Kenney, who in 1931 seemed to Grow to show an interest in tactical support for the Mechanized Force, now viewed the Spanish civil war as offering few, if any, examples of effective modern strategic bombing. Kenney showed more interest in the air transport of Nationalist troops by German aircraft early in the civil war. The cavalry realized from studying the war the importance of defending against air-ground attacks. Now there was a call for a more antiaircraft elements and their training.[114]

Consequently, by the end of the 1930s there was no effective tactical air organization for the Mechanized Force at Fort Knox other than air reconnaissance observation. Not until Patton crossed the ocean to engage the Germans in World War II was serious consideration given to the issue of close tactical air support for armored divisions. When the Third Army became operational under General Patton on August 1, 1944, Brigadier General Otto P. Weyland's XIX Tactical Air Command provided close air support for Patton's fast-moving armored divisions. Air force historian David N. Spires, in his telling history, called their combined success an anomaly because during the interwar period and early in the war, army-air relationships were confrontational and besieged with interservice infighting.[115] The relationship between Patton and Weyland permeated all the way down the chain of command. Grow, who commanded the 6th Armored Division under Patton as the Third Army raced across France and into Germany, praised the excellent close air support his division received from the Tactical Air Command.[116] Major General Elwood R. Quesada of the IX Tactical Air Command, however, established a relationship that molded an effective air-ground team. He set the standard for close air support during the invasion and early weeks in Normandy before the Third Army became operational.

During the 1930s, Van Voorhis and Chaffee, plus many open-minded cavalry officers such as Palmer, Scott, Grow, Crittenberger, Heiberg, Boudinot, White, and Harmon, to mention a few, believed their branch had no choice but to develop a better horse. The belief at Fort Knox was that combat cars offered the critical means for the soldier to fight mounted, but that was not enough, since cavalrymen were expected to fight both mounted and dismounted if necessary. Other vehicles and arms were needed to support the combat cars that enabled cavalry to carry out all its missions. The cavalry leaders at Fort Knox augmented self-propelled artillery and then eventually came to realize that the army also needed armored infantry that could fight mounted and dismounted. This basic thinking and execution during the evolution of mechanized combined arms nurtured at Fort Knox during the 1930s finally prevailed, influencing many of the armored division commanders in World War II.

Just who were the real driving forces in the 1930s that realized change was essential and provided new opportunities for the cavalry?

Earlier Van Voorhis had been instrumental in keeping the Mechanized Force independent of branch control. With the creation of the Armored Force in 1940, Van Voorhis's intentions of 1931 were proved correct. He had argued unsuccessfully for a mechanized cavalry division in 1936–1938. In 1938, he demonstrated the effectiveness of maneuver-oriented self-contained deep operations with the deployment of two striking forces. This impressive operation set the tone for a World War II armor doctrine of combat commands. He verified that rapidity of action with new mounts with agility was superior to current infantry linear tactics and more combat effective than the existing horse cavalry. Chaffee inherited this tactical idea when Van Voorhis was given command of V Corps in October 1938. Grow maintained that Van Voorhis planted the seeds that gave rise to the Armored Force, whereas Chaffee strove to develop a better horse for the cavalry. Grow, however, gave credit to Chaffee rather than Van Voorhis for generating the idea of a flexible organization consisting of combat commands.[117] During the 1938 spring maneuvers Van Voorhis demonstrated considerable skill and flexibility in command and control both on the ground and in the air by deploying his two regiments separately and then consolidating them after a deep offensive deployment against a simulated invasion force. Chaffee learned from this operational concept of deploying a brigade, since his only previous command experience had been deploying the 1st Cavalry during the 1934 maneuvers before being assigned for four years to the War Department's Budget and Legislative Planning Branch.

It cannot be forgotten that it was Van Voorhis who took the initiative. He tactically molded the mission of the Mechanized Force at Fort Eustis into a new organization using obsolete equipment and a few experimental tanks, and armored cars. As he stated, "We got the cart before the horse and made the mission conform to the equipment at hand." Once at Fort Knox, he redefined the mission of cavalry, believing doctrine preceded equipment. His goal was to build up the mechanized cavalry to a brigade with characteristics that had the capability of fighting mounted with new mounts despite delays in receiving new equipment. Under his command, radio communications were developed more extensively than in the other arms so "mobility of the mind equals the mobility of the machine." He believed strength lay in the power of maneuver and firepower, stating, "The very characteristics which make it mechanized cavalry, mobility and firepower, particularly qualify it for combat missions."[118] In addition, Van Voorhis and his two regimental commanders, Colonels Palmer and Scott, were responsible for developing combat command training and tactics for a mechanized maneuver–oriented brigade. During his last maneuvers in May 1938 with the 7th Cavalry Brigade, Van Voorhis again became innovative when at times he commanded the deployment of two cavalry columns in a deep operation from the air. The early organizational concept of deploying two combat commands was later expanded with the creation of the Armored Force. He groomed his operations officer, Willis Crittenberger, to become one of the outstanding generals of the war, ending it as commanding general of IV Corps, in the Italian campaign. Van Voorhis described Crittenberger in May 1938 as "superior in performance and exhibiting leadership skills, both mental and physical. He is of superior value to the service and should be given consideration for promotion to grade of general officer."[119]

During the 1930s, Van Voorhis's influence and drive provided Chaffee, Palmer, Scott, and other cavalrymen the opportunity to experiment with a flexible rather than a fixed organization. In spite of budgetary problems, they understood and experimented with limited combined arms crossover and technology in deep maneuver–oriented warfighting that caught the attention of Patton. However, even with all the efforts of Van Voorhis and Chaffee, they had not produced any new doctrine except ideas on how to work out a solution regarding mechanization. The cavalrymen at Fort Knox were trying to see what kind of agility and mobility could be achieved with the equipment they had on hand and what they could acquire. The ideas germinated at Fort Knox, and the concepts were later converted to doctrine as the Armored Force and Mechanized Cavalry made their own as they engaged in combat.

It could be argued that Van Voorhis more than any other cavalryman at Fort Knox was determined to make sure that the mechanized cavalry was tactically oriented to deal with the modern battlefield in time and space. He was a prime example of a cavalryman accepting command of the Mechanized Force and then adjusting to change by introducing tactical innovations during the early days at Fort Knox. Surely, he was an agent of change and astute in adjusting to many adverse conditions during the 1930s. Van Voorhis performed every task with determination, imagination, and efficiency. By the time Chaffee replaced him in late 1938 as brigade commander, Van Voorhis had taken a mixed collection of soldiers and obsolete equipment and created a disciplined force of cavalrymen in a fighting unit with esprit de corps, the 7th Cavalry Brigade. Surely, he took the first step in the development of American armor. He did not believe in replacing the infantry and cavalry with a mechanized branch. However, he supported a changing mission for cavalry that required replacing horses with fast, hard-hitting vehicles in deep maneuver–oriented operations.[120] Later Van Voorhis proposed a mechanized combat branch for the cavalry because he saw no movement in the Office of the Chief of Cavalry or the Army Staff to take seriously a role for all-out mechanization.

Certainly, Van Voorhis and Chaffee were both most resilient in bringing the cavalry into the future from horse to horsepower. Nevertheless, Van Voorhis deserves more credit than he is given. This is not to take away from Chaffee's accomplishments. Perhaps more than Chaffee, Van Voorhis expanded the meaning of ever-present change that provided new opportunities for the cavalry beginning with the Mechanized Force in 1930 and then with the establishment of a mechanized cavalry force at Fort Knox. Eventually Van Voorhis believed that to bring the cavalry into the future required rethinking the combat arms branch system by the War Department and its removal from under the arm of the chief of cavalry, an opinion Chaffee took up especially when Herr became chief of cavalry. Both Van Voorhis and Chaffee were also determined to bring the mounted arm into the future with new mounts. Both were major marketeers for change.

Colonel Pritchard recalled Van Voorhis's having a sense of humor. In addition, he remembered, "He never asked anyone to do anything that he didn't feel he was willing to do himself." As noted, young Boudinot was impressed with Van Voorhis's demeanor, and so was Pritchard. "He was very particular about people's dress," Pritchard recollected, continuing, "If you didn't keep your equipment clean, you didn't dress well and keep yourself in good shape, he felt you lacked pride. And if you lacked pride, you wouldn't be very much of a soldier."[121]

Grow, on the other hand, viewed Van Voorhis as "stubborn and opinionated" yet a "magnificent disciplinarian and soldier of the old school who saw the need but was bewildered by technical problems. I fought with him and loved him," whereas Chaffee, Grow believed, had "a keener tactical mind."[122] Grow gave credit to Van Voorhis and Chaffee for their organizational and administrative skills and utilization of whatever vehicles and means they had at their disposal to move the cavalry into the future.

I. D. White said of Chaffee, "He was one of the nicest, most rewarding characters I ever met, military and otherwise. He was one of the real fathers of the Armored Force." White gave this accolade to Chaffee because he was instrumental in selling the mechanized force to the War Department during the four years he was stationed in Washington. White also considered Bruce Palmer one of the fathers of the Armored Force.[123]

In spite of efforts by Van Voorhis and Chaffee, little was accomplished in moving from towed artillery to self-propelled artillery and a supporting infantry doctrine suitable for mobile mechanized cavalry operations. Before the Armored Force absorbed the 7th Cavalry Brigade, it was low in reconnaissance strength and lacked infantry assets. When the Armored Force was created, it became tank-heavy, lacking a balance of armor infantry and armored artillery. Major General Jacob L. Devers finally established a balance. He brought self-propelled artillery into the Armored Force, thus rounding out the combined arms team.

There were, however, some disputes over who was the real driving force in bringing mechanization to the army during the 1930s. General Palmer's son recalled his father mentioning that he and Crittenberger were the "real guys who fathered the Armored Cavalry."[124] However, it should be noted that Chaffee had no use for Palmer when the Armored Force was created. And Palmer felt Craig was so politically oriented that his influence was limited and that he did not have the intelligence to be chief of staff. As Chaffee told Marshall in December 1940 when he was chief of staff, Palmer had written a long letter to Craig criticizing his decision and asserting that he, Palmer, should have been appointed. Twice Palmer came to Craig's office, but Craig refused to see him. This attitude disturbed Chaffee, who, like Van Voorhis, expected loyalty, reliability, and trustworthiness from subordinates.[125] Perhaps the other concern Chaffee had was that Palmer was inclined to view the combination of horse and mechanized cavalry as a force twice "as strong as either one alone but immeasurably stronger," a comment he made at the Army War College in October 1936.

One of the heroes in securing the Remagen bridgehead and later the commander of the 4th Armored Division during World War II, Major General William M. Hoge, claimed General Devers was the real father of armor and that Chaffee was more of a theorist. He also thought Chaffee was a pure cavalryman, who during one of his lectures "talked more about horses than he did about armor."[126] Major General Charles L. Scott, who was acting chief of the Armored Force during Chaffee's illness, was more supportive. He claimed the leadership at Fort Knox was not taken by the employment of German panzer divisions because "it came as no surprise and the success, as a matter of fact, was expected." It was Crittenberger, Van Voorhis's operations and training officer, Palmer, and Scott who had met with a prominent German staff officer in 1937 and discussed mechanized maneuver–oriented operations at an operational level.[127] Chaffee believed that, since 1929, Van Voorhis, Palmer, Scott, Henry, and he had "all played some part in starting or forwarding the development of mobile mechanization." More so, he added, "I have always thought that each one contributed his part."[128]

The problem Van Voorhis and Chaffee had faced during the 1930s was dealing with the inertia of army elites that supported the antiquated 1923 doctrine, the branch system, and especially the attitude of the combat arms chiefs who failed to understand that the nature of warfare had changed. The elites prevented the army from thinking about an operational level of warfare and the relationship between strategy and tactics by adhering steadfastly to the traditional autonomy of the branch system that reigned supreme. Because of a fixation on their compartmental system and the desire to defend it against challenges of modernity during a period of budget restrictions, the chiefs became indifferent to significant ideas that could have moved the army to change doctrine. Their perception of how to fight the impending war remained archaic and embedded in the last war of attrition. Added to the situation in the 1930s was a lack of national interest in military affairs both in Congress and among the public, which financially affected the army's ability to prepare for and win the first battle. Charles Scott even went further; he blamed the series of Geneva disarmament treaties during the 1930s that attempted to reduce tanks and artillery or all weapons of offensive warfare.[129]

It was Frank Parker and his foresighted ideas that influenced Chaffee in the late 1920s. Chaffee then integrated them with Van Voorhis's ideas on cavalry mechanization and mounted warfare. Subsequently, Chaffee was instrumental in carrying their ideas even further. Chaffee's understanding of history, especially his study of the Civil War cavalryman General James H. Wilson, convinced him that change was inevitable. Chaffee absorbed these lessons, which provided him with the driving force to deal with new opportunities and finally carry them to completion in 1940. The Armored Force was finally established and eventually became a powerful mobile force strong in firepower and tanks. Consequently, Chaffee was called the "father of the Armored Force." Even though he was successful in implementing his ideas with the creation of the Armored Force, he had not gotten them across broadly enough to change the army's focus from the traditional way of attritional warfare, in the style of Grant's Overland Campaigns in Virginia. The question one has to deal with is, what would have happened if Chaffee had lived, knowing his close relationship with General Marshall?

Van Voorhis, sometimes referred to as the "grandfather of the Armored Force," chose not to recount his experiences other than what appeared in official government documents and journals. John Broom confirmed Van Voorhis's reluctance in an interview with Colonel Pritchard.[130] Van Voorhis, however, did supply some historical information for Chaffee's biographer, Mildred Gillie.

Van Voorhis retired from the army in 1942 as a lieutenant general and died in 1956. A year after his retirement, he commented that the German blitzkrieg coincided with the employment of U.S. Armored Forces. He believed the Wehr-macht learned its doctrine of warfighting with mobile combined arms units from the mechanized cavalry at Fort Knox.[131] How much of this is true, and did the Germans learn something from Parsons's idea of a tank division in the 1930s? What did the German staff officers learn in 1933? What did the Wehrmacht absorb from von Schell's visit in 1937 and Guderian's observations? What American ideas observed at Fort Knox made it to the Wehrmacht's notion of mechanized mobile warfare? One can only speculate. Unfortunately, German intelligence documents regarding these issues were destroyed during World War II. The German military, nevertheless, vindicated the tactical ideas developed and practiced by the mechanized cavalry at Fort Knox and moved them to an operational level of warfare. More so, the Wehrmacht perfected the lightning war doctrine integrated with tactical airpower, an element that the mechanized cavalry failed to develop even though General Parker talked about it as early as 1917.

Brigadier General Sidney R. Hinds, one of the maverick infantry lieutenants during the 1930s who gained considerable experience with tanks, argued, "It was hardly fair to credit Guderian and the Germans or even General Patton himself alone with inventing the blitzkrieg." He claimed, "The idea had been germinating for a long time and only when the vehicles to fertilize it became available did it come to full fruition" in the U.S. Army.[132] Hinds became the commander of Combat Command B, 2nd Armored Division in 1945.

As for von Schell, he became a major general in command of the 25th Panzer Division in Norway, where he planned for a possible invasion of Sweden in case the Allies made an overture to that country. After a short period in Normandy, von Schell took the division to the eastern front in the fall of 1943. He retired from the Wehrmacht as a lieutenant general in December 1944 because of Hitler's usual erratic annoyances with his senior officers on the eastern front. Von Schell was arrested by American forces in April 1945 and shortly appointed, as were many senior German officers, a historical consultant to the U.S Army until 1948. He died in September 1967.[133]

What he brought to military leaders in America during the 1930s was a voice of change in the Wehrmacht, which was formulating a new mobile doctrine, the engine of change called *Die Truppenfuhrung.* In addition, he attempted to explain without success the German tactical philosophy of *Auftragstaktik.* At Fort Knox during the period, the marketeers were there, but there was no engine of change due to the autonomous branch system that hindered the implementation of a new doctrine based on the concepts of all-arms mechanized warfare, in spite of what was being learned from von Schell and reported from Berlin by Truman Smith, who wrote General Marshall in August 1939, "I have grown very tired mentally this year. We have lived more or less on a volcano here and the strain on one's nerves has been tremendous."[134] Many of the army elites, such as Johnny Herr and George Lynch, had not taken heed of the serious military developments in Germany until the blitzkrieg occurred. Certainly, years of changing international dynamics should have been a wake-up call.

6

"So He Lost It All?"

It was evident that the conservative attitude of the line branch chiefs and their separate compartmentalized organizations was far from adequate in preparing the United States for war. During the 1930s, the U.S. Army was aware of military innovations in other countries, especially in Germany, England, and the Soviet Union, through the reporting of their own and other foreign military attachés. Reports were flowing into the War Department about a new form of military art being developed by the Soviet Red Army with large mechanized units that included fast Christie-type tanks deployed at an operational level. These large concentrations of massed mechanized forces were training to fight a major campaign. The reports on foreign military developments were routinely passed on by the military attachés to the War Department, Military Intelligence Division, the War College and the War Plans Division, the Command and General Staff School, and then for distribution to the combat arms schools. Ideas, especially from the Soviet Union and Germany, were advanced regarding operational mobility with a combined arms force necessary to conduct a campaign that flowed from military strategy.[1] The autonomy of the combat arms, however, prevented these ideas from maturing across branch lines.

Even after the United States entered the war, the chiefs, especially of the cavalry, field artillery, and infantry, refused to relinquish their autonomy or even share in crossover cooperation. They were bound to the antiquated 1920 National Defense Act and *1923 Field Service Regulations: Operations* that reflected the experiences of World War I. The *1939 Field Service Regulations (Tentative): Operations* was a reiteration of the *1923 Field Service Regulations* that had the input of Major George A. Lynch, a graduate of West Point and a future chief of infantry.[2] Consequently, the army for decades failed to achieve modern relevance and readiness to deal with the future battlefield that embraced mechanization. Rather than focusing on mechanized maneuver–oriented warfare that offered flexibility, speed, and surprise, the army was comfortable with traditional attrition warfare based on the infantry-artillery team. This doctrine was reinforced in the Protective Mobilization Plan of the late 1930s, which initially emphasized manpower over research and modern equipment.

As prescribed by the *1923 Field Service Regulations* and the *1939 Field Service Regulations (Tentative),* the chief of infantry still controlled the tactical level of engagement. It was a doctrine with a strong French influence that mirrored the infantry-artillery team as the centerpiece of attrition warfare. This branch mind-set precluded any effort to establish a viable combined arms force necessary to bring the army into the future. Adna Chaffee, on the other hand, argued for an all-tank force that stressed mobility as the centerpiece of pursuit and exploitation rather than relying on support from the infantry-artillery team. His problem was that he overstressed cavalry mobility at the expense of an appropriate level of combined arms balance. Chaffee was also stymied by his own branch autonomous organization that prevented the establishment of equilibrium between doctrine and a tank policy. This situation precluded developing the necessary means to achieve operational maneuverability with other combatant arms. Combat arms branch chiefs, especially the cavalry, field artillery, and infantry, stressed procedural changes only as they pertained to their doctrine. There were no attempts to question or challenge branch autonomy except at Fort Knox. There, beginning in 1931, efforts were made to establish a new mounted mobile force. However, during the 1930s army doctrine and equipment development remained tied to traditional branch autonomy. This attitude, along with budgetary restraints, prevented the army school systems from thinking through a combined arms doctrine at an operational level. In addition, the military branch system, with the specific schools, encouraged autonomy; accordingly, the army and its elites by 1940 were not prepared to fight and win the first campaign with a mechanized maneuver–oriented combined arms team.

Shortly after the beginning of the German blitzkrieg against Poland in September 1939 and President Franklin D. Roosevelt's announcement of a limited national emergency, the U.S. Army implemented its Protective Mobilization Plan. Days earlier, *Newsweek* had reported that the U.S. Army was ranked seventeenth in effectiveness among the armies of the world.[3] By December 1939, the *Army and Navy Register* reported that a reorganization of the cavalry was planned. The chief of cavalry had in mind increasing each horse-mounted regiment from 790 to 1,275 horses and increasing the number of scout cars from seven to ten for each regiment. At the time there were fourteen regiments; twelve were horse-mounted and two were mechanized. Clearly, Johnny Herr was doggedly determined that mechanization would not take over his branch.

With the world again changing to a war footing, the army began to grow at a rapid pace due to the mobilization of conscripts, reservists, and the National Guard. On September 16, 1940, President Franklin D. Roosevelt signed the Selective Service Act into law. Unfortunately, the regular army and even the National Guard divisions were far from equipped to absorb the manpower onslaught. Certainly, the economic conditions caused by the Depression produced congressional budget restraints along with a public attitude consumed by isolationism and pacifism. This contributed to the army's unpreparedness as war shadows were rapidly consuming the country.

After Poland was absorbed by Nazi Germany and the Soviet Union, the future general Bruce C. Clarke, who was commissioned from the Military Academy, recalled his displeasure with the curriculum while a student at the Command and General Staff School at Fort Leavenworth. They still were assigned problems "with the horse cavalry after the successful use of panzers in Poland," he complained. Then at Christmas 1939, three or four students went in to see the commandant, Brigadier General Lesley J. McNair. They told him, "The panzers are overrunning Europe and we have been studying World War I only up until today. Now are you interested in armored warfare?" McNair commented, "We will do something about that." However, his action was meager, to say the least. According to Clarke, he directed that "all division problems from then on would have one company of tanks attached to them."[4] This was hardly a way to deal with training officers for armor warfare.

By then the existing issue of another conflict affecting the United States could no longer be avoided. How was the small army, which finally was rapidly expanding, going to start preparing for war? In early August 1940, General Marshall appointed McNair chief of staff of the newly activated War Department General Headquarters of the Army (GHQ). He now had the responsibility of preparing the ground forces for war. Marshall picked the wiry McNair for his special gifts in organization and training. The job, however, would be far from easy. It became a herculean struggle to transform the rapidly growing army into a major fighting force. McNair faced a monumental organizational challenge; he had to deal with internecine conflicts between the combat arms chiefs and their compartmentalization. McNair declared the situation was "the blind leading the blind."[5]

McNair had been commissioned from West Point in the Field Artillery. While commandant at the Command and General Staff School, he reviewed antitank history and the recent effective use of antitank guns during the Spanish civil war. Earlier in 1934, the chief of infantry, Major General Edward Croft, maintained that new armor-piercing ammunition, especially the .50 calibers, were the antidote for tanks and low-flying aircraft. This infantry antitank attitude continued as an antidote to armored fighting vehicles, and by the late 1930s, McNair had brought it to the forefront. In addition, congressional budget restraints regarding weapons development and procurement, plus reports from the Spanish civil war, prompted a greater interest in expanding an antitank doctrine. In 1939, McNair published a text for instructional purposes, *Antimechanized Defense*, which emphasized that antitank elements needed to be pooled at division level to defeat massed attacking tanks. This passive doctrine would become McNair's solution to the German blitzkrieg and the breakthrough of massed panzers. He brought this concept to GHQ.[6]

The Wehrmacht had another take on McNair's antitank idea. The year before publication of *Antimechanized Defense*, Adolf von Schell recommended defensive dispositions well forward in order to repulse hostile tanks "along the forward edge of the infantry position," meaning a conception of linear distribution as opposed to distribution in depth. His solution was to deploy a force of "tank chasers" called JagerPanzers. Other than this, he noted, improvisation was apt to be too late because "speed was essential." Von Schell was pessimistic about positioned antitank defensives, a lesson the French would learn a few years later.[7]

Lieutenant Colonel Mark W. Clark, who impressed Marshall, was appointed McNair's chief of staff. In this position, according to Marshall's biographer Forrest C. Pogue, Clark became overconfident, which alienated many senior officers. Later in World War II, during the Italian campaign, Clark became a controversial figure.[8] McNair, with Marshall's support, became the driving force for an aggressive antitank philosophy that was soon organized and gained momentum. McNair, along with his chief of staff, now had the authority to also deal with the branch chiefs and the new protobranches, especially the Armored Force and its tanks, which he felt was too expensive. Also confronting him was how the future of horse-mechanized cavalry would maintain itself as a combat arm.

McNair had more faith in the central importance of infantry backed by artillery. Though McNair had numerous tasks to accomplish, he became known for changing the heavy square infantry division of four regiments to a light triangular division of three regiments. His purpose was to increase the flexibility and mobility of infantry divisions by reducing numbers and eliminating brigade headquarters.

When McNair took over GHQ, the army as a whole was embedded in a hemisphere defense–only policy. To be sure, national policy in times of war dictated that a small army reinforced by the National Guard would be sufficient for defense of the homeland. Thus, the mechanized cavalry was limited in executing an aggressive offensive combined arms doctrine with an appropriate infantry, air, and artillery balance. To avoid cross-coordination, the chief of field artillery refused to introduce self-propelled artillery, instead preferring motor-drawn and horse-drawn, while the Air Corps was focused on long-range strategic bombing. This kept the mechanized cavalry from thinking about a combined force at a larger unit level capable of fighting the deep battle at an operational level beyond the realm of tactics. By 1940, this attitude began to change, especially after the May 1940 German blitzkrieg in France. No longer were tactics driven by a hemisphere defense–only national policy, but by the possibility of again fighting on foreign soil.

Willis Crittenberger advised General Herr that the U.S. military attaché in Paris noted, "Mobility has been restored to the battlefield by combining airplane, tank and portee infantrymen in unified action. To date, the basis of the German success in the breakthrough and exploitation which has followed has been due to close liaison between combat aviation and mechanized elements of the Panzer Divisions supported by portee infantry." What more information did Herr need than that an entirely new efficient mount now replaced the role of horse cavalry?[9]

Even before the Wehrmacht overran most of France in May 1940, the U.S. Army issued a training directive to prepare units to take the field on short notice. This was incredible, since the ground forces were far from being prepared and still were governed by a doctrine originated after World War I and the 1920 National Defense Act. The army was lacking a doctrine to deal with armored warfare or equipment to carry out combined arms warfare as demonstrated by the German forces in Poland and then in western Europe. There were widespread problems regarding a suitable doctrine, organization, and equipment necessary to fight a mechanized war.

Crittenberger's longtime friend and strong military ally, Senator Henry Cabot Lodge Jr., told his colleagues, "Our army today is not what it ought to be." Furthermore, "We have weapons for only an army of 75,000 for a Regular Army and National Guard of 432,000. At present we are suffering from apathy which has been a general condition regarding our national defense since the World War." On the issue of preparedness he chastised his colleagues, saying in essence that the country was in a terrible dilemma when things were left to the last minute.[10] Lodge eventually gave up politics; he resigned his seat in the Senate and joined the mechanized cavalry, serving with distinction under Crittenberger during the Italian campaign of 1944 and participating in the liberation of France and occupation of Germany.

The regular army and National Guard divisions were unable to reach full strength during the interwar period or have the opportunity to engage in large-scale maneuvers to test doctrine until 1940. After all, the country was not at war. First to receive funding for expansion and modernization was the regular army. At battalion and company levels, infantrymen were receiving more firepower, such as machine guns and mortars. The field artillery was finally receiving towed 105mm howitzers. The mechanized cavalry at Fort Knox was receiving more light tanks for Brigadier General Adna R. Chaffee's proposed mass armor formation. Yet, in spite of the increase in mobilization and equipment, the autonomy of the combat branches precluded any attempt to establish a crossover combined arms force. In the branches, new equipment was being issued without any changes in doctrine that reflected World War I thinking and the 1920 National Defense Act.

To deal with the impending crisis, maneuvers were held in Louisiana in the spring of 1940. One author wrote in the *Cavalry Journal*, "Our cavalry leaders, too, are seriously handicapped by too infrequent opportunities for combined training with infantry, divisions and higher echelons. Intimate association with other arms and services insures team play."[11] The showdown Willis Crittenberger had predicted now finally occurred. The event was the first corps-versus-corps maneuver and the largest held in peacetime, involving seventy thousand soldiers. The maneuvers were designed to experiment with doctrine, leadership, organization, and new equipment. Previously, when funds were available, each branch usually conducted its own training and maneuvers independent of other combat arms. This situation again added to the intensification of making army warfighting doctrine obsolete. Now for the first time, combined arms maneuvers were held with large units, providing higher commanders with the opportunity to test their skills in command and control.

While planning for the Louisiana maneuvers, Chaffee wrote Colonel Charles L. Scott that he was going to Third Army headquarters in Atlanta to do some "missionary work," meaning seeking more combat elements for his brigade. As a result, the War Department gave him the 6th Infantry Regiment (Motorized) to be attached to his 7th Cavalry Brigade. "So little by little we are getting the troops if not the name and dignity of a mechanized division," he wrote Scott. The 6th Infantry, whose history went back to 1789, moved to Fort Knox to train with the 7th Cavalry Brigade for the upcoming maneuvers.[12]

Chaffee's brigade was the center of the corps maneuvers and was employed in a number of combinations with Brigadier General Bruce Magruder's Provisional Tank Brigade, which was pooled from infantry resources made up of light and medium tanks. The chiefs of cavalry and infantry attended, but only as observers. Chaffee was not content to stand by like his branch chief. According to his aide, Captain Gustavus West, he was all over the maneuver area, many times not sleeping or eating for days. Chaffee was a hands-on leader unlike many of the commanders who were influenced by the position warfare of World War I, meaning they thought they could manage the maneuvers from their headquarters by relying on maps and telephones.[13] However, this was not Chaffee's idea of how to manage mechanized warfare.

What became evident was that American armor elements needed to be upgraded and consolidated. Independent branch tank policy for the cavalry and infantry was no longer appropriate for modern warfighting. It was clear to the maneuver participants that the concentration of motorized and mechanized elements had demonstrated that unity of effort was now essential.[14] It was apparent Chaffee's reinforced mechanized cavalry had to be raised to a division, an argument both he and Van Voorhis had been presenting for years. Traditional branch intransigence was moving toward a showdown over how to prepare the army for war.

While the Louisiana maneuvers were under way, the German military launched another blitzkrieg through the Low Countries and northern France in May 1940. This created even more pressure in the United States to unify efforts to establish a separate mechanized branch. By June, France, which had more tanks than the Germans, was totally defeated, as Adolf von Schell had predicted at Fort Knox in the summer of 1937. Meanwhile, the British Expeditionary Force and some French troops were evacuated from France at Dunkirk. The Wehrmacht's remarkable success in western Europe did not escape the notice of the country. It had a traumatic effect. The success of German panzer units supported by tactical airpower was a wake-up call that made it imperative that the organization and equipment issue of mechanization must be resolved.

Now Chaffee had the catalyst he needed. It now became apparent on the importance of rapidly raising mechanized warfare to a maneuver-oriented operational level with a means of fighting in deep operations with a self-contained force. General George C. Marshall's assistant chief of staff, G-3, Brigadier General Frank M. Andrews, a West Point classmate of Chaffee, took serious notice of the maneuvers and events in Europe. In late May 1940, Andrews called an unscheduled meeting at an Alexandria, Louisiana, schoolhouse to critique the maneuvers with leading members of Chaffee's 7th Cavalry Brigade (Reinforced) and Magruder's Provisional Tank Brigade. In attendance were General Andrews's executive, Lieutenant T. J. Camp, Generals Chaffee and Magruder and his executive, Colonel Alvan C. Gillem Jr., and Major General Walter Krueger. It was unanimously determined that the unification of all armor elements "could no longer be delayed." The consensus was that it was no longer possible to continue to arm on an individual cavalry and infantry branch basis.[15]

Also present at the conference was Colonel George S. Patton Jr., who before the Louisiana maneuvers was still a professed horseman. He was now convinced otherwise. Before the maneuvers, however, Patton continued his opportunistic balancing act. At the time he was commanding the 3rd Cavalry at Fort Meyer, Virginia. In February 1940, he wrote a presumptuous letter to one of his mentors and a die-hard horseman, Major General Kenyon A. Joyce, on information he had received from a friend, Colonel John S. "P" Wood, chief of staff of the Third Army. Patton wished Joyce success for the horse cavalry in the upcoming maneuvers. He suggested to Joyce that his horsemen engage in radio interference to blunt the movement of Chaffee's brigade. Patton believed radios in M3 scout cars were strong enough to be used for this purpose if they were tuned in on the same band. "It would be a great joke if our friendly enemy on wheels and tracks could be totally deafened." He concluded in longhand, "Please keep the dope I gave from Col. Wood SECRET."[16] Ironically, Chaffee later heard from a bored Patton, who was now motivated by a resurrected interest in tank warfare. Chaffee then saw to it that the flamboyant cavalryman would return to tanks.

General Herr and the chief of infantry, Major General George A. Lynch, who attended the maneuvers, were not invited to the schoolhouse conference for the critique. Without the branch chiefs' influence, cavalry and infantry leaders participating in the maneuvers were free to share their views on consolidating mechanization in a separate combat arm. General Andrews concurred and immediately discussed the conclusions reached at the schoolhouse meeting with General Marshall. Andrews, the strategic-minded heavy bomber airman, became instrumental in convincing his boss, General Marshall, to approve a move toward establishing an Armored Force. Herr and Lynch immediately disagreed with Andrews's proposal. Lynch believed panzer units were armored cavalry. Furthermore, "The fallacy of shaping an organization on the basis of weapons, rather than adapting weapons to the mission, is a constantly recurring one." MacArthur's 1931 memorandum, Lynch stated, "was one of the soundest documents ever promulgated by the War Department. It lays down the principle that development of weapons and units be based on mission."[17] From Lynch's position, this meant that infantry tanks were designed to be accompanying weapons. Nonetheless, it was now clear to many officers attending the maneuvers that separate mechanization established by General MacArthur in 1931 for each combat branch was too conservative for a rearming of America.

On June 5, Andrews sent a memorandum to Marshall requesting Chaffee, Magruder, Scott, and Brett be called to Washington for consultation regarding the issue of army mechanization. Herr now began to point fingers, claiming the bulk of agitation came from noncavalrymen, such as Sereno E. Brett, who declared for all to hear at Fort Leavenworth before he left that "he was going to Washington to advocate a separate armored corps."[18]

Lieutenant Robert M. Lee, Chaffee's aide and personal pilot, recalled during this period that the two visited Herr in a last effort to present a rationale for mechanizing horse regiments and forming the first mechanized cavalry division at Fort Knox. Herr sat straight in his chair, looked at Chaffee, and said, "Adna, not one more horse will I give up for a tank," and Herr banged his hand on the desk. "Johnny," Chaffee said, "if that is the way you feel about it all right." He saluted and left, commenting to Lee on the way out, "I'm through with him. Let's get on our way, crank up the airplane and get back" to Fort Knox. There for years he had people who gave him strength, meaning he surrounded himself with like-minded people, mentioning Crittenberger, Grow, and Harmon. These were the type of cavalrymen that Chaffee expected to be reliable, loyal, and trustworthy in efforts to mechanize, according to Lee. He also added that part of Chaffee's strength was the West Point honor code, "Duty, honor of country and the honor code thou shall not lie, cheat or tolerate anyone who does."[19]

On June 10, a conference with chiefs of the general staff, arms, and services was held, and this time Herr and Lynch were invited. Herr had objected to the word "tank," and Lynch had objected to the word "mechanization." In spite of their objections, Andrews announced that two armored divisions would be formed; other than that discussions were wide open. What disturbed Herr was what he called all "the secret conclaves" that were held by Andrews without his and Lynch's participation. Again, the chiefs strongly defended their branch system, arguing that infantry and cavalry should develop mechanization within their respective combat arms as directed by General MacArthur's 1931 memorandum. Herr reiterated Lynch's position that panzer divisions performed cavalry missions, concluding that there was no need for a separate combat arm in the army.[20]

Herr and Lynch did not let go. They were in a highly defensive mode to save their branches. They still continued to promote their argument that the 1931 memorandum ordering each combat arm to mechanize separately should be upheld, thus allowing the chiefs individually to deal with their own prescribed mission as outlined in the *1939 Field Service Regulation (Tentative): Operations*. This acceptance of a virtually unchanged *1923 Field Service Regulation* was in contradiction to Andrews's efforts in establishing a combined arms force based on a new doctrine that depended on a self-contained armored force capable of deep independent operations. During one of the long and bitter debates, Chaffee finally blurted out, "Speed is essential. We must [not] stop and haggle over a lot of detail and figure out a lot of things that have been studied over by boards and by commanding officers in the field and tested in maneuvers time and again."[21]

The decision was made. It became a historical watershed in army history. On July 10, 1940, fifteen days after the fall of France, the War Department decided to create a separate Armored Force at Fort Knox "for the purpose of service tests" because the 1920 National Defense Act had not provided for a separate branch. Chaffee was designated its first chief. General Van Voorhis's comments to Grow in 1937 that it would take a directive from the War Department to establish a mechanized force that had been validated. Charles Scott, who commanded the 13th Cavalry (Mech), claimed Van Voorhis as early as January 1931, read a paper to the chief of staff, Douglas MacArthur, "urging the organization of an Armored Force."[22] Finally, it happened. The following units were employed:[23]

7th Cavalry Bridge (Mech), 1st and 13th Cavalry (Mech)
Separate Combat Car Squadron from Fort Riley, Kansas
2nd Battalion, 68th Infantry (Light Tanks)
66th Infantry (Light Tanks)
67th Infantry (Light Tanks)
6th Infantry (Motorized)
68th Field Artillery (Mech–Towed)
7th Signal Troop (Mech)
47th Engineer Troop (Mech)
4th Medical Troop (Mech)
17th and 19th Ordnance
Quartermaster and heavy maintenance companies.

On July 15, 1940, Chaffee officially took command. The I Armored Corps with the 1st and 2nd Armored Divisions was activated. Chaffee became the corps commander, with Lieutenant Colonel Sereno E. Brett (Infantry) as his chief of staff. The 1st Armored Division at Fort Knox was commanded by General Magruder, with Lieutenant Colonel Willis D. Crittenberger (Cavalry) as his chief of staff. The 2nd Armored Division at Fort Benning was commanded by General Scott, with Lieutenant Colonel Geoffrey Keyes (Cavalry) as his chief of staff.

Other cavalry officers who assumed staff positions with the Armored Force were Major Charles H. Unger and Major Ernest N. Harmon, with Headquarters, 1st Armored Corps at Fort Knox. Other general staff officers assigned to the 1st Armored Division were Lieutenant Colonel C. M. Daly and Major William Barriger. Lieutenant Colonel Robert W. Grow and Major R. F. Perry were assigned as general staff officers with the 2nd Armored Division.

Shortly after the Armored Force was created, Chaffee addressed many of the officers who were to command in his new organization. He told them that his duties included the development of tactical and training doctrine for all units, and engaging in research and advisory functions pertaining to armament and equipment to be used by the Armored Force. He reminded the officers that in the War Department Directive of June 10 the wording "for service tests" was necessary because the 1920 National Defense Act had not provided for a separate armored branch, adding that Congress specifically prohibited the creation of a new branch without its authorization. He outlined the organization he had in mind, noting that it should be designed to be offensive and aggressive, with the tank as "the backbone and the principal arm." Next he outlined his tactical intent. The Armored Force would be employed "to attack through hostile weakness. It creates surprise; which is essential to successful combat, by the sustained celerity and power of tactical movement. It uses its mobility to choose the most favorable direction for attack to reach vital enemy rear areas." Remember this, he concluded, "Success with modern armored troops demands the very utmost of every one from top to bottom and back again." Chaffee's tactical philosophy established a new American military culture that embraced all-around combat arms coordination with leaders from top to bottom and back again who demonstrated reliability, trustworthiness, and loyalty. Chaffee felt these leadership traits were necessary if the Armored Force was to eventually become a successful mechanized maneuver–oriented combat arms branch.[24]

Thereafter the Armored Board was created at Fort Knox to conduct research and tests under the direction of Lieutenant Colonel C. C. Benson, an earlier enthusiastic endorser of the Christie M1928 chassis. In October, the Armor Force School and the Armored Force Replacement Training Center were authorized as separate organizations under the command of Colonel Stephen G. Henry, an infantry tanker. Chaffee picked Henry not only because of his knowledge of tanks and mechanics but more because of his progressive ideas and talent as an instructor.[25] By these moves, Chaffee finally obtained all the responsibilities of an arms branch chief. Certainly, the decision had not sat well with old-order thinking and the branch prerogatives of Herr and Lynch.

Another one of Chaffee's pilots recalled about this time that Ernie Harmon, whom Chaffee selected as a staff officer, used to ride as a copilot when Chaffee was required to extend his travels as the Armored Force was being formed. Harmon was not a good pilot, claimed Lieutenant Harvey Alness (later lieutenant general, USAF). When Harmon took over the controls, the plane, a C-45, weaved back and forth. According to Alness, Chaffee grinned and said, "You having a little trouble up there, Ernie?" Harmon replied, "Take this goddamn thing and straighten it out." Alness also recalled another aircraft incident involving Chaffee and Patton. Though Patton had experience as a single-engine pilot, he had problems with a multiengine aircraft. As the copilot, he managed to redirect the course of the plane by bouncing and weaving. As it got worse, so did Patton's language. Chaffee "came to the cockpit and said quietly, 'Calm down George.'" He did, and the rest of the flight improved, according to Alness.[26]

Early on the Armored Force moved to fuse ideas as a light mobile force capable of deep operations that ensured sustaining shock action, pursuit, and exploitation for all its mechanized combat elements. At first the basic tactical organization for the combined arms team was the armored division with an armored brigade organized into one heavy tank regiment and two light tank regiments. The Armored Force division became a tank-heavy regimental organization. Two combat command headquarters, known as Combat Command A and Combat Command B, were established for each regiment, an arrangement close to what Van Voorhis had worked out in the 1938 maneuvers when he divided the 7th Cavalry Brigade (Mech) into two separate columns under the command of Palmer and Scott in a self-contained maneuver–oriented deep operation. The table of organization and equipment for the Armored Force divisions listed two armored regiments, each with four medium tank battalions and two light tank battalions of three companies each.[27]

All other combat elements were organized around the tank brigade, such as a towed field artillery regiment, a ground reconnaissance battalion, and an attached aviation observation squadron. A motorized infantry regiment, an engineer battalion, and service units provided support. The defined tactical doctrine expected the brigade to take advantage of weak points and strike deep, engaging the enemy's vital rear areas or enveloping flanks. In the beginning, however, the Armored Force lacked the capabilities to establish a balanced combined arms team for unity of action that ensured sustaining driving power for all its combat elements. The expressed tactical doctrine expected the armored division to take advantage of weak points and strike deep, engaging the enemy's vital rear areas or enveloping flanks. It was anticipated that the Armored Force mission of deep exploitation and pursuit would take advantage of the initial breakthrough by the infantry-artillery team. Flexibility and mobility were projected as critical to the new doctrine, reflecting the cavalry's traditional mission of fighting mounted.[28]

The distinctive triangular insignia worn by the World War I Tank Corps was adopted by the Armored Force. It contained equal amounts of yellow for cavalry, blue for armored infantry, and red for field artillery. Selected from the 7th Cavalry Brigade (Mech) insignia and superimposed on the new triangular insignia were a tank track, which stood for mobility and armor protection, a gun, which stood for firepower, and a lightning bolt, which stood for shock effect. This symbolized that the Armored Force was moving to a balanced, self-contained combined arms force.

During the organizational dispute, one issue had been resolved. The Armored Force was beginning to train for combat. Bob Grow commented that two forms of cavalry were apparent. The Armored Force, which was training as a heavy striking force, was to engage in exploitation and pursuit. Meanwhile, the mechanized cavalry was training as a light covering force whose mission was to engage in reconnaissance and security. Seemingly, the basic principles of warfare remained. Nevertheless, mobility and mechanized equipment gave a new tactical capability to armor and cavalry.[29]

With the branch debates becoming more passionate and cavalrymen moving over to the Armored Force, Third Army maneuvers continued, as did the instinct to maintain some semblance of the tactical value of horse cavalry. In the month of August 1940, Captain Bruce Palmer Jr., son of Colonel Palmer who had commanded the 1st Cavalry at Fort Knox, described in the *Cavalry Journal* a day's action and adjustments made by a platoon from F Troop, 6th Cavalry. Palmer's experience at the tactical level demonstrated the problems of connecting the machine with the horse. The platoon had four M3 scout cars for a reconnaissance mission. Motorcycles were moved to the rear because their noisy engines revealed the position of the platoon, which was split into two sections "to contact the [hostile force], hang on, and follow up every move they make." Palmer found the M3s too road-bound in the bayou country, which was laden with swamps and thick vegetation, and both were filled with hungry mosquitoes. Some of the terrain was wild and open, with charred stumps littering both sides of the roads, he recalled. Within this terrain, hostile forces blocked all passable roads, making it difficult for scout cars to maintain contact with the enemy's main force. The solution was dismounted reconnaissance, mainly at night to move around the enemy's security forces. Palmer concluded that he often "wished to high heavens for a horse" because he believed they would make night reconnaissance more feasible. He suggested that once they moved close to the hostile force outposts, four horses be detrucked from a small trailer for further scouting and the scout cars were to be hidden. Herr certainly would have approved Palmer's final comment: "Here is the mechanized, the motorized, and the horse working together in small units in their quest for information."[30]

At first, the War Department seriously considered establishing the Armored Force as a separate branch on a par with the cavalry, infantry, field artillery, and other arms and services. This action, however, would require an act of Congress. In a November 19, 1940, memorandum to General Marshall, Andrews concluded that Chaffee's Armored Force had now fulfilled its mission of initial organization. Therefore, he stated, "In the interest of efficiency, it should be legally established as a separate arm." Unsurprisingly, both Herr and Lynch stridently objected. Herr claimed Andrews's recommendation for a legal branch status was a "petty effort." Besides, the chief of cavalry again complained that conspirators had launched a whispering campaign against him because he was a staunch horseman.[31] Herr again expressed alarm over the developments of the past months. He was annoyed by the conferences held by General Marshall with Generals Andrews, Chaffee, Magruder, Scott, and Colonel Brett over the issue of future mechanization. His agitation was even more pronounced because he and Lynch were usually ignored. Both chiefs strongly believed in the continuation of compartmentalization of their branch system, arguing that infantry and cavalry should develop mechanization within their respective combat arms. In a long, sour review of the proceedings regarding mechanization, Herr declared that there was nothing in the success of the Armored Force that "could not be accomplished equally well or better through established agencies of the War Department." He pressed his legal argument that the Armored Force violated the terms of the 1920 National Defense Act that created the established line branches and not "non-infantry and non-cavalry armored units."[32]

The chief of infantry in turn argued, "The Armored Force had only asked for a field force headquarters, not a separate arm." Furthermore, Lynch continued, "the infantry and tank battalions under units of the Armored Force were suffering from a lack of combined arms training and these units should be returned to infantry control."[33]

The proposal for a separate branch raised other important issues that confronted the Armored Force because it had not followed the developmental route, as in the Army Air Corps. It could be argued that a more effective defense, such as the antitank guns so effectively used against light tanks during the Spanish civil war, was more of a concern than were aircraft. In addition, the army had yet to work out a successful coordination plan between the combat arms and supporting services. Finally, a tank destroyer arm to neutralize massive tank attacks was being developed, which had an impact on cavalry elements that were not absorbed into the Armored Force. This came about under the leadership of General McNair.

Meanwhile, as the Armored Force was forming up, the 1st Armored Division chief of staff, Willis Crittenberger, provided Van Voorhis in the Canal Zone a detailed report early in December 1940 of major events happening at Fort Knox. He told Van Voorhis, who certainly was pleased, that as the force structure grew it could expect the arrival of large numbers of scout cars, light tanks, and the first arrival of medium tanks. However, he wrote Van Voorhis, there still was a lack of half-tracks. George Patton, he noted, would probably replace Scott, who temporarily moved back to Fort Knox to command the Armored Force after commanding the 2nd Armored Division. In his report, Crittenberger let it be known that the temporary loss of Chaffee was "somewhat of a handicap" to the Armored Force. Earlier Crittenberger wrote a friend in Headquarters, Panama Canal Department, that Chaffee was "run down and tired out by long hours and fatiguing work incident to the organization of the Armored Force." Those who were acting in his place, though "working assiduously," just did not quite match up to Chaffee's ability. Crittenberger regretfully advised Van Voorhis that Chaffee was diagnosed with a "small spot, which has showed up continuously in pictures of his lungs, as a tumor." On a more amusing note,

Crittenberger told his former chief that Fort Knox's "Forest Hill is almost covered with buildings and will no longer be the objective of our vigorous and repeated assaults" being conducted by the mechanized brigade in the late 1930s. Crittenberger concluded that he only went into detail "because of your [Van Voorhis] interest in this new and lusty arm."[34]

Crittenberger also wrote Senator Lodge about Armored Force Headquarters efforts to create a separate branch. He was not so sure this was a good idea because "older branches have much they can still contribute to our development." It was obvious in his letter to the senator that he was not in favor of rushing into separation: "We should continue to utilize some of the facilities of the older branches, rather to hurry into a separation, which, for example, would immediately deprive us of a reservoir of trained combat officers—infantry and cavalry." Though he had discussions with Lodge over this issue, he understood that for the time being it benefited him to remain discreet. However, Crittenberger felt he and Lodge "had ringside seats in the altercation, and therefore are able to see both sides."[35]

The *1941 Field Service Regulations: Operations*, published after the creation of the proto–Armored Force and influenced by the success of German blitzkrieg, was more aggressive and attack-oriented than the *1939 Field Service Regulations (Tentative)*. The antiwar attitude of Americans and a government bent on isolationism and budget controls led to a defense of the Western Hemisphere that produced an inadequate ground force, which since 1920 had never risen above 138,000 officers and enlisted men. Even though the new regulation emphasized the American version of attrition warfare, where the infantry was to be the essential arm of close combat, the 1941 regulation created a new debate over a maneuver-oriented doctrine. More attention was directed to the enemy's rear to degrade his center of communication and other vital areas. This also meant dislocation and demoralization as an important elements for defeating the enemy. The new manual emphasized a turning movement with the objective of envelopment and not engaging the mass of the enemy's frontline defenses. This operation was suitable for the newly created Armored Force. The *FM 100-5 Field Service Regulations, 1941: Operations* surprisingly noted that cavalry was "capable of offensive combat exploitation and pursuit." Nevertheless, the manual prescribed the primary missions of the cavalry as reconnaissance and counterreconnaissance. The manual also warned of diluting the strength of cavalry by deploying it in "indiscriminate detachments."[36] This warning, as will be detailed later, was ignored during World War II, when cavalry elements were detached to support higher units on various missions.

In addition, the manual emphasized a defense-offense tank destroyer doc-

Light tank M2A4 mounting a 37mm cannon. Courtesy of the Patton Museum.

trine that weakened the emphasis on an offensive tank maneuver–oriented doctrine. It was first Frank Parker, then Van Voorhis, and finally Chaffee, Palmer, and Scott who developed the mechanized maneuver–oriented operational concept. Now the concept was contending with a defensive mind-set, which challenged cavalrymen and infantrymen in the Armored Force who argued for greater offensive mobility with a balanced combined arms team.[37]

The assistant chief of staff, G-2, Brigadier General Sherman Miles, suggested in March 1941 another War Department evaluation for organizing a new army. He proposed supplanting the infantry-artillery team with a tank-air team and motorized infantry divisions. Chaffee, however, did not exactly agree with the War Department's evaluation of the modern battlefield as being dominated primarily by the tank-air team.[38] McNair also responded to the War Department's question on the future role of armor by disagreeing. He believed that "the tank is superior to all other ground weapons; as a corollary, that troops in general have no place on the battlefield of today unless behind armor."[39] This response supported his conviction that tanks should be used as infantry-accompanying weapons.

Nonetheless, strong resistance by the chiefs of cavalry and infantry continued, thus stalling a legislative solution authorizing a new branch organization for the Armored Force. General Marshall, for the time being, placed it on hold. On March 24, 1941, McNair wrote to Scott, who at the time was temporary chief of the Armored Force due to Chaffee's recurring illness, that "the unfavorable action on a separate arm was to be expected, since the proposal is a bit brusque." McNair advised Scott that the setback was not the last word on the subject.[40]

Even though war clouds were rapidly beginning to shroud the dogmatic thinking of the chief of cavalry, Herr continued preaching that horse cavalry could be used en masse. This attitude was illustrated in *Life* magazine in April 1941, which featured horse cavalry maneuvers and the reorganization of a partially horse-motorized division at Fort Bliss, Texas. The 1st Cavalry Division, the article noted, "could lick its weight in enemy tanks," and unlike road-bound trucks and cars, was at its best over rough terrain, such as hills, rocks, and deserts. By mid-1939, the horse cavalry division contained two brigades of two regiments each. In support, the division had one horse-drawn 75mm field howitzer regiment along with reconnaissance elements and eighteen light tanks. By 1941, the 1st Cavalry Division—as during the American Civil War—was expected to ride into action and then fight dismounted. Hundreds of trucks, scout cars, armored cars, and motorcycles and more than a dozen light tanks were anticipated to support the horse units. It was expected that cavalry with more than eight thousand riding horses and supporting elements might be deployed in western defensive operations, such as in Mexico and the Panama Canal, or possibly along the eastern South American coast, where enemy landings might occur. Large cavalry trailers called portees would deploy riding horses in long lines (eight men and eight horses per truck-trailer combination) to terrain that favored cavalry tactics. The division also had hundreds of pack and draft horses, and mules. M3 scout cars augmented the horse team by extending reconnaissance and acting as troop carriers, plus providing antitank and antiaircraft support when necessary. In addition, scout cars were provided with radio receivers for command and control. A ¼-ton 4 x 4, four-wheeled vehicle capable of carrying four riflemen, the Bantam car, was introduced during the maneuvers.[41] Designed for the cavalry, the Bantam was considered more efficient than motorcycles, which Herr earlier had supported. The vehicle, the article noted, was expected to be the main rival of the horse. In fact, it would replace the horse.

The year before the *Life* article, the Bantam Car Company of Butler, Pennsylvania, won an invitation-only bid in September 1940 over the Willys-Overland Company from Toledo, Ohio, to deliver to the army a prototype vehicle similar to the German military version of the Volkswagen. The four-seater vehicles were used during the blitzkriegs as small personnel carriers. After some remarkable resourcefulness, the army received the Bantam prototype for tests. The

Bantam truck ¼-ton 4 x 4: the army's oldest jeep and preferred by the cavalry. Courtesy of the Patton Museum.

vehicle was called a Bantam Reconnaissance Car (BRC). Its arrival introduced a new wheeled vehicle for warfare, which would have a lasting impact on ground force mobility, especially for mechanized cavalry operations. A number were built for evaluation; however, the company did not have the mass-production facilities required for wartime industrial mobilization. The army opted to give the contract to Willys-Overland, which had the Bantam plans and the ability to handle improvements and alterations required for a new lightweight reconnaissance car. Willys-Overland and the Ford Motor Company, using the Willys design, eventually produced hundreds of thousands of what was referred to throughout the war as the "jeep." The famous military correspondent Ernie Pyle commented later that the jeep was "faithful as a dog . . . strong as a mule . . . and agile as a goat." The Bantam Car Company subsequently received government orders for trailers.[42] For years, many cavalrymen still referred to the jeep as the Bantam.

The *Life* article also took a swipe at cavalrymen who joined the Armored Force by calling them "cushion pounders." The featured article depicted a rough, crusty cavalryman, Major General Innis P. Swift, saluting his division as he reviewed his command on Fort Bliss's dusty parade ground.[43] His look foretold what was coming. The 1st Cavalry Division would no longer hear the squeak of saddle harnesses or see troopers saddle up and lead their horses out of corrals to formation. By August 1943, horses had long disappeared. Swift would take the division, consisting of the 5th, 7th, 8th, and 12th Regiments, to the Pacific theater, where the cavalrymen fought dismounted.[44]

The broadly distributed popular magazine article reflected the struggle to reconcile traditional cavalry with modern mechanized warfare. Even with the creation of the Armored Force, the army elites displayed ambiguity and uncertainty over a future cavalry organization and its tactical doctrine. As noted, at the heart of the issue was the adamant Herr and his emphasis on the importance of horse cavalry alone or in combination with mechanized units.

In March 1941, McNair let it be known that "the infantry division will continue to be the backbone of an army." More specialized mission weapons and troops, he maintained, needed to be pooled at higher headquarters, including nondivisional armor battalions and mechanized cavalry. His attitude was that there would be no self-sufficient Armored Force at army level, instead it would be deployed for the time being at corps level as part of the army.[45]

The terminally ill Chaffee challenged McNair. He began by stating that the tactical procedures of World War I were obsolete. Infantry doctrine of open warfare using firepower and maneuver was too geared to a battle of annihilation rather than mobility and deep operations. Tactical actions of armored divisions, he stated, "were based on fire and movement of the component armor units of the combined arms." Chaffee, however, placed emphasis on massed armor for pursuit and exploitation at the expense of self-propelled artillery and infantry. Regarding the past, he stated, "It cannot be comparable to the cavalry charge as so many erroneously believed." Chaffee viewed aviation as support elements for the Armored Force to augment its power and extend the range of artillery, but there were problems in solidifying a tank and tactical air team. There were no developments in the United States similar to the German Stukas tactical dive-bombing developments with the mechanized cavalry before the creation of the Armored Force.[46] Chaffee was now convinced of the importance of tactical air support as the Armored Force expanded. Yet it took time and wartime experience for the coordination to finally materialize.

Finally, by the summer of 1941, the army had grown large enough for McNair to engage it in maneuvers. The GHQ 1941 Louisiana and Texas maneuvers were the first large-scale exercises held by the U.S. Army.[47] Approximately

The horse versus the combat car. Courtesy of the Military History Institute, Crittenberger Papers.

350,000 troops gathered in central Louisiana that rainy September to stage what *Life* magazine described as the "greatest sham battle in history."[48]

Nonetheless, during the GHQ maneuvers, Herr attempted to validate his horse portee–mechanized units. By now it was evident the mounted cavalry had lost its prestige to a new arm, the Armored Force. Herr was not optimistic; he even expressed these feelings before the maneuvers. However, there was some satisfaction for the cavalry branch. To some observers, the mixed horse portee–mechanized cavalry regiment functioned best as a corps reconnaissance element. Another observer from the Command and General Staff School noted that the cavalry mix functioned best when the horse-portee squadron deployed as a firebase while the mechanized squadron maneuvered.[49] There were the usual debates. General Bruce Palmer Jr., then a cavalry captain then, recalled that regimental control of cavalry favored mechanized elements rather than the horse.[50] In spite of mixed feelings over the role of cavalry, the maneuvers gave Herr some fleeting satisfaction, but it was short-lived.[51] He persistently believed his branch would avoid decimation in favor of armor. This was not the case; the maneuvers demonstrated that horse cavalry had lost its mobility. McNair found the cavalry no longer useful.[52]

The 107th Cavalry (Horse Mechanized) Louisiana, 1941. The photograph demonstrates the unsuitable organization that constituted General John K. Herr's efforts to retain his horse units. Courtesy of the Patton Museum.

A Louisiana native at the time, B. D. (Bud) Robertson, recalled the calamity of a beautiful cavalry horse galloping past his family's farm. It was obvious the riderless horse, with full gear, was frightened and in a hurry. Shortly a dusty and confused cavalryman came by looking for his noble companion, which had gone astray. The bewildered trooper explained that he had tied his horse to bushes-that were covered with "little yellow flies." The so-called flies that launched the horse past the family farm were in fact stinging yellow jackets. Later the embarrassed cavalryman came past the farm with the horse in tow. Bud thought for sure this was a "Yankee" who had no knowledge of the South or its insects. Bud also recalled a few cavalrymen spotting chinaberry trees on the farm and deciding they were fruit trees. After eating the chinaberries, they commented to Bud's father that they were the worst-tasting wild grapes they had ever eaten. The cavalrymen were told that the berries were supposedly poisonous. Immediately their facial expressions changed and they began "spitting, sputtering, and coughing." The cavalrymen also had the habit of liberating the locals' corncribs when on long day and night marches, either to feed their horses or to utilize a new form of toilet paper. It was apparent to some Louisiana locals that many of

Major General George S. Patton Jr. and Lieutenant Colonel Robert W. Grow at Manchester, Tennessee, June 1941, in M3 scout car. Note Patton's trademark air horns on hood of vehicle. Courtesy of the Patton Museum.

the horses were being turned in for mechanized equipment, such as scout cars.[53] Times were changing. Herr's attempt to marry the horse with the machine had failed, a point even the locals could sense.

General McNair was impressed with the deployment of tank destroyers during the maneuvers. He believed these new mobile self-propelled weapons would increase the pace of battle. More so, he still was convinced that combined arms meant the traditional infantry-artillery team.[54] The tank destroyer doctrine was designed as a reactive defense-mobile force to counterattack armor breakthroughs. The antitank doctrine benefited from the 1941 Louisiana maneuvers, when Lieutenant General Ben Lear, a testy old cavalryman who commanded the Red Army, misemployed his two armored divisions. Consequently, the I Armored Corps commander, Charles Scott, lacked firm control of the corps' 1st and 2nd Armored Divisions. The tankers were embarrassed, but Patton, who had commanded the 2nd Armored Division since November 1940, was complemented for displaying "high morale, technical proficiency, and devotion to duty. Your leadership has produced a fighting organization," according to Lieutenant General Walter Krueger, who commanded the Blue Army.[55] Patton's G-3 was Lieutenant Colonel Grow until he was named commanding officer of the 34th Armored Regiment, 2nd Armored Division, in September 1941.

General Krueger's motorized infantry and antitank guns outmaneuvered the tankers, who were confined to a main highway and were not able to make flank attacks. Lear was concerned that the swampy terrain would swallow up his tanks. *Life* reported, "Swamps, airplanes and anti-tank guns stop advances of the tanks."[56] Lear and his armored divisions were not the only ones embarrassed. The 1941 maneuvers also saw the temporary decline of large armored units because McNair began to favor independent tank battalions being assigned to infantry divisions. Nevertheless, as historian Christopher R. Gabel concluded, with the move from the Protective Mobilization Plan to the experiences of the 1941 maneuvers, the army was able to respond to Pearl Harbor and prepare the nation for war. Confidence had changed in one year.[57]

Sid Hinds, who commanded an armored infantry regiment during the maneuvers, also received some "personal attention" from General Krueger for not being realistic because he ran his regiment horizontally over the "doughboy'" linear positions and establishing ambushes. This mobile initiative put one of Krueger's regimental teams out of one of the maneuver problems, so Hinds "scoured the countryside to find them." He declared this was representative of the "infantry hardliners" who reflected their branch doctrine of linear thinking. Hinds added that the maneuvers proved "beyond a doubt the necessity for a combined arms team." One issue that bothered Hinds during the maneuvers was the lack of close-in tactical air support. He maintained "there was no direct communication between the troops on the ground and the supporting pilots in the air." This problem was solved later when Tactical Air put one of its forward observers (pilots) in a tank with an air radio to direct ground assaults. It was interesting to recall, he claimed, "that only one air-ground observer ever volunteered for a second tour. He was determined to win a Silver Star, which he did darn well earn, along with a Purple Heart."[58]

As a result of the maneuvers a new quasi arm was activated on December 1, 1941, and designated Tank Destroyer Center; in March 1942 it was redesignated Tank Destroyer Command; and in August it again was redesignated as Tank Destroyer Center. At this time, the army again considered the Christie suspension system for a full-tracked fast-moving tank destroyer. The Buick Division, employing the long helical independent Christie suspension system, built a few models for the Ordnance Department. Eventually, this design gave way to a new type, torsion bars that ran under the chassis, eliminating the side spaces required for the helical springs.[59]

Other quasi arms were also established; the Antiaircraft Command was established in March 1942 and the Airborne Command was activated in March 1942.

Unfortunately, Chaffee's bout with cancer and exhaustion took their toll. In December 1940 General Marshall was made aware of how seriously ill his friend was when he received a letter from Dr. Edward D. Churchill of Massachusetts General Hospital. Marshall was advised in a confidential report that it was not reasonable to expect an arrest of the tumor. Surgery was not considered because the diagnosis of primary carcinoma of the right lung revealed it to be inoperable because of extension to the lymph nodes. Deep X-ray with a million-volt machine was employed. However, the physicians were not optimistic, giving Chaffee not much longer than a year to live.[60] He could no longer carry on his fight for a separate branch or deal with the stubborn Herr. Perhaps his growing illness prevented the mechanized maneuver–oriented principle from being articulated throughout the army's elites, for many of them failed to absorb his revolutionary doctrine before they entered combat. Patton, however, understood this when he commanded the 2nd Armored Division during the 1941 maneuvers, as did many other cavalrymen who moved over to armor.[61]

Chaffee's last enlightening message on the issue of mechanization was explained to Congress in May 1941. The prewar situation that created considerable uncertainty over an independent armor doctrine he blamed on the combat branch chiefs' competition over funding, resulting in their lack of responsiveness to modern warfighting and differences of opinion. He also took issue with the country's pacifist tendencies that made the nation unprepared. This was the same observation Van Voorhis had made when he took over command of the 7th Cavalry Brigade in 1936. He and many senior cavalrymen at Fort Knox believed branch-specific doctrine and tactics drove separate developmental policies when it came to organization and equipment, thus making funding for new weapons and armored fighting vehicles much too competitive for Depression-squeezed defense dollars. Chaffee added that the U.S. military "failed to evaluate properly the importance of combined arms in armored units."[62] This was a reiteration of what General Parker argued for in the late 1920s.

Shortly after Chaffee's message to Congress, he died on August 22, 1941. His biographer wrote, "At last came the day when the last ounce of the general's once phenomenal energy was spent, and he returned to his bed, never to rise again."[63] One could assume that the excessive stress and anxiety caused by years of struggling over cavalry mechanization and the creation of an Armored Force added to his illness and accelerated his death. Later Ernest Harmon wrote, "There can be no tangible record of Chaffee's brilliance, but there were to be times in Africa, Italy, and Germany when the Allied Command could have well used his judgment and shrewd intelligence."[64]

Bob Grow was also distraught over the loss of Chaffee. Later he recalled Chaffee's attempts to deal with the unsettled questions of organization before he died. Grow professed Chaffee wanted the combat commands and visualized how they would be used. His idea was for a flexible tactical headquarters to which could be attached any number of battalions or other units required for a particular mission. Grow claimed part of Chaffee's tactical legacy was his idea of combat commands and that they were never to be used for administrative control or become a fixed organization. Chaffee's organizational concept provided mobile flexibility. "It was possible to reinforce either combat command at any time with any unit and get completely satisfactory results," Bob Grow stated. He later claimed Chaffee had quite an influence on him when he commanded the 6th Armored Division as it spearheaded the advance through Europe.[65]

Meanwhile, General Marshall was looking for an expert in firepower. He chose a prominent polo player, Major General Jacob L. Devers, an artilleryman and classmate of Patton, to take command of the Armored Force in August 1941, after the death of Chaffee. What was missing from the force was a balanced, self-contained combined arms organization of mechanized cavalry, armor, armored infantry, and armored self-propelled artillery. And there were unresolved equipment issues about armored cars, half-tracks, self-propelled artillery versus towed artillery, and light tanks versus medium tanks.

Devers was also told by Marshall to report directly to him and not to McNair's GHQ, thus avoiding the red tape that Devers so despised. Soon major

A very focused PFC driving three smiling generals. General George C. Marshall is in the front seat, with Generals Stephen O. Henry, commandant of the Armor School, and Jacob L. Devers, commandant of the Armored Force, sitting in the back seat. Fort Knox, Kentucky, May 1942. From the Stephen O. Henry Files, Patton Museum.

changes at Fort Knox began to take place. Earlier Devers had been chief of staff to General Van Voorhis, who was the commanding general of the Panama Canal Department after he left command of V Corps. Chaffee's biographer claimed Van Voorhis and Devers discussed the role of the 7th Cavalry Brigade (Mech) in mechanized warfare and analyzed the successful execution of the German blitzkrieg. Though Devers was not convinced at the time of the utility of an Armored Force, he did digest Van Voorhis's ideas and subsequently introduced new organizational changes and operations far beyond those suggested by Chaffee.[66] When the new Armored Force commander arrived at Fort Knox, he immediately began crushing the remaining mechanized cavalry and infantry-tank branch jealousies. Devers began to move the Armored Force from a light to a heavy combined arms ground force.

In addition, Devers took steps to expand the firepower of the Armored Force. Unfortunately, Chaffee had earlier tried to emphasize the importance of self-propelled field artillery but failed. Part of Chaffee's dilemma was a lack of support from the chief of field artillery, who saw no need to develop self-propelled artillery for the mechanized cavalry. Devers, however, changed this. He added self-propelled artillery to increase firepower, thus enhancing the operational mobility of the tank-dominated Armored Force. Devers, along with his chief of staff, Lieutenant Colonel Edward H. Brooks, a future armored division commander during World War II, played a major role in acquiring the M7 105mm self-propelled howitzer and the M8 75mm self-propelled howitzer assault gun so critical for mobile armored and mechanized cavalry tactics. Another one of Devers's innovations was adding light spotter aircraft and ground forward observers for battalions. His actions provided a new way of applying self-propelled artillery and aviation with Chaffee's tank-heavy force.[67] Earlier semitracked vehicles called half-tracks had gone through a series of developments. Now they became standard for armored infantry regiments and battalions. Finally the Armored Force had a balance of armored fighting vehicles—tanks, self-propelled artillery, and half-tracks for the armored infantry.

The struggle, however, for branch supremacy still impeded development of a well-thought-out armor and cavalry doctrine. Chaffee's cavalryman ideas dominated early Armored Force doctrine, and now Devers moved to innovate doctrine with his reorganization. General McNair responded caustically. He believed the Armored Force was embracing branch autonomy and suggested it "should join the Army." Meanwhile, McNair's tank destroyer idea began to move forward under the command of an infantryman, Colonel Andrew D. Bruce. Consequently, the autonomy and status of the Armored Force began to decline. McNair's doctrine was implanted, meaning, "The counter-attack long has been termed the soul of defense." He did accept the possibility of a tank-versus-tank counterattack; however, he felt "the procedure would be too costly."[68]

While Devers was taking command of the Armored Force and beginning to implement changes, Herr's negative influence continued to plague Marshall's attempts at reorganization. With the Armored Force splitting from the cavalry branch and creating its own identity, the question arose about how separate or remaining mechanized and horse-mechanized cavalry units fit in during this period of organizational turbulence and the movement to centralized control that McNair was striving for.

Herr's response was explained in the U.S. Army's 1941 *Cavalry Field Manual 2-15*. The traditional combat mission of cavalry was outlined under his direction. It reiterated the call for greater mobility and firepower, meaning both the horse and mechanized elements. At the regimental level, the mission was reconnaissance and security for army corps. At squadron level, the emphasis was on "sneak and peek" reconnaissance, and how to avoid combat when necessary.

Horse-motorized cavalry. Note scout car on the left and the motorcycle along with the disembarkiing horses. Courtesy of the Patton Museum.

The mechanized cavalry squadron was also assigned security or delaying missions. The manual added that horse-portee units from cavalry regiments could, if necessary, reinforce the mechanized squadron. Furthermore, it emphasized Herr's dogma that various combinations of mechanized elements and horses allowed for greater flexibility. The manual also defined the role of mechanized reconnaissance troops assigned to infantry divisions, which were to engage in pursuit and exploitation after a breakthrough. In addition to reconnaissance, the mechanized cavalry troop was to perform special operations, such as gap fillers, a mobile reserve and liaison between larger units. Mechanized cavalry elements, the manual stated, "were to be equipped with armored and self-propelled motor vehicles designed for combat purposes and in which weapons are mounted."[69]

The War Powers Acts of 1941 and 1942 gave President Franklin D. Roosevelt, as wartime commander in chief, the legal right to exert authority over war agencies, including the War Department. This authority was passed down to Marshall, giving him the legal authority to reshuffle the antiquated branch system. In March 1942, Herr and the other branch chiefs' influence finally ended with the elimination of the Offices of the Chief of Cavalry, Infantry, Field Artillery, and Coast Artillery. While these events were unfolding, General Marshall made reference to the branch chiefs, believing them bitterly opposed to the reorganization. He was referring to Herr and Major General Robert M. Danford, the chief of field artillery. General Lynch had stepped down the year before and was succeeded by Major General Courtney H. Hodges, considered a loyal supporter by Marshall.[70] The old-order chiefs were not perceived as loyal members of the army's new team. General Herr was gone, as were his horses. After the elimination of the line branch chief offices, the Army Ground Forces began to mechanize the remaining regular army and National Guard horse-mechanized and horse-motorized units.

Establishing a new combat organization and its redesignation was not an easy task. Though at first McNair seemed to favor an ad hoc Armored Force status, he later placed his faith in the infantry-artillery team supported by an antitank force to deal with a tank breakthrough. For greater centralized control, he continued to advocate that specialized units be massed and placed at higher headquarters, especially mechanized elements. This action was designed to centralize control over the decentralized compartmentalized organization of the branch system, especially the new specialized ones. Certainly, to the delight of Herr, this action by McNair precluded any effort to legalize an armor branch.

McNair's decision to emphasize the infantry-artillery team and tank destroyers was at odds with the Armored Force concept of combined arms with the tank as the main maneuver element for exploitation. The tank destroyer doctrine was reactive rather than proactive like the Wehrmacht, with its offensive-minded panzer elements where field artillery, motorized infantry, and tactical airpower were all integrated in deep maneuver–oriented operations. The blitzkrieg worked for the first few years; it was a revolution in mobilized warfare. Nevertheless, bear in mind von Schell's interest in the Mechanized Force in 1931, Van Voorhis's and Grow's comments on German Staff officers' visits to Fort Knox in 1933, and von Schell's conversations with Van Voorhis, Palmer, and Scott in 1937. Also recall that Van Voorhis, Chaffee, Scott, and Grow claimed the Germans learned from cavalry mechanization at Fort Knox, where the combat car was the main element for deep maneuver–oriented operations. To the Germans this was the solution to static trench warfare. In May 1941, Chaffee, in a statement to the House Committee on a military appropriations bill for 1942, noted, "It is true that the organization of our armored divisions is very similar in principle to that of the German panzer divisions but it is not a question of us blindly following the latter. Rather it proves the soundness of our own development since the Germans in general followed the principal of our own mechanized cavalry organization which existed prior to the development of the present German organization."[71] Many cavalrymen were convinced that the blitzkrieg was built on the doctrine and organization that were worked out at Fort Knox.

Another problem that confronted the reorganization of the army for war was mechanizing the remaining horse-portee elements. To simplify this situation and the army's mission, General Marshall established the Army Ground Forces (AGF) in March 1942, replacing GHQ as the element tasked to raise and train a rapidly expanding army. The new organization created three field commands: the Army Ground Forces (AGF) commanded by McNair, the Army Air Forces (AAF), and the Army Service Forces or supply system (ASF). The AAF and ASF reported to the War Department; however, AGF under McNair, as one historian wrote, "remained a field command separate from the War Department."[72]

Restrictions were imposed when the AGF was established, providing McNair greater control over the newly created specialized proto-branches. The Armored Force command structure was no longer directly linked to the War Department and General Staff but channeled through the AGF. With the departure of General Devers in March 1943 and the arrival of Major General Alvan C. Gillem Jr.—a product of infantry tradition—closer ties were established with AGF and McNair's influence. In July 1943 the Armored Force was redesignated the Armored Command and placed under the control of AGF and McNair, who believed in the traditional importance of the infantry-artillery team as the central fighting force, supported by antitank elements. McNair eliminated the Armored Corps, assigning armored divisions to predominantly infantry corps. The number of armored divisions was reduced to free up assets needed to create separate tank battalions, which were then assigned to infantry divisions. This move rejected the role of armor as a concentrated force.

The new reorganization in some respects must have pleased Herr and Lynch. Both could claim some solace over their efforts to block the creation of a new branch. Once AGF was established, uncertainty over the use of the name "Force" became an issue because of confusion with AGF, ASF, and the AAF. It was determined that "Armored Command" was a more representative designation that placed it on the same organizational level as the newly created Tank Destroyer, Antiaircraft, and Airborne Commands. On July 2, 1943, the Armored Force was redesignated as the Armored Command with Gillem commanding. Nevertheless, for years the struggle for armor organizational identity continued. It was a legal issue until 1950 when, by congressional action, the cavalry branch was absorbed into the armor branch.

Even though mechanization passed him by, Herr continued preaching that horse cavalry could be used en masse or horse-mechanized. He continued advocating his ideas during the war, and even after the conflict he lamented the loss of his beloved horse soldiers. In his 1953 book written with Edward S. Wallace on the history of U.S. Cavalry, he called for one remounted cavalry division and noted that the 1st Cavalry Division still mourned the loss of its horses.[73] Historian Edward M. Coffman recalled one occasion during the Korean War when Herr told former World War I army chief of staff, Peyton C. March, that the army "betrayed the horse." Furthermore, Herr stated to the retired March that with the aid of one horse regiment the United States would win the war in Korea. March just chuckled, Coffman remembered.[74]

The chief of cavalry, however, would also have some satisfaction when Japan invaded the Philippines. Little did General Herr realize that Lieutenant Edwin P. Ramsey, who commanded two horse-mounted columns from the 26th Cavalry (Philippine Scouts), would lead his unit in the last horse-mounted cavalry action.[75] According to General Jonathan M. "Skinny" Wainwright, who surrendered to the Japanese in 1942, Ramsey effectively used his horse cavalry to cover the retreat to Bataan. After the withdrawal, the horses were butchered and eaten. Years after the war, Wainwright advocated the retention of one horse cavalry division equipped with 75mm mountain howitzers for deployment "in any weather, in any country, in any terrain." He wanted to remount the 1st Cavalry Division. According to Herr, Wainwright's opinion on the successful use of mounted cavalry had no influence on the War Department.[76]

Meanwhile, Devers was reshaping the Armored Force as a combined arms organization with specialized equipment for heavy fighting. The question was raised over the future mission of cavalry, which now had a combat role separate from the Armored Force. On the eve of the United States' entry into the world conflict, three cavalry organizations were finally established at the tactical level, the horse-motorized, the combined horse-mechanized, and the mechanized cavalry regiments from Fort Knox. Manpower for the cavalry was also becoming a problem due to allocations for other troop demands, especially for armored and infantry divisions and the Army Air Corps. The divisions had their own organic reconnaissance elements.

As for the 1st Cavalry Division, it was reorganized as an infantry unit. It retained, however, its cavalry designation. In World War II, the division fought in four major campaigns in the Southwest Pacific. Eventually the 1st Cavalry Division, then commanded by Major General Verne D. Mudge, was the first Allied force to reenter the capital city of Manila. This was accomplished by a typical cavalry maneuver. After the division assembled on Luzon Island, it cut around the western flank of friendly forces and through Japanese-held terrain. In three days, it had moved 120 miles to the heart of the capital.

At Fort Riley, the 2nd Cavalry Division was activated on April 1, 1941, as a racially mixed formation. Due to a reduced need for horse units, the division was deactivated in July 1942 and reactivated as the 2nd Cavalry Division (Colored) in February 1943. In March 1944, the division was sent to Oran in North Africa and deactivated a few months later. Subsequently, the colored cavalrymen were assigned to service units.[77]

By September 1943, the Armored Command again had been reorganized based on lessons learned in North Africa. The important experience was that the ratio of infantry to tank battalions had not provided adequate infantry, and the regimental organization and the combat command concepts resulted in a redundancy of headquarters. Consequently, the solution was to eliminate the regimental structure and decrease the number of tank battalions in the armored division from six to three. The new organization consisted of three tank battalions, each having three medium tank companies and one light tank company, three armored infantry battalions, three armored field artillery battalions, one armored engineer battalion, combat support, including an ordnance and medical battalions and an armored signal company. The armored reconnaissance battalions were relabeled cavalry reconnaissance squadrons, being assigned to each armored division forming up in the United States. The new reorganization consisted of fourteen light armored divisions. The two heavy armored divisions, the 2nd and 3rd, retained their regiments because they were already overseas, the former in the Mediterranean theater and the latter in England. Added to Combat Commands A and B was a reserve command, Combat Command R.[78] This reorganization to light divisions provided more flexibility and balance for the combined arms team for exploitation and pursuit deep in the enemy's vital rear areas. Chaffee, according to Bob Grow, would have approved this reorganization.

For all his efforts, Chaffee gained the reputation as the "father of the Armored Force." His unrelenting drive to support a mechanized cavalry, especially when he served four years in the Office of the Chief of the Budget Branch in the War Department, unfortunately took a toll on his health. His reputation as a major agent of change was considerably enhanced with the publication of Mildred H. Gillie's *Forging the Thunderbolt* in 1947. The book was and remains the standard reference on Chaffee's role in creating the Armored Force. Unfortunately, all of Gillie's important primary sources were either lost or destroyed. Those sources she mentioned in the book were meager and not revealing.[79] Kenneth Heckler, an army historian, also used the Chaffee papers as a source for *History of the Armored Forces Command and Center*. While a history instructor at the Armor School at Fort Knox during the early 1990s, Dr. John T. Broom began an extensive search for the Chaffee papers. Broom, in concert with Dr. Timothy

K. Nenninger, from the National Archives, and John Purdy, curator of the Patton Museum of Cavalry and Armor at Fort Knox, finally determined that the papers were "simply sold as bulk paper to a pulp manufacturer." Other portions of Chaffee's belongings were auctioned off to collectors and antique dealers due to the family's dire economic situation. Some collectors were identified but unfortunately would not come forward and contact the Patton Museum.[80]

Undoubtedly, the advent of new functional developments, such as armor, antiaircraft, airborne, and tank destroyer commands challenged the traditional way combat arms were organized. This resulted in tactical reorganizations to deal with modern warfighting. McNair's mission was to reshape stateside training and organization for the ground army and end the interwar combat arms branch autonomy. McNair made sure Marshall's orders were carried through regarding branch autonomy.

General McNair believed there were too many echelons and planned on reducing a number of the functions of auxiliary units, such as cavalry. Nonethe-

Saluting the standards from an M3 scout car, Major General Willis D. Crittenberger and Colonel William S. Biddle review the 113th Cavalry, Iowa National Guard, Camp Hood, 1943. Courtesy of the U.S. Army Military History Institute.

less, within McNair's reorganization mechanized cavalry as an arm had some purpose, a traditional purpose, meaning reconnaissance and security. There soon were changes. Reconnaissance was to be conducted not only by mechanized cavalry elements but also by the recon platoons from infantry regiments and forward infantry and intelligence patrols.[81] Finally it was determined by AGF to virtually eliminate horse-motorized and horse-mechanized cavalry and send the reorganized mechanized cavalry units to North Africa and Europe.

As for McNair's idea of a Tank Destroyer Command, it never worked out as the war progressed. Most wartime commanders took issue with the doctrine. Generals Devers and Patton believed tank destroyers should be tanks. Major General Omar N. Bradley in turn objected to the high silhouette of self-propelled tank destroyers, noting the effective use of dug-in antitank guns by Germans in North Africa. He leaned more toward towed antitank guns that could be dug-in, thus presenting little exposure. Already the doctrine and the deployment of tank destroyers were in doubt.[82]

Because of his death, Chaffee was never able to move mechanized cavalry to a balanced combined arms doctrine with appropriate infantry, air, and artillery support. Furthermore, branch tactical dominance and congressional budget restraints kept the mechanized cavalry from developing a large force capable of fighting a deep battle at an operational level beyond the realm of tactics. Only a few farsighted officers understood the possibility of this and moved to establish a new proto-branch. Armor won out through the untiring efforts of Van Voorhis and Chaffee. The cavalry, however, still existed as a branch, but new wartime missions meant it had to give up its horses.

Concluding, one could easily ask, What was one of the major fallout effects of Herr's reluctance to totally mechanize the cavalry? Many cavalrymen saw the inevitability of change and rejected the horse cavalry and Herr's adamant attitude.[83] They believed change was ever present and provided new opportunities to integrate doctrine and technology. These cavalry officers left their branch and grew to be dominant leaders in the new Armored Force and influential in molding a new doctrine of combined arms. They made their own doctrine as the war went on. By December 31, 1941, the cavalry was only at 59,804, with army strength of 1,686,403.[84] Virtually all who joined the Armored Force were influenced by what Van Voorhis and Chaffee had accomplished and demonstrated during the 1930s at Fort Knox. Certainly, General Patton stands at the top of the list of opportunists, because he kept himself constantly aware of events that affected mechanization of the army. The histories of Martin Blumenson's *Patton Papers*, Carlo D'Este's *Patton: A Genius for War*, and Roger H. Nye's *The Patton Mind: The Professional Development of an Extraordinary Leader* demonstrated this.

Major Ernest N. Harmon became one of a number of distinguished cavalrymen who moved over to armor. Harmon, who eventually commanded armored divisions and a corps during World War II, remembered a painful meeting with Herr. The chief of cavalry told Harmon that if he went to tanks, their friendship was ended. It was evident to Harmon that Herr was angry and distressed over the events that would replace his beloved horse cavalry. Harmon went to tanks because he wanted to learn more about a new method of warfighting. He was assigned to the 1st Cavalry (Mech) and became acquainted with Chaffee.[85] This launched his legendary career in armor, where he gained the reputation of a hot-tempered "Old Gravel Voice."

Another one of General Dwight D. Eisenhower's renowned cavalrymen, who eventually commanded an army during the war, was Lucian K. Truscott Jr. He came into the army from civilian life and spent most of his long military career in the cavalry, which he dearly valued because of its historical tradition. He recalled Herr's possessing great charm and welcomed him as an honest and forthright chief of cavalry. Yet Truscott noted Herr was impatient with cavalrymen who held contrary views. As he had with Harmon, Herr made his opinions known clearly.[86]

After his "gorgeous years" at Fort Riley, James Polk was assigned to the famed 3rd Cavalry, which had provided the escort for the burial of the Unknown Soldier on November 11, 1921. Until 1941, the regiment also provided the guard detail at the tomb. Polk admitted later that "old horse cavalry" opposed armor. The disturbing part of this attitude, Polk recalled, was that it affected a cavalryman's efficiency rating in his branch if he went over to armor. Herr, who Polk also said was too horse-bound, let it be known that if cavalrymen went to Fort Knox, their standing in the cavalry branch was in jeopardy. Yet Polk believed after the war that "dedicated horse people were stupid in not going to the mountains and the tough terrain where cavalry could hold its own against any enemy." Though he supported the use of horse cavalry under certain conditions, he believed it was the tank actions in the western desert maneuvers before the United States entered World War II that killed the cavalry as the traditional mounted arm.[87]

Grow, who served in the Office of the Chief of Cavalry during Herr's tour, also began to think armor. In 1940, he ended up with the 2nd Armored Division and the beginning of a long, admirable association with Patton. However, he and Patton had different personalities; whereas Patton was flamboyant and theatrical, Grow was methodical and businesslike. Yet they got along because of their common belief that armor was the new form of mounted warfare. Later, when Grow commanded the 6th Armored Division under Patton's Third Army in Europe, the colorful general referred to Grow as one of his top armored division commanders.[88] Grow felt that Patton exemplified the principles of leadership with his actions, "but most were good and some were not so good."[89]

Regrettably, Herr had no use for Grow's opinions on using new mounts for exploitation and pursuit. After World War II ended, Herr let his displeasure be known again. Herr openly criticized Grow, especially for supporting a separate armored branch because of his short article in the *Cavalry Journal* at the end of 1945. He wrote that the bulk of armored divisions in Europe executed the mission of cavalry, thus establishing the fact that a separate arm had to be defined by law.[90] Herr still believed, even after the successful deployment of armored divisions and mechanized cavalry elements during the war, that his position on the value of horse cavalry was correct. Herr just failed to understand the significance of mechanization as the future of cavalry, which caused the loss of many cavalrymen to armor. It was Grow's belief that if Herr had taken a strong position for complete mechanization of the cavalry, there would not have been an Armored Force but instead mechanized cavalry divisions. Grow's well-known comment was, "He lost it all."[91]

Parker, Van Voorhis, and Chaffee were the major spokesmen for change, the marketeers for a new vision of mechanized warfighting. At Fort Knox, other cavalrymen embraced these new operational concepts and were willing to subject their ideas to trials. Unfortunately, the chief of cavalry was not convinced regarding the relevancy of mechanizing. His cultural bias prevented the establishment of mechanized cavalry divisions. Herr valued tradition and stability more than change and progress. This conflict ignored the need to establish at least a conceptual framework of doctrine to serve as the engine of change, meaning establishing a doctrine from what was written, taught, and worked out at Fort Knox reflecting concepts, which emphasized depth of attack, initiative, synchronization, and agility with equipment on hand. In December 1941, the army went to war with an Armored Force and armored cavalry elements, all in preliminary stages of transitions to mechanization. As before, the army entered the war substantially unprepared. As General Donn A. Starry mentioned, the army was "in search of but not in command of approved, tested, validated doctrine; and not equipped, organized, or trained to execute whatever they thought doctrine might be."[92] This was the legacy of the intertwar period as a result of the establishment of new combat arms branches by the 1920 National Defense Act, the infantry-dominated *1923 Field Service Regulations* based on World War I experiences, the little-changed *1939 Field Service Regulation (Tentative)*, and MacArthur's 1931 decision to have each of the combat arms branch chiefs decide how the branch should mechanize. Consequently, no one was in charge of doctrine development.

7

"Sneak and Peek" or Fight

Doctrine was being rapidly developed around equipment on hand while the 1st and 2nd Armored Divisions were being organized. The arrival of new platforms for the mechanized force was anticipated as industry moved to a war footing. At Fort Knox an officer school was established to deal with leadership and tactical issues. One of the instructors, Major H. H. D. Heiberg, the former Christie combat car T1 troop commander at Fort Knox in the early 1930s, taught that the reconnaissance battalion's primary role was to obtain a continuous, complete, and coordinated picture of terrain and the enemy's situation in cooperation with attached observation aircraft. Once the main enemy force was identified, reconnaissance units pulled back and closed up with an approaching main force, the heavy armored division. The cavalry then moved to a flank protection and security mission. Though the reconnaissance battalion was a force of considerable combat strength, Heiberg lectured, "It is primarily trained, organized, and equipped for reconnaissance." Combat, he noted, "can be offensive, delaying action or pursuit." Offensive combat, however, was necessary only when essential information was required for higher headquarters. What Heiberg was teaching was traditional cavalry doctrine of stealth "sneak and peek." He emphasized that it was important to cooperate with aviation, combat engineers, and supply units. On logistics, Heiberg emphasized that supply, maintenance, and evacuation problems were more difficult for mechanized cavalry than for any other unit in the armored division. This was because the cavalry reconnaissance battalion was expected to operate on a wider front than any other unit in the division. He did not forget to mention the importance of a medical detachment for field aid and evacuation. He concluded by telling the officers in attendance that "reporting information consumes time; it is hard work, and scattered over a wide territory. So, it is important for cavalrymen to be constantly on the alert." Heiberg held that recon cavalrymen possessed the exceptional qualities of "initiative, courage, and common sense."[1]

Heiberg was expressing the standardized *April 1941 Mechanized Cavalry*

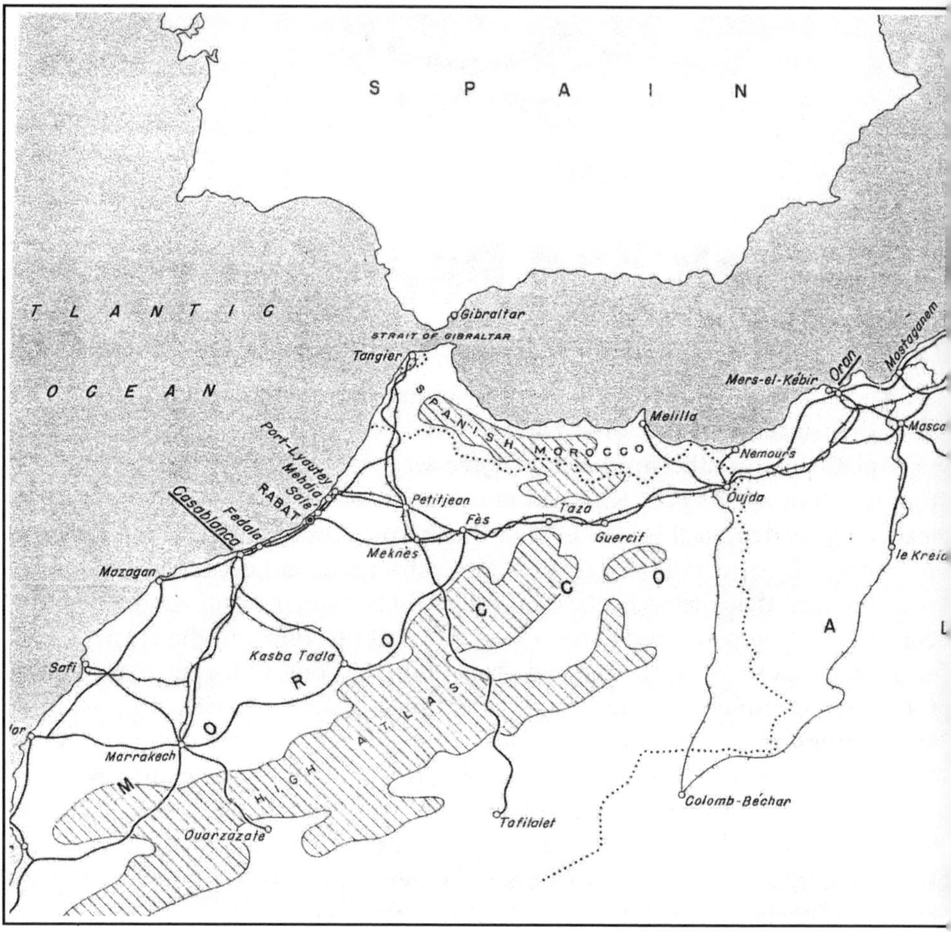

North Africa. Printed by the U.S. Army Topographic Command.

FM 2–10 that recognized the changes since 1938 and subsequent maneuvers that conceded the departure of tanks to the Armored Force, whose mission was exploitation, pursuit, and shock action, a horse cavalry tradition. However, General Herr made sure that the manual also stated the same mission for the horse cavalry. The mechanized cavalry was expected to carry out its mission of stealth reconnaissance and avoid or bypass combat whenever possible. Security was not mentioned. At the platoon level, reconnaissance elements had a scout section with Bantams, some armed with mortars, an armored car section, and for fire support an assault gun, the half-track 75mm gun motor carriage.

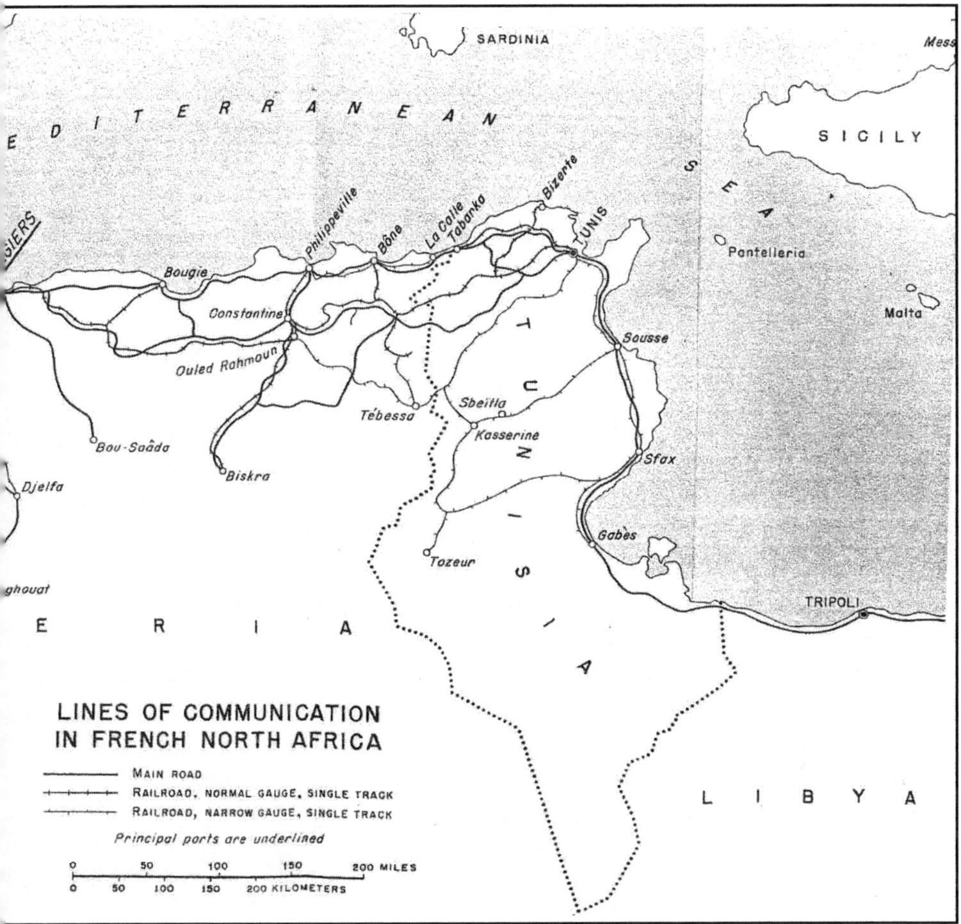

Like Major Heiberg, Major General Charles L. Scott had been with the mechanized cavalry during the 1930s as a regimental commander. Because of Chaffee's illness in 1940, Scott became the temporary Armored Force commander. Afterward, in 1942, he was sent to observe British Eighth Army operations in North Africa. He wrote about his experiences. His position on the future of cavalry was different than Heiberg's. It was widely known, he reported, "that adequate security begins with long distance air and ground reconnaissance." This was important in all phases of modern mobile ground warfare because "there was no line or well delineated front," he observed. As a result, combat elements, which he called groups, needed protection for their front, flank, and rear. To accomplish this mission, reconnaissance elements required not only air observation but also more "punch." Continuing, "In this day and age, long distance reconnaissance must be organized to fight in execution of its mission for time to send formations in, and to fight for time for the main body to properly utilize the information sent in." Once this mission was completed, he stated, the reconnaissance elements would revert to a battle security mission guarding flanks and the rear of the main force. He was critical of current cavalry reconnaissance regiments because of too much "overhead." A battalion organization was favored because it was best suited to perform both reconnaissance and battle security for U.S. Armored Divisions. However, for both organizations he suggested more firepower, such as organic self-propelled antitank guns and more light tanks. He was critical of the cavalry's failure to tie its ground elements to air support, for both long-range air reconnaissance and combat. For air defense, the cavalry was relying too much on its mounted .30- and .50-caliber machine guns, which Scott favored. His article stressed cavalry offensive, meaning fighting with greater firepower, both mounted and dismounted, along with air support.[2]

The first real test of what Scott suggested and what Heiberg was teaching occurred when Allied forces encountered the Axis powers in North Africa during Operation Torch at the end of 1942. The Center Task Force landed at Oran, Algeria, in November 1942 against sporadic resistance. On January 5, 1943, Lieutenant General Dwight D. Eisenhower, the Allied commander, appointed Major General Mark W. Clark, the commander of the II Corps in England, to command the newly organized Fifth Army. Opinions began to emerge over the question of leadership when the chief of staff, General George C. Marshall, selected the officers to command large army units. Major General George S. Patton Jr., the commander of the Western Task Force that landed in Morocco in November 1942, wrote before the landings at Oran that he did not yet trust Clark, and that Clark was more interested in his own future than in winning the war. After the landings, he felt that Clark was "too damned slick" and did not show "the least interest in troops and installations."[3] Meanwhile, on January 6, the American II Corps was assigned to the Fifth Army. Subsequently, Clark turned over the field command of II Corps for deployment with the British to an infantryman, Major General Lloyd L. Fredendall. He was handpicked by General Marshall, who had a high opinion of Fredendall's command ability, as did General McNair, commander of Army Ground Forces. At the end of January, Eisenhower attached the II Corps and some French forces to the British First Army for the planned operation to engage the Axis forces in Libya.

Fredendall's guide as a corps commander was the current 1942 *Field Manual 100–15, Field Service Regulations: Larger Units*. Recall that General Parker in

1930 prepared a provisional manual for larger units for his chief of staff, General Summerall. It was designed as an instructional manual for commanders of units, at corps, and higher. At the time it ran into trouble because it was considered too cumbersome. Some army leaders claimed it was plagiarized from the French manual for larger units that prescribed a rigidly controlled operation called the "methodical battle."[4] After a review, Summerall supported Parker. The manual was released in May 1930 and became the first effort to articulate a doctrine for functions and operations of larger units in the field until the United States entered the war. The corps commander's mission was to exert the tactical will that would propel his force at all combined arms levels through the shock and friction of battle. In 1942 General Marshall ordered the publication of a new manual for larger units, describing new doctrines and functions, avoiding the methodical battle, and emphasizing combined large theater operations with joint land, sea, and air units. However, the tank-artillery team and attrition warfare still existed as the centerpiece of doctrine.

The 1942 *Field Manual 100–15* did explain how the corps was to operate in regard to prescribed doctrine and functions. It was Fredendall's task to use tactical events of execution and maneuver at an operational level between strategy and tactics with two or more divisions under his control. The corps consisted of a headquarters and certain organic elements designated as corps troops depending on the assigned mission. In addition, he was expected to allocate reinforcing units such as tank destroyers, tanks, artillery, and special troops as the combat situation evolved. Fredendall was also expected to leave the details of execution of his plan to his division commanders, while at the same time communicating to his superior strategic commander. As corps commander he was expected to realize the importance of looking deeper in time, space, and events. In order to carry out his mission, Fredendall was to maintain close contact with the leading divisions and be in a position to reinforce or change dispositions to affect the battlefield outcome. Most important was timely intelligence, which should flow up from tactical reconnaissance units. For a corps commander to attain unity of effort, he must clearly identify objectives and reinforce the relationship among subordinate forces, such as the 1st Armored Division and its units, including the mechanized cavalry. Since the static and attritional nature of World War I, mechanization changed the character of warfighting, placing greater emphasis on maneuverability.[5] Fredendall, a staff officer and infantryman, had problems dealing with change; thus, he failed to plan accordingly, especially when dealing with the mobility-minded legendary tactician Field Marshal Erwin Rommel and his Afrika Korps.

The problem with Fredendall was that he did not have corps cavalry, an essential requirement for corps-level maneuver and as an economy of force to be used discriminately in defining the enemy force. The North African theater provided opportunities for maneuver warfare. Nevertheless, Fredendall refused to go along with corps cavalry because he was inclined to the methodical battle even though the corps cavalry group of two mechanized squadrons had been established in 1942 to provide for the follow-on reconnaissance force.

General Fredendall was also expected to cooperate with British forces while organizing his corps around Major General Orlando Ward's 1st Armored Division. Ward, a cavalryman, was a respected and highly intelligent decorated World War I officer. His division was tasked to be the main striking force supported by artillery and units from the 1st Infantry Division commanded by Major General Terry de la Mesa Allen, a former instructor at the Cavalry School. The first cavalry unit deployed to hard combat was the 81st Armored Reconnaissance Battalion (ARB), organized with three reconnaissance companies and an M3 light tank company, which was attached to the 1st Armored Division. Lieutenant Colonel Charles J. Hoy commanded the battalion.

On January 15, 1943, the 81st ARB entered the combat zone. The unit was originally designated 2nd Reconnaissance Battalion and cadred from the 7th Cavalry Brigade (Mech) and assigned to the 1st Armored Division. On May 12, 1941, the unit was redesignated 81st Armored Reconnaissance Battalion. Hoy commanded the battalion at the time the cavalrymen landed in North Africa. His philosophy stressed the traditional role of reconnaissance, meaning, "to look far afield, to be inquisitive, and to take nothing for granted." Hoy's approach was to engage in stealth and avoid combat, the "sneak and peek" approach. The 81st ARB was expected to engage in all types of reconnaissance and security and furnish information to higher headquarters.[6] The unit's basic equipment was the ¼-ton Bantam or jeep, which was the primary recon vehicle, and the open-topped, lightly armored motor vehicle M3 scout car armed with .30- and .50-caliber machine guns. The wheeled M3 scout car was designed primarily for recon support, providing an overall link to the rear. For security screening ahead of a division, the M3 scout car did not have the firepower or off-road mobility to engage the enemy for information when necessary, until a heavier force arrived. The battalion also had ¾-ton wheeled M6 37mm gun motor carriages and M3 Stuart light tanks. In addition, lightly armored M3 half-tracks were used as personnel carriers for transporting troops and supplies to the combat zone and as mobile combat mounts for light weapons, such as machine guns, 75mm howitzers, and 60mm and 81mm mortars. These weapons were the best that the cavalry units could muster at the time. However, they were not a match for German tanks and antitank guns when the cavalry had to fight for information or engage in defensive combat.

A 1st Armored Division patrol in Tunisia. Note the jeep, the primary recon vehicle. Courtesy of the Patton Museum.

At the time, cavalry reconnaissance units were tasked to engage in stealth reconnaissance and security rather than dismounted combat action. This reflected what was being taught at the Cavalry School and later proved a detriment to the reconnaissance battalions. While engaging in reconnaissance, the cavalrymen were expected to brush aside small enemy groups or pass around and peek to determine their movements. Most important was the unit's ability to report timely information to higher headquarters. The recon's organization was unusual because it was originally a battalion and not a squadron. The cavalry reconnaissance battalion or squadron's table of organization and equipment consisted of M3 wheeled scout cars, supported by wheeled gun motors and infantry carriages, half-tracks, and light tanks. An air element was expected to provide long-distance reconnaissance.

During the months after the landing in Oran, the 81st ARB was not engaged as an economy force, meaning a reciprocal of mass at a critical time to complete reconnaissance and security screening for the 1st Armored Division. However, security was not discussed in the *1941 FM 2–10*, so the battalion had to improvise. The battlefield in Tunisia was too fluid for high commanders who were embedded in the doctrine of attrition warfare and had no corps cavalry. As a result, the 81st ARB lacked the judicious employment and distribution to execute all its unit's combat elements for the mission of reconnaissance and security screening. The 1st Armored Division split up the recon battalion and deployed it as separate elements in various diversified missions, such as controlling high ground, blocking action, dismounted combat, and defense of a minefield. The reluctance to use the 81st ARB as an economy force came about as a result of command leadership from above to corps level. The 81st ARB commander, Colonel Hoy, however, attempted at best to engage in stealth reconnaissance and security as prescribed in training manuals and taught at the Cavalry School.[7]

Shortly after commitment the II Corps experienced its first severe test of doctrine, organization, leadership, combat, and equipment. To say the least, it became an exercise in poor leadership that rippled down to tactical levels. It became more evident as the II Corps deployed to Tunisia to take a defensive position south of the British First Army in a sector known as the Eastern Dorsal, a mountain range running north-south. These leadership deficiences proved disastrous and led to one of the U.S. Army's worst defeats and a rout in some cases. The army was far from ready to fight and win its first major battle. The problem was command failure at an operational level, where major operations are planned to achieve strategic objectives. Though operational level implies a broader dimension of time and space than do tactics, it ensures support for logistical and tactical means to achieve strategic objectives.

The subsequent defeats of the II Corps at Sidi-Bou-Zid and Kasserine by Rommel's Afrika Korps in February 1943 demonstrated major flaws in U.S. Army leadership at an operational level. No doubt Fredendall was intimidated by Rommel's success in earlier, spectacular, large-scale armor battles against the British. The problem also at Sidi-Bou-Zid, an oasis in the desert, and the Kasserine Pass was the inability to verify intelligence at the tactical level. This can be attributed to Fredendall and Brigadier General R. A. McQuillin, a cavalryman and commander of Combat Command A, 1st Armored Division. The corps commander prescribed detailed deployment for Combat Command A, which had not taken advantage of the terrain around Sidi-Bou-Zid for the proposed attack to take the Faid Pass just northeast of the oasis. To make the situation even more tenuous, the Germans as usual easily dispatched the French forces operating in the area.

Proper deployment of cavalry reconnaissance units by higher headquarters might have led to the discovery of German troops maneuvering into this area. Security measures were also lacking that could have prevented a surprise attack by Rommel's forces. In addition, Fredendall's corps was dispersed, with Combat Command B under the command of Brigadier General P. M. Robinett, who preferred not to go forward during actual combat because it could compromise his situation. At the time his command was tied up with General Kenneth A. N. Anderson's First British Army for an anticipated German attack. Faulty air and ground intelligence convinced Anderson that the main German attack would strike north of II Corps' area of operation. Meanwhile, a counterattack by McQuillin led to an ambush destroying an entire tank battalion. This again pointed out the deficiencies in reconnaissance, security, and command leadership, which had not understood the use of cavalry as an economy force. Too often cavalry recon units were detached as orphans.

The result was a disproportionately complicated chain of organizational command and control, poor unit and troop management, a lack of coordinating air cover and deep reconnaissance, and the execution of timely intelligence. Units felt these deficiencies in command and control leadership. The 81st ARB, which was expected to be deployed as a stealth reconnaissance force, instead became more of a spread-out fire brigade, acting as a defending and delaying force while American forces retreated. The 81st ARB, like the rest of the 1st Armored Division, was widely dispersed. Thus there was no effort to establish economy of force or unit integrity to perform a prescribed cavalry mission. There was no problem of division cavalry being split up. This was understood in 1943 with the addition of an additional troop. However, this assumed corps cavalry would become part of the fight.

Thus, the 1st Armored Division became a casualty of one of the most inept corps commanders of World War II.[8] Patton commented from his viewpoint in Morocco about his old nemesis, "Fredendall lost control of the fighting."[9]

What caused Fredendall to lose control of the battlefield? Elements of his corps were badly intermingled west of the Eastern Dorsal, and as a result, the 1st Armored Division and its units were widely dispersed. Fredendall's method of coordinating strategic objectives with the British and executing the corps' mission—as the primary tactical unit of execution and maneuver—was to attempt it from the safety of his buried command post deep in a ravine located many miles to the rear at Tebessa. At times Fredendall had no idea what to do. He was a "chateau general," a term used to describe high-echelon commanders in World War I who rarely ventured outside their headquarters. The corps commander was fighting the Tunisian campaign on a World War I model. He avoided the front lines, thus losing close contact with the 1st Armored Division, his major striking force and reconnaissance unit. In addition, his persistent anti-British remarks and lack of coordination with his allies was well known. This attitude compromised operational coordination with the British. General Eisenhower found this disturbing, since he was trying to build a coalition force with the British First Army operating on II Corps' left flank.[10]

In a diplomatic way Eisenhower earlier had tried to warn Fredendall. However, his behavior did not change, and eventually lead to the debacles at Sidi-Bou-Zid and at Kasserine Pass. This resulted in a major rout of the 1st Armored Division along with its 81st ARB, now involved in defensive combat, such as security and delaying actions. The gloomy Fredendall was content to sit in his deep bunker far from the action just like the French and British generals sat in their bunkers on the western front during most of World War I.

Alarmed at this situation, General Eisenhower sent to Tunisia a dynamic and profane cavalryman, Major General Ernest N. Harmon, who commanded the 2nd Armored Division during the Morocco landings. He was temporarily tasked to take charge of the deteriorating battlefield situation, without replacing Fredendall or Ward. Receiving no support from the II Corps commander, Harmon, nevertheless, was able to stabilize the worsening situation with Ward's division and other American units. In a few days he returned to his division in Morocco. Before leaving Tebessa, Harmon recalled that he reported to the corps command post and found Fredendall in bed displaying the effects of alcohol consumption. Reporting back to Eisenhower, Harmon indicated that the apathetic Fredendall was not a team player. Harmon suggested Patton take command of the II Corps.[11] He then told Patton, "Fredendall is a physical and moral coward."[12] Eisenhower took note of the command problems and advised General Marshall that Fredendall lacked the capability to coordinate plans expected of a corps commander.[13] It was the responsibility of corps command leadership to implement coordinated tactical plans for execution and maneuver at an operational level so as to push the Germans and Italians out of North Africa. In this, Fredendall failed. He could not shake the influence of the post–World War I branch system that did not allow cross-coordination and the use of combined arms in mechanized operations. The II Corps commander had a bunker mentality.

By now the 1st Armored Division had a major morale problem and had lost its confidence. Fredendall, from his underground command post, wasted no time in blaming Ward. General Hamilton H. Howze, who at the time was a lieutenant colonel and Ward's operations officer (G3), had some interesting things to say about Frendendall. Howze graduated from West Point in 1930 and was commissioned in the cavalry, serving his first four years with the 7th Cavalry at Fort Bliss, Texas. Like most cavalry officers, he was fond of polo during the 1930s, before his sudden transition to armor. He was an ardent reader of Karl von Clausewitz's famous treatise *On War* and years later became a serious sky cavalry supporter.

Howze claimed Fredendall was not as mobility-minded as Ward. The problem with Fredendall was that he was imbued with the doctrine of attrition warfare and not able to execute mobile warfare with a combined arms team. He was unwilling to change or adjust to mechanized warfare. Howze claimed the II Corps commander dictated to Ward how he should maneuver his units and was too far from the battlefield to evaluate military operations as they unfolded. On one occasion, Fredendall bypassed Ward and gave orders to a combat command.[14] This action was a clear violation of the chain of command. The open animosity that developed between corps and division commanders filtered down to all 1st Armored Division units, including the 81st ARB, which found it difficult to coordinate its assigned mission of reconnaissance and security screening. To make morale worse, General Eisenhower reported that the British First Army considered the 1st Armored Division "noncombat worthy."[15] Ward's units felt this negative perception, as Patton soon discovered.

During the Tunisian operation, Colonel Hoy and his men were placed many times in precarious positions because of the changing nature of mechanized cavalry doctrine from reconnaissance and security screening to a fire brigade mission. The African experience also caused deterioration in morale, especially after the envelopment of Companies A and B. Many cavalrymen were captured, and those who were not had to fight a costly delaying action. Now the realization set in. Cavalrymen were victims of a questionable doctrine and poor high command leadership. Later that year Hoy wrote, "The *esprit* of an outfit is one of the greatest contributing factors to its success in battle." He went on to say—in almost a reference to leadership—that pride and reputation were synonymous with unit identity. He chastised those "men who believed undisciplined and unmilitary troops are the best for battle."[16] "They did not look like soldiers; they did not fight like soldiers. How do you expect them to act like soldiers?" Actor George C. Scott told this to Major General Omar N. Bradley in the 1970 film *Patton*.[17] This was also one of Fredendall's legacies.

Meanwhile, Fredendall was sent back to the United States, where he received a hero's welcome and a third star. Even before the II Corps commander left for Oran, Eisenhower had doubts about his indifferent behavior. In hindsight, Eisenhower should have acted sooner in dealing with Fredendall. Perhaps the delay was due to Marshall and McNair's support of the lethargic II Corps commander.

Eventually the command situation and morale problems were resolved when Fredendall was sacked and Patton took temporary command of the II Corps on March 6. British general Harold Alexander then became commander of the Eighteenth Army Group, consisting of the British First Army, some French units, and II Corps. He had considerable combat experience, and his demeanor was quiet and unassuming, with a strong sense of terrain.

The feisty and profane Patton put a renewed fighting spirit into the II Corps. Soon Ward was sacked because Patton thought he was not aggressive enough. Another reason for the action was a lingering atmosphere of distrust that permeated the relationship between the former corps commander and his armored division commander. Eisenhower wrote Marshall, "Ward was too sensitive, both to criticism from his immediate superiors and to the loss of his friends and subordinates. . . . He cannot develop the necessary veneer of callousness."[18] Ward went back to the United States and took command of McNair's pet Tank Destroyer Center. General Harmon returned to Tunisia and took command of Ward's division. Harmon used the 81st ARB in its traditional role rather than the confusing one the recon battalion was saddled with during the Tunisian operation. He attempted to move the unit to its proper cavalry role as the eyes and ears of his division. Unfortunately, the 81st ARB again was subjected to a collapse in leadership at higher levels when the 1st Armored Division became part of General Clark's Fifth Army in Italy.

Meanwhile, another prominent cavalryman, Colonel C. C. Benson, a strong proponent of mechanization and the Christie vehicle during the interwar period, was also sacked. In 1929 Benson wrote an influential article that praised the Christie armored car as the M1940, meaning it was ten years ahead of its time. At the time Benson was president of the Armored Force Board, the testing agency serving the chief of the Armored Force on development of training and tactical doctrine. Benson was given command of Combat Command B, 1st Armored Division. Shortly General Harmon, the new division commander, relieved Benson for his lack of aggressiveness during an attack in early May, when for the first time the division fought under a unified command. When Benson's command failed to reach an objective, Harmon asked what caused them to fail. Benson replied that his men were worn out after two to three hours of fighting. Harmon gasped at Benson and caustically said, "You must be worn out." Not one to mince words, Harmon told Benson he was finished and to "go and sleep it off." Howze was placed in temporary command and carried through the attack only after Harmon realized that the Germans had moved up antitank guns after Benson had begun his attack. With extra artillery placed on German positions, Howze was able to move his armor around the flank and roll up the antitank defense.[19]

Another example of one of the three echelons of mechanized cavalry employment during the North African campaign was the 91st Cavalry Reconnaissance Squadron (CRS), whose original mission was intended to fill the corps' cavalry gap. The 91st was previously the CRS of the 1st Cavalry Division and subsequently was detached and deployed to II Corps. In April 1943, it was attached to the 9th Infantry Division, as the Allies squeezed the Germans and Italians closer to Tunis. Earlier, during Kasserine, the 91st could have been really important but was not available. Just what was Fredendall thinking? During this phase the squadron fought both as cavalry and as dismounted infantry. Later the squadron was attached to the 1st Armored Division, where it performed its normal role of reconnaissance and screening in advance of the division's Combat Command A.

Eventually the Germans and Italians realized they could not stop the Allied advance and began to surrender. The end was at hand. They were now being pressured by the British Eighth Army moving west toward Tunis to link up with the British First Army, some French units, and the American II Corps, now composed of the 1st Armored Division, and the 1st, 9th, and 34th Infantry Divisions under the command of Lieutenant General Omar N. Bradley. Patton had been reassigned in preparation for the upcoming Allied invasion of Sicily, where he was to command the American Seventh Army.

The Axis retreat and surrender of Rommel's famed Afrika Korps and General Juergen von Arnim's Fifth Panzer Army was due to supplies and equipment being diverted to the eastern front, and increasing Allied manpower and equipment and control of the Mediterranean Sea. Also contributing to the defeat was the poor cooperation between two German armies at a critical time as they merged to face British and American allies in Tunisia. There was no unity of command. This was a result of Hitler's initial interference and his insistence on a divided command between Rommel's Afrika Korps retreating from the British Eighth Army and Arnim's Fifth Panzer Army in Tunisia. Finally on May 13, 1943, the German Fifth Panzer Army and Rommel's Afrika Korps, along with their Italian allies, surrendered. The Tunisian campaign taught the U.S. Army

An array of early mechanized equipment at Fort Knox in 1942. *Left to right*: wheeled 37mm gun carriage, M3 75mm gun half-track, M3 medium tank, and M3 light tank. All but the half-track would see their demise in North Africa. The half-track eventually became a staple in armored infantry. Courtesy of the Patton Museum.

valuable combat lessons and brought unity with the British in the first trial of engaging in combined and joint operations.

It was now time for the 1st Armored Division, which had lost most of its heavy equipment during the Tunisian campaign, to spend time refitting and training for an invasion of Italy. The division did not participate in the Sicilian invasion. It was also time to reevaluate doctrine, equipment, and the major shortcomings at the operational level of leadership that filtered down to all tactical units, including the 81st ARB and its mission of reconnaissance and security.

Discussing the equipment used by the American forces in Tunisia, Bradley was far from complimentary. He found half-tracks overrated because they offered little crew and troop protection from enemy fire. Regarding gasoline-driven tanks, he reported crews called them firetraps because they blazed too easily when hit. One tanker told Bradley, "These gasoline engines go up like torches on the first or second hit. Then you've got to bail out and leave 'em burning."[20] Gasoline engines powered all the M3 light tanks used by mechanized cavalry reconnaissance units.

Army Group Commander Alexander had the opportunity to meet both Patton and Bradley in Tunisia. "Both were two completely contrasting military characters," he later wrote. Patton was impatient, and "Bradley was unwilling to commit himself to active operations unless he could clearly see their purpose." The British general found Bradley more cautious. However, he considered Patton "a thruster, prepared to take risks. Patton should have lived during the Napoleonic wars—he would have been a splendid Marshal under Napoleon."[21] Alexander visualized Patton leading a dashing charge for La Grande Armée of Napoleon, whose maxims on cavalry can be summed up as "Cavalry needs audacity and practice; above all it must not be dominated by the spirit of conservatism or avarice." For cavalry to be successful, Napoleon believed in a combination of speed with a shock force brandishing the *arme blanche*. However, Napoleon placed discipline first.[22] As Alexander implied, Patton had studied Napoleon's maxims. His extensive study of history enhanced Patton's cavalry instincts for mobility of the mind and thinking mounted. He possessed this attitude through most of World War II, rather than the methodical approach of infantry linearism and attrition warfare. These doctrinal attitudes would be at odds as the Allies invaded Fortress Europe.

After the end of the African campaign, the commanders of the 81st ARB and the 91st CRS claimed they had usually deployed their units in the traditional reconnaissance and security missions. However, these deployments were affected by four imponderables: leadership, terrain, weather, and the enemy. In addition, there was a tendency for the cavalrymen to disperse when fired on, and as a result they were reluctant to fight for information as required by higher headquarters. Both Hoy and Chandler, however, stood fast in believing in the traditional stealth "sneak and peek" and in the security mission for cavalry. At times both units were forced to deviate from prescribed cavalry missions due to the exigencies of combat and poor operational leadership from higher-echelon commanders. The confused deployment was felt more by the 81st ARB, since it was the first mechanized cavalry to experience a reversal in combat. In addition, both reconnaissance units had to perform their missions with the cavalry's main scout vehicle: the wheeled M3, usually armed with a .50-caliber and two .30-caliber machine guns. The lightly armored M3 personnel half-tracks and howitzer motor carriage half-tracks, both of which were more terrain capable than the wheeled motor carriages and M3 scout cars, supported the scouts.

The experience in North Africa soon demonstrated that the long series of developments of reconnaissance wheeled scout cars beginning with the cavalry in 1929 had ended because they were too lightly armored, the open top too vulnerable, and the vehicle too restricted by terrain. By 1944, production of the M3 White scout car series ended. The M3 scout car was replaced by the M8 armored car, with a 37mm cannon mounted in an open turret. The M8 had its origins in the Tank Destroyer Force as an antitank weapon for defensive action. Fully tracked M8 75mm self-propelled howitzers replaced the M3 75mm half-track assault guns. The small, rugged, four-wheel drive jeep became the scout car of choice. It had a fittingly low silhouette; however, the vehicle was vulnerable to small arms fire and mines. The M3 light tank was upgraded to the M5 light tank by adding more armor, increasing the speed, and making driving improvements. Nevertheless, it was no match for heavier German armor. The 37mm tank gun was retained to deal with soft targets. However, the new twin gas-driven Cadillac engines installed in the M5s still made the tanks vulnerable. The M3 lightly armored half-track—originally designed as a cavalry personnel carrier—was retained and upgraded in numerous configurations as a specialty vehicle, becoming the workhorse of the army during the rest of the war.

After the surrender, an assessment was made of the 91st CRS operations in Tunisia. The principal reconnaissance vehicle, according to Lieutenant Colonel Harry W. Candler, the squadron commander, was the M3 scout car. The scout cars, he maintained, were ideal for command and control because of their ability to transport radio communication equipment used between platoons and troops and then between troops and squadrons. He noted that his cavalry squadron had more crew-served weapons but not enough men to serve them. He praised the timely use of the squadron's principal weapon, the 81mm mortar, which he claimed could be effectively mounted on gun carriages. Candler found the high protective shield for the 37mm gun on the 4 x 4 Farago ¾-ton M6 motor carriage impractical because of its high silhouette. The four-wheel drive jeep, which he called a Bantam, was ideal for traditional "sneak and peek" patrolling.[23]

The experiences in North Africa made it evident that reconnaissance and security were just two of the many missions assigned to mechanized cavalry elements on a chaotic battlefield. It was evident that the arrangement and maneuver of cavalry units in relationship to their primary mission of stealth reconnaissance was not synchronized in a unified operation at an operational level as an economy of force element, since it was not expressed in doctrine. Not until new leadership took command and understood that the doctrine of stealth reconnaissance was not workable given the dynamics of a diverse battlefield would equipment requirements change. Light tanks more and more took on reconnaissance roles. The 75mm M3 gun motor carriage mounted on a standard M3 personnel carrier proved to be useless as an antitank weapon. Nonetheless, the M3 motor carriage was at the time considered the most effective system when used in battery by E Troop, 91st CRS for artillery support.

While both reconnaissance units were reworking doctrine, organization, equipment, and personnel retraining, it was evident as a result of the African campaign and the high intensity of mobile warfare occurring in Sicily, that as in all wars psychological issues emerged, compromising unit integrity and tactical mobility. It brings out the human dimension in warfare. What Alexander had stated about Patton's impatience was demonstrated in August 1943 when the general cursed and struck two soldiers with his gloves. Patton's perception of what constitutes nonwounded battle casualties was blurred. The articulate cavalryman had no patience for what he called "cowards." The first soldier he struck was diagnosed as having a history of diarrhea and battle exhaustion (a classification of battle neurosis or fatigue), leading to a "psychoneurosis, anxiety state, moderately severe." The second soldier, who had a commendable record in the field artillery, suffered from sleeplessness, followed by persistent nervousness. All his symptoms were brought on when a shell fragment wounded his best friend.[24] One of the most popular forms of treatment throughout the war for all combat units, armor, cavalry, field artillery, infantry, mechanized, and combat engineers, was narcosynthesis, meaning soldiers received intravenous sodium pentothal to induce a state of seminarcosis.[25]

Apparently, Patton had not recognized this aspect of mental health management when he read Adolf von Schell's book *Battle Leadership*, published in the early 1930s. The German graduate from the advance course at the Infantry School at Fort Benning wrote, "Soldiers can be brave one day and afraid the next. Soldiers are not machines but human beings who must be led in war. Each one of them reacts differently, therefore each must be handled differently." He goes on to say, "To sense this and arrive at a correct psychological solution is part of the art of leadership."[26]

In September 1943, armored divisions were reorganized for the last time to achieve greater balance. All armored divisions, except the 2nd and 3rd, became light for greater flexibility and command and control. Three combat commands ,A, B, and R (reserve command), were established, each with a tank battalion, a self-propelled armored artillery battalion, and an armored infantry battalion. The light armored divisions were now organized more for deep exploitation and pursuit rather than tank-versus-tank combat. The cavalry armored reconnaissance battalion was relabeled cavalry reconnaissance squadron, and a standard organization was set for division and corps cavalry. Support units were retained. By end of 1943, the army was well on its way to becoming the most mechanized ground force in the world.

With the slapping incidents over and with the invasion of Italy, another campaign involving the 1st Armored Division began. The division was again subjected to questionable leadership from high command. Before I discuss the issue, I will examine the events leading up to the problem in command.

The Allied drive for the liberation of Italy and the subsequent brutal campaign up the Italian peninsula. It was a long, exhausting, and depressing military campaign. Even today, controversy surrounds the strategic decision to engage the Germans in the Mediterranean theater because of this question: Did the strategic decision to invade Italy contribute to the Allied overall effort to defeat Germany?

The strategic military plan was the brainchild of Prime Minister Winston Churchill with President Roosevelt's reluctant approval. Their military staffs strategically planned first to provide the Allies unrestricted use of the Mediterranean region. Next, Churchill's "soft underbelly of Europe" grand strategy was to drive up the Italian boot and penetrate the eastern Alps toward Austria. It was Churchill's reasoning that it was imperative as the Allies moved up the boot to establish sites for air bases, establish control of the Adriatic for possible operations in the Balkans, provide access to Near East oil, secure supply lines to the Soviet Union, and finally attack Austria. Both leaders agreed that it was important to force Germany's ally, Italy, out of the war. Doing so, the operation was expected to draw needed German troops and resources from other fronts and occupied territories to fill the void. Further, the strategic operation in Italy was expected to alleviate hostile military pressure from the cross-Channel invasion of France scheduled for 1944.

On September 3, 1943, 15th Army Group began the invasion of the mainland with Field Marshal Bernard Montgomery's experienced Eighth Army landing on the toe of the Italian boot. Six days later the inexperienced U.S. Fifth Army landed at Salerno with the objective of seizing the vital port of Naples, thus preventing Luftwaffe Field Marshal Albert Kesselring, the German commander in southern and central Italy, from establishing a cohesive front. Kesselring was a determined and charming senior German officer who had the reputation of understanding the importance that airpower played in neutralizing the battlefield. He and his senior commanders now became skillful practitioners of the art of defense, fighting delaying actions by utilizing mountainous terrain and river barriers not suitable for Allied armored operations. The American Fifth Army was under the command of Lieutenant General Mark W. Clark, who Patton described as being self-centered and creepy. Clark had been McNair's protégé when he was picked as chief of staff of the AGF in May 1942. Shortly he was appointed commander of ground forces in England and began forming the American II Corps.

Quick German reaction placed Clark's Fifth Army in a precarious position, facing reembarkation before Montgomery's Eighth Army reached Salerno from the south. Clark almost panicked. This was his first major command after demonstrating abilities in staff work and administration. Was he to face another Dunkirk-like extraction? To avoid this situation, 15th Army Group commander Alexander ordered in reinforcements, including a heavy concentration of naval gunfire, and aerial bombing. The crisis for Clark finally ended on September 16 with the linkup with Montgomery's army. Like Fredendall, Clark had problems dealing with his British allies as the forces moved up the Italian boot.

The speed of Kesselring's reaction and the determined resistance of the German Tenth Army commanded by Generaloberst Heinrich von Vietinghoff apparently surprised the Allies. The mountains ringing Salerno strengthened von Vietinghoff's position. The Germans were also able to funnel forces more quickly by land than the Allies could by sea. Clark sent in the VI Corps' inexperienced Texas National Guard 36th Infantry Division without naval and air support. One of the division's most mobile units was the 36th Cavalry Reconnaissance Troop, commanded by Captain Rush S. Wells Jr. a professional soldier from Mississippi. The troop's mission was to be the eyes and ears of the division. Scouts were deployed to forward positions by regimental commanders, and at times acquired intelligence was reported to the divisional commander, Major General Fred L. Walker. The troop was usually equipped with radio jeeps and machine guns, each crewed by three men, plus M8 armored cars. During World War II, infantry divisions in the Mediterranean and European theaters of operation included a mechanized cavalry reconnaissance troop. Usually the troop contained one command-intelligence platoon and three scout reconnaissance platoons, one for each regiment as an employment option depending on the combat situation. The scout platoons were used for static observation, forward scouting and reconnaissance, and flank liaison. During one forward scouting mission in September, after the breakout from the beachhead Wells was wounded. Observing German tanks ready to counterattack, he then scrambled to call in artillery fire. The attack was warded off.[27]

What soon became evident to the cavalry recon squadrons was that the Italian campaign in general and the terrain in particular were unsuitable for mobile operations. This situation greatly affected the employment of the 1st Armored Division and the two reconnaissance squadrons, 81st and 91st.

By October 1, the vital port of Naples had fallen to the Allies, and the large airfield complex around Foggio was secured for use in extended strategic bombing. During the drive to Naples the 36th Infantry Division experienced difficult fighting. It was their baptism of fire and foretold what was to come as the Allies moved up the Italian boot, with mountains and rivers ideal for defensive operations. Meanwhile, Kesselring planned to retreat slowly northward to link up with Field Marshal Erwin Rommel's Army Group in northern Italy in time to establish a line in the northern Apennines. However, the two German field marshals were at odds over the defense of Italy. Rommel had not felt it was practical to support Kesselring at Salerno or send reinforcements to southern Italy. This was fortunate for Clark. Now Clark and Montgomery had to deal with a classic defensive campaign of delay and attrition warfare brought about by the skill and initiative of Kesselring and his troops.

The Allies grimly fought their way up the deadly Italian boot, with Clark's Fifth Army on the Mediterranean side and Montgomery's Eighth Army on the Adriatic side. Rather than a series of coordinated tactical events at an operational level, Clark instead moved his army up the boot in separate tactical engagements. Hitler, as usual, tried to out-general his generals. He ordered a change in Kesselring's plan for a delaying defense, directing him to stop the Allies south of Rome. Kesselring set up a series of defensive lines, first with the Winter Line and then the Gustav Line. It worked, resulting in stalemate and attrition warfare comparable to the slaughter of World War I. By December the Allies had reached the formidable German barriers shielding the road to Rome.

By January 1, Field Marshal Kesselring (now commander in chief of all German forces in Italy) had his forces deployed in a series of defensive positions providing registered interlocking fire. In addition, river obstacles and the Apennine Mountains overlooking the Cassino bastion augmented German defensive positions. The Gustav Line stretched from the Adriatic to the Mediterranean, with the town of Cassino providing the key pivotal defensive position for the Germans. Clark and his allies were well aware of the German strongpoints as a result of aerial reconnaissance and interrogating prisoners of war. Conjecturally, the cavalry, which provided troopers with a better sense of terrain training than the infantry, could have provided Clark with better intelligence than aerial reconnaissance, which was affected by weather conditions. Here was a lost opportunity. As one general officer later stated, horse cavalry easily could have reconnoitered the mountains and rough terrain. In this environment horse cavalry could have held "its own against any enemy," he believed.[28] Apparently, Clark did not take this into consideration during this stage of the war.

Major General Lucian K. Truscott Jr., a cavalryman, shared this opinion. Truscott was a battle-hardened veteran with considerable experience. He commanded the 3rd Infantry Division in Sicily and then up the rugged terrain from Naples-Foggio to the landing at Anzio. Throughout his career, Truscott lamented the loss of cavalry divisions and their horses. As he was quoted in the *Cavalry Journal*, "I am firmly convinced that if one squadron of horse cavalry and one pack troop of 200 mules had been available to me on August 1 [1943], they would have enabled me to cut off and capture the entire German force opposing me along the north coast road and would have permitted my entry into Messina at least 48 hours earlier." He also believed that had Patton's Seventh Army had a cavalry division, it could have prevented the German escape across the Straits of Messina. Furthermore, he noted, "one good cavalry regiment deployed after the Fifth Army crossed the Volturno River in October 1943 may have prevented the debacle at Cassino and Anzio." As it turned out, Truscott created an improvised provisional cavalry troop organized with pack mules and pack artillery. Though the troopers lacked cavalry training, they provided valuable service and reduced losses as the 3rd Infantry Division moved up the formidable Italian boot. The provisional cavalry troop was disbanded during Lucas's defensive posture at Anzio. According to Truscott, Patton agreed with his views over the need for a cavalry division.[29]

Meanwhile, the 36th Cavalry Reconnaissance Troop moved with the 36th Infantry Division as it struggled through the Winter Line toward the dreaded Gustav Line. During this period the recon troops at times filled a gap in the lines and attacked with the infantry. The troop commander complained about this mission because the cavalrymen had little training as infantrymen. Unfortunately, they had the wrong equipment to deal with German firepower and entrenched positions because cavalrymen were not issued entrenching tools. By the time the cavalrymen reached the Cassino bastion, they were called upon again to perform a new mission that was not included in their training. Though the troopers set up a mobile observation post for the division, the men were now required to establish an observation post line of resistance. This meant the cavalry had to obtain defensive equipment, such as large, bulky entrenching tools to deal with the difficult mountainous terrain. The granite terrain provided no easy task for the mechanized cavalrymen, who were now grudgingly required to dig in. The "sneak and peek" mission now encompassed static defense, resulting in the troopers losing their mobility.[30]

At the operational campaign level General Alexander now set in motion an ambitious plan to deal with the stalemate along the new German defenses. He had, however, the tendency to tactfully suggest rather than command. Alexander proposed to outflank the Gustav Line with a frontal attack along the Rapido-Cassino Front with Clark's multinational Fifth Army and a secondary effort by the British X Corps on his left. At the same time, the British Eighth Army was to put pressure on the Germans on the Adriatic coast in the east. Concurrently, the U.S. VI Corps, under the command of Major General John Lucas, was to deploy to Anzio-Nettuno from the sea behind the Gustav Line and move inland northward toward the Alban Hills south of Rome. The key strategic objective for Clark was crossing the Rapido River and controlling the mountainous Cassino bastion for a move to secure the Liri Valley. This provided Clark the opportunity to break out with his armor over Highway 6. Once the Liri Valley was secured, it provided the gateway to Rome and less compartmentalized terrain, allowing commanders to utilize cavalry elements to monitor German movements and concentrations along the main roads leading to Rome. The valley also offered a suitable route for linking up with Lucas's VI Corps after breaking out of the Anzio-Nettuno beachhead and gave the 1st Armored Division an opportunity to engage retreating Germans. The British X Corps on Clark's left was to attack and cross over the lower Garigliano River. However, this plan turned into a disaster for the Allies because of poor coordination. They were again stymied. Clark's actions, as part of Alexander's proposed campaign, demonstrated that he had not understood the nature of the whole battlefield at an operational level. In spite of what Clark claimed, he also had not understood the importance of coalition warfare. He called it a "hodgepodge army as compared to the tightly knit German armies," thus complicating the whole issue of command and control, communication, and intelligence.[31] Though Alexander was more tactful and unassuming, Clark was more arrogant and publicity motivated. Who would get to Rome first, the British or Americans?

The Liri Valley offered a suitable route for deploying armor and linking up with the forces breaking out of the Anzio-Nettuno beachhead, thus opening the way over Route 6, the most practical road to Rome. Now 1st Armored Division could have the opportunity to exploit and pursue retreating Germans with the possibility of cutting their escape route north. The 13th Armored Regiment under the command of Colonel Howze was to be the leading strike force for exploitation once the infantry established a breakthrough.

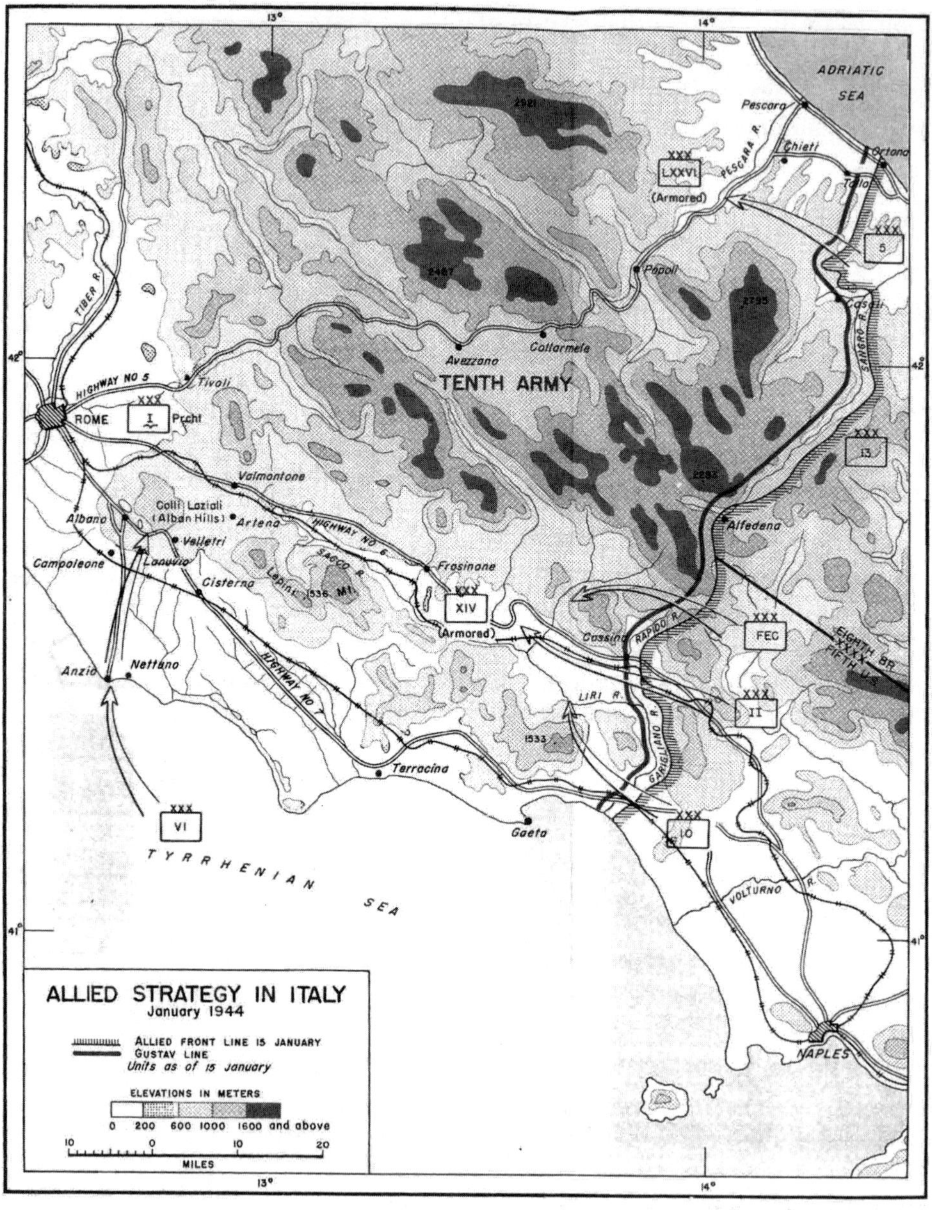

Allied dispositions in Italy, January 1944. Printed by the U.S. Army Topographic Command.

Clark's plan for the frontal assault over the Rapido during January 1944 provided an example of controversial decisions at an operational level that reminded many of World War I's static attrition warfare. The most debated act over the misleading notion of the suitability of attrition warfare occurred when Clark and Major General Geoffrey Keyes, U.S. II Corps commander, ordered the 36th Infantry Division to ford one of the strongest sectors of the Gustav Line, the marshy Rapido River. The problem with Clark's plan was that he used too small a force to engage such a formidable objective. In addition, the 36th Infantry Division had been battered as it moved up the deadly Italian boot.

General Walker insisted that he had good recon intelligence. He tried to persuade Clark and Keyes not to send his division across the Rapido on an impossible mission. Walker was aware the river was swift and forty to fifty feet wide, about twelve feet deep, and with high banks.[32] On the other side Kesselring held the high ground with the veteran 15th Panzer Grenadiers, supported by considerable artillery and interlocking firepower. The Grenadiers were anxiously waiting.

Field Marshal Kesselring later commented that Clark's decision to attack on a narrow front was a mistake because it made it easier to concentrate German artillery. A broad frontal attack, he maintained, possibly could have been more successful.[33] What bothered Walker was that Clark and Keyes had not seen the difficulties in ordering the 36th Infantry Division against such a strong defensive position held by German veterans, especially when considering the topography of the area for the proposed attack. What caused Walker to challenge Clark and Keyes was based not only on his reconnaissance unit but on the experience he gained during World War I when commanding a battalion of the 30th Infantry Regiment, 3rd Division. On one occasion his battalion had dispatched thousands of Germans attempting to cross the Marne River. Now Walker felt that 1918 was about to be repeated, unfortunately this time with his division.[34]

The attack to penetrate the Gustav Line began on January 20 and within a few days had culminated in a disheartening stalemate. The attack was bogged down by the tenacious German defenses and formidable terrain combined with severe weather conditions. The Allies had been severely bruised and were stymied by the determined resistance. The endless cycle of death continued around the Cassino bastion.

Clark finally called off the operation, but not before many of the "T" Patchers had been cut to shreds. Walker wrote in his diary, "Battles are not won by wishing, while ignoring the facts. The stupidity of some higher commanders seems to be profound."[35]

Elements from the 36th Division were flanked by the British X Corps, which successfully crossed the lower Garigliano River near the coast and established a beachhead. The German corps commander became so concerned that he asked Kesselring for reinforcements. He claimed that if the British Corps was reinforced, it could achieve a breakthrough and the entire German front would be rolled up.[36] The Gustav Line was cracked, but Clark and the British Army Group commander had not comprehended the success. The Allied commanders failed to take advantage of cavalry reconnaissance elements that were trained in stealth operations. Later in May when the Gustav Line was finally breached, it was through the same area that the British almost exploited in January and Clark had ignored.

The Rapido affair eventually led to a postwar congressional investigation. Clark argued that it was imperative to attack across the Rapido at the town of Sant' Angelo to pin down the defending Germans so the landing at Anzio-Nettuno could succeed. The chairman of the House Committee on Military Affairs, A. J. May, wrote Secretary of War Robert P. Patterson, "The action to which the 36th Division was committed was necessary . . . General Clark exercised sound judgment in planning. . . . undoubtedly drew the Germans away from our landings at Anzio." Clark responded, "I can only say that under the same circumstances I would have to do it over again." He referred to the 36th Infantry Division as guards and tackles of a football team because they were called to attack and attack hard in conjunction with other divisions on the left and right.[37]

The 13th Armored Regiment commander of the 1st Armored Division, Colonel Howze, viewed the situation at the Rapido differently than Walker. Howze, who in the 1950s emerged as one of a few perceptive officers recognizing the potential of air cavalry, claimed Walker had not deployed his supporting units correctly. Howze's interest in the operation stemmed from his mission to break out with the 13th Armored Regiment and spearhead the move around the Cassino bastion and up the Liri Valley. He had blamed not Clark but the division's execution of its mission. The initial setback at an attempted crossing by the 141st and 143rd Infantry Regiments, with the 142nd Regiment in reserve, was due to the division's reconnaissance patrols and engineers. They had not sufficiently checked for German mines. As a result, the first pontoon for bridging was blown up at the river site before a crossing was attempted. Howze's reconnaissance troops, who were at one of the crossing sites, called one of the bridges "Chocolate Alley" because of the bloody bodies and remains surrounding the site. Many torn bodies quietly drifted down the Rapido. In addition, Howze claimed Walker had not used his tank battalion, the 760th, effectively. Walker had the idle tankers deployed back in a wooded area rather than providing supporting direct fire in a defilade position close to the river crossing. The

36th also had the 636th Tank Destroyer Battalion for support. He believed the Americans had the advantage of supporting artillery and airpower. However, Howze observed that there was a lack of artillery coordination. Since he observed the operation, Howze concluded the attack by the 36th Infantry Division could have succeeded if tactical planning at division staff level had been better executed and fire support coordinated.[38]

General Truscott found Walker intelligent, well trained, and a professional who had a strong attachment for his National Guard Texans, which was reciprocated throughout the division. Truscott, however, had not considered Walker an outstanding battle leader.[39]

Field Marshal Kesselring and other senior German officers who served during the Italian campaign commented years later that the Allies were too preoccupied with limited objectives such as capturing a particular hill. This, Kesselring claimed, gave his forces time to coordinate defenses for the next push. He found American troops predictable because they never advanced beyond their air cover. Regarding the Cassino bastion and the Rapido River crossing, the terrain, he felt, benefited both combatants. The Americans, however, failed to exploit their terrain advantages. After the war, the Germans who opposed the attackers believed the Allied offensive was a failure due to an insufficient concentration of forces in a vital sector, such as the lower Garigliano River. German commanders felt surprised that only one division carried out the Rapido offensive. These comments from the German senior officers indicated that ground intelligence was poorly executed on the U.S. side, and Clark had not understood that a single thrust over a narrow strip gave the German defenders the ability to concentrate all available firepower on one objective.[40] The "sneak and peek" intelligence expected from a cavalry reconnaissance troop failed, as did the leadership, which had not understood the deployment of cavalry as the eyes and ears for higher echelons. What the cavalry could claim was the extensive use of mules to deal with the rough terrain.

What was left of the 36th Infantry Division moved back from the north bank of the river, while the 36th Cavalry Reconnaissance Troop manned a thin outpost. One troop of the 91st Reconnaissance Squadron was detached and added to the division's defenses.

Meanwhile, on January 22 the IV Corps landed on Anzio-Nettuno. The plan was to outflank the Gustav Line and take the pressure off Clark's war of frontal attrition allowing a quick breakout toward the Alban Hills that would hasten the fall of Rome. The corps commander, Major General John P. Lucas, an artilleryman, consolidated his beachhead in spite of the fact that he had landed most of his forces in forty-eight hours without opposition. The Fifth Army commander's reminder of his slugfest agony and near reembarkation at Salerno

Allied situation in Italy, May 1944. Printed by the U.S. Army Topographic Command.

may have swayed Lucas to inaction. Clark advised Lucas not to take reckless chances. Certainly this was due to Clark's "Salerno complex." It gave Kesselring the opportunity to react quickly to deal with the critical situation, thus providing the Germans with valuable time to redirect their tactical situation and prepare counterattacks.

Lucas's hesitation allowed for a continued but frantic enemy buildup, with reinforcements coming from Germany and its occupied countries. One Wehrmacht artillery officer, Siegfried Knappe, who in haste had just arrived at Anzio from Germany, wrote later that early in the landings the German forces would not have had much of a chance if Lucas's corps had attacked with audacity. His reason was that at the time German leadership had created ad hoc units that lacked initial organizational cohesion, heavy weapons, artillery, and tanks to deal with the landings at the beachhead. Knappe was astonished that the Americans just dug in as he looked out to the Tyrrhenian Sea at the Allied ships. He believed an aggressive commander such as Patton could have taken Rome with little effort early in the landings.[41]

The expanded German Fourteenth Army pinned the VI Corps to a narrow beachhead, attempting to push the Allied forces into the sea. In spite of heavy German counterattacks in deplorable weather conditions, the Allied forces held their positions due to the tenacity of the soldiers, timely naval gunfire, and air support. Nevertheless, the 81st ARB from the 1st Armored Division managed to engage in some activities less stressful than reconnaissance and security, which was becoming difficult due to extensive German counterattacks. Company B, after setting eight M8 75mm self-propelled howitzers in defilade along a stream bank and dispatching one to an outpost to serve as an artillery battery supporting the 1st Special Service Force, decided to buy a cow from an Italian farmer and butcher it. Fortunately, the area was somewhat quiet at the time. Eventually the farmer sold all his cows when Italians in the area were relocated to Naples. The 1st Armored Division historian, George F. Howe, related a story about a soldier who was plowing a field using abandoned horses. The German observers close by took a benevolent attitude and let the soldier continue his pastoral endeavor.[42]

Nevertheless, the beachhead was not full of the creature comforts. A detached unit from the 81st became part of a combined arms team that engaged in a small dawn raid. Attached to the 1st Special Service Force along with assault guns, tank destroyers, and tanks, the unit surprised the Germans and returned with sixty-one prisoners but not before losing two tanks. A few days later the 81st was again involved in a night raid with a combined arms team. A blockhouse was destroyed. However, the team and its vehicles ran into a minefield and suffered a number of casualties, including all their armored vehicles, which were eventually retrieved under smoke cover.[43]

The Special Service Force was a combination of U.S. and Canadian forces trained for mountainous operations in all weather conditions. In addition, Ranger battalions were established. They were the brainchild of the hard-fighting cavalryman Lucian K. Truscott Jr. In mid-1942, as American soldiers arrived in Ireland, he modeled an elite force patterned after British commandos, calling them "Rangers." Since they were trained for raiding, the Rangers were lightly equipped. However, as with the mechanized cavalry, they were also used at times as conventional infantry. Unfortunately, two Ranger battalions were ambushed spearheading an attack on Cisterna on January 29 and almost totally wiped out.

Cisterna, which had one major road to Anzio, was a communication center for the Germans. VI Corps decided to send the Rangers, followed by Truscott's 3rd Infantry Division, to secure the town. This move could deprive the enemy of one of the most important logistics routes to the Gustav Line, Germany's primary line of resistance. One historian who analyzed the operation argued that the timid Lucas failed to plan properly by not taking the initiative and seizing critical terrain, such as the small town of Cisterna. In addition, Truscott's division was not concentrated but spread too thin. This situation limited firepower and mass for a concentrated effort. The plan that doomed the Rangers was Truscott's idea to send two battalions of Darby's Rangers, the main attacking force, along a single infiltration route rather than infiltrating along multiple routes. The Germans were aware of the operation before the attack was launched, and thus had the opportunity to set up various killing zones. Many Rangers died and others were marched off into captivity and paraded through Rome.[44]

General Lucas's faintheartedness brought into question his qualifications for continued command. As a result, he was relieved and replaced by the more forceful General Truscott. In fairness to the departing Lucas, the Anzio landing was conceived on too small a scale and was limited by logistical demands for sea transportation and accompanying air support for the upcoming amphibious landings in Normandy, France. In addition, Allied air support was overextended because of the intensity of a battle of attrition at the Cassino bastion and Garigliano bridgehead. As it turned out, none of the primary objectives were achieved. In retrospect, the landing at Anzio might have been more effective after a breakthrough on the southern front had it been achieved.

In the meantime, Alexander, the Allied Army Group commander, attempted to relieve the pressure at Anzio by again mounting a series of costly frontal battles along the Gustav Line around Cassino, where the medieval Benedictine abbey was the pivotal point for making progress toward Rome. Alexander was pressured by the New Zealand division commander, Lieutenant General Sir Bernard Freyberg, to obliterate the monastery. Doing so had not neutralized resistance because German troops did not occupy Monte Cassino, only the ridge below it. Alexander, with Clark's reluctant concurrence, authorized its saturation bombing beginning on February 15. The bombing turned the abbey into rubble. Though the enemy's defenses around the Cassino bastion were temporarily ruptured, the Allied ground forces hesitated and did not immediately attack their objectives. The Germans now had time and excellent cover due to the prevalence of rubble. They were again able to repel the attackers. Once more intelligence operations on German dispositions had failed.

The tactical stalemate at the Cassino bastion continued until May. With sufficient reinforcements and benefiting from an increase in intensive Allied air interdiction, Alexander was able to concentrate most of his forces on his left flank for an overwhelming offensive designed to destroy the right wing of the German Tenth Army and the recently formed Fourteenth Army. This was the same area where in January the British X Corps cracked the Gustav Line, but it was not reinforced. Now it was mass versus velocity. Alexander's full-scale frontal attack, supported by massive artillery, struck between Cassino and the sea, surprised the defenders, in part because the Allies continued to gain intelligence derived from decrypted signals from the German's Enigma. In addition, the Germans lacked normal air and ground reconnaissance. After many grueling and bloody months, a linkage with the VI Corps at Anzio had finally occurred. As the Allies broke out of the Anzio beachhead, the 81st ARB, as the 1st Armored Division's reconnaissance team, continued to make combat sorties. On one occasion, A Company cut a road ahead of the division's advance, hauling in a number of prisoners. The door to Rome was opened, presenting the opportunity for the capitulation of German forces, especially the Tenth Army. The advance to the Eternal City, however, led to another operational controversy that entangled coalition leadership with Clark and Alexander at the center stage.

This controversy began after the 3rd Infantry Division spearheaded the breakout from the beachhead. After linking up with the southern force, General Alexander ordered Clark's Fifth Army to turn aggressively against the German bridgehead at Valmontone on Highway 6 just east of the strategic Alban Hills. Instead, Clark shifted his axis at Velletri on Highway 7 just south of the Alban Hills, directing his main body northwest toward Rome. This change in direction gave the Germans time to check the advance. After a tenacious defense by the Germans at the towns of Valmontone and Velletri, they broke contact and withdrew north. Clark's reason for changing the axis was that the Germans had too many roads parallel to Highway 6, the main road to Rome. This situation, he claimed, benefited the Germans, and he did not have the forces to deal with encircling the Germans at the Valmontone axis. With this excuse one could ask Clark: What were your ground-air intelligence elements doing?

Clark's problem of dealing with coalition warfare also showed its ugly head. Clark asserted, "He [Alexander] always had to be careful about making it possible for the British Eighth Army to get an easy victory" at the expense of the Fifth Army. He chastised the British for poor leadership and a lack of aggressiveness. "Everybody was anxious to get to Rome or take part in its capture," he stated. Calling Alexander one of his best friends, he nevertheless told him that if the British Eighth Army wanted to take part in the capture of Rome, he would have his troops fire on the Eighth Army.[45]

General Truscott viewed Alexander as an outstanding leader among the Allies, believing he had the personality of Montgomery and Patton but not their flamboyancy. Although Truscott felt both Clark and Alexander lacked experience in high command, he respected Alexander's intellect and diplomatic skill.[46] Truscott considered Clark an able tactical commander. However, he disagreed with Clark's decision to change the main attack from the Valmontone axis and wanted to personally discuss the change of direction with Clark. At the time, the Fifth Army commander was unavailable. Truscott believed the momentum toward Valmontone was sufficient for exploitation. Since he had received an order from Clark, he changed the axis of his corps. General Alexander's plan to block the escape of the German Tenth Army north was doomed on May 26 when Clark made his fateful decision. For convenient political reasons, he avoided the battle of annihilation of the Tenth Army for a geopolitical objective, Rome. Patton, who was also a consummate actor, had a point when he earlier called Clark self-serving. Clark's Rome decision allowed most of the German forces to retreat and avoid annihilation only to fight another day. Prime Minister Churchill called Clark's political actions unfortunate. Alexander later chastised Clark for not executing his military plan for destroying German forces south of Rome. A German panzer corps commander at the time was also critical of Clark for not sticking to Alexander's plan. There was nothing at the time, he noted, to prevent defeat of the retreating German forces.

On June 4 Clark, along with his usual entourage of correspondents and photographers, entered Rome in a jeep, two days ahead of the massive cross-channel Normandy landings. France now became more important in the strategic effort to defeat Germany. Truscott recalled that Clark's triumphal entry into the Eternal City was his typical method of operation, which was fueled by a desire for publicity. Truscott noted that it "was his greatest weakness." He later commented, "The quality of Clark which everybody knew about was his love for publicity." This, Truscott recalled, may have hampered the debonair commander's ability to see or feel the whole battlefield. When Clark visited Truscott's command post, he usually arrived with correspondents and photographers, not to mention his key public relations officer.[47]

Clark, however, praised the 36th Infantry Division as one of his best, and reasoned that was why he used it in the Anzio breakout. Again this praise was self-serving rather than courageous in light of the Rapido debacle. In the meantime, the 36th Cavalry Reconnaissance Troop finally had the opportunity to function as a recon unit for the 36th Infantry Division, which no longer retained its Texas identity due to casualties and replacements. Attached to an engineer combat regiment, the cavalrymen moved through the key town of Velletri along the Appian Way (Highway 7) toward Rome. Captain George K. Fell, the troop commander, radioed that the town was cleared of Germans. Fell reported that for the first time since the landing at Salerno the cavalrymen were used for mechanized reconnaissance, which "we have been thoroughly trained for and qualified in." As the division raced toward the Eternal City in early June, the cavalrymen set up a prisoner of war camp and mopped up stray Germans.[48]

In the meantime, the 1st Armored Division with supporting artillery advanced toward Rome. The 81st ARB deployed as a security force on the flanks. When the division reached the Eternal City, reconnaissance elements were the first to enter, capturing the chief of police. The elated chief told the cavalrymen that there were no other Allied troops in the city.[49] That was some consolation for the 36th Infantry Division. The 36th Cavalry Reconnaissance Troop led Walker's Texans through Rome, naturally after Clark's triumphal entry in a jeep accompanied by a film crew to capture the event.

In July the 1st Armored Division was reorganized as a light force, eliminating the regimental organization in favor of battalions for greater flexibly. General Harmon left the division to again take command of the heavy 2nd Armored Division. Major General Vernon E. Pritchard, a cavalryman, arrived from the United States and took command. On July 24, 1944, the 81st ARB was redesignated the 81st Cavalry Reconnaissance Squadron (CRS). The reorganization was more in line with *1943 Field Manual 2-30, Cavalry Field Manual: Cavalry Mechanized Reconnaissance Squadron* that replaced the *1941 Field Manual 2-14*. The mission of cavalry was redefined as reconnaissance, security, screening, and fire and maneuver. The manual specified that the cavalry was to engage in combat only when necessary. The division CRS was assigned as an organic element of the division, and the corps CRS was attached from army to corps.

For the next three months the cavalry squadron, however, was employed in two conspicuously different missions compared with other reconnaissance elements of armored divisions and not as prescribed in *1943 Field Manual 2-30*. First, for the remainder of the operations in Italy, the squadron was deployed as an economy of force element. As before, the cavalrymen were not employed in the usual piecemeal operations, as became the practice of other armored divisions. The mechanized cavalry was deployed as reconnaissance and security elements with combat teams or task forces as determined by the division commander. General Pritchard employed his cavalry squadron as a unit, assigning it specific areas of responsibility to reconnoiter in front of or on the flanks of his division. Second, the squadron deployed as dismounted infantry during the drive toward the rugged mountain barriers of the Apennines. To get into the Po Valley it was necessary to control the unbroken mountainous chain of ridges and peaks. The new deployment was determined by these conditions along with determined German defensive operations. As the Germans retreated, they used a wide range of demolitions to block their withdrawal routes, thus preventing the cavalry from using its vehicles. The cavalrymen resorted to aggressive dismounted patrolling, securing prisoners and ambushing retreating Germans. On another occasion when engaged in firefights, the troopers relied on their assault guns. Toward the last months of the campaign the squadron again, in accordance with an armored reconnaissance squadron, mounted their light armored vehicles and engaged in mobile tactics pursuing the enemy.[50]

During this period of three months, the 81st CRS scrambled over mountains and rivers dismounted. At times the squadron used pack mule detachments to carry heavy loads over mountainous terrain. One study noted "that over the trails used, a man could carry a load of approximately fifty pounds while a mule was capable of carrying approximately two hundred pounds." The problem with dismounted action was that the squadron had equipment that was not adequate, such as the .30-caliber carbines. The cavalrymen would have preferred the M1 Garand used by the infantrymen.[51]

Meanwhile, back on August 15 and over the objections of Alexander and Clark, the campaign in southern France began. It was designed to protect the Allies' flank as they raced through France toward Germany. For the landings to be successful and to prevent Hitler from withdrawing troops to shore up his crumbling western and eastern fronts, it was necessary to contain as many of their troops and supplies as possible in Italy.

After the landings in southern France, the Italian campaign was reduced to a major diversion and a challenge for the remaining experienced Allied troops. Even Clark, in his book *Calculated Risk*, wrote that after the landings in France, Italy became the "forgotten front."[52] He saw no need for an invasion in southern France. It was Clark's strategic plan after the fall of Rome to strike through northeast Italy at Austria and Hungary, thus providing flank protection for Josip Broz Tito, the Yugoslavian partisan resistance leader against the Hitler regime. To do so he had to keep all the divisions in Italy marked for the invasion of southern France, including the 36th Infantry Division. What was really on Clark's mind? In an interview after the war he admitted he longed for a new command. If it was southern France, then it fit conveniently with his ambitions. When involved in the planning for the invasion, his reaction was "What the hell? Here I am up to my neck in a tight situation in Italy and this Southern France situation comes up."[53] He was in a tight situation, and it was his campaign to resolve. In December 1944, the flamboyant Clark replaced Alexander as commander of the Fifteenth Army Group. General Truscott took over command of the Fifth Army. The Germans continued to take the situation in Italy seriously, putting up strong resistance by establishing a series of defenses along the Apennine Mountains and the Po Valley, which had the finest road nets in northern Italy. Nevertheless, it still was a deadly struggle up the Italian boot until the war ended. General Vietinghoff, depleted of heavy tanks and vehicles and ammunition, made sure German defenses were significant enough to salvage his forces. He had to deal not only with this situation and advancing Allied forces but also with a major eruption of Italian partisan activities in his rear areas. Shortly, however, the war in Italy ended. Even today, the decision to accept Churchill's "soft underbelly" strategy by moving up the Italian boot is controversial.

Field Marshal Alexander (promoted July 4, 1944) had no doubt that the plan to invade Italy was a wise strategic decision. He believed it forced Hitler to fight on three fronts, thus adding to Germany's logistical problems and diverting manpower strength needed on other military fronts, in addition to drawing off troops essential for security duties in occupied countries. The Italian campaign also retained some of the Wehrmacht's most gifted combat commanders. When the French invasion was launched, Alexander claimed fifty-five German divisions and their required logistical resources were tied down in the Mediterranean theater, thus causing fewer Allied casualties when western Europe was assaulted. By the end of April 1945, a million Germans had marched into captivity. Both Clark and Alexander continued to support the Italian campaign because it could have placed Allied forces in the Balkans, thus avoiding a Red Army takeover. Alexander wrote in his memoirs, "The Mediterranean might have been the most decisive of all in post war history."[54] The Italian campaign, however, had little impact on the outcome on the eastern front.

The 36th Cavalry Reconnaissance Troop was deployed through the rest of the war more for reconnaissance and security, and with its elements employed intact. For example, on September 9, 1944, the 36th Infantry Division assaulted southern France, and the 36th Cavalry Reconnaissance Troop again became the eyes and ears of the division. One proud cavalryman recalled liberating a .50-caliber machine gun from an American B25 shot down in Italy and placing it on his M8 armored car because it had a higher cyclic rate of fire than the .50 calibers issued to armored cars. The cavalrymen were looking for more firepower. The weapon was mounted on the turret and saw action in France, where the cavalrymen were deployed to find and fix the enemy for the follow-on main force. On one reconnaissance mission the troopers ran into an ambush on a curve in the road and suffered a number of casualties when confronted by a 20mm German antiaircraft gun. It was becoming more and more evident that M8 armored cars were not suitable for mechanized combat at troop level. They did not have the combat firepower to fight for information. [55]

Colonel H. J. Samsel, the commander of the 117th Cavalry Reconnaissance

Squadron attached to the VI Corps, also found the .50 calibers useful. His cavalrymen used incendiaries because the rounds were more effective than the .30-caliber machine guns and 60mm mortars in dislodging the enemy. When the rounds hit a hard surface or hard ground, they made a sizable explosion. When used in underbrush or tall grass, the colonel noted, the incendiary rounds caused small fires and smoke, "which seem to have a definite psychological effect." The 117th found the M8 75mm howitzer unsuitable for the squadron because of its limited bursting radius. Before the squadron moved to southern France, the M8s were replaced with the M7 105mm howitzers. This was not a table of organization change but a local adjustment. The squadron now had greater firepower and bursting radius, plus the advantage of massing firepower when supporting artillery was not available. The cavalrymen were happy to receive the new M24 light tanks replacing the M5s. The M24s with their 75mm tank cannons were used as direct assault weapons when forward reconnaissance units needed to deal with enemy resistance. However, it was noted that dismounted action was more often than not the rule. The squadron commander stated the cavalrymen operated more effectively as an economy force and did not like the idea of his troops being broken up. He commented, "The most satisfactory results have been obtained when the Troop operates as a unit."[56]

The 117th Cavalry Reconnaissance Squadron also had an interesting experience working with the Office of Strategic Services (OSS), the principal intelligence organization serving in all theaters of military operations. During the Italian campaign, OSS elements worked with the military and conducted missions throughout central and northern Italy, supporting resistance activities. A number of OSS intelligence personnel speaking native languages were attached to the 117th Cavalry Reconnaissance Squadron and worked closely with the squadron's S-2, providing information about enemy activities. They operated with forward units and melted into the native population, recruiting local Italian patriots and securing valuable information on enemy dispositions, strength, and possible intentions. The OSS operatives were also able to identify pro-German fascists. Captain R. G. Melragon, the S-2, found the OSS working in small groups was able to gain considerably more enemy intelligence than forward cavalry reconnaissance elements. Melragon noted that the relationship with OSS was valuable because the squadron was able to provide higher headquarters with a greater amount of key intelligence.[57]

The 91st Cavalry Reconnaissance Troop from the 91st Infantry Division, which entered combat as a whole on July 13, 1944, north of Rome, adopted the unofficial motto "Sneak, Peek, and Retreat." Trooper M. Carl Ramga recalled that his unit's mission was based on the old cavalry idea of "Go lightly armed, move fast, probe deep into enemy territory, and get out as best you can." He

M8 armored car with the 37mm gun serving perhaps its most useful combat purpose: a portable clothes line. Courtesy of M. Carl Ramga, 91st Cavalry Troop, 91st Infantry Division.

noted that recon troopers were ordered not to fight or purposely engage enemy units. Ramga claimed his unit was specially trained and good at "prisoner acquisition." Even though the cavalrymen worked closely with the infantry, they complained that they were denied the coveted Combat Infantry Badge. The 91st Recon Troop found the M8 armored car's 37mm gun to be a useful portable clothesline as well as an effective "paint scratcher." Toward the end of the Italian campaign, three 91st recon platoons with M8 armored cars were used to work the point, leading a task force consisting of an infantry battalion, a dozen tanks, six tank destroyers, and medical personnel. When the mounted troopers drew heavy enemy fire, they identified the source and let the tanks and tank destroyers clear the way for the infantry. The recon platoons captured more than twenty-seven hundred German soldiers in five days.[58]

Mechanized cavalry at troop level was critical in finding and fixing the enemy for follow-on combined arms task forces. When it came to deployment of cavalry units higher than troop level as an economy force, they required more firepower and additional attached units for forming a task force. No doubt, the problem during the Italian campaign was that the highly compartmentalized maneuver corridors prevented mobile forces from spreading out. Many high commanders usually attached reconnaissance elements to combat commands, whereas cavalry reconnaissance troops usually remained with their assigned infantry division. For more firepower, troops were reinforced with other units as the division commander saw fit due to the exigencies of combat.

Perhaps higher-echelon commanders, such as Fredendall, Lucas, and Clark, had not comprehended that leadership meant the art of command, and that their leadership affected the spirit of the army down to the lowest unit. As the dynamics of the battlefield changed, they had not planned sufficiently for adjusting at an operational level as Kesselring noted. This, in part, was due to a lack of critical and timely reconnaissance and intelligence that a stealth cavalry unit may have provided. Clark and Lucas and a few divisional commanders displayed examples of questionable leadership that exhibited reluctance regarding when and how to make adjustments. Furthermore, they had not understood or heard the tactical needs of subordinate units. Neither had they understood the mission of mechanized cavalry as an economy force for high command to use discriminately so as to provide timely information ahead of the bulk of the main force, especially after the Cassino bastion was neutralized. Only through proper ground intelligence provided by mobile cavalry could high-echelon commanders, such as Clark and Lucas, set the stage for major operations to determine the enemy's center of gravity or to act at a decisive point. The cavalry had the ability to find the enemy, develop the situation, and provide commanders with reaction time and security. It was evident that there was no excuse for attempting to use cavalry as an economy of force element discriminately to find and fix the enemy. Perhaps Clark and Lucas viewed the terrain as too difficult for such cavalry operations. Nevertheless, good leaders know how to change and adjust. Since the beginning of operations in North Africa and in the Mediterranean theater, the Cavalry School dogma of stealth reconnaissance was challenged as not workable under the current table of organization and equipment.

Historian Louis D. DiMarco, in his extensive study of mechanized cavalry in World War II, convincingly argued that doctrine of reconnaissance and security, as defined in manuals, focused on the tactical level rather than the operational level. He maintained that at the operational level doctrine had not articulated the role mechanized cavalry played as an economy force.[59] Battlefield experience also brought to the forefront the dispute over missions: the stealth "sneak and peek" method of reconnaissance or the mission to fight mounted or dismounted when important information was to be gained.

The cavalry reconnaissance squadrons or regiments deployed in the Mediterranean theater and Italy were frequently detached for service either with combat commands or with infantry regiments. Only later, after the Anzio breakout, was cavalry deployment changed emphasis to fighting dismounted instead of being used as an ad hoc stepchild for higher-echelon commanders. Previously, cavalry missions primarily emphasized stealth reconnaissance and security, and less often offensive and defensive combat. Most important, the emphasis was placed on "sneak and peek," as Hoy explained in his article for the *Cavalry Journal* in November–December 1943. After the North African campaign, Hoy was assigned to the staff and faculty at the Cavalry School, where he taught the importance of stealth through speed and maneuver. This traditional pre–World War II approach taught at the Cavalry School conflicted with the views in General Scott's article that appeared in the same journal the year before. Hoy's ideas were in keeping with the glorious cavalry maneuvers of the Civil War. Conversely, Scott believed in a more extended mission for cavalry based on his relationship with Van Voorhis and Chaffee, and his experience as commander of the 13th Cavalry (Mech) at Fort Knox, and later with the Armored Force. Scott stressed "offensive action," meaning to fight for information. This reflected the mechanized force's attitude during the 1930s at Fort Knox, an outlook that carried over to the Armored Force. This meant that cavalry must be trained to fight and equipped to carry out combat missions, especially when conducting long-distance reconnaissance against strong opposition or engaging in exploitation and pursuit.[60]

Colonel Hoy saw Scott's position differently. He agreed that the cavalry squadron required sufficient firepower. However, he was concerned that weighing down the squadron with too many tanks, 81mm mortars, and 37mm assault guns would make it "too unwieldy." This, he argued, compromised command and control. "If I get the armored car, then I don't want the tank," he added.[61] Hoy also stated that reconnaissance by fire should not be used ungoverned. He believed that a reconnaissance unit should not fight for information but acquire information through the technique of "sneak and peek." His experience, he claimed, was confirmed by actions of Major J. Lawson's squadron from the 11th Hussars, a British armored car cavalry regiment.[62] During the North Africa campaign, the 11th Hussars' mobile patrols were noted for penetrating behind enemy lines and reconnoitering the rear, thus providing timely intelligence for future operations by the main force.[63] It was interesting that the 11th Hussars, whom Hoy admired, went to war with Rolls-Royce armored cars of World War I vintage.[64]

Thus, the deployment of cavalry in North Africa and Italy at battalion or squadron and troop levels brought out the disagreement over the new meaning of cavalry. It was apparent that the mechanized cavalry doctrine of basic reconnaissance had not fit the battle conditions experienced during the North African and Italian campaigns. Mechanized cavalry experiences were inconsistent with established doctrine. One school, the "eyes and ears" traditionalists, adhered to stealth reconnaissance, security, and avoidance of combat if possible. The other school, the "fighters," maintained that the mission of cavalry was much more than just reconnaissance and security. They believed mechanized cavalry needed to be more versatile and equipped to fight both mounted and dismounted. The two-school debate also brought to the forefront the issue of light versus heavy, and the appropriate reinforcements and mix of equipment for mechanized cavalry. Most important was the issue of certain high command leadership, which was never able to reconcile the nature of cavalry as a pure reconnaissance and security force or an economy force capable of fighting both mounted and dismounted.

8

Mechanized Cavalry from Normandy to the End of the War

In this chapter the focus will be on leadership and on several organizational levels of mechanized cavalry. It will also emphasize changing doctrine, technique, and missions caused by adjustments made as a result of the chaos, tensions, and demands of warfighting. It is not intended to be a chronological history of mechanized cavalry operations in the European theater of operations (ETO) but an examination of the demise of traditional reconnaissance stealth "sneak and peek" doctrine and its replacement by one placing emphasis on defensive and offensive combat, special operations, security, and fighting for information with combined arms elements called task forces.

The armored divisions that entered the ETO were designed to spearhead the infantry and support the accomplishment of its fundamental mission to destroy the enemy's land forces and seize and hold ground. The armored divisions were the basic element of the Armored Force created by Major General Adna R. Chaffee Jr. to provide high mobility with a combined arms combat command and to provide protective firepower during rapid offensive action in deep operations. Once committed to combat, armored divisions took on the cavalry mission of pursuit and exploitation. However, the establishment of the Armored Force had created perplexing problems in organization and tactical doctrine for future cavalry and its role in mechanized warfare.

Major General Robert W. Grow, who commanded the 6th Armored Division during the war, commented that cavalrymen faced two obstacles. First, from time immemorial cavalrymen have contended that for reconnaissance to be effective, units must fight and survive to accomplish their missions on the battlefield. Second, he noted, with "the physical limitations of machines it became difficult to build greater mobility into a reconnaissance vehicle than a combat vehicle and at the same time give the reconnaissance vehicle the fighting power it needs." During World War II reconnaissance units had problems operating freely in front of the main body unless given sufficient armor and armament to fight for information. This situation, Bob Grow concluded, left commanders of higher units advancing with their heaviest fighting power in the lead and relegating reconnaissance elements to flank and liaison duties on many occasions.[1] Grow's basic philosophy was that the mission of cavalry was to fight, and as the dynamics of warfare change, the organization and equipment for mechanized cavalry required changes to meet the evolving realities of the modern battlefield. However, he noted, the role of cavalry remained constant, only the mechanism had changed.

Major General John Herr's *1941 Cavalry Field Manual 2-15 (Employment of Cavalry)*, *1943 Cavalry Field Manual 2-30 (Cavalry Mechanized Reconnaissance Squadron)*, the *1943 Cavalry Training Circular 107 Field Service Regulation,* and the *1944 FM 100-5* did not take into consideration changes in cavalry operations as a result of mechanized warfare. The idea was there, but traditional doctrine hindered the adoption of new ideas. Only in response to the experience of modern warfare were changes made in doctrine.

In 1943 reconnaissance was emphasized as the basic mission of all mechanized cavalry in *Training Circular 107*. By the time of the invasion of Normandy in 1944, the bulk of mechanized cavalry was deployed in the ETO and trained as a light, mobile force. The prescribed mission, tactical doctrine, and technique for the mechanized cavalry involved reconnaissance and security with a minimum of fighting. The *Field Service Regulations 100-5: Operations (1944)* stated: "Mechanized cavalry units are organized, equipped, and trained to perform reconnaissance missions employing infiltration tactics, fire, and maneuver. They engage in combat only to the extent necessary to accomplish the assigned missions."[2]

The 1944 *FM 100-5* also outlined the mission for horse cavalry, which "obtains its best results by the rapidity and flexibility of its methods in attack. It should be assigned missions which permit exploitation of its mobility." The manual—no doubt to the delight of General Herr—pointed out "horse cavalry habitually maneuvers mounted, but ordinarily fights on foot." This was a reflection of the traditional nineteenth-century way of warfighting for the horse cavalry. Expected to complement the horse cavalry were "weapons similar to those of the infantry and armored cars for reconnaissance, and motor transport to supplement its animals transportation for supply."[3] No doubt this 1944 regulation pleased the deposed Herr. However, with the creation of armored divisions and the mechanization of horse cavalry, the utility of the horse was lost, as was its historical tradition. It no longer met the requirements of the battlefield. The mission of armor was exploitation, pursuit, and shock action. Army Ground Forces' tactical manuals expected mechanized cavalry to fulfill its mission of reconnaissance and security ahead of the main force with a minimal amount of fighting. As the war progressed in Europe, this tactical role, however, did not find favor with corps and division command leadership.

The preferred role of mechanized cavalry, which goes back to the birthing period at Fort Knox during the 1930s, was to fight mounted with new mechanical mounts. This doctrine questioned the *FM 100-5* tactical mission of "sneak and peek." The rationale to fight with a self-contained mounted force was established at Fort Knox. It was demonstrated that the inherent mobility of the mechanized cavalry made it capable of executing missions on the fringe of battle better than any other ground force. The cavalry, Grow explained, needed to be equipped and trained to carry out all types of combat missions. He indicated that during the formative years the basic training mission of cavalry at Fort Knox was to fight and not engage in what cavalrymen called "sneak and peek" reconnaissance, which was emphasized in training directives. Nevertheless, for cavalry reconnaissance to be effective, it must include flexibility, fighting power, and the ability to survive on the modern battlefield, Grow contended.[4]

During World War II, the mechanized cavalry was not organized, equipped, and trained to carry out all types of combat missions, as Grow had advocated. Doctrine, he noted, set forth a narrowly defined cavalry mission, which was stealth reconnaissance and security. No sooner was mechanized cavalry deployed to France than it became obvious that doctrine was not suitable for the European battlefields. The problem was that mechanized cavalry was dispersed to three levels of command: a separate cavalry group of two squadrons usually assigned from army command to corps, mechanized cavalry reconnaissance squadrons and armored reconnaissance battalions assigned to armored divisions, and finally a mechanized cavalry reconnaissance troop assigned to infantry divisions.

When cavalry squadrons landed in Normandy, they consisted of four reconnaissance troops lettered A, B, C, and D mounted in M8 armored cars referred to as Greyhounds and jeeps, or what the cavalrymen liked to call Bantams. The letter D is skipped in corps cavalry. Next were the more heavily armed cavalrymen, lettered E and F Troop. E Troops, the assault gun troop, was mounted on 75mm self-propelled howitzers, while F Troop, the light tank troop, was mounted in M5 light tanks armed with a 37mm cannon and machine guns. The mounted reconnaissance troops, the tank troop, and the assault gun troop were trained to function as a reconnaissance team to gather information for higher units. The remaining troops were service and medical detachments traveling in jeeps, half-tracks, and trucks to provide logistical support.

The M8 armored cars wind their way through houses and streets blasted by artillery fire. Courtesy of the Patton Museum.

The mission statement for a mechanized cavalry squadron initially prescribed reconnaissance and security, flank protection, maintaining contact with adjacent units, and engaging in active patrolling both mounted and, if necessary, dismounted. Generally, all cavalry reconnaissance squadrons began operations after the landing in Normandy with fifty-two M8 and M20 utility armored cars, 106 ¼-ton jeeps (Bantams) as the basic vehicles for reconnaissance and command and control. In support of the armored cars and jeeps were M8 75mm self-propelled howitzers on light tank chassis, six for corps cavalry and eight for division cavalry, as well as thirty-two M3A1 half-tracks, seventeen M5 light tanks, mortars, and mounted cavalrymen armed with M1 carbines, .45-caliber submachine guns and .30- and .50-caliber machine guns, .45-caliber pistols, antitank rocket launchers, and, for logistic support, twenty 2½-ton trucks.[5]

The cavalrymen were expected to fight mounted in either jeeps or armored cars. Jeeps were equipped with a single .30-caliber machine gun mounted on the dashboard and operated by a trooper sitting in the passenger seat. At times .50-caliber machine guns were mounted in a four-jeep mortar platoon consisting of two recon jeeps and two 60mm mortars. In order to maximize mobility and versatility, jeeps compromised on armor. The M8 armored car mounted a 37mm gun in a movable turret and a .30-caliber machine gun. The M20 armored car was designed as a utility vehicle armed with a .50-caliber machine gun on a ring mounted over the open-top crew compartment. Both jeeps and armored cars could obtain speeds over fifty miles per hour on suitable roads. For heavier support each mechanized cavalry squadron had a light tank troop with M5 tanks, which had speed and maneuverability but lacked sufficient armor and firepower to engage heavier German tanks and assault guns. Each squadron had an assault gun troop armed with snub-nosed 75mm howitzers mounted on a light tank chassis. With this type of equipment the three cavalry elements—reconnaissance troops, the assault gun troops, and the light tank troops—operated as a team. The equipment authorized for the mechanized cavalry was designed for reconnaissance, such as to seek and observe the enemy, and not to engage heavy opposition when encountered. This table of organization and equipment reflected light cavalry. Once the mechanized cavalry entered the war in western Europe, however, it found out through the chaos of uncertainty of warfighting that neither the 1944 *FM 100-5* nor *Training Circular 107* was appropriate for an ever-changing battlefield environment, nor was their organization and equipment.

Soon after the landings in Normandy, V Corps commander, Major General Leonard T. Gerow, in Lieutenant General Omar N. Bradley's First Army, tried to organize his mechanized cavalry as outlined in the 1944 *FM 100-5*. He believed his cavalry, unlike armored divisions with more firepower, was not equipped for strong attacks and its vehicles, such as Greyhounds and M5 light tanks, were too vulnerable to antitank weapons. It was expected that once the enemy was found and fixed, the cavalry mission was over and higher units with more firepower and mass would then engage the enemy. The cavalry was then expected to probe for gaps through which it could operate on the enemy's flanks. This was a textbook-prescribed mission. However, the Normandy Bocage country, with its hedgerows and chopped up terrain, brought into question the prescribed cavalry mission. It proved to be a difficult and costly mission to overcome.

The 106th Cavalry Group commanded by Colonel Vennard Wilson soon discovered that its organization, equipment, and mission were not suitable in

The 113th Cavalry Reconnaissance Squadron ready to fire its M8s against retreating Germans near Heure la Reman, close to the Dutch border, September 9, 1944. Note the hedgerow cutter on the nearest M8. Courtesy of the Patton Museum.

the dense hedgerows of Normandy. The unit's lineage went back to July 1897, when it was constituted in the Illinois National Guard as a cavalry squadron. In September 1940 the 106th was horse-mechanized in the Illinois National Guard. In March 1944 elements of the 106th Cavalry were reorganized as the 106th Cavalry Group (Mech) with the 106th and 121st Cavalry Reconnaissance Squadrons attached.

The initial missions of the 106th Cavalry Group were not difficult, providing contacts between infantry divisions moving forward, and then closing the neck of the Normandy peninsula. These missions were typical, following the applicable manuals, meaning reconnaissance and security, and flank protection for higher echelons, usually corps and army, as prescribed in an operation memorandum. After encountering little resistance, the mechanized cavalrymen engaged mainly in patrolling and providing artillery support with their M8 75mm self-propelled howitzers. Once this mission was completed, information was passed to higher headquarters, such as the division to which the squadron was attached or to the corps commander through the cavalry group commander. The main means of communication was the radio-equipped M8 armored cars stationed near the division's command post. It was planned that certain cavalry groups were to be used by the corps commander to establish contact between the separate divisions, protect exposed flanks, provide security, and defend if necessary the corps' command post.[6] Soon the cavalry group found itself engaged in dismounted night patrolling to find and fix an elusive enemy in the Bocage. Before long the cavalrymen from Illinois began to place hedgerow cutters on their M5 light tanks.

Suitable hedgerow cutters were invented by Sergeant Curtis G. Culin Jr. of the 102nd Cavalry Reconnaissance Squadron, 102nd Cavalry Group. He successfully welded steel scrap from a destroyed enemy roadblock to a M5 light tank. Upon inspection by General Bradley and his staff, orders were issued to ordnance units to mass-produce the hedgerow cutters. Meanwhile, a young lieutenant colonel, Cyrus A. Dolf III, took command of the 102nd Cavalry Group after its poor performance in Normandy. Dolf, a graduate of West Point with a commission in the infantry, was one of Fort Benning's tankers and the only noncavalry officer to command a cavalry group during the war. It did not take long before Dolf's remarkable leadership had an effect on the cavalry group.[7]

As the war became more of a mobile pursuit after the breakout in Normandy and the race to Paris and the Seine River, the mission for mechanized cavalry began to change. This brought about a refining of doctrine. To understand how mechanized cavalry was employed and underwent changes inconsistent with *1944 FM 100-5*, it is important to understand doctrine, technique, deployment, and leadership at three levels: cavalry group, cavalry reconnaissance squadron, and cavalry reconnaissance troop. First a few cavalry groups will be examined to illustrate how they were originally employed, how they underwent changes, and the effects of leadership adjustments to change. Five cavalry mission types will be analyzed:

Reconnaissance: The "sneak and peek," or observation and information,
Security: Blocking actions, screening both moving and stationary, flank protection, and liaison contact between friendly units,
Special operations: Operating as information service for army level, acting as mobile reserves, security, and control of rear areas,
Offensive combat: Attack and pursuit and exploitation, and
Defensive combat: Delaying action, holding key terrain for follow-on units.

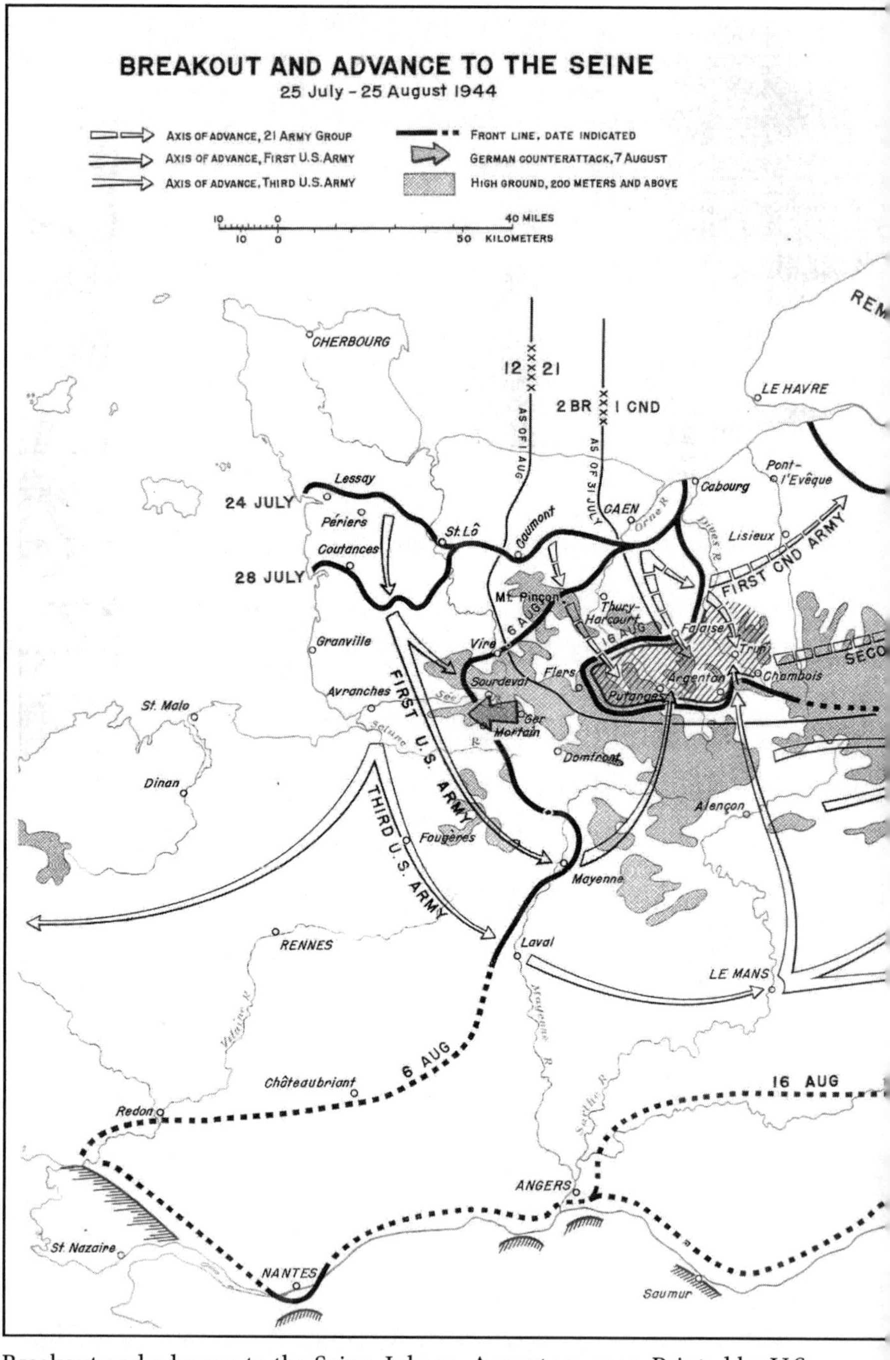

Breakout and advance to the Seine. July 25–August 25, 1944. Printed by U.S. Army Topographic Command.

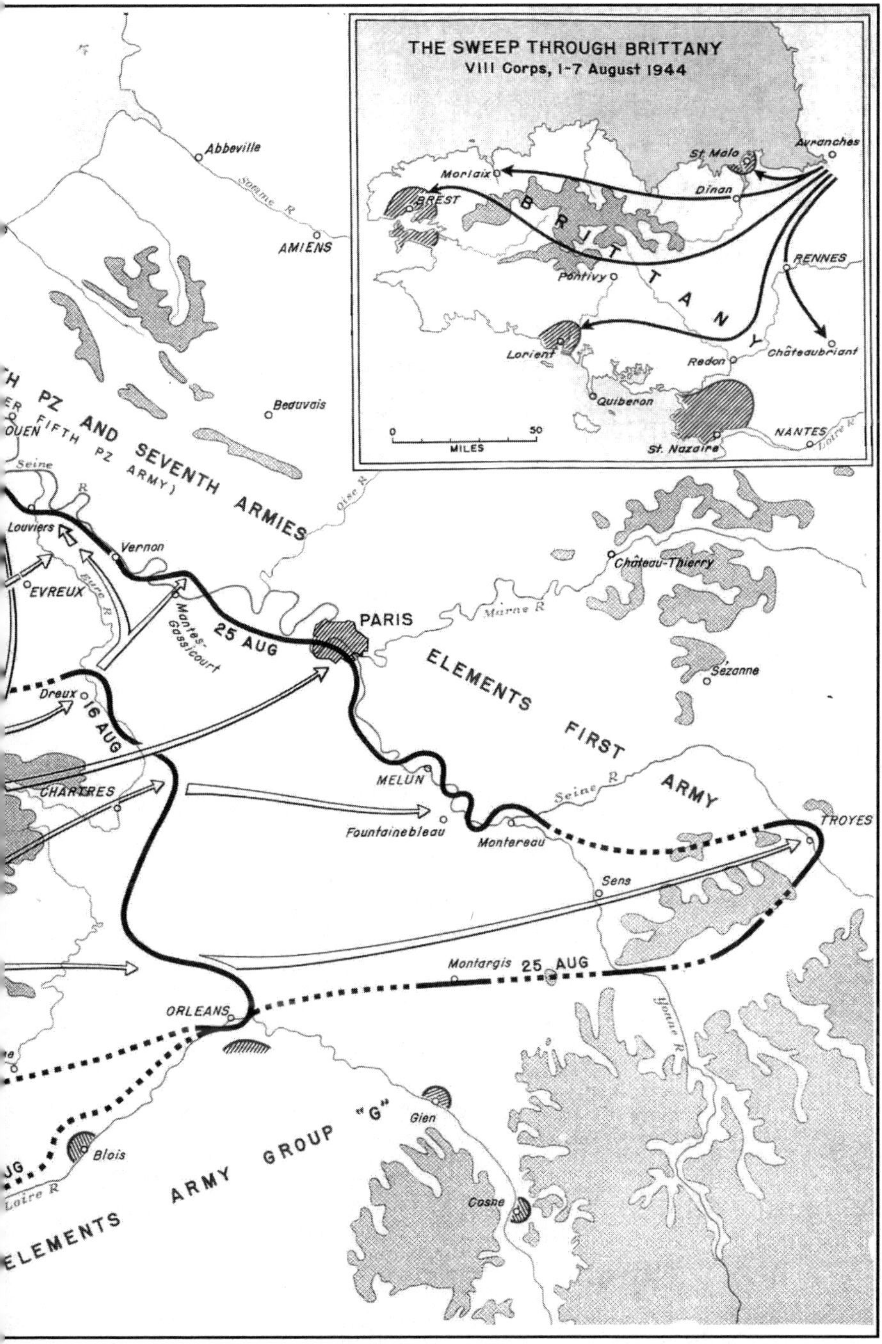

THE SWEEP THROUGH BRITTANY
VIII Corps, 1-7 August 1944
Avranches
St Malo
Morlaix
Dinan
BREST
BRITTANY
RENNES
Pontivy
Châteaubriant
Lorient
Redon
Quiberon
St. Nazaire
NANTES
Loire R
0
50
MILES
Abbeville
Somme R
AMIENS
Beauvais
PZ AND SEVENTH ARMIES
FIFTH PZ ARMY)
Seine
Louviers
Vernon
EVREUX
Eure R
Mantes-Gassicourt
25 AUG
PARIS
Oise R
Château-Thierry
Marne R
Sézanne
ELEMENTS FIRST ARMY
Dreux
16 AUG
CHARTRES
MELUN
Seine R
Fountainebleau
Montereau
TROYES
Sens
Montargis
25 AUG
Yonne R
ORLEANS
Gien
Blois
Cosne
Loire R
ARMY GROUP "G"
ELEMENTS

Cavalry Groups

In the beginning each cavalry group had a headquarters and a headquarters troop and usually two attached cavalry reconnaissance squadrons. Generally the cavalry group was assigned to an army command, which attached it to a corps. The cavalry group was designed to serve as a tool for corps or army commanders, plus providing information for higher commanders on friendly troops during fluid operations. Thirteen mechanized cavalry groups saw action in Europe: the 2nd, 3rd, 4th 6th, 11th, 14th, 15th, 16th, 101st, 102nd, 106th, 113th, and 115th.[8] Emphasis will be on changing doctrine and leadership rather than a detailed battle history of each cavalry group. A selected number of groups and their commanders and how they adjusted to the ever-changing battlefield environment will be examined.

One of the most interesting cavalry groups was the 6th Cavalry Group (Mech), known as the "Fighting Sixth." By an act of Congress in August 1861, the 3rd Cavalry was designated the 6th Cavalry Regiment, whose motto became

A 6th Cavalry Group jeep, followed by a M8 light armored car, leaves the main entrance to the Tanderagee Castle Estate, North Ireland, March 1944. Courtesy of the National Archives and Records Administration.

"Led by Love of Country." Since participating in the Civil War, the regiment had served in the Indian wars, the war with Spain, and then the China Relief Expedition, where the future major general Adna R. Chaffee Jr. served with the 6th Cavalry as a young lieutenant. In addition, the 6th Cavalry served in the Philippine insurrection and the Mexican expedition and engaged in remount and military police duty in World War I. In December 1938, the 6th Cavalry underwent a complete reorganization from an all-horse to a horse-mechanized regiment. Horses were to be deployed in portee trailers, and M3 scout cars and motorcycles were added. In addition, the number of machine guns was increased to provide more firepower. There was apprehension at the time that the regiment might become fully mechanized. One old trooper wrote, "Personally, I hope it will never lose its identity as the U.S. 6th Cavalry."[9] Much to the disappointment of the old-timers, the unit was eventually mechanized. On January 1, 1944, the 6th Cavalry (Mech) was redesignated 6th Cavalry Group (Mech) with the 6th and 28th Cavalry Reconnaissance Squadrons assigned as its constituent elements.

On April 3, 1944, Lieutenant General George S. Patton Jr. sent a "Letter of Instructions" to his Third Army's corps, division, and separate unit commanders specifying that infantry reconnaissance be stressed both day and night and that junior officers be inquisitive when seeking precise information. Furthermore, he instructed reconnaissance elements to maintain continuous contact with the enemy. Patton questioned the use of M5 light tanks for reconnaissance because their noise at night alerted the enemy, causing them to fire on the tanks' positions.[10] However, the following month, the 6th Cavalry Group, under the command of Colonel Edward M. "Joe" Fickett, received a special mission directive from Patton that deviated from his "Letter of Instructions."

The group departed from its prescribed mission of reconnaissance and security and became a unique Army Information Service (AIS) channel for General Patton. The AIS was referred to as his "Household Cavalry." Joe Fickett, who looked like a hard-bitten, mustached cavalryman of the frontier breed, had come up through the ranks beginning as a cavalry private in the Illinois National Guard. Soon he was commissioned as a second lieutenant in the cavalry, and in 1917 accepted in the regular army. Earlier, in 1923, he wrote an award-winning article in the *Cavalry Journal* on the importance of air reconnaissance to support mounted operations. In 1926, he attended the cavalry troop officer's course and then attended the Command and General Staff School in 1939. In May 1944, Fickett's cavalry group was given a special mission to establish a channel, both physical and technical, directly to the army advance command post. This arrangement bypassed normal communications for monitoring friendly frontline intelligence and operations when the Third Army became operational in France. Patton's "Household Cavalry" became active on August 1 when his

Third Army was officially deployed as a breakout and pursuit and exploitation force.[11] According to Patton's G-2 (Intelligence), cavalryman Colonel Oscar W. Koch, the "Household Cavalry" was designed to keep the Third Army commander apprised of the location of friendly units and their enemy contacts in a timely fashion. Koch was fluent in German and during the interwar period he translated German military manuals. The Household Cavalry allowed him to gather intelligence on the enemy's intentions.[12] During August, while the Third Army drove south to the Loire and then east to the Seine, the cavalry group kept Patton informed of the location and disposition of his frontline elements and provided flank liaison and security between friendly combat units.

No sooner had the 6th Cavalry Group engaged in the mission of being an information service than the commander was wounded. Lieutenant Colonel James H. Polk, who briefly commanded the 14th Cavalry Group, replaced him. Polk commanded the group for about a month until Fickett returned. He called the 6th Cavalry Group "a most unusual outfit" because "it was an old line Regular Army unit that went from horse cavalry to portee cavalry to armored cavalry with the same organization and personnel." He added, "They were sharp and snappy with beautiful equipment, experienced men and officers." Their radio operators, he noted, were "the best in our rapidly expanding army because they operated their nets from troop level up in rapid flawless transmissions." Polk had a front-row seat at Third Army Headquarters observing the 6th Cavalry's Army Information Service working directly with Patton and Koch. Polk noted that the AIS was important for command, control, and communication with friendly units as Patton's army engaged in exploitation and pursuit. It did, however, cause some perceptions by division commanders that the AIS was being used as spies.[13]

Once the Seine River was crossed at the end of August, the Allied forces changed strategy and continued pursuit to the German frontier rather than regroup to deal with emerging logistic problems. Supplies were still channeled in from the landing beaches in Normandy. This situation compromised General Eisenhower's planned broad-front advance toward the German frontier with all American, British, and Canadian forces moving forward like an outstretched hand, thus stretching out the Wehrmacht's defenses. Eisenhower focused on routes directing the Allies toward the Ruhr and Saar industrial regions, neutralizing them, and then moving on to Berlin. Eisenhower's broad-front approach came in conflict with General Bernard L. Montgomery's idea of a narrow-front approach, which he hoped would end the war by Christmas. Thus, when Patton's Lorraine campaign began early in September, the Allies' strategy had changed and as a result, Third Army, whose mission was focused on the Saar industrial region and crossing the Rhine River, began to lose its momentum.

General Eisenhower, who assumed control of ground operations on September 1, had reluctantly accepted now Field Marshal Montgomery's "single concentrated thrust" called Market (air assault) Garden (ground assault). His operational strategy after securing the ports of Le Havre and Antwerp was to drive a spearhead beyond the lower Rhine at Arnhem in Holland to outflank German fortifications. This was expected to put British forces in a position to strike the Ruhr industrial region from the north. Supporting Montgomery's right flank was Lieutenant General Courtney H. Hodges's U.S. First Army, which was driving toward a band of concrete German frontier fortifications called the Westwall, or Seigfried Line. These formidable fortifications ran from the Swiss border to a point north of Aachen to the Dutch border. The Allies referred to these fortifications as the Siegfried Line supposedly after the popular British tune "We're Going to Hang Out the Washing on the Siegfried Line," which was in reference to the German Siegfriedstellung of World War I.

Meanwhile, south of the First Army, Patton's Third Army began the Lorraine campaign, initially designed to establish bridgeheads over the Moselle River and then take Nancy and the old fortress of Metz, the capital of Lorraine. Once it was secured, Patton planned to push his army into Germany and strike at the Saar industrial region. The problem was that Patton was obsessed with the siege of Metz and its surrounding fortifications that blocked his army's move into southern Germany. Rather than continuing exploitation and pursuit, he bogged down his army in a morass of siege warfare. The Metz operation, Market Garden, and General Eisenhower's broad-front strategy gave the Wehrmacht a reprieve to consolidate its defenses along the Siegfried Line. It now became a war of attrition. This resulted in cavalry reconnaissance squadrons having to change their traditional mission of pure recon and security to defensive and offensive actions, and dismounted combat. Meanwhile, the main concentrated thrust was given to Montgomery at the logistical expense of Hodges's and Patton's armies.

Days before Montgomery set in motion Market Garden a reconnaissance patrol from the 113th Cavalry Group (Mech) attached to Major General Charles H. Corlett's XIX Corps became the first Allied troops to cross the border into Holland in the vicinity of Maastricht. This achievement began when Corlett was given the mission to protect Hodges's First Army's left flank. Corlett in turn ordered the 113th Cavalry to secure a crossing of the Meuse at Liege and clear the east bank along with the area between the river and Albert Canal. The offensive mission was accomplished but not until mines were cleared and German roadblocks destroyed. The cavalrymen had to fight to achieve their mission. As with all the cavalry groups, they were now beginning to realize the importance of refining an outdated concept on how cavalry should fight a future war. Once the area was reconnoitered and secured by the cavalrymen, the 30th Infantry

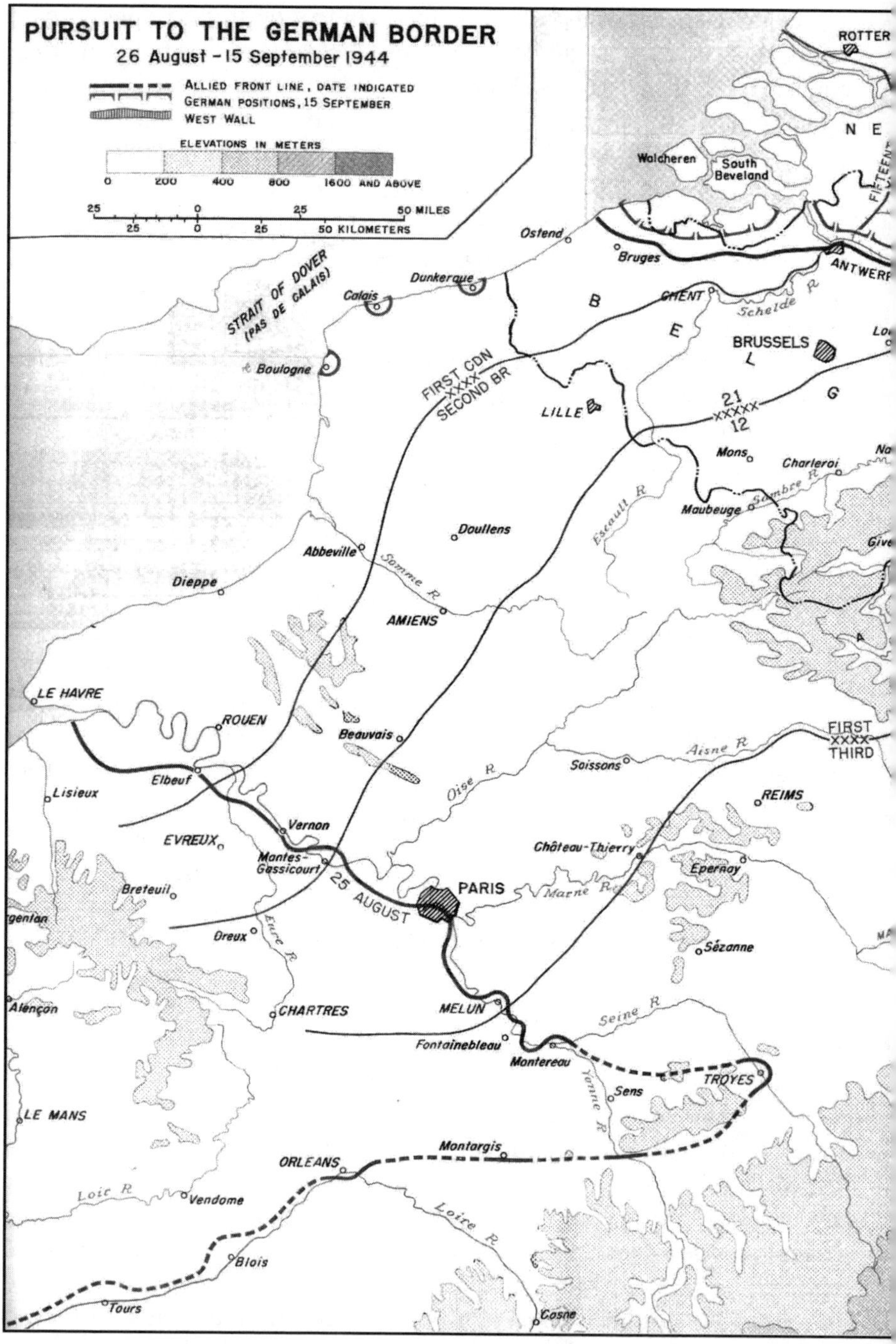

Pursuit to the German border. August 26–September 15, 1944. Printed by U.S. Army Topographic Command.

Arnhem
MÜNSTER
Nijmegen
Grave
Cleve
Wesel
Hamm
Lippe R
Paderborn
ANDS
Maashees
Niers R
RUHR
Eindhoven
Venlo
Mörs
Nuttlar
Ruhr R
KASSEL
HARZ
Weser R
Werra R
Roermond
München-Gladbach
DÜSSELDORF
Fulda R
MEUSE
CANAL
Roer R
Erft R
COLOGNE
Hersfeld
Düren
FIRST PRCHT
XXXX
SEVENTH
Sieg R
Siegen
Maastricht
AACHEN
Bonn
Lahn R
Giessen
LIEGE
Remagen
Sinzig
VOGELS BERG
Fulda
HOHE RHÖN
Huy
Malmédy
Ahr R
Koblenz
Mayen
St. Vith
Prüm
EIFEL
FRANKFURT-am-Main
Hanau
Wiesbaden
TAUNUS
Main R
Bastogne
Bitburg
Wittlich
MAINZ
Bad Kreuznach
Oppenheim
WÜRZBURG
15 SEPTEMBER
TRIER
HUNSRÜCK
Worms
ODEN WALD
LUXEMBOURG
MANNHEIM
Kaiserslautern
HAARDT
SAAR
Heidelberg
Thionville
Saarlautern
SAARBRÜCKEN
Conflans
METZ
Sarreguemines
Wissembourg
Lauter R
KARLSRUHE
Pont-à-Mousson
Hagenau
Pforzheim
STUTTGART
Moder R
Sarrebourg
FIRST
XXXX
FIFTH PZ
NANCY
Savern
STRASBOURG
Neckar R
Moselle R
Meurthe R
Danube R
FIFTH PZ
XXXX
NINETEENTH
Neufchâteau
St. Dié
Mirecourt
Epinal
Colmar
Neuf-Brisach
THIRD
XXXX
SEVENTH
Luxeuil
MULHOUSE
Belfort
Vesoul
Altkirch
Saône R
BASEL
SWITZERLAND
ZÜRICH
Pont-de-Roide
VOSGES
SCHWARZWALD
GERMANY

Division was able to push through and enter the southern part of the country. In mid-September, the cavalry group demonstrated an example of defensive action. Reinforced, the group provided fire support, defended over a nine-mile line for six weeks, and engaged mainly in dismounted patrolling. After this action the suggestion, which would be repeated, was to authorize more ground troops for patrolling and combat vehicle protection while the cavalrymen were dismounted. One method of firepower the group adopted for defensive combat was to use 75mm self-propelled howitzer assault guns, which were employed in an indirect fire role.[14]

At the end of September, the 113th Cavalry Group moved from a defensive mission and went on the attack in Holland toward Roermond in the Peel marshes. Corlett's XIX Corps was to attack and reduce a salient between the two army groups as they were poised for an attack into Germany. The cavalrymen, reinforced by tanks, tank destroyers, and combat engineers, ran into stiff German resistance. The attack made little headway after six days of fighting. The combat experience of the 113th Cavalry Group was typical of all cavalry groups. They began to adapt to the friction of battle, which, as the war progressed, caused considerable adjustments to existing cavalry doctrine and equipment. As a result, doctrine that emphasized "sneak and peek" infiltration tactics without engaging in combat was disregarded.[15] By the end of the war the 113th was still on the offensive and engaged in bitter fighting in the Teutobueger Forest and taking part in sealing off a large German pocket in the Hartz Mountains.[16]

The 113th Cavalry Group also had an interesting commander, Colonel William S. Biddle III, a noted polo player and an acclaimed equitation instructor. He was commissioned a second lieutenant in the cavalry after graduating from West Point in 1923. Biddle spent most of the interwar years in the horse cavalry. His future unit, the 113th Cavalry Group, was originally an Iowa National Guard horse-mechanized regiment. When federalized, the 113th "Red Horse" was completely mechanized and eventually organized as a cavalry group with two squadrons assigned, the 113th and 125th Cavalry Reconnaissance Squadrons. When the group landed in France in early July and began operations, Biddle was in command and stayed in this position with his beloved 113th until the end of the war. He maintained that throughout his career the time spent with the cavalrymen from the Iowa National Guard was the most gratifying. After the war, he became one of the key officers appointed to analyze cavalry operations during the conflict. Subsequently he moved up the ranks, commanding a division and then a corps.

Meanwhile Montgomery's Market Garden proved to be a "bridge too far." The egocentric and contentious Montgomery failed to focus on the banks of

Reviewing the 113th Cavalry (Mech), General Willis D. Crittenberger and Colonel William S. Biddle, Camp Hood, 1943. Courtesy of the U.S. Army Military History Institute.

the Schelde estuary connecting Antwerp to the sea. Before the launch of Market Garden on September 17, Montgomery's Twenty-first Army Group had surprised the Germans. The important port city of Antwerp and its strategic docks were taken intact. However, he failed to secure the Schelde River and its estuary that branched out to the ocean. The clearing of the Schelde estuary was critical. Eisenhower felt it was of vital importance to alleviate the burgeoning logistic problem as his armies moved toward the German frontier. Finally, months after Montgomery's Twenty-first Army Group captured the port city, the Canadians and British finally cleared the Schelde estuary, and the channel was swept of mines. At the end of November, critical logistics finally began to flow. As a result of the supply demands for Market Garden and Montgomery's failure to neutralize the Schelde estuary, Hodges and Patton were deprived of critical logistics required for their armies to continue exploitation and pursuit to the German frontier. It was no longer a pursuit; Patton had to halt his drive on September 25. Eisenhower's compromise with Montgomery's Market Garden did not sit well with Patton or Hodges, whose left flank was exposed. General Bradley, U.S.

12th Army Group commander, was not brought into the plan. When notified by Eisenhower, he objected strenuously. Bradley believed that Montgomery first should devote his efforts to securing the Schelde.[17]

Meanwhile, Joe Fickett had returned and concluded that his unit was diminishing in value as the Third Army paused. The 6th Cavalry Group, due to its expert management of communication and speed of M8 Greyhounds, was able to keep the Third Army commander aware of his advancing units many miles away from the command post. By October, however, the 6th Cavalry Group's information mission had changed from the one prescribed by Patton. October was a difficult month for Fickett's cavalrymen because for the first time the Third Army had to fight for rough terrain, thus yielding less information returns. Extremely wet conditions early in October hampered vehicle operations, especially the group's motorcycles that were used to carry the bulk of messages during this period. The cavalry group, however, continued to furnish Patton information on his frontline troops and their disposition, as well as his flank units. The group also deployed a mobile monitoring team to monitor radio nets to determine if there were discrepancies in signal security. The group also used pigeons for communications, a ridiculous holdover from World War I. Because of weather and the static situation Patton was confronted with in Lorraine, it was finally determined that this type of primitive communication was not as realistic as radio communication. Fickett decided to rely less and less on motorcycles and pigeons for communication because of the cold rain and mud.[18]

In November, a major change had occurred for the cavalry groups. No longer were the cavalrymen and infantry wearing old canvas leggings that were a descendant of the puttees of World War I; they now were issued combat boots to deal with the uncommonly cold weather, which would favor the Wehrmacht as Hitler prepared his major counteroffensive.

In November, Joe Fickett began outlining a new mission for his cavalry group. The general plan was for the 6th Cavalry Group (Mech) (Reinforced) to become a mobile all-arms force or what would come to be called a task force, capable of executing strong, aggressive offensive action as directed by the army or corps commander.[19] Usually, to form a task force for offensive and defensive combat, security, and special operations, cavalry groups were reinforced with a combined arms team consisting of field artillery, tank destroyers, and combat engineers. Fickett's Task Force, or what he called a "mobile force," consisted of a battalion of infantry (rangers), tank destroyers, assault guns, and combat engineer elements. The force also had the organizational capability to absorb a tank battalion and self-propelled artillery.[20] The training order for November directed that "the advancement of tactical knowledge of mechanized warfare, with particular emphasis on cavalry as an offensive combat arm," would be implemented. This major change in mission was to prepare the cavalry group to "function within one or more types of employment."[21] This meant establishing a moving or standing counterreconnaissance screen along with aggressive reconnaissance to determine the enemy's situation as well as fighting for information. Fickett's reinforced group, with heavy firepower and light mobility, had all the characteristics to execute strong and limited offensive action associated with exploitation and pursuit executed by the major maneuver element, the corps. The most important function was providing all commanders with a flexible organization to carry out assigned offensive missions. In addition, Fickett's combined arms mobile force was to take advantage of the ability of the cavalry reconnaissance element to "find, see, and report," and then employ aggressive tactics to maintain the forward momentum of the mobile force.[22] Along with the reorganization, the 6th Cavalry Group's mission as Patton's Army Information Service continued on a small scale.

The 6th Cavalry Group was reconstituted to form an all-arms cavalry force, an innovative tactical move that deviated from *1944 Field Service Regulation FM 100-5*. The purpose of Fickett's all-arms cavalry force was to find the enemy; the engineers were to clear and help; the Rangers were to maneuver and assault; the assault guns and tank destroyers were to support; the tanks were to strike; and the reserves were to close and destroy. Concluding his idea for a mobile force, Fickett stated, "Intelligence, initiative, aggressiveness, and the desire to close with and destroy the enemy must be the motivating force in the mind of every commander employing an element of this force."[23] The new organization that Fickett proposed made his group too large for pure reconnaissance. Again, it was evident that cavalry doctrine as taught before the invasion was not adequate for the ever-changing battlefield.

For the rest of the campaign the 6th Cavalry Group was employed as a light combat command totally indifferent to the principles of traditional cavalry and modified by new means of fighting using a mix of equipment and units with greater firepower. Fickett's Mobile Force or Task Force was to be formed under a temporary cavalry commander to carry out separate missions or a specific operation, such as acting as the forward spearhead for the main maneuvering element, the corps. Because of the employment and extended combat actions with the use of heavy firepower and light mobility, the 6th Cavalry Group's emerging doctrine as a mobile task force was inconsistent with the *1943 Cavalry Training Circular No. 107* and the 1944 *FM 100-5*. It was evident that reserving cavalry exclusively for reconnaissance and not participating in combat was brought into question. By January 1945, Fickett and other cavalry group commanders were beginning to challenge the name "cavalry reconnaissance squadrons (mech)."

They were now suggesting the description be changed to "cavalry squadron (Mech)." By now it was evident that combat actions demonstrated that cavalry squadrons were a light mobile force, well suited to perform all cavalry missions and capable of all types of offensive actions. Battle conditions in western Europe did not allow cavalry to just "sneak" its way through German resistance. With this being the case, the task forces were required to take cavalry elements and fight through enemy lines. Another important development was that tactical air reconnaissance became more efficient in supplementing the need for distant ground reconnaissance.[24]

The terms "combat team" and "task force" were derived from the German *Kampfgruppe,* which was an ad hoc organization for a special combat mission. For example, in the 6th Armored Division a combat team was fashioned by the combat commands to accomplish different missions, usually with a mix of organic units marked by the absence of reconnaissance troops. Beginning in September, cavalry groups were designated as maneuverable task forces for specific missions established by the corps commander. Like combat teams, task forces were a mix of units, but they were not organic to cavalry groups or squadrons. The U.S. Army, in a vague effort to describe a *Kampfgruppe,* many times used the two terms interchangeably, depending on corps and division commanders and prescribed missions.

On December 1 and until the end of the war, Fickett's group, while attached to a number of corps and divisions in the Third Army, began a long series of successful assignments as a task force. One operation was significant due to its offensive nature. Before Metz was captured, Task Force Fickett—reinforced by the 5th Ranger Battalion, tank destroyers, and engineers—was given the mission of screening the right flank of XX Corps, then breaching the German line of defense at L'Hopital, and clearing the Karlsbrunn Forest. L'Hopital became a problem due to successful enemy counterattacks and artillery fire. The initial problem was insufficient time to conduct reconnaissance to determine the nature of German dispositions. The problems were resolved when heavy artillery from the 19th Field Artillery Battalion of the 5th Infantry Division made its capture possible. A combined vigorous attack by the Rangers and Fickett's cavalrymen captured L'Hopital, with the attack carrying through to Karlsbrunn Forest, causing the enemy's artillery to displace. The forest was quickly cleared. Task Force Fickett's offensive action was almost entirely dismounted. Their M8 Greyhounds, tanks, and tank destroyers, were unable to support the attack due to heavy mud, and the steep approaches to a railway embankment needed to be crossed before engaging L'Hopital. The result of this offensive operation by a reinforced cavalry group indicated some of the problems fighting dismounted without mobile fire support. The combat report mentioned that the action also illustrated that mechanized cavalry reinforced by infantry and aggressively led can overcome handicaps.[25]

By May 1945, Task Force Fickett had become part of the campaign in the Ardennes. The cavalrymen reinforced, screened, and maintained security for III Corps in the Bastogne area and, after the breakout, defended portions of the Our River. Then the group breached the Siegfried Line and went over the Rhine River and into the heart of Germany. By the end of the war, the "Fighting Sixth" was employed in mopping up dejected enemy stragglers.[26] Other outstanding task forces that were formed by group leaders and singled out by Fickett included Polk's 3rd Cavalry Group, MacDonald's 4th Cavalry Group, and Biddle's 113th Cavalry Group. This is not to say other groups did not have interesting histories. They all had to make adjustments, setting aside established doctrine and changing to conform to the ever-changing needs on the battlefield.

One of the extraordinary cavalry group task forces, which Fickett discussed, was the 4th Cavalry Group (Mech). The 4th Cavalry Group, whose motto was "Prepared and Loyal," was constituted on March 3, 1855, as the 1st Cavalry Regiment. On August 3, 1861, it was redesignated the 4th Cavalry. In 1940 the regiment was reorganized as horse-mechanized, and in April 1942 it was again reorganized as 4th Cavalry (Mech), commanded by Colonel Joseph M. Tully, who took command in November 1941 when the unit was still horse cavalry. In 1943 the regiment was again reorganized and redesignated the 4th Cavalry Group (Mech), with the 4th and 24th Cavalry Reconnaissance Squadrons (Mech). The 4th Cavalry had participated in the Indian wars, the Civil War, and the Philippine insurrection before entering World War II.[27]

On June 6, 1944, the VII Corps, under the command of Major General J. Lawton (Lighting Joe) Collins, landed on Utah Beach and slugged its way across the Cotentin Peninsula, capturing Cherbourg on June 27. Collins commanded the corps until Germany surrendered in May 1945. Colonel Tully commanded the 4th Cavalry Group assigned to Collins's corps. Collins called the group his "Light Armored Brigade."[28]

In November, Colonel John C. Macdonald, Cavalry, became commander of the 4th Cavalry Group, succeeding Colonel Joseph M. Tully. After mechanizing, Tully led the 4th Cavalry Group through France. MacDonald, whom General Harmon called a "dour and imperturbable Scot,"[29] was from Nova Scotia and became a U.S. citizen in 1917. That same year he received a commission as a second lieutenant of cavalry in the Officers' Reserve Corps. However, he served as an infantry unit commander in France during World War I. In 1920, he graduated from Norwich University, where he subsequently became commandant of cadets and professor of military science and tactics. In 1923, MacDonald was returned to the cavalry, graduating from the Troop Officers' Course in 1928 and the Special Advanced Equitation Course in 1931. He soon joined the 1st Cavalry (Horse) at Marfa, Texas. From 1933 to 1938, MacDonald commanded one of the first armored cavalry units in existence, Troop A, 1st Armored Cavalry Squadron, 1st Cavalry Division at Fort Bliss, Texas. When he sailed for Europe in 1944, he was the G-2 in what was to become the Third Army.

The idea of using the 4th Cavalry Group strictly as a reconnaissance element prescribed in the 1944 *FM 100-5* was soon abandoned. As battle missions changed, the common practice of the 4th Cavalry Group was to attach tanks, tank destroyers, artillery, and combat engineers to provide a self-sufficient task force, which in many cases Macdonald used until the end of the war.[30] Until the time Hodges's First Army paused at the border and prepared to invade Germany, VII Corps' 4th Cavalry Group was responsible for security and screening a large gap between VII Corps and V Corps. By October Hodges's army had penetrated the Siegfried Line at Aachen and east of Roetgen.

When First Army and VII Corps started their drive into Germany, Collins used the reinforced 4th Cavalry Group as an economy of force to save on the strength of other divisions in his corps. Its new mission began in mid-September when the decision was made to penetrate the Westwall at Aachen, a German city located on the Dutch and Belgian borders. The area became the scene of one of the bloodiest battles confronting the American forces. Aachen was also part of the triangle of Dueren and Monschau that surrounded the densely evergreen Hurtgen Forest. Efforts to take the city in mid-September had failed. Early in October another attempt was made, this time from the north. Again, it failed. Eventually, the city was encircled and after bitter street fighting surrendered on October 21.

Before moving to the critical Roer dams east of the forest, General Hodges decided to send his army headlong into what turned out to be a mess. This became one of the bloodiest campaigns of the war in western Europe, the Battle for Hurtgen Forest. The potential nightmare of the forest environment apparently did not bring history to mind. Bradley, and especially Hodges, forgot the lessons of attrition warfare engaged by Ulysses S. Grant's Battle of the Wilderness in 1864 and General John J. Pershing's driving the doughboys into the Argonne in World War I. The forest clogged all the mobility, firepower, airpower, and technology that had propelled Hodges and his corps across France and Belgium. The perceived problem was that Hurtgen Forest threatened First Army's right flank and rear. The forest consumed an area of about fifty miles covered with towering, thick fir trees. This haunting environment produced a gloomy and mysterious muted light that filtered down onto a rainy, muddy, snowy, and frozen ground littered with dead trees and undergrowth. Unsuspecting infantrymen had to deal with "Bouncing Bettys," land mines that caught them between their legs, resulting in the loss of sex organs if not lives. There was also the constant odor of rotten logs. Hodges pushed division after division into the forest. This was the beginning of a three-month struggle in which thousands of Americans were killed, wounded, captured, or reported missing. Many emerged from the eerie forest with trench foot, respiratory diseases, and combat fatigue. General Hodges—demonstrating a World War I general's mentality—continued to feed unit after unit into the grinding horrors that the forest possessed. Noted army historian Charles B. MacDonald wrote that Hodges's action showed a shocking disregard for the German-held Roer dams and their importance in controlling the lower Rhine. With the dams in their hands, the enemy controlled flooding of the lower Rhine. If the dams were destroyed, the lower Rhine could become flooded, thus closing a route into northern Germany. The question the combatants had to deal with was whether to control the flooding of the lower Rhine or to control the lower Rhine from flooding.

MacDonald argued that if the dams had been seized early in the campaign, the Germans might have been forced to withdraw from the forest, thus losing their defensive advantage of flooding downstream and potentially stalling the Allied drive.[31] Finally, by January, the misery ended when the worn-out and demoralized Americans emerged from the forest.

Historian Russell F. Weigley argued that the Battle for the Hurtgen Forest was worse than the appalling casualties suffered during U. S. Grant's Wilderness campaign and Pershing's Argonne offensive. Weigley maintained that higher commanders lost the feel of the whole battlefield, meaning the significance of the Roer dams, which Weigley believed should have been the objective rather than the forest, contained only immobile German formations.[32] Hodges had cavalry groups, which he could have reinforced, to provide security and control of the rear and flanks as the First Army moved to the important Roer dams. However, he was too cautious and methodical. He failed to see that the decisive point was the Roer dams.

The densely wooded environment itself was not conducive to cavalry operations. For example, the 28th Cavalry Reconnaissance Troop assigned to Major General Norman D. Cota's 28th Infantry "Keystone" Division—one of the divisions Hodges fed into Hurtgen—was given a screening mission. The recon troop was to retain contact with the 4th Cavalry Group, located south in defensive positions. Cota considered this mission appropriate, since it was a normal one for a cavalry troop. It was expected that in conducting this mission the recon troop would avoid as much enemy contact as possible. It was to be the traditional "sneak and peek" operation. More so, the difficult terrain in the Hurtgen Forest made the recon troop ineffective because its Greyhounds and half-tracks, on which the troop depended, were not usable under these conditions. The infantry regiments usually provided their own reconnaissance using dismounted scouts. The 28th Division typically used its recon troops for traffic control and route reconnaissance, such as bridge classification and locating suitable terrain containing roads.[33]

During the early weeks of the Hurtgen campaign, the 4th Cavalry Group was given a security mission, screening Collins's VII Corps' right flank, south of Aachen. On one particular mission, the group was to secure the high ground and flanks while Collins's corps made an effort to push through the Hurtgen Forest. In the process, the cavalrymen along with engineers removed mines, especially the dreaded "Bouncing Bettys." On a number of occasions, the cavalrymen repulsed German counterattacks, located enemy defensive positions, and engaged in roving patrols. The latter created an illusion that the group had greater strength than it actually did. On October 1, the entire V Corps took over the 4th Cavalry Group's sector.[34] Between December 19 and 21, the 4th Cavalry Group was again heavily engaged as Collins's "Light Armored Brigade." The cavalrymen were employed as dismounted infantry with the mission to seize the high ground and secure a gap that existed between two infantry units moving toward the Roer River. MacDonald had to group all his vehicles in the rear area with the necessary personnel to deal with maintenance and security. He found that this situation left him with an insufficient number of riflemen for his mission. While the 4th Cavalry Group was engaged in dismounted fighting, all hell broke loose farther south when the Germans launched their major counteroffensive through the Ardennes area. Collins's VII Corps was relieved from its mission and rushed south to help check the German drive designed to split the Allies in half and capture the critical port of Antwerp. The offensive seriously endangered the entire Allied front. The 4th Cavalry Group was tasked with screening the movement south and then given an active defense mission. The 4th, however, was given limited attack missions to secure VII Corps' movements as it deployed to deal with the German offensive. It was becoming more fluid and confusing as the Germans persisted in their attack, making identifying friendly troop locations difficult.

As the Allied forces struggled to contain the German counteroffensive, the 4th Cavalry Group was detached from VII Corps and attached to General Harmon's 2nd Armored Division. MacDonald sent a troop into Humain, Belgium, and against little opposition secured the town. The "dour and imperturbable Scot," however, felt the attack on Humain was deployed in piecemeal fashion against what he perceived as an organized German position. His argument was that with just the deployment of a troop, it left his cavalry group in an overextended position. This situation, he maintained, denied the commander the ability to retain a sizable reserve to deal with any combat contingency. He claimed it was a violation of an economy of force.[35]

In analyzing his group's combat performance during the war, MacDonald referred to its organization as comparable to a combat command of an armored division without the firepower. The 4th Cavalry Group's organization was inadequate to perform most of the missions assigned, according to MacDonald. Early in February 1945, MacDonald complained that his group was attached to fourteen different units within a thirty-day period. He mentioned to an observer from the First Army's Combat Lessons Branch, G-3, "There is generally an attitude of indifference toward attached units. Many feel, upon receiving a new unit, that they have a fresh force available to them while as a matter of fact, my forces are often depleted from an operation just completed."[36]

Though the 4th Cavalry Group attempted to be a self-sufficient task force comprised of tank destroyers, light tanks, artillery, and engineers, it also operated with at least a minimum of combined arms necessary to engage in heavy combat. There was never a sufficient number of infantry for dismounted action. The M8 armored car proved to be a disappointment as a reconnaissance vehicle because it lacked cross-country maneuverability and its 37mm gun was of insufficient caliber to deal with the combat conditions that usually confronted the cavalry troop. MacDonald preferred the 105mm self-propelled howitzer to the 75mm self-propelled howitzer as a supporting weapon for his cavalry group. The towed antitank guns assigned to the group, MacDonald believed, were not useful for mobile cavalry operations but suitable only for defensive action. In addition, the M1 carbines proved to be unsuitable for dismounted action, as were the SCR 300 radios required for command and control when fighting dismounted. Instead, liaison aircraft was preferred for command and control.

The cavalrymen from the 4th developed unique means of night protection, using trip wires and grenades. They placed a grenade with the pin removed in a tin can and then secured the lever inside. Separate trip wires were attached to both the can and the grenade; a pull on either caused the grenade to arm. On one occasion an explosion alerted a machine gun team, which succeeded in killing five Germans attempting to enter the cavalrymen's perimeter at night.[37]

Colonel MacDonald's mission-type orders for the 4th Cavalry Group meant deploying it as a task force with cavalry reconnaissance capabilities and as an economy of force with elements possessing high mobility and adequate firepower. A subsequent study at the Armor School in May 1950 concurred, describing the problems of moving from a reconnaissance and security role to that of a mobile combined arms task force using the 4th Cavalry's combat experience. Representatives from the 4th, 15th, 16th, 101st, and 113th Cavalry

Groups supported and concurred with the study after analyzing their after-action reports.

Another interesting cavalry group commander who rose to fame was Jimmy Polk. As noted earlier, he accompanied one of the first cavalry units to arrive in England, the 106th Cavalry Group (Mech). Polk had a southern aristocratic heritage. He was a descendant of an officer who served in the Continental Army and was a member of the Society of Cincinnatus. His great-great-uncle Lieutenant General Leonidas K. Polk was one of the few Confederate corps commanders to die in combat. Young Polk, while at the Cavalry School before the war, sympathized with General Herr over the loss of his beloved horse cavalry. At times he did not show similar feelings for Patton, especially after his "blood and guts" speech that was filled with profanity and phrases like "I didn't come to get killed, I came to kill Germans." Polk felt the speech alienated about 20 percent of the command because it was distasteful to see a senior officer using such profanity.[38]

With the invasion under way, Polk was anxious to leave England and was relieved when his cavalry group landed on Utah Beach at the end of June. How-

Troop E, 106th Cavalry Reconnaissance Squadron, shelling enemy positions in Geislautern, Germany, February 8, 1945. Courtesy of the Patton Museum.

ever, Polk was stunned at seeing the sea and beach littered with sunken ships, burned-out tanks, vehicles, German bodies, and dead animals. While awaiting new mission orders Polk took temporary command of the 106th Cavalry Group.[39] Shortly he took command of the 14th Cavalry Group, which he thought was a screwed-up outfit and "awfully ragged."[40] After the Third Army became operational, Patton took notice of Polk, writing in his diary on July 22 that he had put Jimmy Polk in command of a cavalry group: "He is 32 years old and seems awfully young, but I was a colonel at 31 in command of a much larger force."[41]

The 3rd Cavalry Group (Mech) was activated in June 1943. It was not the old and legendary 3rd Horse Cavalry of prewar days. The "Brave Rifles" Regiment had been unhorsed and turned into tank battalions. When Polk finally took command of the "Brave Rifles" in September, it had no relation to the old 3rd Horse Cavalry, although it had inherited its colors. It had a large number of draftees from northern New York and western Pennsylvania, farmers and factory workers descended from eastern Europe, especially Poland. Polk found them tough and resourceful but lacking discipline. Their origins made them ideal for intelligence gathering because most knew German and central European languages. When the group landed in France in August 1944, it had been commanded by Colonel Frederick W. Drury, a 1919 graduate of West Point. Before Polk took command, Drury's group patrolled and engaged in forward reconnaissance as the XX Corps moved toward the fortress city of Metz. Major General Walton H. Walker was looking for a XX Corps crossing site and sent Drury's cavalry on a probing mission along the river to locate an intact bridge. No luck. Drury's patrol was ambushed in a defilade, and his jeep was later discovered riddled with bullets. The colonel was reported missing. Later it was confirmed that he was wounded and a prisoner of war. He was liberated in Poland by Soviet forces in spring 1945.[42]

After the announcement of Polk's appointment to the new command, his mother received a complimentary letter from Patton's wife, Beatrice, reflecting what Patton had written in his diary and adding what the general thought at the time, stating, "[Polk] earned it having done a wonderful job." Polk remembers reporting to Patton for his official orders and the general telling him, "I will not put up with a beaten commander. So if you're ever overrun or your outfit is ever overrun and badly beaten up, you're going to be relieved." He continued, if bad command is your fault, "I never want to see you again, you understand?" Polk said, "Yes sir," saluted, and left without a pause. He then reported to General Walker, whom Patton called "my fightingest son-of-a-bitch," also known for his love of Johnnie Walker scotch. For security reasons the media called the XX the "Ghost Corps" as it raced across France. What Polk saw, however, was not a happy corps commander. Walker stated that he did not know what the 3rd Cavalry Group was doing or where it was. He claimed he never received messages from the cavalrymen and that overall communication was terrible. "Polk," said Walker, go "straighten it out at once and good luck. . . . That's all." The young cavalry group commander left abruptly to get acquainted with his troops. What Polk found was a battle-wise outfit, although it had not reached the expertise of Joe Fickett's 6th Cavalry Group.[43]

Colonel Polk discovered that his two reconnaissance squadrons, the 3rd and 43rd, were separated north and south of the 5th Infantry Division along the Moselle River and not deployed as an economy of force element. Some of the cavalry troops deployed as infantry were hunkered down in a wooded area. Both squadrons were spread too thin for effective patrolling and reconnaissance. It was apparent that the corps commander, General Walker, who usually believed in aggressive reconnaissance and the value of information, had no understanding of prescribing a mission for the cavalry group, believing the unit was a tool for higher headquarters. Shortly this changed. Polk began training his men to use armored cavalry in reconnaissance, as a covering force, and in economy of force missions for XX Corps.

Meanwhile, right after taking command of the 3rd Cavalry Group, Polk received his first surprise, a "bidet." He wrote his wife, "I've never seen anything quite like it but the consensus of opinion is that it is intended for females only." At the time of his discovery, the group was deployed in a defensive screen between the Meuse and Moselle rivers. Meanwhile, "Uncle George" (meaning General Patton) visited Polk's headquarters and "caused quite a turmoil among the troops." Nevertheless, Patton was satisfied with what he saw. At the time of Patton's visit on September 21, the 3rd Cavalry was officially designated Task Force Polk.[44]

The reason for the reorganization was that the group was no longer made up of just two cavalry squadrons but had attached a battalion of engineers, two tank companies and two assault gun troops from the 6th Cavalry Group, and a field artillery battalion. Task Force Polk, or what Polk called his private army, was reorganized for the purpose of carrying out special missions as directed by higher authority. In addition, the task force had attached a French battalion from the 1st Regiment of Paris, which Polk called "a bunch of crazy communists" that were ill equipped and ill trained. He believed General Charles de Gaulle wanted the "crazies" out of Paris. Nobody wanted them, so they were forced on Task Force Polk. Later Polk credited them with "ambushing, guerrilla and underground activities, but not at all qualified to do well in a front line position."[45]

The mission for Task Force Polk was to secure some twenty miles of the Moselle riverfront and act as a link between the 90th and 5th Infantry Divisions. Still he complained that some of his cavalrymen were deployed as infantry. Polk now realized the reason why General Walker did not know what the 3rd Cavalry Group was doing was because it was deployed some distance from the corps command post and widely dispersed. In addition, the common radio nets and the landlines were unsatisfactory. The landlines, as was the case throughout the war, had a tendency to be stolen, or tanks just rolled over them. Polk's solution was two high-powered radios mounted on trucks with connections between his Headquarters and XX Corps Headquarters. As a result, relays improved considerably.

Meanwhile, Patton's Third Army finally took Metz on November 22, and by December 13 the outer forts were neutralized. The XII and XX Corps were now up against the Westwall and by mid-December had established small bridgeheads across the Saar River. During this period, Task Force Polk engaged in defensive action in front of the Siegfried Line and was the first unit of the Third Army to cross the German border. Through these actions, Polk's Task Force gained the nickname "3rd Cavalry Division" in both American publications and German intelligence reports.[46]

When the Third Army finally came up against the defenses of Metz and the east bank of the Moselle, the terrain became a major obstacle. Patton had to deal not only with this situation but also with changing weather conditions, giving the German army time to strengthen its defenses. In November the Third Army finally reached the Westwall. Meanwhile, thousands of American infantrymen were still locked up in the dense Hurtgen Forest. The defenses into Germany were now degraded, but not as much as was expected. The German army was still a dangerous foe determined to defend the Fatherland. They were falling back relatively intact and fighting for time. For months, it would be a battle of attrition along the Westwall.

At the end of November, Task Force Polk without warning bumped up against the Siegfried Line. The task force was taken by surprise because intelligence failed to provide information that such a line existed at what Polk called the "Switchline," which ran up the Moselle River and lay across the base of the Saar-Moselle triangle. Jimmy Polk vividly remembered his experience on the river: "We can't ever seem to get away from the damn thing: that twisting, turning, steep-sided, stinking Moselle that continues to haunt me." The task force took a number of casualties when confronted by a large number of concrete pillboxes, mines, and rows of concrete called "dragons' teeth." The force found it challenging, had to regroup, and was eventually taken out of the line. Polk moved his command post into the "rat house," a Germanized old hotel over the border in France.[47]

General Grow wrote later that Patton's Lorraine campaign was "mobility unused" because the principle of the objective was compromised. Walker's XX

General Robert W. Grow at Greencock, Scotland Harbor, February 22, 1944. Courtesy of *The Super Sixer*, January 1986. Author's collection.

Corps' mission was to secure the crossing over the Saar River and then strike on to Germany's major natural defensive barrier, the Rhine. However, the defended Moselle intervened, and thus Metz became an intermediate or secondary objective. Grow asked: Was the capture of Metz a compelling requirement, or should it have been isolated or ignored? Its road and rail net and its historical and political importance made it significant. However, he believed the German forces defending Metz were relatively unimportant even though the city was important to Patton's prestige and to his strategic, tactical, and logistical operations. Grow thought a thin screening force would have been sufficient to contain the immobile Germans. North of Metz was Thionville on the Moselle. Grow believed a crossing there might have served XX Corps' purpose of providing a wide encirclement of Metz and exploitation to the Rhine. The primary objective was now abandoned for an intermediate objective. XX Corps, which had, according to Grow, unlimited mobility, was now pinned to Metz, whose German defenders were relatively immobile.[48]

Grow made another interesting observation, stating that security was closely allied to mobility. He wrote, "A mobile force in relatively unrestricted territory has little difficulty maintaining security against a less mobile enemy." Patton's actions in changing the plan led to "the abandonment of the primary objective in favor of an intermediate objective not essential to the former." This not only violated the principles of movement and economy of force, but also the principle of the objective, which remains fixed, Grow argued. Regarding the principle of economy of force being violated, Grow meant a the lack of allocating the means, both physical and material, to the objective, which stays constant.[49] All of Patton's subsidiary actions compromised the principle of the objective. Yet the principle of economy of force was difficult to apply because of its close interdependence with the ever-changing conditions of warfighting that the Lorraine campaign presented.

Concerning Patton's Lorraine operations Major General F. W. von Mellenthin, who saw action in every German theater during the war and at the time was Chief of General Staff, Army Group G in the West, wrote after the war that the Americans made a mistake in coupling their armored divisions too closely with the infantry. U.S. corps commanders who fought the battles usually had two infantry divisions, one armored division, and one cavalry group. As the Wehrmacht was in the process of rushing in replacements and organizing the defense of the Westwall, he believed that three armored divisions formed into a tank corps could have achieved a decisive breakthrough. He singled out the 4th and 6th Armored Divisions reinforced by the French 2nd Armored Division because he and his staff felt that these two American divisions were "experienced formations and were ably commanded."[50] Both the 4th and 6th Armored

Divisions had aggressive cavalry reconnaissance squadrons, the 25th and 86th, respectively. Wehrmacht senior officers expected U.S. armored divisions to be deployed as self-sufficient task forces possessing mobility, firepower, and armor, striking deep into German territory. However, the broad-front strategy precluded any such approach, which meant quickly selecting decisive points to overcome the enemy's center of gravity. At times, the American high command was more inclined to plan for a methodical battle of annihilation that would maintain Allied cooperation.

In the meantime, the major German counterattack began. Regarding the surprise attack, Jimmy Polk commented, "It illustrates what a fool's paradise everyone lived in for a while." Many thought the war would be over by Christmas, however, this was not the attitude at Third Army Headquarters. Lieutenant Colonel Robert S. Allen, before the war a Washington syndicated columnist and author, was a deputy G-2 reporting to Colonel Koch. Allen was one of a select few who had access to ULTRA, the intelligence derived from decrypts enciphered from the German Enigma cipher machines. On December 10, he and Koch reported to Patton a major German buildup north of the Third Army, mentioned at an ULTRA briefing. Koch's deputy was concerned that Walker's XX Corps might be hit on its northern flank by a powerful German force. Allen also reported that intelligence noted a heavy concentration of enemy artillery gathering in the area since August.[51]

Also on December 10, the 9th Air Force warned that there was a substantial German military buildup, not only of men and equipment but also air elements. This massing, the warning noted, would probably hit the First and Ninth Army areas within two weeks. Air commanders were advised that if a counteroffensive occurred, maximum effort possible should be made to disperse equipment and supplies, and that vital installations needed to dig in and fortify. In addition, Lieutenant Colonel George Ruhlen—later a major general—who commanded the 3rd Armored Artillery Battalion, 9th Armored Division, reported increased German patrols and artillery registration. An escaped Polish laborer reported that tanks were dispersed and hidden under haystacks and in almost every barn west of the Rhine River facing the Ardennes. This information was passed up to the 9th Armored Division's G-2. Ruhlen later answered no when asked if the G-2 had received the information. "It may be still in the file," he noted sarcastically.[52]

What transpired next was Patton's Third Army's most brilliant operational maneuver, made possible by outstanding staff planning and intelligence from Koch's G-2. Patton's corps and division commanders began an immediate ninety-degree turn, moving over one hundred treacherous miles on icy and snowbound roads to launch an attack to relieve the encircled American forces at

Bastogne. On December 26, the 37th Tank Battalion, led by Lieutenant Colonel Creighton Abrams, broke through and linked up with the encircled Americans. Before Christmas, Patton had written his wife that Walker's front was quiet and that Task Force Polk was facing "low grade" troops. Polk, itching for combat and affronted by his cavalry's deployment, told Patton, "They are nothing but Poles with ulcers" in front of his cavalrymen.[53]

Supported by massive air strikes and supplies, the Allies were finally able to break the Germans' grip around Bastogne. Then, for three weeks, Patton's forces fought a meat-grinding battle to break the back of determined German forces. By mid-January, at Houffalize, Third Army met up with Lieutenant General Courtney H. Hodges's First Army, which had taken the brunt of the German attack. During this period Task Force Polk mainly held defensive positions for flank security and engaged in gathering prisoners for intelligence.

On January 10, Allen wrote that the alibis for the Ardennes fiasco were beginning to form up. Koch's outspoken deputy claimed Bradley, who commanded the 12th Army Group, and Hodges were "criminally negligent." Allen claimed both British and American intelligence "muffed the ball badly" because ample information on the Germans' marshaling for a major attack was available and made known. Allen predicted that for years to come books would be written over and over about the Battle of the Bulge, but it can be summed up in a few words, "We were surprised and got one hell of a beating. . . . They [the Germans] took the initiative away from us." He argued that Eisenhower and his British G-2 also should bear the responsibility for not reacting to the German buildup.[54] Allen had a debatable argument, because Bradley failed to place reserves behind the two corps that bore the brunt of the German counteroffensive.

Meanwhile, Task Force Polk was on the move again, operating from forward positions within close proximity of the retreating Germans. Again, Polk was referring to his unit as a cavalry group even though it still fought as a task force. The mission was defense along the Westwall by containing the enemy in what Polk called deploying in an economy of force role as cavalry assets were maximized for essential combat actions. Beginning February 28, 1945, the cavalry group operated in front of and on the flanks of XX Corps until March 9, when Polk had the 16th Cavalry Group (Mech), commanded by Colonel William Busbey, attached. As a joke, the 3rd and 14th Cavalry Groups without authorization began labeling themselves the 316th Provisional Cavalry Brigade (Armored). The name caught on. Soon the designation appeared on operations maps around the theater. It was a presumptuous move that might have caused trouble, but it did not. The 16th Cavalry Group was new to the theater, and Polk referred to its commander as a "bird." Polk told Busbey he had better "establish a reputation for bravery to direct your command." Busbey, who graduated from West Point in 1924, years before Polk, excitedly concurred. During his first attack he rode on the deck of the leading tank firing a .50-caliber machine gun. Later Polk told Busbey he was overdoing it a bit.[55]

In early March, the mission of the provisional brigade was to attack in conjunction with XX Corps and protect the north flank in an area south of the Moselle River east of Trier. Supporting the brigade's move were a field artillery battalion, an engineer battalion, a tank destroyer company, and an air tactical support element. Meanwhile, elements of the combined cavalry groups maintained contact with the 94th Infantry Division on the south as that division crossed the Ruwer River for a breakout. For eight days the 316th fought as a combined arms force on a seven-mile front, against stiff German opposition and over difficult terrain, which caused many cavalrymen to fight dismounted. The action of the provisional brigade cleared the way for a combat command from the 12th Armored Division to exploit the 94th Infantry Division's attack.[56] Polk's brief deployment of the 316th Provisional Brigade was an example of offensive action that employed a combined arms team bringing all elements of mechanized cavalry combat power into play.

After the 316th Provisional Brigade was reorganized to its original cavalry group status, the 3rd Cavalry Group continued to engage in some extensive fighting, crossing the Rhine on March 29 and linking up with the 6th Armored Division, which had swept east of Frankfurt and isolated the city. Polk's group also bypassed the city, leaving the following infantry to deal with its surrender.

By the end of March, Jimmy Polk's cavalrymen were usually changing direction per orders from the corps commander. Polk had praise for the way Walker was directing his mobile forces. He was impressed with Walker's employment of a box formation with the 4th and 6th Armored Divisions leading in exploitation and pursuit while two infantry divisions where cleaning up the rear. The 3rd Cavalry Group deployed on the flanks. This to Polk was an excellent example of mobile operations, which he gloated over as an example of operational warfare, meaning how to fight a battle at a campaign level.[57] For a cavalryman like Polk, it was exciting to be part of one of Patton's best teams. It could be argued that Third Army had the only commander who understood the operational level of warfare.

All did not, however, go well for Polk's 3rd Cavalry Group. On April 10, Koch's deputy, Robert Allen, was reported missing after he had gone through the group's front lines. Naturally Patton and Koch were more than concerned, to say the least. An investigation determined that the intelligence-minded Allen and several members of Third Army staff were in a jeep followed by a jeepload of MPs. They had driven through Polk's area without checking in or being challenged at an outpost. Polk assumed Allen knew what he was doing and was aware of the situation when his jeeps entered enemy-controlled territory. He was badly wounded when his team was ambushed and captured by German civilians. Allen lost half of his arm but recovered and was liberated several weeks later. Polk was reprimanded.[58]

Eventually, 3rd Cavalry Group ended up in Austria and ran into the back of a Waffen SS division facing the oncoming Red Army. The Germans had orders to let the cavalrymen through. Polk recalled, "It certainly was a spooky feeling to be moving between Krauts armed to the teeth and no truce or peace or anything, just sort of an unwritten agreement. To tell the truth, I was scared sick."[59] The cavalrymen never met the Red Army, leaving that up to the infantry. For the entire 261 combat days, the 3rd Cavalry Group remained attached to Walker's XX Corps.[60]

With the war over, the 3rd Cavalry Group planned the ninety-ninth anniversary of its organization in mid-May 1945 at Seewalchen, Austria. All former 3rd cavalrymen in the area were also invited, including General Patton and his staff. Patton handed out meritorious awards to Polk and praised his leadership and the cavalrymen for their outstanding contributions to the Third Army. Patton, as usual, was outspoken, and the troops loved it. During lunch Patton described how he thought American forces ought to attack the Red Army and drive it back to prewar boundaries. He claimed that he had prepared objectives for the XX Corps and the 3rd Cavalry Group and expected to get support from the German army to operate in the rear areas. There was a feeling of surprise and concern because "the war was over and many of the attendees were so naïve as to believe the Russians were our gallant allies and not our long term enemies," Jimmy Polk recalled. After lunch, which was placed to overlook a lake and the Austrian Alps, the cavalrymen—now supported by generous amounts of Benedictine, cognac punch, and cigars—decided to take a boat ride. Fortunately, the celebrants did not sink the boat. Afterward a nostalgic visit was made to the stables to see horses that had been liberated by the cavalrymen. After Patton and his staff left, Polk staggered to his quarters. He wrote his wife that he was in such a dazed state—no doubt helped by the Benedictine and cognac—that he just collapsed in bed.[61]

General Walker, who was unable to attend, sent a letter praising the 3rd Cavalry Group with "all the dash and daring of Stuart's famous 'raiders' themselves. . . . Today I call you XX Corps' finest." As Task Force Polk "your efforts contributed as much as any single unit in the reduction of Metz . . . . [and] you were the first of XX Corps to enter Germany." He also complimented the cavalrymen because in the concluding days of the campaign they acted as a special combat team for his corps by performing important and significant tactical operations.

The 99th anniversary of the 3rd Cavalry, May 19, 1945. Generals James H. Polk and George S. Patton taking a cruise on a lake after a rewarding lunch! Courtesy of the U.S. Army Military History Institute.

In an emotional tribute Walker echoed the words of General Winfield Scott addressing the 3rd Cavalry after the national flag was raised at Chapultapec during the war with Mexico one hundred years earlier: "Brave Rifles! Veterans! You have been baptized in fire and blood and have come out steel."[62]

Task Force Polk, however, received one last mission order that had not pleased General Mark W. Clark, the Fifteenth Army Group commander in Italy. In Patton's book released after the war, *War as I Knew It*, he commented, "Little is known or has ever been said of the Third Army's duplication of Hannibal's feat—crossing over the Alps." With the situation in Yugoslavia uneasy after the surrender of Nazi Germany, Patton ordered three divisions to move over the Alps and north of Trieste. Polk was ordered to reconstitute his task force on May 20. Within twelve hours Polk had it hitched up and on the move, crossing the Alps and thoroughly mixing in with British troops in northern Italy. Polk was a little concerned because the war was over and he might have to deal with grumbling and desertions. This was not the case; his cavalrymen were still in a fighting mood. Naturally, redeployment had not set well with the surprised Clark, who had not been informed of Patton's move. No sooner had Task Force

Polk made contact with the British than Clark advised General Eisenhower that Patton's troops were not wanted. No doubt with tongue in cheek, Clark reluctantly but sarcastically congratulated Patton on his "alacrity and boldness" in employing his cavalry troops.[63] Surely Clark did not want Patton to steal any more of his glory as the triumphant liberator of Rome.

Polk recalled years later that the commander of XX Corps, General Walker, and General Patton as commander of the Third Army "set the example in personal drive, in mission type orders and in timely decisions. If they left you alone, you knew you were doing your job." The 3rd Cavalry Group under Polk's leadership received high praise for carrying on the tradition of horse cavalry. Patton added, as "a mechanized cavalry I have never seen a better regiment."[64] What Polk learned, as did most of the cavalry group commanders, was to use mechanized cavalry more aggressively in reconnaissance, such as a covering force, meaning providing all-around security that required advance, flank, and rearguard action. He also learned the necessity of functioning as an economy of force element and fighting when necessary. Polk was the first cavalry group commander to deploy a provisional cavalry brigade composed of an all-arms force. Again, as with the other cavalry groups, the combat experience of the 3rd Cavalry Group indicated the inadequacy of earlier doctrine. Polk's World War II experience, however, had an effect. He developed a philosophy of command, meaning sound cavalry tactics are sound armor tactics; sound infantry tactics are sound armored infantry tactics. Once this is understood, he advised, learn equipment and maintenance. Thus, it was important to understand doctrine first and then apply equipment.[65]

The war had moved the organization of cavalry groups from two squadrons to combined arms task forces for special operations, a change that many traditional cavalry leaders began to understand. No longer was mechanized cavalry's mission simply reconnaissance and security with a minimum of fighting.

At the end of the war, Polk suggested that *FM 2-15 (Employment of Cavalry)* be entirely rewritten. He stated, "Horse cavalry is strictly a special purpose unit . . . ; mechanized cavalry has assumed the traditional role of cavalry. It has become the light, fast, hard hitting unit of the corps." For flexibility, he noted, "Mechanized cavalry is ideally suited to the task force principle" of being formed up for special missions. The commander of the 6th Cavalry Group, Colonel Fickett, added that cavalry groups were employed many times as light combat commands. This being the case, he noted, "There is no occasion, no opportunity, and no justification for the maintenance in large commands of such an extremely costly, highly trained organization simply for the purpose of reconnaissance." Force, however, was definitely required, adding, "Stealth will not take the squadron as a unit through enemy lines." His solution for reconnaissance was tactical airpower and the ability of ground cavalry to fight for information. He concluded that *FM 2-15* should reflect the normal mission of cavalry, although not to the point where it became too heavy as a task force. Furthermore, he noted, the tactical concepts in *FM 2-15* for horse cavalry missions "are, with certain modifications because of differences in transportation and arms, equally applicable to mechanized cavalry."[66]

By May 1945, thirteen mechanized cavalry groups and one unattached mechanized reconnaissance squadron had seen action in the ETO. These units were assigned to armies and routinely attached to corps. In most cases these attachments were permanent. Corps commanders further attached mechanized cavalry for operations to divisions, especially infantry divisions. Thirteen mechanized cavalry reconnaissance squadrons fought as organic units of light armored divisions, and two as organic to heavy armored divisions. Forty-two mechanized cavalry reconnaissance troops fought as organic units of infantry divisions.

The missions for the mechanized cavalry groups noted earlier were classified first as reconnaissance and security. Their training was based on this mission, as was their equipment. The mission of cavalry groups had been that of traditional cavalry; however, the chaos and tension of battle required numerous adjustments in doctrine and equipment to meet the demands of an ever-changing murkiness and friction in war-ravaged Europe. As the war progressed, combat, both offensive and defensive, and special operations became more frequent for cavalry groups.

By now cavalry missions were continuously altered due to the fluidity of combat. For example, mechanized cavalry groups rarely engaged in pure reconnaissance missions as prescribed by the 1944 *FM 100-5*. Defensive and delaying missions were becoming more common. For example, there was Colonel William S. Biddle's 113th Cavalry Group defending the north flank of Bradley's Twelfth Army Group while it struggled to reach the Westwall.

An example of defensive action was the 106th Cavalry Group during the battle for Luneville. After weeks of fighting in most disagreeable weather and extensive minefields on the XV Corps' left flank, the cavalrymen slugged their way through the Foret de Parroy north as dismounted infantry near the contested town of Luneville. The dismounted cavalrymen fought a defensive battle alongside the 44th and 79th Infantry Divisions. It turned out to be one of their toughest campaigns.[67]

Another example of defensive combat was Colonel Charles "Hank" Reed's 2nd Cavalry Group, 2nd Squadron, engaging in an outpost defense on September 18 during an enemy attack on Luneville.

Like many group commanders, Reed had excelled in equestrian sports at

A M5 light tank from the 2nd Cavalry Group passes a French woman leading her cow through the battle-scarred town of Beauzemont, France, November 1944. Courtesy of the National Archives and Records Administration.

West Point, especially polo. He spent his whole interwar career in the cavalry. The 2nd Cavalry Group's lineage goes back to 1836, when it was constituted in the regular army as the 2nd Regiment of Dragoons. When the Civil War broke out, it was designated as 2nd Cavalry. The regiment fought in all American wars and by World War II it had lost its horses. The unit was then redesignated as the 2nd Cavalry (Mech) and finally as the 2nd Cavalry Group (Mech) in December 1943 with the 2nd and 42nd Cavalry Squadrons (Mech) assigned. Early in its combat operations the group gained a reputation—supposedly thanks to the German High Command—as "Patton's Ghosts" because of the speed at which Reed drove his cavalry group through France while attached to Major General Manton S. Eddy's XII Corps and protecting the 4th Armored Division's left flank. Reed reported after moving some two hundred miles since landing in France on July 26, "We are washing our tanks in the Loire River."[68] Reed was severely wounded at Luneville while his troops unsuccessfully attempted to contain an attack by enemy tanks supported by infantry. The cavalrymen came up against the German 111th Panzer Brigade and were outgunned, as their cavalry vehicles were easily destroyed. American 75mm shells from M8 assault guns just bounced off the Panthers. Nevertheless, the cavalrymen took the initiative and were able to deal with dismounted German infantrymen long enough to escape an encirclement. Reed and his cavalrymen's defensive actions prevented the enemy from occupying Luneville.[69]

Months following the battle of Luneville, the 2nd Cavalry Group found itself dismounted and fighting as infantry in the French mud north of the Foret de Parroy. After recovering from dealing with superior German armor at Luneville, the 2nd, as a reinforced task force, moved to defend the west bank of the Moselle during the Battle of the Bulge and in 1945 raced through Germany defending Eddy's XII Corps' flanks. By May 1945, the 2nd Cavalry Group had penetrated all the way into Czechoslovakia, where the cavalrymen met up with Red Army elements. Reed eventually returned to command his group during these operations and became one of the most decorated cavalry group commanders of the war.[70]

After emerging from the forest in November, the 106th Cavalry Group mounted its vehicles for the next mission, which was to protect the left flank of XV Corps west of the Vosges Mountains. While screening the corps, the group detected the advance of the elite 130th Panzer Lehr Division moving to interdict Allied supply lines. After reporting the movement to higher headquarters, Colonel Vennard Wilson's Group then engaged in delaying Panzer Lehr's advance south against the 44th Infantry Division. During this action enemy tanks counterattacked and broke into a town occupied by the cavalrymen and began firing rounds into basements where the troopers had retreated. One report noted that the 37mm shells and bazookas just ricocheted off "the Mark V Panthers like hail bouncing off a tin roof." An after-action assessment of the operation indicated that the delaying mission worked; however, it was evident that greater organic firepower was needed as well as more dismounted infantry, especially when engaged in urban warfighting.[71] In March 1945, the 106th was given an offensive mission and spearheaded the advance of XV Corps through the Siegfried Line and into the heart of Germany.

A cavalry group again deviated from its customary mission and engaged in special operations when Joe Fickett's 6th Cavalry Group became an army information service for Patton's Third Army. An example of a security mission found MacDonald's 4th Cavalry Group screening General Collins's VII Corps' right flank south of Aachen in September and early October 1944, and the 11th Cavalry Group protecting the north flank of XIII Corps during the advance from the Rhine River to the Elbe River during the last months of the war. When engaging in offensive missions, a battalion of field artillery and tank destroyers, and a company of engineers usually accompanied the cavalry groups. An example of cavalry groups employed in offensive combat found Task Force Polk engaged in offensive action to seize a town and clear a forest early in December, and Task Force Fickett engaged in pursuit at the end of April 1945. Finally, a rare example of reconnaissance was Biddle's 113th Cavalry Group operating in the zone of action of the XIX Corps from September 5 to 8, 1944.

By the time the war ended, mechanized cavalry groups had spent 33 percent of their time in defensive combat (delaying actions, holding terrain for arrival of main forces); 29 percent in special operations (mobile reserves, information service, and rear area security); 25 percent in security (blocking, screening, protecting flanks, filling in gaps, and maintaining contact with larger units); 10 percent in offensive combat (attack, pursuit, and exploitation); and 3 percent in reconnaissance ("sneak and peek"). As the war progressed, it became apparent that defensive missions had become prominent and the ratio for mounted employment was one in every eight days. The problem was that high commanders had not employed cavalry groups as an economy of force to shape operations for the bulk of combat power.[72]

In summarizing the missions of cavalry groups, it is evident reconnaissance was rare and that defensive missions were common. The days in dismounted combat were slightly more numerous than days mounted. Finally, for defensive, offensive, and security deployment it was typical to reinforce the cavalry group with a battalion of field artillery and tank destroyers, and a company of combat engineers. It became clear that cavalry was no longer deployed as prescribed in the 1944 *FM 100-5*. The war had changed the mission of cavalry, which was reconstructed at times as combined arms task forces designed for special missions by corps commanders who required light elements that had mobility and adequate firepower. Doctrine was constantly being redefined.

Cavalry Reconnaissance Squadrons

The next level to consider is the deployment of mechanized cavalry reconnaissance squadrons and armored reconnaissance battalions in armored divisions. Again, five cavalry mission types will be analyzed: reconnaissance, security, special operations, and offensive and defensive combat.

Their employment and types of missions were different than those of cavalry groups. A number of cavalry squadrons will be focused on to understand how they were originally employed, how they underwent changes, and the effects of leadership. As with cavalry groups, these squadrons and battalions were originally trained under the guidelines of General Herr's *1941 Cavalry Field Manual 2-15 (Employment of Cavalry)*, the *1943 Cavalry Field Manual 2-30 (Cavalry Mechanized Reconnaissance Squadron)*, the *1943 Cavalry Training Circular 107*, and *1944 FSR: Operations (FM 100-5)*.

There are a few distinct differences, however, in deployment of mechanized cavalry reconnaissance squadrons and armored reconnaissance battalions as compared with cavalry groups because they performed more recon missions. In addition, cavalry elements were often detached for service with division combat

Biddle Report, MHI

UNIT	OFFENSIVE		DEFENSIVE		RCN		SECURITY		SPEC OPNS		DAYS CONSIDERED
———	Days	%	Days	%	Days	%	Days	%	Days	%	———
2d Cav Gp, Mecz	29	9.9	112	38.1	8	2.7	118	40.1	27	9.2	294
3d Cav Gp, Mecz	22	10.5	119	56.4	8	3.8	44	20.8	18	8.5	211
4th Cav Gp, Mecz	66	21.6	65	21.2	10	3.3	96	31.4	69	22.5	306
6th Cav Gp, Mecz	53	17.4	52	17.0	0	0	36	11.8	164	53.8	305
11th Cav Gp, Mecz	1	0.7	70	47.9	0	0	34	23.3	41	28.1	146
14th Cav Gp, Mecz	14	9.4	19	12.8	1	0.6	14	9.4	101	67.8	149
15th Cav Gp, Mecz	14	4.5	15	4.9	22	7.2	152	49.3	105	34.1	308
16th Cav Gp, Mecz	7	11.9	5	8.6	0	0	3	5.1	44	74.4	59
101st Cav Gp, Mecz	16	18.4	33	37.9	9	10.4	16	18.4	13	14.9	87
102d Cav Gp, Mecz	14	4.2	110	35.5	6	1.8	71	21.2	125	37.3	335
113th Cav Gp, Mecz	35	11.2	160	51.3	16	5.1	39	12.5	62	19.9	312
115th Cav Gp, Mecz	0	0	40	69.0	2	3.4	2	3.4	14	24.2	58
117th Cav Gp, Mecz	118	6.7	120	44.9	12	4.5	82	30.1	35	13.1	267
TOTAL (No. of days)	289		929		94		707		818		2837
GRAND TOTAL No. of Days	———		———		———		———		———		
Percentages	10.2%		32.8%		3.3%		24.9%		28.8%		100%

NOTE: Offensive combat includes attack, and pursuit and exploitation. Defensive combat includes defense, delaying action, and holding key terrain. The accomplishment of reconnaissance missions normally involved offensive combat. Security includes screening (both moving and stationary), blocking, flank protection, maintaining contact, and filling gaps. The performance of security missions frequently involved offensive combat. Special operations includes mobile reserve, rear area security, and Third Army Information Service.

SOURCE: Report of the General Board, United States Forces, European Theater, "Tactics, Employment, Technique, Organization, and Equipment of Mechanized Cavalry," Armor Section, ETO, Study No. 49, dated November 20, 1945, MHI.

commands and task forces. Another difference was that once elements of cavalry units were detached, the remainder were held in reserve. This compromised the recon's teamwork and unit integrity to find and fix the enemy so the high command could allocate the bulk of its forces for decisive operations.

Cavalry reconnaissance squadrons integral to armored divisions trace their history to the 7th Cavalry Brigade (Mech), which was an outgrowth of Daniel Van Voorhis's Mechanized Force established at Fort Eustis, Virginia, in 1930. If one looks back at the history of the U.S. Cavalry, it is arguable that cavalry reconnaissance squadrons indirectly traced their tradition to the maneuvers of horse cavalry during the Civil War. However, as the Civil War progressed, cavalry units began to function more as mounted infantry using horses for mobility and then engaging the enemy in dismounted action. The post–World War I armored car and combat car squadrons led to improvements and changes in cavalry equipment and organization. By the time the U.S. Army was engaged in combat in World War II, squadrons had taken on a nature all their own. Like the cavalry groups, they rarely were used in economy of force roles to shape operations for the main force. Instead, they were split up and attached to combat commands in response to the division commander's reaction to a mission-type order from higher headquarters. To explain the changes and leadership, a number of mechanized reconnaissance cavalry squadrons organic to armored divisions will be selectively examined.

One of the first examples is route reconnaissance and security, a pure cavalry mission executed by the 86th Cavalry Reconnaissance Squadron, 6th Armored Division, during exploitation of the Brittany Peninsula from August 1 until the division closed in on Brest on August 8, 1944. After attending the Command and General Staff College, Lieutenant Colonel Albert E. Harris, who graduated from West Point in 1930, took command of the 86th from the original commander, Lieutenant Colonel James B. Quill, in September 1943.[73] One advantage the squadron had was that the division was commanded by General Grow, who believed in the importance of traditional cavalry screening of the front and fighting for terrain and information if necessary. During the Brittany campaign, Grow made sure that the 86th was under division control and operated as an economy of force element.

When the 6th Armored Division entered combat and before entering the Brittany Peninsula, Grow had an interesting experience with Patton. Earlier, Patton had written his wife, "Bob Grow has come and so has P. Wood," commander of the 4th Armored Division. "The clans are gathering." Later, on July 29 at Coutances, Patton's confidence was disturbed because of a lack of movement by Grow's division, which was stalled on a road along a stream. Grow and one of his combat commanders, Brigadier General James Taylor, and staff were engrossed in examining maps looking for a ford to cross. Irritated and impatient, Patton told Grow he had checked the river and noticed just a few Germans on a distant hill and an erratic machine gun. He told the startled Grow that the river was not deep, and "unless he did something, he would be out of a job." He then preceded to repeat a Japanese proverb: "One look is worth one hundred reports." Patton added, "why in hell" did you not go down to the river personally and reconnoiter it for a crossing? Apparently Grow had not taken advantage of his cavalry squadron. Also Grow had a mission-type philosophy, which was to tell his combat commanders what to do but not how to do it. In record time the division crossed the river and began a rapid exploitation through the Brittany Peninsula. After the Brittany campaign, Grow relieved Taylor because he did not possess the leadership qualities of an aggressive commander. Patton later commented that Grow "learned the lesson and from then on they were a very great division."[74]

On July 31, 1944, the Third Army became operational, and within a few hours Patton ordered Grow to "capture Brest." The 86th Cavalry Reconnaissance Squadron covered the front and maintained contact with the division's two combat commands as they raced toward the port city. An example of cavalry techniques occurred before the initial breakout and exploitation by the 6th Armored Division. Troop B, under the command of Captain Donald L. Tilleman, 86th Cavalry Reconnaissance Squadron, was reconnoitering ahead of one of Grow's leading combat commands. A company of M5 light tanks, commanded by Captain Walter C. Smith, supported Tilleman's cavalry. Approaching the French town of Brehal, the recon force and tanks ran into an enemy roadblock. Smith pulled back his M5s so tactical air support could neutralize the obstacle. Smith then moved his tanks through the roadblock, securing an opening for the advancing armored infantry, which shortly came under sporadic mortar and artillery fire from Brehal. Smith then ran two tanks through the town, pausing long enough to throw a few rounds from their 37mm tank guns into the church steeple and its wooden front doors. Weeks later a staff officer from VIII Corps reported, "There were a tremendous amount of firing because they thought there was a sniper in the church steeple. Actually, they killed the French bell ringer who was up there tolling for victory." After the report, the 86th devised a habit of dropping off a trooper to calm the town's enthusiastic bell ringer, because the cavalrymen found on a number of occasions that the bells started ringing as forward elements came abreast of the church and the Germans responded with previously registered artillery fire.[75]

During the initial exploitation to Brest, the 86th Cavalry Reconnaissance Squadron was used as an independent economy of force element and usually deployed several hours in front of the advance elements of the combat commands.

However, the move from reconnaissance in force to security in force requires some explanation. After the Brittany campaign in August 1944, the squadron commander Colonel Harris complained that the cavalrymen were not allowed to operate as an independent economy of force. The squadron's troops instead were attached to combat commands, thus degrading the ability of the squadron commander to facilitate or control reconnaissance. Grow at first employed the squadron in an economy of force role. However, as the division approached the port city of Brest, he attached cavalry recon units to combat commands. Grow used the 86th during the exploitation phase, but when the division reached the outskirts of Brest, he decided that it was necessary to attach recon units to combat commands rather than under division control.[76] This change eventually became standard for all armored division commanders.

The drive to Brest by the 86th was not all recon-associated with combat action. Lieutenant John Hughes found a large cache of liquor—at least 250 cases of champagne, cognac, and numerous bottles of alcohol—left behind in the stalls of an old cavalry building at Pontivy by retreating Germans. The French mayor was reluctant to part with the cache but changed his mind when he was told that it belonged to the U.S. Army. He then took Hughes to a warehouse where one hundred thousand gallons of red wine were stored in casks. Hughes proceeded to liberate some of the cache, and as the 86th passed through the town he handed out some of the spoils. A number of jeeps then sped away with the liquor. Hughes was later told that the supply of alcohol he gave his squadron members gave them the courage they needed to carry on the exploitation. Decades later, when Hughes was in his late seventies, he finally shared with his veteran friends one of the three remaining bottles of cognac he had sent home in August 1944 to Rushville, Indiana. Hughes commented, "I wanted to catch just a moment of my youth in my seventy-eighth year with my friends, the other old soldiers, as we slowly sipped that bottle of cognac and remembered those days in Pontivy."[77]

Another example of effective combat reconnaissance under difficult circumstances was action of the 25th Cavalry Reconnaissance Squadron and its commander, Lieutenant Colonel Leslie D. Goodall. The 25th was assigned to the 4th Armored Division then under the command of Major General Hugh J. Gaffey, who replaced John S. Wood early in December 1944. Gaffey had been Patton's chief of staff. This was at a time when the Allies were engaging in a battle of attrition along the whole Siegfried Line. Eisenhower's broad-front strategy was designed to put continuous pressure on the German army, which now, due to weather and terrain conditions, was able to put up a determined defense. General Bradley, the army group commander, still had faith in a World War I style of attrition warfare, especially in the First Army sector, where the methodical Courtney Hodges was sucked in for months into the grueling battle for the

Hurtgen Forest. During this period the Third Army took the fortress town of Metz and eventually established a beachhead over the Saar River and up against the Siegfried Line.

The 4th Armored Division received what the tankers referred to as the "fire call" after the major German counteroffensive in the Ardennes that began on December 16.[78] The area was located in the center of General Bradley's 12th Army Group and was considered a quiet sector, referred to as a "ghost front" or, as a company commander said, "a nursery and old folk's home."[79] Now it became imperative that Patton's Third Army stop its Lorraine campaign and rapidly move nearly one hundred miles north. Through outstanding staff work, the Third Army hastily executed a ninety-degree turn with the 4th Armored Division leading the way and performed one of the most remarkable operational-level movements in modern military history. Advancing quickly and under severe weather conditions, the Third Army prepared to attack the southern edge of the German counteroffensive and relieve the encircled city of Bastogne.

Colonel Goodall's cavalrymen from the 25th began screening the flanks of the division as it advanced toward Bastogne. His mission was to maintain contact with the 26th Infantry Division, which was on the 4th Armored Division's right flank. Troop A was detached from the squadron and assigned the mission of screening the east flank. Throughout the operation toward Bastogne, Colonel Goodall's 25th, minus some of his cavalry elements, was not operating as an intact force role. The squadron's troops, as with most armored divisions, were habitually assigned or detached to combat commands and sometimes to the CCR, the Reserve Command.

As the 25th cavalrymen moved through mud, cold, and snowy weather conditions, they employed firepower, mobility, communication, and flexibility combined with security to obtain information by aggressive combat. Throughout the operation toward Bastogne, Goodall's 25th was not operating in an economy of force under division control. The 25th never engaged in "sneak and peek" but was a fighting unit, operating with thin-skinned vehicles. The cavalrymen's aggressiveness demonstrated again and again that the "sneak and peek" prewar doctrine they were taught was inconsistent with current combat conditions. This evolving ad hoc doctrine meant the cavalrymen from the 25th had to adjust and then fight through the enemy's outer screen to determine the location and disposition of major German forces encircling Bastogne.

During the drive to Bastogne, Combat Command B and Troop B were hit on the right flank by German patrols armed with dreaded *Panzerfausts* while moving toward the Belgian town of Chaumont. German soldiers, like the cavalrymen from the 25th, also engaged in security and penetration reconnaissance. Lieutenant Colonel Albin F. Irzyk, commanding the 8th Tank Battalion, 4th

Armored Division, recalled that the German fire from the woods was so heavy he had to send in the cavalry. After heavy fighting Troop B prevailed; however, the troop commander was missing and never heard from again. The slogan "snoop and poop" gave way to "hoot and shoot." It was obvious to Goodall that the prewar doctrine of "sneak and peek" was not paying off in obtaining information. Al Irzyk—later brigadier general—commented that Goodall was always mission-oriented and dependable, which made him an outstanding cavalry leader. Goodall, who made use of his unit's mobility and flexibility, remained commander of the squadron throughout the war.[80]

One example of offensive combat began when the 87th Cavalry Reconnaissance Squadron assigned to the 7th Armored Division was reinforced as a light combat command and assigned a separate mission. The background for the 87th Cavalry Reconnaissance Squadron's plans for offensive employment began on August 31, 1944, when Walker's XX Corps reached Verdun and occupied a bridgehead east of the Meuse River. Patton originally ordered Walker, whose "Ghost Corps" was spearheading the rapid advance across France, to cross over the Meuse and head for Metz and then through the Siegfried Line, Germany's western front fortifications. Once the line was penetrated, Patton planned to move toward the Rhine River, where a crossing could be forged at Mainz, Germany. The final objective was Frankfurt. Patton bragged that his Third Army was planning to go through the Siegfried Line "like shit through a goose."[81] The Allies, however, moved too fast, something the logistic planners had not anticipated. Problems began when bad weather curtailed unloading at the Normandy invasion beaches in addition to inadequate supply routes that affected the mobility of the Third Army. On September 25, Patton's drive was finally halted due to a lack of supplies and gasoline, which had been a problem since the end of August. Even though logistic problems piled up, General Eisenhower reluctantly decided to approve Montgomery's Operation Market Garden to gain a bridgehead across the lower Rhine through Arnhem in the Netherlands. Patton was piqued, to say the least, at Eisenhower's support for Montgomery. The enemy now became more tenacious in defending the Fatherland. In addition, the German army had the advantage of a series of impressive terrain obstacles, such as rivers, mountains, and forests.

What operations were maintained during the initial period of gasoline shortage? Walker's corps was not idle. Early in September, the 87th Cavalry Reconnaissance Squadron, commanded by Lieutenant Colonel Robert J. Stewart, was tasked for a special operation to force an opening toward the Moselle River and attempt to seize bridges for Major General Lindsay McD. Silvester's 7th Armored Division so his armor could establish a bridgehead beyond the river. This was the first time that the squadron had the opportunity to function as an economy of force element. Previously, it had engaged in reconnaissance roles where troops, supported by assault platoons, were assigned to various combat commands. To facilitate the squadron's mission, Walker reinforced it with infantry, tanks, tank destroyers, field artillery, and combat engineers. The cavalry task force was to move over four routes with four unbalanced forces. The squadron took on the shape of a light combat command and was given a mission-type order usually reserved for armor units.

Why was this mission-type order given to a cavalry squadron? There are two reasons to consider. First, it was determined that the squadron was the easiest force to assemble and provide with gasoline when the C 47 cargo planes arrived. There was also the enemy situation, which was perceived as degraded, unorganized, and in retreat. It was not realized that the German army was now able to reorganize a defense as a result of the gasoline shortage that stopped the Third Army. However, Walker and Silvester had not taken this into consideration when tasking the cavalry task force. Though the reinforced squadron advanced in four columns, it failed in its mission because it could not find crossing sites. The problem was that the cavalry task force moving simultaneously in four columns was unbalanced. Shortly the infantry attachment was relieved, and in a last attack the cavalry task force was confronted with determined German artillery and mortars, which were organized and thus able to concentrate fire. Stewart, recognizing the futility of continuing the operation, canceled the attack. The squadron now returned to division control to rest, reorganize, and pull maintenance. It was time for the cavalrymen to wait for a new mission.[82]

Another offensive operation, aside from a special mission, by the 87th Cavalry Reconnaissance Squadron occurred on April 11, 1945. The mission was to capture the town of Meinkenbrackt as part of the 7th Armored Division's mopping-up operations in the Ruhr pocket. The operation called for dismounted action, as three-plus troops advanced rapidly through the woods to assault the town. Meanwhile, one troop provided flank security and was given the mission of taking the high ground north of the town. The action resulted in a rapid and successful dismounted assault against the defended Meinkenbrackt, which had been hammered by the 434th Armored Field Artillery Battalion in direct support. This operation by the squadron demonstrated the effectiveness of dismounted mechanized cavalry supported by field artillery.[83]

Meanwhile, back to the situation along the Siegfried Line. Beginning in October 1944, the Allies had to deal with the worst weather in decades as they approached Germany. As a result, in October and November a continuous cold and drenching rain fell on the combatants. Because of Eisenhower's broad-front strategy, XX Corps' main body was now immobilized at Verdun. Lieutenant General Courtney Hodges's First Army found itself up against the cold and muddy Hurtgen Forest, thickly wooded, with mine-filled narrow trails and steep defiles that favored the enemy. The forest was no place for armor due to a lack of maneuverability. Again, German artillery made life miserable by firing tree-high bursts, spreading lethal splinters over American infantrymen as they struggled through the forest in anticipation of reaching the Roer River. For two months the struggling infantry had to endure bloody fighting in the cold, snow, and mud that had a devastating impact on morale.

Defensive combat for mechanized cavalry squadrons occurred in many situations during the attempted breakout of the Hurtgen Forest, which resulted in dismounted action. One example was the 85th Cavalry Reconnaissance Squadron commanded by Lieutenant Colonel George C. Benjamin. The squadron was assigned to the 5th Armored Division, which was maneuvering as part of the V Corps. For the ensuing roadblock combat, the 85th was under the command of MacDonald's 4th Cavalry Group for tactical employment, such as screening the division's southern right flank. Again the issues of weather, terrain, and the enemy were the imponderables. Troop A from the squadron was given an infantry mission on December 20, 1944, to set up a roadblock and cut off a main German logistics route. The mission was made more difficult because the troop had to travel to its defensive position without vehicles and hand-carry additional ammunition for its rifles, grenades, and bazookas, which were the only antitank weapons available. The movement was under arduous conditions as they moved to occupy key but uninviting terrain. For days the cavalrymen were confronted with the task of dismounted action. The pressure was intense. The Wehrmacht was pushed against the Roer, hemmed in and in need of reinforcements to be moved over critical roads. The spirited cavalrymen held their position against determined enemy attacks. With the help of logistic support—but without supporting troops—the cavalrymen demonstrated versatility and flexibility by adjusting back to their traditional mission and fighting dismounted until relieved on December 23 by an infantry-tank team.[84]

Colonel John T. Cole, the commander of Combat Command B, 5th Armored Division, had realized, as did Colonel Benjamin, that the undertaking was unusual because it was an infantry mission. Cole felt he had to order it. He recalled that he had "only a vague recollection of this particular situation, as the whole week was a nightmare in all respects." Continuing, "By skillful use of terrain and through digging in, they [the cavalrymen] held their position despite repeated attacks by German armor and infantry, . . . some of whose vehicles and tanks were not stopped until put out by hand within 30 yards of their position."[85]

Another example of defensive action by a mechanized cavalry reconnaissance squadron occurred during the initial stages of the Battle of the Bulge. Troops from the 89th Cavalry Reconnaissance Squadron assigned to the 9th Armored Division were bruised in attempts to counterattack the oncoming Germans. Finally, pressure forced the cavalrymen to engage in a delaying action and then organize a defensive position. Once it was secured, the recon troopers started active patrols. On December 21, a task force from the 10th Armored Division attacked north of the 89th and secured its positions, thus allowing the cavalrymen to maintain active patrolling along the pressure points on each of their flanks until December 26, when the squadron was withdrawn for a rest.[86]

The next example is a security mission by a mechanized cavalry squadron. By the end of March 1945, the 6th Armored Division was given the mission to "seize Kassel," which contained the huge Henschel tank works and fixed antiaircraft batteries capable of firing deadly flak not only against planes but also against advancing U.S. ground troops. The city also offered a barrier to crossing the Fulda River. It was Grow's philosophy to advance in multiple columns toward Kassel and cross the river on a broad front, thus seizing an opportunity to secure undamaged bridges and establish a bridgehead. However, Walker changed the mission of the 6th Armored Division to cross the Fulda south of Kassel and continue the attack toward Muhlhausen. To make a rapid crossing, Grow broke up his division into numerous reconnaissance units to search for a place to cross. When the division reached the Fulda, all the bridges had been destroyed except for one railroad bridge a few miles north of the planned crossing in the vicinity of Malsfeld. Engineers built a bridge during the night, and a combat team fought dismounted across the river. Meanwhile, the 86th Cavalry Reconnaissance Squadron screened the movement by picking its way unsteadily across the railroad bridge. Once across the Fulda, the division came up against the Werra River and determined enemy resistance. Two bridges were intact; one was heavily defended on the opposite side, but advancing forces secured the other. Grow then decided to send three of his armored columns over the secured bridge.

Eventually the entire division crossed and again began fanning out toward Muhlhausen, a city of about fifty thousand inhabitants that included important German officials and served as a critical supply and transportation center. Bob Grow, thinking as a cavalryman, noted that just west of the city was the Thuringian plain, "as fine a maneuver ground as one can imagine." The mission was to move in a crescent formation to encircle and secure the city and prevent the escape of these important officials.[87]

The 86th cavalrymen were given the mission of screening the exposed north or left flank of the division while six combat teams poured out of the hills west of Muhlhausen. The combat teams quickly deployed, knocking out all opposition. The city was completely surrounded and secured. Grow was pleased with this tactical maneuver. To him it was cavalry actions complete with mounted speed. During the encirclement, the 86th was involved in a few concentrated attacks by enemy forces that were repulsed by the cavalrymen. Later it was determined that because of the screen, the enemies' plans were disturbed and they were forced to move east rather then south as originally planned.[88]

Since the Brittany campaign, the 86th Cavalry Reconnaissance Squadron had not operated in an economy of force role. Its unit integrity was split. As the division approached the port city of Brest, Grow placed units of the 86th under control of his combat commands rather than under division control. The former commander of the squadron, Colonel Harris, believed cavalrymen might have been more effective if allowed to operate as a unit. It was the feeling that squadron leaders could have given better support to reconnaissance troops had they been under their control rather than separated and under the control of combat command commanders. The 86th screening in an economy of force role at Muhlhausen was successful. It raised the issue of an economy of force mission in mechanized cavalry operations, meaning that squadron commanders could have been more effective had the cavalrymen worked together as a unit.

Finally, an example of reconnaissance as a combined arms economy of force element was the action of the 92nd Cavalry Reconnaissance Squadron assigned to the 12th Armored Division in the Neustadt area. Supported by one light tank platoon and two medium tank platoons and the 342nd Armored Field Artillery Battalion, the squadron began closing in on the retreating enemy. On April 8, 1945, the cavalrymen with tanks moved out. Soon they encountered stiff German resistance. For days the squadron maintained contact and pressured the retreating Germans. By April 15, the 92nd had completed its reconnaissance in force, fighting for information necessary to maintain the exploitation and pursuit by the 12th Armored Division.[89]

The selected historical examples noted here provide some insight into the employment of cavalry reconnaissance squadrons. When examining all cavalry squadrons in western Europe during World War II, one has to look at missions performed. The most frequent at 48 percent was special operations, followed by security at 24 percent, reconnaissance at 13 percent, defensive combat at 11 percent, and offensive combat at 4 percent. Mechanized cavalry reconnaissance squadrons assigned to light armored divisions and armored reconnaissance battalions assigned to heavy armored divisions performed considerably more reconnaissance missions than did cavalry groups. Frequently, cavalry squadron troops were detached for use with the division's combat commands and task forces.[90]

Biddle Report, MHI

UNIT	OFFENSIVE		DEFENSIVE		RCN		SECURITY		SPEC OPNS		DAYS CONSIDERED
——	Days	%	Days	%	Days	%	Days	%	Days	%	——
25th Cav Rcn Sq, Mecz	4	2	10	5	25	12	110	51	74	30	223
82d Armd Rcn Bn	21	7	35	11	75	23	15	5	188	54	334
85th Cav Rcn Sq, Mecz	10	4	52	21	20	8	56	22	113	45	251
86th Cav Rcn Sq, Mecz	2	1	47	17	76	27	59	20	102	35	286
87th Cav Rcn Sq, Mecz	6	3	21	11	6	3	9	5	148	78	190
89th Cav Rcn Sq, Mecz	0	0	16	10	0	0	44	26	107	64	167
90th Cav Rcn Sq, Mecz	6	4	1	1	15	10	66	44	62	41	150
92th Cav Rcn Sq, Mecz	19	12	6	4	23	14	55	35	57	35	160
TOTAL (No. of days)	68	—	188	—	240	—	414	—	851	—	1761
GRAND TOTAL No. of Days	—	—	—	—	—	—	—	—	—	—	
Percentages		4		11		13		24		48	100

NOTE: Offensive combat includes attack, and pursuit and exploitation. Defensive combat includes defense, delaying action, and holding key terrain. The accomplishment of reconnaissance missions normally involved offensive combat. Security includes screening (both moving and stationary), blocking, flank protection, maintaining contact, and filling gaps. The performance of security missions frequently involved offensive combat. Special operations includes mobile reserve, rear area security, and Third Army Information Service.

SOURCE: Report of the General Board, United States Forces, European Theater, "Tactics, Employment, Technique, Organization, and Equipment of Mechanized Cavalry," Armor Section, ETO, Study No. 49, dated November 20, 1945, MHI.

Mechanized Cavalry Reconnaissance Troops, Infantry Divisions

Regarding infantry divisions, mechanized cavalry reconnaissance troops were normally employed intact and used for security, special operations, and, to a lesser extent, reconnaissance and defensive and offensive combat. Two uncommon examples of the employment of mechanized cavalry reconnaissance troops in infantry divisions will be examined.

The first example at a troop level was an infrequent reconnaissance mission and cooperation with an armored division by the 80th Cavalry Reconnaissance Troop during the advance of the 80th Infantry Division commanded by Major General Horace L. McBride. The reconnaissance mission had its origins late in March 1945, when the XX Corps commander, General Walker, ordered the 6th Armored Division to Leipzig. But first it had to take Kassel. The 80th Infantry Division was to follow Grow's division. Meanwhile, Lieutenant Colonel Harry Brindle's 86th Cavalry Reconnaissance Squadron from the 6th Armored Division, followed by the 80th Infantry Division, was leading two combat commands moving abreast. During this period D Troop (supported) from the 86th moved rapidly to secure a bridge over the Eder River in close proximity to the Fulda River. This move was critical to XX Corps' attack against Kassel. Moving quickly over the intact bridge, M8 armored cars ran over a few German soldiers on the east bank. Taking positions, the reinforced troop proceeded to repulse enemy attempts to destroy the bridge and bridgehead.[91] At the end of March, the 6th Armored Division was advancing toward the Eder River south of Kassel. By early April, the "Super Sixth" had crossed the Fulda River. At first, the outer defenses presented some problems, stopping one of the division's armored infantry battalions. Meanwhile, the 80th Cavalry Reconnaissance Troop reconnoitered in advance of McBride's division move toward Kassel. When the 80th Division reached the outskirts of the city, it met unyielding German resistance.

General McBride believed the city could be taken by an armor thrust. He told Grow that there seemed to be little in the way of taking Kassel itself. McBride encouraged Grow to push through his armor. He was wrong. Grow remembered all too well his division's experience with the fortified port city of Brest and Patton's "mobility unused" during the efforts to take Metz. Kassel's defenses were still organized, making an armor operation difficult. However, the German defenders were relatively immobile and pinned to the city. The next day Walker halted Grow's division six miles from the city. Walker then ordered Grow to move his division east, jump the Werra River, and take Muhlhausen, an important communication center, while McBride's infantry was dealing with Kassel. Mission orders went out from McBride to the 80th recon troops to reconnoiter the division's zone of attack. For three days the cavalrymen engaged in reconnaissance. Enemy resistance was light and scattered at first, but by the time the 80th Infantry Division entered the city it had a fierce fight on its hands—often house to house. For four days the infantrymen had to deal with counterattacks, including tanks, until the German garrison finally surrendered on April 4. Meanwhile, recon troops continued to reconnoiter and secure positions with adjoining units. Shortly before the garrison surrendered, McBride ordered the cavalry troop to reconnoiter and secure the division's right flank south of the city.[92]

A second example is a rare offensive mission by a reinforced cavalry troop. During the last month of the war, the 104th Cavalry Reconnaissance Troop was assigned to the 104th Infantry Division commanded by Major General Terry de la Mesa Allen and part of General Collins's VII Corps. Allen was the former 1st Squadron commander, 6th Cavalry. General Bradley did not like Allen, known as "Terrible Terry" when he commanded the 1st Infantry Division in North Africa. In spite of Allen's leadership qualities, Bradley found that the rambunctious former cavalryman and his infantry division lacked discipline. Allen disapproved of the maxim that discipline makes good soldiers. Earlier, Bradley had viewed Allen as a rebel and sent him back to the United States before the invasion of Sicily.[93] In October 1943, Allen, unlike other leaders from the North African debacle, was back in the good graces of General Marshall, who decided to give him command of the "Timberwolves" division. He began turning the division into an effective fighting force.

After participating in the encirclement of the industrial Ruhr, which was completed early in April 1945, plans were set to continue the move east. By mid-April the encirclement had led to the capture of more than three hundred thousand worn-out and dejected German soldiers. This operation now completed, General Collins, as part of Hodges's First Army, began positioning his corps to meet the converging Red Army. With the dramatic success of the Allied armies, it became only a matter of time until the end of the war. The thousand-year Reich was no longer organized to stop the Allies. Soon the collapsing Nazi regime lay in ruins as a result of total war. Meanwhile, the 1st and 104th Infantry Divisions were given a mission order to follow the 3rd Armored Division as it spearheaded the drive to the east. The objective was to advance "along the axis Kassel-Leipzig to gain contact with the Soviet Forces." Hodges's army was to exploit any opportunity for seizing a bridgehead over the Elbe River and be prepared to continue the advance east.[94]

On April 4, the 104th Cavalry Reconnaissance Troop moved out as a reinforced task force, protecting the division's left flank and then moving ahead of the division. The cavalry troop was reinforced with a platoon of infantry, an assault gun platoon, and the 750th Tank Battalion attached. The reinforced cavalry speedily fought its way toward the Weser River in advance of the division's regimental combat teams. Though it faced a lightly defended front, the 104th Cavalry Recon Troop Task Force captured succeeding key enemy positions before reaching its final objective, the Weser River, with all its units intact. Unfortunately, all the bridges over the river had been destroyed. Nevertheless, the mission of the reinforced cavalry troop for offensive action made it possible for "the bulk of the 104th Infantry Division to advance without loss of time." Even though the bridges were blown, the division—using ferries and a bridge in another corps area—established a foothold on the east bank. Once it was established, engineers built bridges across the Weser, allowing the 1st Infantry Division with both tanks and artillery to cross. The 104th Cavalry Recon Troop was given a new mission protecting the division's flank as it moved toward the Leine River, east of the Weser.[95]

The exploits of the 80th and 104th Cavalry Reconnaissance Troops demonstrated two rare occurrences in deployment. Overall reconnaissance in the European theater comprised 6 percent of missions performed by cavalry troops assigned to infantry divisions, and offensive combat was at 1 percent. After the war a study determined that security missions consumed one-half of the time cavalry troops spent while in combat. Special operations were at 39 percent, and defensive combat at 4 percent. In most cases cavalry troops assigned to infantry divisions were employed intact. During the drive to Kassel, D Troop (Reinforced) from the 86th Cavalry Reconnaissance Squadron also engaged in rare offensive combat to secure a key bridge over the Eder River.

In conclusion, it can be said that again the mechanized cavalry went through changes caused by the turmoil and tensions of warfighting. With the horse gone, cavalrymen had to accept the fact that doctrine was evolving, and existing techniques and equipment most of the time had not met mission expectations. To be sure, the preinvasion doctrine bore no relation to events and demands of the European battlefields. Mechanized cavalry elements participated in every major campaign in the ETO, compiling a remarkable record by conducting any combat mission assigned by higher headquarters. Daniel Van Voorhis's original ideas developed at Fort Knox during the 1930s and recorded by Grow were confirmed—mechanized cavalry could perform all cavalry missions without horsepower. After the war, cavalry leaders reflected on their changing missions and efforts at managing the chaos of modern warfighting, attempting to find a role for cavalry in the combat arms team.

After the end of hostilities, a General Board, Armored Section, was established "to prepare a factual analysis of the strategy, tactics, and administrative employment by U.S. Forces in the European Theater."[96] One board dealt with ex-

Biddle Board, MHI

UNIT	OFFENSIVE		DEFENSIVE		RCN		SECURITY		SPEC OPNS		DAYS CONSIDERED
———	Days	%	Days	%	Days	%	Days	%	Days	%	———
1st Cav Rcn Troop, Mecz	0	0	5	2	0	0	18	68	82	30	271
3d Cav Rcn Troop, Mecz	Data on types of missions by days not available.										
5th Cav Rcn Troop, Mecz	4	2	0	0	3	2	85	47	87	49	179
30th Cav Rcn Troop, Mecz	1	0.3	26	9	16	5	115	39.7	136	46	294
79th Cav Rcn Troop, Mecz	0	0	7	4	21	11	62	34	93	51	183
80th Cav Rcn Troop, Mecz	2	2	3	3	19	20	29	29	46	46	99
83d Cav Rcn Troop, Mecz	0	0	0	0	12	5	175	73	54	22	241
90th Cav Rcn Troop, Mecz	0	0	7	9	21	26	41	50	13	15	82
104th Cav Rcn Troop, Mecz	11	6	10	5	3	2	70	39	87	48	181
TOTAL (No. of days)	18	—	58	—	95	—	761	—	598	—	1530
GRAND TOTAL No. of Days	—	—	—	—	—	—	—	—	—	—	
Percentages	—	1	—	4	—	6	—	50	—	39	100

NOTE: Offensive combat includes attack, and pursuit and exploitation. Defensive combat includes defense, delaying action, and holding key terrain. The accomplishment of reconnaissance missions normally involved offensive combat. Security includes screening (both moving and stationary), blocking, flank protection, maintaining contact, and filling gaps. The performance of security missions frequently involved offensive combat. Special operations includes mobile reserve, rear area security, and Third Army Information Service.
SOURCE: Report of the General Board, United States Forces, European Theater, "Tactics, Employment, Technique, Organization, and Equipment of Mechanized Cavalry," Armor Section, ETO, Study No. 49, dated November 30, 1945, MHI.

amining various tactical operations conducted by the cavalry during the war. Its mission was to "prepare reports and recommendations on the tactical employment, technique, organization and equipment of mechanized units." The results were listed in the Report of the General Board, United States Forces, European Therter, "Tactics, Employment, Techniques, Organization, and Equipment of Mechanized Cavalry" (1945). Brigadier General Joseph A. Holly chaired the board, whose analysis became known as the "Biddle Report," named after Colonel William S. Biddle, who commanded the 113th Cavalry Group and compiled the extensive document. The board relied on wartime experiences and combat records, such as after action reports and numerous army, corps, division, cavalry group, squadron, and troop commanders' comments on mechanized cavalry operations. This board's extensive report was one of the most forthright reflections on mechanized cavalry operations by senior cavalry officers and high commanders from corps to army level.

One of the most interesting observations came from Fickett's 6th Cavalry Group. The commander of the group's 28th Cavalry Squadron, Major Thomas H. Stewart, sent a postwar review to his superior, Colonel Fickett. At the time Fickett was preparing a larger report for the postwar Cavalry Board, which was to review and analyze all cavalry operations at three levels. The squadron commander's review became controversial. Included was a paragraph recommending armored divisions be disbanded. The argument was that the war demonstrated that they were seldom employed alone and at corps tactical level were usually combined with infantry divisions. In almost all cases, he argued, an infantry regiment or a regimental combat team had to be attached. Furthermore, he noted, infantry divisions normally had been adequately provided with a tank battalion and mechanized equipment. He also suggested that the tank destroyer arm be dissolved. "It is not a tank and should be incorporated in the Field Artillery," said Stewart. The same was said of antiaircraft artillery.[97]

Fickett concurred with Stewart's findings that armored divisions should be disbanded, adding that the campaign in western Europe was a cavalry operation of exploitation and pursuit. His reasoning was that after the Saint-Lô breakout in July 1944, major aspects of campaigns were characterized by a combination of mobility, heavy firepower, and shock action. These actions, he noted, are "certainly the actions expected of highly mobile troops such as cavalry." He cited those cavalry groups that were organized as task forces, where units were intermixed with combined arms for a special mission or operation, commanded by a cavalry officer. As expected, Fickett's suggestions sent disturbing vibrations through the armor community, which still had not achieved formal branch status. In addition to his own task force, Fickett noted three others that he considered outstanding: Task Force Polk (3rd Cavalry Group (Mech) (Reinforced); Task Force MacDonald (4th Cavalry Group (Mech) (Reinforced); and Task Force Biddle (113th Cavalry Group (Mech) (Reinforced).[98]

In fact, all cavalry group commanders were top-quality professionals who adjusted when tested by the demands of combat. It was Hank Reed's leadership that drove the 2nd Cavalry Group. And it was Reed, not Patton, who rescued the famous Lipizzaner stallions of Austria at a German remount depot in Czechoslovakia before the Red Army arrived, thus avoiding possible extinction of the historic breed. He gave the orders before he told Patton what he was going to do. Unlike many other subordinate commanders, Reed, the old horseman, was comfortable around the Third Army commander, and thus able to pull off the rescue that Patton took credit for.[99]

Two West Point graduates, Colonel Joseph M. Tully, who commanded the 4th Cavalry Group, and Colonel Lawrence G. Smith, who commanded the 14th Cavalry Group, were noted for their aggressiveness. Tully eventually commanded the 90th Infantry Division with distinction. Smith drove his cavalrymen with skill during the Bulge breakout and retired after thirty-four years of active service.[100]

Cyrus Dolf, the most junior of the group commanders and the only infantryman, was another outstanding cavalry leader. He shaped the 102nd into a formidable fighting force after its poor performance in Normandy.

A platoon from Colonel Charles B. McClelland's New York National Guard's 101st Cavalry Group claimed they captured Field Marshal Albert Kesselring and his entire staff at Zeller See on May 8. The 101st Cavalry Group also took credit for capturing numerous German government officials and the Japanese ambassador and his diplomatic staff.[101]

Colonel Wilson of the 106th Cavalry Group certainly summed up the feelings of many of his cavalrymen: "We can feel a special reverence for those comrades whom we knew intimately, whom we saw while their blood was still warm, men who did not let us down in a fight."[102]

All these task forces Fickett referred to were combined arms elements built around the cavalry group and used by army or corps commanders as weapons of opportunity. He argued that these successful task forces were mission tasked by higher commanders to perform operations beyond "the powers and capacities of any other type of troops available to them." Fickett complained, however, that cavalry elements of the task forces suffered from equipment, limitations of organization, and inadequate training placed upon them by faulty concepts. He argued that this was caused by the existence of an organization emphasizing reconnaissance. This "has been definitely determined to be unsound," he wrote, especially when many cavalrymen had to fight as infantry.[103]

Colonel Polk wrote to the board that Herr's 1941 Cavalry "*FM 2-15* must be entirely rewritten" because "horse cavalry is strictly a special purpose unit." Mechanized cavalry, he noted, "has assumed the traditional role of cavalry except in the concept of its use in mass." The war had changed the techniques of cavalry; "it has become the light, fast, hard hitting unit of the corps." More so, he stated, mechanized cavalry was a complement to armored and infantry divisions because it could be grouped into brigades or divisions for special operations. Polk believed mechanized cavalry was ideally suited to the task force configuration. Fickett added that cavalry groups usually were employed as light combat commands. Lieutenant Colonel Charles J. Hodge, commander of the 117th Cavalry Reconnaissance Squadron attached to the Seventh Army, made a weighty statement to the board that "the mission of cavalry is to fight." This pronouncement echoed the voices of Van Voorhis, Chaffee, and Grow during the early 1930s, when these pioneers were reshaping the doctrine and techniques of cavalry operations. Hodge added that any teachings, such as those prescribed in cavalry field manuals that limited the mission to "sneak and peek" reconnaissance, were far removed from the actual experiences in the ETO. As the techniques moved away from stealth and observation, "it became a dangerous doctrine to the young cavalry officer," he concluded.[104]

Lieutenant Colonel John J. Rhodes, commander of the 4th Cavalry Reconnaissance Squadron (Mech) attached to the 4th Cavalry Group, told the board that he was convinced the war demonstrated nothing new in tactics. "The doctrine of horse cavalry as has been taught for years are sound, correct, and applicable to modern war," he explained. He commented that his group "was repeatedly committed in every type of role except one—we were not dropped by parachute or glider." He added, "We were committed both mounted and dismounted in attack, penetration, mop-up, pursuit, and defense, in every conceivable type of terrain and weather."[105]

General Simpson, the commander of the Ninth Army, viewed the matter differently. He believed the general mission of mechanized cavalry was reconnaissance, and "it is generally agreed that this will be its future role." For the future, Simpson believed cavalry groups should remain under army control; General Harmon, who commanded the XXII Corps at the end of the war, said the groups should be abolished and the cavalry regimental organization reestablished.[106] General Patton agreed with Harmon, as did First Army Headquarters. A board of officers from Third Army Headquarters, including Colonels Polk, Reed, and Wilson, and Colonel Lawrence G. Smith from the 14th Cavalry Group (Mech), recommended, along with First Army staff, that a cavalry regiment of two squadrons replace the groups.[107]

Major General John W. Leonard, commander of the 9th Armored Division, agreed to some extent with Simpson that mechanized cavalry units should be organized and trained to perform reconnaissance and security, and when necessary trained for combat missions. He did concede that in his experience, the 89th Cavalry Reconnaissance Squadron was rarely used for stealth "sneak and peek." The problem, he maintained, was that the squadron was generally "used on missions involving combat and had to be strongly reinforced." The commander of the 2nd Armored Division, Major General I. D. White, viewed his experience differently. He believed that there should be no limit placed on the type of mission given a mechanized cavalry unit, adding, "We should teach that aggressive action generally requiring combat is the best way to obtain information of the enemy."[108]

The board also took up the issue of cavalry equipment, which became controversial. For example, retaining the M8 armored car was an important necessity according to Polk. The only reason for its existence, he argued, was that it was faster and more mobile in mechanized cavalry operations than the elements of an armored division. He believed, "We have now reached our top limit on tanks." He also argued for improving the widely used half-track and activating a 81mm mortar platoon. Colonel George Busbey, who with Polk was part of the 316th Provisional Cavalry Brigade, stated that the need for a wheeled vehicle was enduring, adding that the cavalry's basic vehicle should not be the tank. The most interesting comment came from Colonel John Rhodes, who claimed that the 4th Cavalry Recon Squadron's experiences demonstrated that wheeled vehicles, such as the jeep and the M8 armored car, were preferred over tracked vehicles. His argument was based on the drive across France and Belgium when his squadron was never delayed or stopped as a result of vehicle failure. The M5 light tanks were a problem, however, because they were repeatedly left behind for maintenance. General White disagreed, stating, "I have no use whatsoever for the M8 armored car." He favored a full-tracked personnel carrier because the current table of organization and equipment "does not lend itself to dismounted combat." This, he believed, was important because modern cavalry "must be trained as riflemen for dismounted action" supported by light tanks and assault guns. Lieutenant General Lucian K. Truscott Jr., who salvaged the VI Corps at Anzio, entered into the debate. Regarding armament and equipment, he preferred not to go "whole hog," meaning settling for either all M8 armored cars or all light tanks. He preferred a more balanced approach, such as the combination of armored cars and M24 light tanks introduced to cavalry units at the beginning of 1945. General Walker wanted a 105mm howitzer mounted on the M24 tank chassis to be used by cavalry troops as an assault weapon. The importance of liaison aircraft and the need for improved radio sets to enhance communication were noted. There was a general agreement that tank destroyers were not suitable for the mechanized cavalry.[109]

The consensus was that mechanized cavalry was not limited to reconnaissance, "nor did such develop to be the primary mission of that arm." Reconnaissance was frequently performed but usually in conjunction with "the execution of other missions rather than as a mission of its own." The information on the enemy's intent and dispositions had to be fought for rather than acquired by stealth and observation. In addition, supporting aviation greatly reduced the need for distant ground reconnaissance. Thus, a commander who had mechanized cavalry elements rarely used them for pure reconnaissance, instead employing them on combat missions. What also was evident was that dismounted patrolling was handicapped by an inadequate number of troops assigned to fulfill a mission.

The report concluded that mechanized cavalry executed numerous traditional missions, such as offensive and defensive combat, special operations, screening, seizing and holding important terrain, security, combat liaison, mobile reserves, and deep operations against designated objectives. This flexibility enabled the mechanized cavalry to accomplish a great variety of missions, especially when assigned to a task force or combat command.

Certain characteristics, such as mobility, were superior as long as mechanized cavalry was confined to roads. Off-road capability and cross-country ability were limited to the cavalry's primary wheeled vehicles, the M8 armored car and jeep. The concern of many cavalrymen was the 37mm gun mounted on the M8 armored cars and the M5 light tanks. The M5 proved to be inadequate in almost all operations, except in light reconnaissance and screening actions. The 37mm gun, however, was effective against soft targets but had little effect against German tanks. There was no contest. With its introduction early in 1945—in time for the Rhine River crossing—the M24 light tank mounting a 75mm gun and torsion-bar suspension system seemed to be an improvement. It was near the end of the war, and the M24s that replaced the M5s in cavalry units never were able to demonstrate their usefulness except on rare occasions. Attached tracked vehicles, such as the M10 and M36 tank destroyers, compromised mobility due to their inadequate speed and excessive weight. The M18 76mm tank destroyer was preferred because of its lightweight and improved suspension system utilizing torsion bars. The assault gun was considered to be of "great value," especially the M7 self-propelled 105mm howitzer. As the war progressed, most cavalry units preferred the M7 105mm howitzer to the snub-nosed M8 75mm howitzer. The cavalry, however, was looking for a 105mm howitzer mounted on a lighter but full-track chassis.

The question of cavalry dismounted firepower became a major issue. For example, when fighting dismounted, cavalrymen did not have available the Browning automatic rifles and M1 Garands used by the infantry. The M1 carbines were not preferred because of the short and lighter cartridge and the non-adjustable "L-type" rear sight. The 60mm mortars proved early in combat to be inadequate. By the end of the war in Europe, 81mm mortars had proved to be more suitable for cavalry support. Though the mechanized cavalry demonstrated its ability to operate in enemy territory, units found that their capability suffered from a shortage of organic vehicles for transporting logistics to support their mission. Another problem was that a number of units were fixated, due to training, on a primary mission of reconnaissance and a minimum of fighting. This mind-set caused some cavalry units to begin operations "with a mental obstacle to offensive combat." Many times, especially in defensive operations, units suffered from a lack of reserves and reinforcements, especially when committed to combat for long periods.

The most negative organizational move, according to the board report, was that before the landings in France mechanized cavalry groups were reorganized with two separate reconnaissance squadrons. Frequently during operations, they were split into various task forces or operated as dismounted infantry. The feeling was that there was a lack of economy of force dispositions, which in turn compromised unit integrity. The board also noted that reorganization before the landings in France had an adverse affect on morale and compromised the efficiency of group squadrons in the field. The consensus of the board was that the principal organizational pattern should be a three-squadron regiment, with the designation of mechanized cavalry changed to "cavalry." This was an effort to reestablish the cavalry branch's identity. The maximum regimental strength suggested was thirty-five hundred officers and men with a complement of 850 fighting vehicles. For the future, the board recommended a mechanized cavalry brigade be formed and tested. As noted earlier in the war, cavalry squadrons consisted of four reconnaissance troops mounted in M8 armored cars and jeeps. Next were the lightly armored cavalrymen mounted on assault guns and light tanks. The remaining troops were service and medical detachments. After the war the board suggested that a dragoon troop or infantry elements be added to the cavalry squadron to provide greater firepower for dismounted action.

Regarding the assigned mission of reconnaissance with a minimum of fighting, the board concluded the war proved it unsound. For the future, the board suggested that, as demonstrated by operations in Europe, mechanized cavalry still perform its traditional role. However, it needed to be highly mobile, heavily armed, and lightly equipped to fight as a combat force. The debate over tactical doctrine expressed in *FM 2-15: Employment of Cavalry* found two different opinions expressed: keep the provisions except those that only apply to horse cavalry or completely revise the manual. The most interesting suggestion made by the cavalry board regarding doctrine and technique was encouraging greater coordination and cooperation between all engaged units from corps down, including improving cooperation with the air forces. One proposal was to place greater emphasis on combined arms maneuvers. This suggestion to change doctrine through coordination with other combat arms indicated that the board was seriously making an effort to move away from pure reconnaissance and establish a mechanized cavalry organization to accomplish a range of missions and fight as the vanguard of the main forces. Also, the board believed that the capabilities of a light, fast, and hard-hitting mechanized cavalry force needed to be explored as nuclei for combined arms task forces. Because the future battlefield would likely see more mobility and thinly spread units, the board noted, "there exists for mechanized cavalry a role in the field for light armor, which is of unique importance to the accomplishment of the mission of the ground forces." Continuing, before these suggestions are implemented, changes in organization and equipment "must be made."

Concerning the general mission, tactical doctrine, technique, and future role of mechanized cavalry, the board made a number of recommendations that *FM 100-5* be revised and *FM 2-15* be carefully examined and revised as necessary. On the question of equipment, the board stated that the tables of organization and equipment should be revised and that efforts should be made to improve the armored car and its range. As controversial as the M8 Greyhound was, it again became critical after the war when cavalry units formed a constabulary force—a new mission that was neither anticipated nor planned. The peacetime army did not adopt all the recommendations made by the board. However, by 1948 many of the recommendations became the basis of the first postwar cavalry organization when the 3rd Cavalry Regiment (Light) at Fort Bliss, Texas, adopted a new table of equipment and organization, which included three reconnaissance battalions, composed of light tanks for each company, infantry elements, and self-propelled 105mm assault howitzers. The reorganization of the constabulary into the 2nd, 6th, and 14th Cavalry Regiments followed. A company of medium tanks was also assigned to each battalion. Most important was that this new organizational structure provided more organic assets, which were a major squadron problem during the war, such as infantry elements and antitank capabilities. A few years later doctrine was updated when the army issued *Field Manual 17-35*, which stated that the new mission of the reconnaissance battalion was "to engage in offensive or defensive combat, either mounted, dismounted, or in combination of both, primarily in execution of security and reconnaissance missions."[110]

While the Cavalry Board was debating the future of the cavalry arm, the War Department ordered the establishment of an equipment board, named after the respected General Joseph W. Stilwell, who was appointed its chairman.

After an extensive examination of tactical employment of the combat arms based on postwar reports from Europe, the Stilwell board recommended that "cavalry be discontinued as a separate arm." It was a doctrinal milestone and again questioned the future role of cavalry in the U.S. Army.[111] Unlike before the war, when the army had to adjust its equipment to a doctrine based on branch compartmentalization, the Stillwell Board attempted to solidify a combined arms doctrine before equipment development. Cavalry, however, did not have an independent role in this postwar doctrine or in equipment development.

9

The Terrible Turmoil of Postwar Germany and the U.S. Constabulary

Unquestionably, postwar Germany was in terrible turmoil. The devastation brought about by total war produced numerous challenges in how to manage the chaos within the context of political, economic, and personnel turbulence. This situation was vividly described by Brigadier General Albin F. Irzyk, a cavalryman and tank battalion commander with the 4th Armored Division during the war. Irzyk wrote that after the defeat of Germany a staggering achievement occurred when a new organization was created mainly from armor and mechanized cavalry elements.[1] This new police-security organization, the U.S. Constabulary, was expected to be lighter, more mobile, and faster and to have the ability to cover considerable territory. Its motto was "Mobility, Vigilance, Justice," and the men referred to themselves as the "Circle C Cowboys." Even some mechanized cavalry elements again became horse-mounted. It was evident that change and adjustment were ever present, thus providing new opportunities for cavalry elements soon to become part of a new and unprecedented organization, the Constabulary. This new military force, however, would never see its homeland.

The history of the U.S. Constabulary is in part another story of the mechanized cavalry, which was originally structured on its World War II organization that in itself went through changes as the war progressed. A constabulary force was not new to the army. It was used for the first time in the Philippines in the early 1900s. As the war was ending, changes again were imminent because of the emerging hostile atmosphere brought on by international dynamics that led to new antagonistic debates after the wartime Grand Alliance defeated Nazi Germany in a total war. This growing hostility between the Western Allies and the Soviet Union led to a long and drawn-out period just short of war. It became known as the cold war. Even before World War II ended, the Allies had conflicting ideas about how to create a peacetime Germany. Should the defeated country be restored as a pluralistic capitalistic democracy or transformed into a communist state? Within the setting of attempted new national peacetime goals and the rapidly unfolding international events, a U.S. Constabulary came into being as a new breed of U.S. Army cold war troopers.

What were the early cold war events that had a major impact on the need for a Constabulary, and what brought it into existence as an elite force based on the cavalry organizational model of World War II?

At the Yalta Conference in February 1945, the three major power representatives—President Franklin D. Roosevelt for the United States, Prime Minister Winston Churchill for Great Britain, and Generalissimo Joseph V. Stalin for the Soviet Union—reaffirmed their aims of destroying all Nazi institutions and bringing major war criminals to justice. The three powers agreed to carve up Europe, thus leading to Soviet domination over the whole of Eastern Europe. One of the crucial issues that became a dividing factor of the wartime alliance was German reparations. This meant all heavy manufacturing equipment, rolling stock, machine tools, foreign investments, ships, and all economic means that were used to support the Nazi war machine were to be used as payment for wartime losses caused by German aggression. The Big Three also decided that $20 billion, half of which was expected to go to the Soviet Union, was to be worked out by a Reparation Commission. Germany was also to be divided into zones of occupation to be managed by the quadripartite Allied Control Council.

To deal with the anticipated international and personnel instability, and to allow military personnel to maintain border control over occupied territories, U.S. Military Government Law 161 (MGL 161) was put into effect in March 1945 at the request of the State Department. The purpose of the law was to seal off Germany so former Nazi officials, intelligence operatives, and wanted persons did not escape. The chief weapon to deal with Nazism was the *Fragebogen,* a questionnaire used by the military government to determine Nazi and militaristic associations of Germans. MGL 161 was amended to provide for a "prohibited frontier zone," controlled by army commanders so security nets could be established. By the end of the year, this effort was weakened because of a critical shortage of military personnel who were rapidly being redeployed for demobilization.[2]

In April 1945 the Joint Chiefs of Staff directive Number 1067 (JSC 1067) laid down a harsh and punitive occupation policy. General Dwight D. Eisenhower seemed to favor the talented administrator Lieutenant General Lucius D. Clay, whom he appointed his deputy for military government of Germany. The directive established U.S. policies to be followed by American military representatives on the Allied Control Council, which began functioning in June 1945. It was the mission of Eisenhower and Clay to implement this occupation policy. In May, Operation Overlord ended when Nazi Germany surrendered and plan Eclipse was implemented immediately. This plan called for General George S. Patton's Third Army to advance to the German border and seal it off, already accomplished when XII Corps was in Czechoslovakia. On July 1, U.S. troops withdrew to secure the German state or *Land* of Bavaria. Roadblocks were set up, and all avenues leading into the American Zone were monitored to deal with the onrush of millions of displaced persons, civilian refugees, and former German soldiers trying to escape from the Red Army. At this time, millions of U.S. troops were still on the ground, supported by aerial reconnaissance flights in the European theater to maintain order and secure the communication avenues in Bavaria.

In the meantime, on June 5 the victorious powers, the United States, the United Kingdom and its dominions, and the Soviet Union, assumed supreme authority over the devastated Third Reich. France, no longer a major power after German military forces disgraced the country in 1940, was not at first given a part in the reorganization. The basic objective of the three victorious powers was to exercise control over Germany according to the London Protocol of September 1944 and follow-up agreements. The country was divided into three occupation zones, and the Berlin capital into three sectors. The Allied Control Council was established and made up of three senior military commanders from Great Britain, the Soviet Union, and the United States. The immediate goals were to curb the German appetite for conquest, destroy Germany and especially Prussia as the stronghold of militarism, and hold accountable those responsible for genocidal war crimes. While sorting out these goals, there were already conflicting efforts among the victorious powers. All agreed at first that Germany was to be demilitarized and all Nazi institutions eliminated. However, there was disagreement about the future political and economic goals for a peacetime Germany. The dividing issue became a major event of the cold war. Was the future Germany to be reeducated in the democratic spirit of self-determination or established as a communist one-state system?

Meanwhile, the three powers agreed that France should also be given an occupation zone, although not at the expense of the Soviet Union. The French zone of occupation was cut out of American and British zones, and France became the fourth controlling power. The resurrection of France as a power was due to the efforts of the United States and England and its dominions, which retook the country for France and sent the German occupiers retreating to the Siegfried Line. The four powers were intent on keeping Germany intact until reconstruction was agreed upon. As the Soviet Red Army gained control of eastern Germany, Stalin preferred to transform what was left of Germany's industrial means to his advantage. His motive was to demand huge economic concessions in the form of reparations as compensation for the Soviet Union's terrible sacrifices and losses during the war. He demanded not only billions of dollars but also 80 percent of all German factories. Stalin also wanted a say in a four-power management of the industrial Ruhr while at the same time he began sealing off his eastern zone and rapidly extracting East Germany's wartime heavy equipment. For years, the issues of management and reparations remained a divisive problem that the powers were unable to resolve.

Before he passed away on April 12, 1945, President Franklin D. Roosevelt was more for maintaining a stable central Europe as part of a global equilibrium to be secured by a United Nations. Roosevelt and Churchill both held that a future German democracy and its economic base were indispensable. Before a diplomatic solution could be reached in this area, the defeated Third Reich had to disarm and demilitarize. Once this was achieved, Germany's economic unity was to be restored. The critical questions were under what political system and how? Unfortunately, the original common aim of partitioning the country became no more than an idea, because when the war ended, the United States and England began to view with alarm the Stalinization of Poland and eastern Europe. Winston Churchill told President Truman on May 12, 1945, "An iron curtain is drawn down upon their [Soviet Union] front. We do not know what is going on behind."[3] The curtain indeed had fallen in front of the occupying Red Army, which began to secure the occupied territories and its zone in Germany and to establish provisional governments based on the communist model one-party state.

Now serious border problems began to mount, especially regarding travel between the Soviet and American Zones. Liaison was almost impossible due to the Red Army's political intransigence and the occasional sport of shooting into the American Zone. The Soviet forces, unlike the British and French occupiers, began to funnel a massive number of refugees and displaced persons across the border into the American Zone.

On July 17, 1945, representatives of the three victorious powers, President Truman for the United States, Prime Minister Churchill for Great Britain, and Stalin for the Soviet Union, met in Potsdam to confer on capital peace treaties. Eleven days later the British electorate voted Churchill out of office, replacing him with the Labour candidate Clement R. Atlee. When the conference ended on August 2, guidelines had been established regarding the occupation of Germany. The country was to be disarmed and demilitarized, all Nazi institutions were to be dissolved, a trial of war criminals was to be convened, and, what became a dividing issue, there was to be an effort to encourage democratic ideals and the restoration of local self-governments. This was to be based on the freedoms of speech, press, and religion, all subject to the requirements of internal security. In addition, economic restrictions were imposed, such as those controlling domestic economic development. All these international events had an impact on the future of Germany. Consequently, they had a bearing on molding a Constabulary. The occupation of a part of Germany by U.S. forces, especially the Constabulary, would be a variation of an earlier cavalry mission. This occurred when the 1st Squadron, 2nd Cavalry, Third Army, became part of the occupation force of the Rhine bridgehead after the armistice of 1918, which ended in 1923.

The idea of a Constabulary began before the war ended when it was realized that some type of police-security plan was needed for the occupied territories in Germany. The origins of such a plan took form late in 1944 when the Fifteenth Army made a study of the potential problems in dealing with a multitude of unorganized displaced people. The Fifteenth Army, under the command of Lieutenant General Leonard T. Gerow, had entered the ETO early in 1945 and was tasked to serve primarily as an occupation force while the Allied forces moved through Germany.[4] Meanwhile, as U.S. forces were overrunning the Rhineland, occupation plans were further crystallized by the end of March. General Eisenhower planned to set up a frontier zone in occupied Germany to regulate the flow of military and civilian personnel in the areas controlled by U.S. and Allied forces. With this in mind, commanders of U.S. forces considered establishing an inter-German boundary, known as the "prohibited frontier zone," and setting up control posts to reestablish communications and an economic network.

Once military operational requirements were met, Eisenhower advised Allied army group commanders that they were to divert troops to provide maintenance of the zones. He left this decision up to the group commanders to establish an interim special police organization to handle internal security in occupied territories.[5] Meanwhile, in mid-April 1945, a Fifteenth Army directive ordered the formation of a "Frontier Command," whose purpose was to provide security by setting up roadblocks and various posts while at the same time mobile elements were to patrol captured areas to control the movement of people across the western German border.[6]

In September a G-2 study at the recently created U.S. Forces European Theater (USFET) under the command of General Eisenhower recommended the formation of a "military district Constabulary" similar to a system employed by highway patrols in the United States. This was the first time the term "Constabulary" had been used in planning for an occupation organization in the American Zone. It was suggested the assets come from mechanized cavalry tactical units because of their wartime experience and a long tradition of being highly mobile. Eisenhower's goal was to establish a specialized superpolice or an elite force equipped with an efficient communication network, modern weapons, and adequate mobile vehicles and liaison planes to support the military governing of occupied Germany. The object was to establish a force mobile enough to patrol among the German population and police numerous prisoners of war, displaced persons, and millions of ethnic Germans expelled from Eastern Europe. All these unfortunate and homeless people were wandering through the American Zone.

At the end of October, Eisenhower announced that a superpolice force or Constabulary could eventually control the turbulent German population in the American Zone.[7] Meanwhile, the commanding generals of the Eastern and Western Military Districts in American-occupied Germany, the Military District in Berlin, and the Bremen enclave established "district constabularies" subordinate to the local military district commanders.[8] The Seventh Army, which controlled the Western Military District, designated the 15th Cavalry Group and its two attached squadrons, the 15th and 17th Cavalry Reconnaissance Squadrons, as District Constabulary. Meanwhile, the Third Army organized the Eastern Military District, designating the 2nd Cavalry Group and its assigned 2nd and 42nd Cavalry Reconnaissance Squadrons and the 6981st Provisional Rifle Company as the mobile security force. In addition, troops from the 6th Cavalry Group (Mech) and its two assigned recon squadrons, the 6th and 28th, were also deployed as a security force. The majority of these security forces engaged in patrolling, dealing with the German population, and checking the border procedures of the remaining infantry division tactical units who engaged in border patrol until the establishment of the U.S. Constabulary in July 1946. Until this time, infantry tactical units controlled the borders by establishing static posts that usually covered roads. The problem with this static arrangement was that border violators easily circumvented them by moving cross country.

By the end of November, U.S. military commanders had completed within their districts the deployment of ad hoc mechanized cavalry elements that had some police training. During this paper planning stage the term "military district Constabulary" was being reconsidered, such as calling it some type of a "zone Constabulary." In spite of looking for a new meaning, the district constabularies' missions became too overwhelming due to the restructuring of Germany into the agreed power zones.[9] This understandably led to emerging postwar military and civilian personnel turmoil. Unlike fluid combat, American units became primarily static, trying to manage millions of displaced and expelled people. Consequently, they were unable nor trained to deal with mounting occupation infractions, such as border violations, thousands of roving displaced persons, refugees, pilferage, and subversive activities. At the same time, the mood of many soldiers had drastically changed. Military malcontents started roaming around Europe in an emotional anticipation of going home. By the end of 1945, redeployment to the United States became almost a surge. Virtually all experienced military personnel were heading home, clamoring for their "ruptured duck," issued to discharged soldiers.

Finally, at the suggestion of General Eisenhower, the plan for a U.S. Constabulary was approved by the War Department and was to be used as a basis for preparing a table of organization and equipment. By the end of the year, it was determined that the Constabulary was to consist of a headquarters similar to a corps headquarters, with three brigade headquarters, one for each *Landr* capital at Baden-Württemberg, Bavaria, and Hesse in the American Zone with Berlin, Frankfurt, and the Bremen enclave excluded. The goal was to establish by July 1, 1946, an elite organization of up to thirty-eight thousand officers and enlisted men to patrol forty-three thousand square miles of the American Zone.

What was the postwar climate in the American Zone during the period when plans were being formulated for establishing a Constabulary? It was obvious that the planning efforts were reactive to wartime events rather than investing in proactive planning to resurrect a peaceful Germany based on capitalistic-democratic principles. Directive Number 1067 was too punitive and not constructive enough for restoring a peacetime Germany. The question arose as to whether Germany should be restored or undergo a transformation. The centrifugal political forces between contesting powers and ideologies made reconstruction of Germany a doubtful proposition.

Allen W. Dulles, an Office of Strategic Services (OSS) station chief during the war and later director of the Central Intelligence Agency, addressed the Council on Foreign Relations on December 3, 1945. Discussing the challenge, he declared, "Germany today is a problem of extraordinary complexity," adding, "It defies a solution." He criticized the choppy way the *Laender* in the American Zone were set up for occupation. Dulles was a close observer of early postwar events. This led him to argue for a new form of occupation. The German economy was virtually nonexistent, industries were idle or destroyed, and roads and bridges were impassable, he acknowledged. Furthermore, competent people not allied to the defunct Nazi Party were hard to find. The management of these *Laender* was unworkable due to the international power struggle over reconstruction. Dulles observed that Stalin's policy was hard to fathom because the Soviets had "torn up all the double tracks, they are keeping all able-bodied German prisoners, and they have taken East a great many industrialists, bankers, scientists and the like." Regarding East Germany, Dulles said, "An iron curtain has descended over the fate of these people and very likely conditions are truly terrible." His solution for the form of occupation in the American Zone was to "use small, highly mechanized units and put our reliance on planes." More so, "Germany ought to be put to work to set up a standard of living," and the refugee problem needed to be solved. Now was the time, he concluded, to establish a viable democratic economic future for Germany not based on "mixed collectivist doctrines." What was left of Germany now rested on how the Americans were to police, secure, manage, and restore order in the country.[10] What Dulles was arguing for was a peaceful Germany, which he believed needed to be restored and not transformed as was occurring in the Soviet Zone.

As Dulles said, the problem of Germany was one of "extraordinary complexity." No sooner had the Third Reich been defeated than the U.S. Army entered into a period of disarray because of rapid demobilization. This had a serious effect on establishing a Constabulary. Equipment used to win the war was stored and fell into disrepair. By December, numerous cases of rape, reckless driving, poor uniform discipline, and a virtual lack of military courtesy were reported throughout the American Zone. Many soldiers were prone to rowdiness, excessive drinking, and beating German civilians without regard for their civil rights. Noted historian Earl F. Ziemke reasoned that this behavior was caused by last-fling urges of men and the tensions associated with being imminently redeployed and discharged. In addition, there was the problem of arriving inexperienced and poorly trained, low-scoring soldiers. Consequently, he wrote, these soldiers now constituted the majority of the occupation force.[11]

In the meantime, Soviet military forces in the east were also demobilizing. However, unlike the United States and its allies, Stalin still kept a formidable force in being, with thousands of first-line tanks and aircraft and more than a million men outside the Soviet Union's borders. Almost immediately after the war, the Soviet High Command began a major reconstruction of the Red Army, soon to be called the Soviet Army. The emphasis was placed on greater mobility, especially transforming infantry divisions into armored formations and mechanized divisions. Within a few years it was conceivable that Stalin's military would have the means to overrun western Germany.[12]

Sergeant William Tevington remembered right after the war when he was on occupation duty some of the games being played by Soviet Red Army soldiers. One of their favorite tricks, he recalled, was moving the designating markers between the American and Soviet Zones to snare unsuspecting American troops patrolling the border and then holding them for ransom. Usually the soldiers were released in exchange for whiskey and cigarettes. Private First Class Julian C. McCalla related the same experience when serving with the Constabulary in Berlin. He and his partner were pursuing a Soviet soldier for theft in the American Zone. As their jeep neared the Soviet Zone, McCalla noticed the warning signs had been removed. Before he could reverse his jeep, the two were surrounded by armed Soviet soldiers and looking down their gun barrels. As soon as whiskey and cigarettes were sent over to the Soviet Zone, the jeep and its occupants were released. During this early period, many American soldiers on border patrols found the Soviet soldiers annoying and had problems understanding their intentions.[13]

While Soviet military forces were repositioning, Major General Ernest N. Harmon returned to Europe after a Christmas vacation stateside with his family. He immediately reported to General Joseph T. McNarney, who in November 1945 had replaced Eisenhower as military governor and head of USFET. Eisenhower returned to Washington, replacing General George C. Marshall as army chief of staff. Clay in turn became McNarney's deputy military governor. At the suggestion of a friend from the Italian campaign, Lieutenant General Lucian K. Truscott Jr., Harmon was recommended to McNarney for the new command. At the time, Truscott commanded the Third Army at Bad Tolz.[14]

General McNarney had serious problems, to say the least. First was the management of a country in total ruins, which raised the question of who would be responsible for the military government of the disintegrated former Third Reich. In November 1945, the decision was made to expel millions of Germans from Austria, Poland, Hungary, Czechoslovakia, Romania, and the German region beyond the Oder-Neisse Line, which was given to Poland because of the Big Three agreement at Yalta. In addition to the expelled Germans, millions of displaced Jews and Poles were wandering throughout the American Zone.

Managing civilian affairs and military governments during the war and shortly after were G-5 detachments that moved with the advancing armies. Their mission was to administer captured German cities, towns, villages, and county governments. The G-5 staff sections were functioning at army, corps, and division levels and reported to the responsible army commander. Following World War I, soldiers took on a greater role in civilian affairs as governors and administrators. Within months, this function began to change when army commanders' responsibilities for military government shifted to the Office of Military Government in Germany (OMGUS). In addition, greater liaison was established with the State Department for civilian involvement in governing postwar Germany. It was Clay's intention, which he mentioned to Eisenhower, that a civilian agency such as the State Department should eventually take over the control of governing Germany. The existing troops in USFET were gradually relieved as governors and were tasked instead to support OMGUS.[15]

Another serious problem facing McNarney was troop discipline and the crisis to "get the boys home." Ernie Harmon recalled that on one occasion there was a near riot when a mob of brawling American troops marched on McNarney's headquarters in Frankfurt, shouting, "We wanna go home."[16] It was evident McNarney was in a dilemma; not only how to deal with German affairs, refugees, and displaced persons, but also improving the deteriorating morale of American troops. Accordingly, plans were drawn up in his headquarters to consolidate some type of a mobile police organization as directed by the War Department and suggested by General Eisenhower to take over the long-anticipated occupation of the American Zone of Germany. Next was the question of who was most qualified to accomplish such a mission for military and civilian security. McNarney's solution was to select a cavalryman and experienced armor division leader.

When Harmon reported to McNarney, he was told to take command of a new force, the U.S. Constabulary. To say the least, Harmon was puzzled, asking, "What's that?"[17] McNarney gave Harmon six months to get the new organization into operation. The rugged, profane cavalryman was soon given an explanation of his new and easier-said-than-done mission. His comment later was that this was the first time in U.S. Army history that the army was "called upon to occupy a modern industrial nation." Accordingly, Harmon wrote, "It presented many new and difficult problems."[18]

General Harmon assumed command of the newly formed U.S. Constabulary in mid-January 1946. Within weeks, he established a planning board at Third Army Headquarters at Bad Tolz to formulate a table of organization and equipment for the proposed elite military force. Harmon expected the Constabulary to take the form of a state militia rather than a state police force. His thinking then was that historically, state militias were called out to handle large civil disorders. The stern commander did not want the Constabulary units designated as cavalry. He expected, however, the new units to take on the organizational structure of cavalry. Harmon also expected the Constabulary to become involved in static, but also mobile operations, such as maintaining border patrol posts, traffic control, and protection for the movement of supplies. This concept of operations shortly changed.[19]

Within a few days, Harmon outlined his proposed mission: "To maintain general military security and to assist in the accomplishment of the objectives of the Military Government in the occupied zones of Germany and Austria by means of an active patrol system prepared to take prompt and effective action to forestall and suppress riots, rebellions, and acts prejudicial to the security of the U.S. occupational forces."[20]

Two key members of Harmon's planning staff were Colonel William S. Biddle, who commanded the 113th Cavalry Group, and Colonel Joseph Edward

General Ernest N. Harmon is welcomed to Stuttgart in May 1946. On his left is Colonel Albert Harris of the 3rd Constabulary. Harris was a distinguished commander of the 86th Cavalry Reconnaissance Squadron, 6th Armored Division, during part of World War II. Courtesy of the National Archives and Records Administration.

M. Fickett, who commanded the 6th Cavalry Group during the war. Both had been important contributors to the postwar 1945 General Board report "Tactics, Employment, Techniques, Organization, and Equipment of Mechanized Cavalry." Fickett notified Biddle that he had been selected to be part of a new army organization called the Constabulary. Biddle replied that he knew little about such an unusual mission. When Biddle reported to Harmon, he was told that the "operational element of the occupational forces were to comprise two major forces—a tactical force, comprising roughly of a corps, and a Constabulary force of an approximate overall strength of 38,000." Both forces were to be under the control of Truscott's Third Army, whose mission was to use the tactical force as the general reserve and Constabulary as the first line of defense.[21]

Headquarters, USFET, tentatively approved the plan. General McNarney, however, wanted the Constabulary to be "branch immaterial" rather than exclusively cavalry. This meant reducing the firepower that had been part of the cavalry's equipment during the war. Biddle, who was joined by Colonel Charles H. Reed of the 2nd Cavalry Group, had to rework the organization. Biddle recalled, "Harmon is a man who wants things done fast." So, the planning team worked all night. The new table of equipment eliminated M7 self-propelled howitzers, M18 tank destroyers, and the M3 half-tracks and their variants. The M24 light tank troop for each proposed squadron was to be moved to regimental reserve. Inserted were horse platoons that were intended for use in negotiating rough terrain. The M8 armored cars armed with 37mm turrets, the turretless M20 armored utility cars armed with .50-caliber machine guns, and jeeps were to become the main means for mobile operations. Replacing the versatile half-tracks were 1½-ton trucks used as personnel carriers and command post vehicles. All the changes were structured around the cavalry brigade organization, down to squadron and then troop levels.

Harmon then ordered that Biddle and the planning team "war game" the new plan. He was pleased, and subsequently Truscott approved it. However, he directed that the words "cavalry" and "Dragoon" not be used and that the word "Constabulary" be the proper designation. In addition, he intended that a small motorcycle element be added for traffic control. No sooner had the new table of organization and equipment been agreed upon than Biddle flew to Washington. The War Department, as he noted, "acted with speed which astonished me." It was approved as a tentative table of organization with minor changes. During the formative period, Biddle mentioned that Harmon stressed, "I want to make clear: and I want also to emphasize that it will be a military, not purely a police force. It is not like a state police organization." McNarney, Harmon, and Biddle strongly believed in a cavalry-type organization because it fostered esprit and would be largely under cavalry leadership.[22]

General Harmon, who last commanded XXII Corps, began to make the force operational, in spite of rapid demobilization, including the Armored Force. One of the first priorities was to implement the approved table of organization and equipment. Headquarters elements from the VI Corps became the headquarters for the Constabulary. The 1st Armored Division, which provided infantry, artillery, and special units, including the 13th Armored Regiment, which was the 13th Cavalry (Mech) before being redesignated in July 1940. The 4th Armored Division as a whole became the basic nucleus for the new and unusual organization.

The 4th Armored Division soon took on a more active role in forming up Harmon's Constabulary. After the war, the division was declared a permanent occupation force and subsequently spread over an extensive zone as tactical troops to maintain security in the defeated Third Reich. Major General Fay B. Prickett commanded the division. This operation began before the Military Government was established. Lieutenant Colonel Irzyk of the 4th Armored Division was at the time commander of a *Landkreis,* a rural administrative area similar to a U.S. county. After he was called up to division headquarters at Regensburg, Prickett assigned Irzyk as his S-3 (Operations) and executive officer. He held these positions for the first year and a half, spending most of the time coordinating, training, and approving and modifying or rejecting plans regarding Constabulary operations. Prickett commanded the 1st Constabulary Brigade from the activation until 1947, when he returned to the United States.

When Prickett, Irzyk, and the division staff received orders about a totally new mission, it did not take long to transform the 4th Armored Division into the new elite police-security force. At the time, the division staff was involved in carrying out assigned tactical missions when, as Irzyk recalled, "we were hit by a thunderbolt." The division was to be deactivated and divest itself of all armored equipment that had propelled it across Europe. Irzyk said that he was "astonished" because "we were forming a unit such as had never before been in the Army. This was a completely new drawing without any precedent." Subsequently, the famous 4th Armored Division became the nucleus of the Constabulary. A new organization was created with leftover wartime equipment.[23]

Unfortunately, the new mission was seriously compromised by redeployment. At the time, the 1st and 4th Armored Divisions were undergoing a change with the continuous redeploying of their personnel based on a point system. The remaining tankers, cavalrymen, armored field artillerymen, armored infantrymen, and combat engineers were expected to become the backbone of this unknown organization that had no previous history. For many veterans it was difficult to understand the breakup of one of the army's most distinguished wartime combat divisions. They remembered how the 4th Armored Division was used to spearhead many of the Third Army's deep maneuver–oriented operations. Now the transformation from wartime maneuver-oriented operation to a police-security operation in peacetime became a challenge. It again was a time for change and new opportunities that caused adjustments for many of the remaining soldiers and the new young recruits.

Additional elements from the 2nd, 6th, 11th, and 15th Cavalry Groups were also expected to join the Constabulary. These units, because of their wartime light combat mobility, were chosen to form the nucleus of the reorganized and redesignated U.S. Constabulary. The headquarters was organized similarly to that of a wartime corps headquarters. A security force was established at headquarters whose mission was criminal investigation and counterintelligence.

Three Constabulary brigades were established with three regiments each. Each regiment had three squadrons of five troops organized like the wartime cavalry troops. However, their equipment was modified to exclude heavy weapons. Three mechanized troops were equipped with jeeps mounting .30-caliber machine guns and M8 armored cars mounting 37mm cannons and .30-caliber machine guns and M20 armored cars minus turrets and with a .50-caliber machine gun on ring mounts. Each troop had ten armored cars. Two troops were motorized with 1½-ton utility trucks each mounting one .30-caliber machine gun. In addition, supporting weapons were provided, such as 57mm recoilless rifles and 81mm mortars. Troops were usually armed with M1 rifles, .45-caliber Thompson submachine guns or .45-caliber M3 "grease guns," and .45-caliber semiautomatic pistols.[24] Provisions were also made for motorcycle and horse cavalry.

The troop was to be the primary unit of the Constabulary. Its organization was based on the wartime mechanized cavalry troop model. However, the troop required a greater number of hand weapons and light vehicles for mobility. The troop was to be organized into team sections with three jeeps and one M8 or M20 armored car for support and as a command vehicle. M24 light tanks were expected to be used as mobile reserves, mainly in major cities when a show of force was required.

Surely, to the delight of the last chief of cavalry, Major General John K. Herr, horse cavalry was again used. The horse platoon usually consisted of one officer and thirty-two enlisted men mounted on former German army horses. The men selected were experienced horse handlers. The guidelines for this new breed of cavalrymen were that they be alert, aggressive, and capable of acting on their own initiative. They were required to be skilled policemen, capable riders, and experts with pistol and rifle. And they had to be able to care for their mounts and equipment. One of the problems they inherited was training German horses to understand American commands. Surely, Herr would have been proud to see mounted cavalrymen in their McClellan saddles with their rifles secured in scabbards, wearing traditional riding britches and boots. They became important and useful elements in border patrol operations, negotiating difficult and mountainous terrain that was unsuitable for motorized and mechanized units. The horse soldiers were also used for riot control in refugee camps and for rounding up cattle stolen for the thriving black market. They were ideal "Circle C Cowboys," chasing down old-fashioned cattle rustlers who engaged in that thieving business, which began to flourish as soon as the war ended because of the huge demand for fresh meat.[25]

Perhaps in a meager way Herr could claim with pride that there was no substitute for cavalry. Hopefully this restrained the quips that once were used

(Above) The M8 armored car, 15th Constabulary, Bavaria. Note the special helmet and insignia on trooper and armored car. Courtesy of the Patton Museum. *(Below)* The 16th Constabulary Horse Cavalry platoon on review, Berlin, May 30, 1946. Courtesy of the National Archives and Records Administration.

to greet cavalrymen on the march, such as "Where are your bows and arrows?" or "What's good for the daily double tomorrow?" Motorcycles that Herr considered before the war for his cavalry were reintroduced to provide traffic control on the autobahns.[26]

When Harmon took command, he immediately left an impression that was not forgotten. One trooper recalled that Harmon came on like a tiger. On one occasion after a detailed squadron inspection, Harmon made "ferocious speeches laced with every known profanity and a few that he must have created. We thought that he was really something." Another newly arrived soldier from the soon to be deactivated mobile 474th Antiaircraft Artillery Automatic (AAA) Weapons Battalion (redesignated as the 64th Constabulary Squadron) remembered Harmon addressing his unit. The soldier never forgot what "the little profane major general" had to say, that "it was time to get off our beer-soaked asses and become soldiers again."[27]

On February 15, the Inspector General's Section began operation, under the command of Colonel C. C. Park. The next day the Judge Advocate Section of the Constabulary was activated. On April 1, a military lawyer, Colonel Robert C. Bard, was assigned as the Staff Judge Advocate General (JAG). The following month General Harmon was granted by the War Department general courts-martial jurisdiction over the Constabulary. Before the year ended, the JAG Section was immersed in military criminal law, claims, real estate law, and general legal assistance.

Also on April 1, the Office of Military Government for Germany revised an earlier Allied Authority Control Council Directive ordering all sport and military athletic organizations to be dissolved by January 1, 1946. This included all youth activities beyond the *Land* level. A new German youth activities program was established to make constructive use of leisure time. This offered an opportunity to instill democratic ideas, practice such political procedures, and encourage responsibility. By mid-April General McNarney directed that an "Army Assistance to German Youth Activities (GYA), U.S. Zone" be established by all major commanders and make facilities available to German youths on a part-time basis.[28] General Harmon expressed a strong interest in the German youth program and expected it to be carried out.

In the meantime, on March 15, General Clay succeeded General McNarney. With this move, Military Government was separated from Army Command. Clay was appointed military governor of the American sector, and shortly USFET was redesignated as the European Command (EUCOM). As soon as he took command, he had to deal with colored troops who were in service units.[29] Clay recalled that their disciplinary record "was not good," and the number of incidents with the German population was causing serious problems. Clay gave credit to Lieutenant General Clarence R. Huebner, his deputy and chief of staff and commander of ground forces, for developing an academic advancement program to instill a sense of pride in colored troops. Once this was achieved, they were transferred out of service units, and later in 1947 formed into three infantry battalions incorporated into the Constabulary.[30] General McNarney early in August 1945 recommended to General Marshall that racial integration be tried on an experimental basis "to offset as much as possible the inevitable reaction of radical elements."[31]

While Harmon was in the process of organizing a distinctive U.S. Constabulary, Secretary of War Robert P. Patterson announced early in April 1946 the intended merger of the Cavalry and Armored Force into a single arm. The War Department stressed that the two arms were to be consolidated as a single "Armored Cavalry." This decision was made "in recognition of World War II developments" and President Truman's initiative for service unification. By then, the army had already prompted some preliminary unification by announcing the abolition of the horse cavalry and its merger with the Armored Forces. General Jacob L. Devers, the commanding general of the Army Ground Forces, favored the abolition, commenting that the only value of horses during the war was as pack animals. At the time, the *Army Navy Register* wrote that plans for

Members of the U.S. Constabulary conducting a routine roadblock check at Hof, Germany, June 27, 1946. National Archives and Records Administration.

the horse cavalry were indefinite. What was desirable, the article stated, "was a use of horse units for certain terrain where motorized elements could not travel, and where mobility greater than that of foot troops is required."[32]

As the Constabulary was organizing, Harmon immediately realized the necessity of retraining arriving troops. In March the Constabulary Training School at Sonthofen was established, located in the southernmost tip of Bavaria against a background of the snow-covered Alps. Sonthofen was suitable because for the most part it had escaped the damages of war, even though a few bombs were dropped on the town in April. The history of the facility went back to January 1937, when Adolf Hitler authorized a male youth school be established on "behalf of and for the Nazi Party to be a preparatory school for the National Socialistic Ordensburgen." In German history, the castles of a religious order or fraternity were known as *Ordensburgen.* Party leaders and not the parents selected students, who were to be guided by the principle "state first, individual second." It was Hitler's strong sense that a force of arms never defeated Germany's military in World War I. The problem, he raged, was a lack of strong uniformed political leadership. This he planned to resolve.

The school was named Die Adolf Hitler Schule. Students, who had to be twelve years of age to be admitted, were expected to complete their studies in six years. During this period, the Nazi Party took complete charge of their lives, creating physical and mental conditioning for a future leadership role in the party. In addition, perfect health was expected, as was an unimpeachable character. The students had to demonstrate a proficiency in sports and an efficiency in school, as well as establishing a visible leadership record while a member of the Hitler Jugend. Most important, candidates were to provide uncontestable and traceable proof of genetic fitness and pure Aryan lineage as defined by the Nazi Party.[33]

Now Die Adolf Hitler Schule was to be taken over by the soon to be organized Constabulary school. Earlier in April 1945, the 2nd Moroccan Division of the French army had been the first to occupy Sonthofen. A few months later several American military elements occupied the region and began occupation duty. In mid-January 1946, the 2nd Cavalry Squadron took over the buildings and tower of Ordensburgen. The cavalrymen were told to start preparations for organizing a Constabulary school. They secured the property and billets in anticipation of the arrival of a permanent staff and instructors. During this period of preparation, the mobile 465th AAA Automatic Weapons Battalion relieved the 2nd Cavalry Squadron, becoming the permanent school troops. Later the 2nd Cavalry Squadron was redesignated as the 2nd Constabulary Squadron.

The first groups of instructors were assembled at the Third Army's Constabulary Section at Bad Tolz under the command of the school's commandant,

Colonel Harold G. Colt. On January 23, General Harmon assembled the group to discuss the mission of the new school at Ordensburgen and related potential problems, especially supplies, troops, and their retraining as a police-security force. The nearest depot was 110 miles from Sonthofen. At the time all the U.S. forces in the zone were engaged in static duties, many dealing with displaced persons while at the same time the army's force level was rapidly downsizing. Harmon's problem was that many of the troops he required could not be released from their static commitment. The mechanized cavalry units assigned to the Constabulary were already patrolling troubled areas. They, for the time being, could not be spared for early retraining and integration into the new force. After Harmon's conference, it was evident that the mission of the Constabulary was still in a state of flux. Nevertheless, the school was finally activated on March 4 when the first class of 650 students reported for duty.[34] No replacement center was established. This decision by Harmon and his planners was based on the assumption that once the Constabulary was stabilized and the problem of long-term enlistments solved, Ordensburgen could function as both a school and a replacement center.[35]

To resolve the issue of arriving inexperienced and poorly trained, low-scoring soldiers that Professor Ziemke wrote about, the following qualifications were established:

> Physical: Soldierly bearing; height not less than five feet, seven inches; weight not less than 140 pounds, stripped; good vision without glasses; well knit, good muscular coordination, strong, erect, well proportioned; hard, tough, enduring, quick reactions.
> Mental and Nervous: Intelligence; stable mentality; good judgment and tact; alert, observant.
> Moral: Excellent character; personal honesty; loyal; cooperative; conscientious; adaptable.[36]

The generation of these qualifications was due primarily to the efforts of the Constabulary school at Sonthofen and the training efforts of the S-3s (Operations), brigade, regimental, squadron, and troop commanders. One of the problems, however, was that "Old Army Non-Coms" were conspicuous by their absence, thus placing a greater burden on postwar young enlisted men and their officers who had to train them. There was also the problem of retaining technical specialists, such as maintenance personnel and radio operators. Nevertheless, leadership and the school system became the heart and soul of the Constabulary that eventually turned these postwar soldiers into a formidable police-security force. By 1947, the average age of the Constabulary trooper was twenty-two.[37]

A four-week curriculum for officers and enlisted men was planned months before the first class reported. Initially four academic departments were established: the Tactics Department, the Police Department, the Geopolitics Department, and the Communications Department. In addition, a supportive Visual Aids Department was created. Supporting the key department heads were numerous instructors. These groups of officers put together a table of organization and curriculum for the school. To assist in the training a *Troopers Handbook* was written under the direction of Lieutenant Colonel Warren D. Haskell, a former state police commissioner of Rhode Island and head of the Constabulary's Police Department. The handbook reflected the considerable experience and practical ideas Haskell had gained while engaged in civilian police work.[38]

While the department heads and faculty members were gathering information on courses to be taught, Colonel Henry C. Newton was assigned as assistant commandant on February 13 and became director of training. He became instrumental in guiding the Constabulary School through its beginnings and early growth. Newton, an educator and graduate of Los Angeles Polytechnic, came from the civilian component of the army and was commissioned in the field artillery after World War I. Shortly thereafter, he served on the faculty of the University of California. Newton ended his civilian career when his unit, the 160th Infantry Regiment, California National Guard, was federalized in 1940. During World War II, he had considerable experience in the Armored Force doing research and tactics on armored infantry. In addition, he organized and commanded the Armored Force Officer School at Fort Knox, which included courses in tactics and techniques of armor. He also served later with the 12th Armored Division. Newton's mission was to manage and coordinate the activities of all academic departments, including directing an administrative agency dealing with all student matters. His associates and graduates called the school "Newton College."

During his tour, he made a number of changes. Lieutenant Colonel James P. Hannigan, who was commissioned from West Point as a field artillery officer, became the Constabulary's school secretary and Newton's assistant.

The departments created were:[39]

Department of Tactics included courses in tactics of raids and searches, border control and patrolling, riot control, intelligence gathering.

Department of Communication taught radio operator courses that were critical, since the Constabulary's mission required patrolling, covering considerable territory.

Department of Motors taught courses in vehicle maintenance and repair.

A trooper from the 28th Constabulary Squadron and policeman check for contraband, June 1946. Courtesy of the National Archives and Records Administration.

Department of Geopolitics provided courses on German history, geography, politics, and characteristics of the German people.

The Police Department became the Department of Public Safety, which became the largest educational function in the school. It was tasked to train troopers in police work, such as laws and techniques of arrest, elements of crime, rules of evidence, traffic control and preparing accident reports, town patrolling, military police duties, interrogation, confessions and statements, and cooperation with other agencies.

A Department of General Subjects was established to teach map reading, geography, report writing, and unarmed defense. Later courses in first aid and the use of key German words were taught. In addition, a Desk and Records Section was established.

A demonstration and guard troop was organized from the mobile 465th AAA Automatic Weapons Battalion, which was reorganized from five batteries into three. Troop A was tasked to provide interior security. A mechanized platoon, motorized platoon, motocycle section, light tank section, and mounted horse squad were organized to provide demonstrations in tactics and in the firing of all weapons assigned to the students soon to become Constabulary troopers. An academic troop was formed to operate as a housekeeping unit for all academic departments.

Colonel Newton made sure that the theoretical side of instruction was held to a minimum. He emphasized the application of theory in such areas as basic tactics, police functions, weapon proficiency, and motor skills and maintenance. Newton made sure that Harmon's directive requiring all troopers to raise their standards of discipline, military appearance and courtesy, and general efficiency was followed.

The main problem Newton had to deal with was personnel instability.[40] The serious problem of redeployment became quite evident early in the school's existence because personnel turmoil affected all academic departments and the quality of their instructors. As Harmon noted, the most difficult problem he incurred was obtaining qualified men for the Constabulary. Once trained, many were redeployed. The turnover rate was high. At times new men were placed in positions of responsibility without adequate training and experience.[41]

General Harmon addressed the first graduating class after it completed its four-week curriculum. General Truscott, the commander of the Third Army, addressed the second class, which started on April 2. By the end of 1946, a total of fifty-seven hundred officers and enlisted men had graduated from Ordensburgen. Changes and modifications were made in the curriculum during the school's existence based on the operational experience of the Constabulary and the reevaluation of anticipated needs as the cold war heated up. For example, a Department of Special Weapons and later intelligence courses were introduced. Added to the school was a Women's Army Corps detachment to alleviate the critical shortage of administrative personnel.

Headquarters VI Corps moved to Bamberg in early March 1946 and on May 1 was officially redesignated as Headquarters, United States Constabulary, with General Harmon commanding. Harmon liked Bamberg because it was near the eastern frontier of the American Zone. To facilitate his movements he liberated former Reichsmarschall Hermann W. Göring's private train. Shortly after he fell heir to Göring's luxurious train he had it painted in Constabulary colors. However, Harmon kept Göring's interior fittings intact because they "suited my purpose just fine."[42]

In the meantime, the Third Army merged with the Seventh Army and moved from Bad Tolz to Heidelberg.

Until the U.S. Constabulary became operational on July 1, district constabularies were established in Bamberg on April 1, consisting of the 2nd, 6th, and 15th Cavalry Groups. These temporary district constabularies received on-the-job training for about eight weeks until the time they were to be absorbed by U.S. Constabulary. The Constabulary *Troopers Handbook* was distributed to all units. Operations consisted primarily of search and seizure, especially in Jewish displacement centers. In addition, the district constabularies patrolled the inland waterways. Border control was given to the district constabulary engineers. At the end of the month, L-5-type aircraft were assigned to Constabulary units. On April 26, the 2nd Constabulary Brigade became operational, effective May 1.[43] The district constabularies ceased to function and were absorbed by the U.S. Constabulary when it became operational. The tactical divisions, which patrolled the borders, were either sent back to the United States or used to maintain security at displaced person camps and prisoner of war installations.

Soon after the move, provisions were made for supplying the distinctive Constabulary uniform, which had a unique shoulder patch to make the new force distinguishable. To accomplish this image Harmon decided on a symbol taken from the Armored Force, which represented the combined arms. The colors blue (infantry), red (artillery), and yellow (cavalry) were selected. Next, a circle unit insignia was designed to incorporate the combined arms colors, including the lightning bolt, which signified the quick striking power of a mounted force. The German population called them *Blitz Polizei.* The distinctive force was required to wear bright yellow scarves, Sam Browne belts across their chests, and combat boots with a smooth outer brown surface. Special helmet liners rimmed with yellow and blue stripes were prescribed bearing the Constabulary insignia. The motto "Mobility, Vigilance, Justice" was adopted and soon became "watchwords throughout the American Zone."[44]

It was also evident in April that there were serious equipment problems. For security missions it was determined that 940 armored cars were required. Constabulary units had to go to widespread depots throughout the American Zone, as well as the liberated areas, to find M8s and M20s. However, the cars were in poor condition, and 200 were in need of complete rewiring. With the assistance of Ordnance and Signal Corps personnel, the M8s and M20s were made operational. Another potential equipment problem was a demand for horses. It was planned to establish a horse platoon with at least forty horses, including replacements for each Constabulary regiment. The solution was to utilize captured German horses, which became available after the end of the war.[45]

On April 14, 1946, Colonel Biddle took command of the 11th Constabulary Regiment at a time when the regiment was overwhelmed. It was a difficult month for organizing. Biddle soon found out there were some serious personnel issues. On April 15, he addressed the officers and men from the regiment. After delivering a short history of the 11th Cavalry and its tradition, he let the men know he was not happy with what he saw: "There are a lot of things I see around here which show me that most of us aren't sufficiently alert yet." For example, he chastised the men for not saluting and for improper uniforms and poor vehicle maintenance; finally, he expected the men to keep the area clean. Biddle expected the troopers to police the area, meaning pick up cigarette butts and trash. Months after he delivered his speech he heard from his brigade commander, Brigadier General Hobart R. Gay, who was angry that the regiment had failed a vehicle maintenance inspection. Gay advised Biddle that of all the regiments in his command, his was the worst. More disturbing was the venereal disease rate. Harmon told Gay he was not at all happy with the high statistics, especially with the 8th and 25th Squadrons.[46]

There was the habit of transporting German women in army vehicles for self-serving gratification. German males resented this. The Constabulary's monthly report for April noted that in Munich 68 women were apprehended, and of this number 29 were reported infected. In Mannheim 2,311 men and women were apprehended, with 571 reported infected with venereal disease. The venereal disease rate was especially a concern for occupying troops. For American soldiers the rate was 1,000 cases per annum, and for Constabulary soldiers it was 140 per annum, according to the Surgeons Section, Medical Affairs.[47]

General Gay was also displeased with Colonel Bard's JAG Section. He criticized the regiment's Trial Judge Advocate for failing to carefully analyze cases before trial. Gay told Biddle in October, "I believe our Court-Martial system is the fairest and the best in the world but the best system is of no value unless the persons who put it into effect know their business." Gay then ordered Biddle to establish a school in each squadron on the Manual of Courts-Martial. Biddle had his hands full, to say the least.[48]

The April 1946 report from Constabulary Headquarters identified other serious problems. Many German and Austrian civilians resented fraternization between American soldiers and German women, who were loathed and insultingly referred to as "Yank brides" and "chocolate girls." The report further stated that even more pronounced was resentment toward colored troops who fraternized with German women, especially in areas where they were stationed. Part of the problem was caused by the frustration of segregation. Colored troops felt, the report noted, that they were entitled to their "share of the *Frauleins.*" One

German university student typified the local attitude: "We don't mind being occupied so much, but why did you bring colored troops to do it?" This policy of open fraternization, the Germans felt, led to serious disturbances. Another part of the problem was the failure to recognize the extensive degree of fraternization that led to prostitution by many fräuleins due to basic human needs and survival. This obviously made it difficult to curb the impossible, the oldest profession. The headquarters' report suggested that this situation led to the birth of "a new nationalism" because the German males resented the attention given to their fräuleins.[49]

General Irzyk, however, felt that the fräulein and fraternization issue was not as much of a problem as was the constant demand to execute successful operations. "We never sat with heads bowed commiserating about disciplinary problems," he recalled.[50]

In all fairness to Biddle and other Constabulary commanders, April 1946 was a most difficult month for all in formulating an innovative organization to meet the July 1 operational deadline. It should be noted that most of the statistics came from that busy month of April. Yet, to Harmon's credit, he managed to stay the course, as did Prickett, Gay, Biddle, Irzyk, and many other dedicated commanders and noncommissioned officers. Nevertheless, it was a trying and difficult month for all in the Constabulary, as more and more American soldiers were demanding to be demobilized.

While Biddle was trying to shape up his regiment, he told a friend that the future cavalry would remain a branch and be renamed Armored Cavalry. Not only did he have to deal with the regiment's current police-security mission and General Gay's anger, he had to think again about uncertainties of another change.[51] In less than a year since it was officially organized, it was evident throughout the Constabulary that change again confronted the U.S. Cavalry.

The activities and problems of the district constabularies were soon reflected in an April report from Harmon's headquarters. First, it was indicated that there might be a delay in the activation of the U.S. Constabulary due to personnel redeployment. The report noted that internal security was at risk, primarily due to the actions of American soldiers. This was caused by the complicated point system, which was designed to send men with the largest number of points based on time in service and combat to the Zone of the Interior for discharge. It noted that liquor and excessive drinking caused many of the incidents. While waiting to be deployed back to the States, many soldiers had too much time on their hands. Some cases of Germans attacking military personnel were reported. This predicament was never completely solved, according to Harmon.

Professor David Brion Davis from Yale University, a former president of the Organization of American Historians, recalled his experiences while serving with the Constabulary in the spring of 1946. A disturbing incident occurred when this teenage trooper had to arrest a Polish soldier for raping a six-year-old girl and giving her gonorrhea. It influenced him so much that he remembered her name was Maria. Davis viewed the problem of segregation as the most upsetting during his tour with the Constabulary. On one occasion his unit was called to the Mannheim railroad station to keep a trainload of colored troops from "acting up." Instead what he found was a conglomerate of uprooted humanity of all descriptions and nationalities. When the colored troops arrived at the station, Davis commented, "They seemed like a good bunch." He became alarmed, however, when a dozen or so American soldiers started shouting racial slurs. Later his Constabulary unit was called out on a night mission to investigate a racial disturbance. Apparently, a group of colored troops had entered an all-white club, resulting in a fight. Before Davis's unit arrived, the military police had already engaged the colored troops who were armed. One MP was shot. The order was then given to the Constabulary troopers to fire on the perpetrators. Fortunately, cooler heads finally prevailed. Davis's limited experience in the Constabulary, he claimed, brought to his attention the arrogance, xenophobia, and insensitivity exhibited by many American soldiers. He eventually concluded that the "contours" of American culture, which had a fear of communism and intended to democratize the world, were contradicted by an unwillingness to deal with racial issues that exploded during the early Constabulary era. He wrote home at the time that the biggest mistake the army made was segregation.[52]

The future professor alleged Harmon's "racial hostility was outrageously exploited." It was quite evident that Harmon's demeanor offended Davis. Davis mistakenly wrote in the *Journal of American History* that Harmon was "trounced by the German Afrika Korps at the Battle of Kasserine Pass." His contact with the general, which seemed to irritate him, occurred when Harmon came to Stuttgart to inspect his unit. "Ernie Harmon came down today," he wrote home. "Our cavalry boots were shining like a pool of oil in the sunlight and the Colt .45s were bone dry." Sirens blaring and in drove Harmon in his long Packard all decked out in "dazzling cavalry boots" and a chest full of ribbons. He was actively engaging in one of his many inspection tours. Somewhat self-indulgent, Davis wrote that it reminded him of another god who rode into Stuttgart and was greeted by a roaring crowd during the war years. Apparently, he was referring to the notorious racist Adolf Hitler. Davis arrogantly wrote home that he was the only one in his unit who understood this comparison. He professed to have quoted Harmon, who said, "The Constabulary is the best organization in the ETO. I'm not so goddamn sure it's not the best in the whole army, and I'll lick the son-of-a-bitch who says it isn't." This attitude offended Davis, as did the issue of segregation, which he blamed on Harmon and the army. He did admit he saw no torture of the Germans, "despite all the arrogance, xenophobia and insensitivity."[53]

Davis's personal grudge was resolved when President Truman issued his historic executive order in 1948 desegregating the military, the first American institution to do so. Davis was a teenage trooper at the time and had at best a limited view of the whole picture. Surely, when he wrote his article in 1990, he was still inflamed by memories of his youthful idealism and influenced by the military negativism of the antiwar movement of the Vietnam and postwar era. His article ignored the real accomplishments of the Constabulary. It did not do justice to the difficulties the Constabulary had to overcome during the cold war period, when youth and inexperience were transformed into an elite police-security force. However, when Davis returned to the States, he was not aware that plans were being made to integrate colored units into the Constabulary and that General Huebner was introducing an academic advancement program for these troops. Once trained, they were formed into three infantry battalions and incorporated into the Constabulary late in 1947.[54]

As General Irzyk later noted regarding the period when the young Davis was indulging in his critical letters home, there were no governing bodies, such as ßürgermeisters, *Kreis* commanders, or *Laender* leaders in control of a devastated country, which was flooded with thousands of displaced persons and refugees all fighting for subsistence. Under these circumstances, personnel instability and crime became rampant. If not for Harmon's leadership, Irzyk maintained, the postwar chaos could not have been controlled through the police-security activities of the Constabulary.[55]

By May 1946, it was more than evident that the primary security threats the Constabulary had to deal with continued to be local and infiltrating communists, former Nazis, and a restless German population. In addition, there were two divergent and problematic groups of displaced persons, the Jews and Poles. Historically, the Poles hated the Germans for past territory violations and invasions, and the Jews wanted revenge for the Nazi Holocaust. With this horror behind them, the Jews were anxious to move to their traditional home in Palestine.[56] Even more so, they had no opportunity to acclimate to the German population because of the anti-Semitism left over from the Nazi regime. As a result, they were not receptive to any outside authority.

The Jewish problems were many-sided, all quite different than the Polish situation. Since the war ended, there had been a steady increase in the number of Jewish displaced persons arriving from Poland because of the anti-Semitism that swept over the Soviet-occupied countries. The United Nations Relief and Rehabilitation Agency (UNRRA) administered the displaced Jews, as it did the Poles. By September 1946, the general security in the American Zone was threatened by a huge number of black-market activities connected to the large influx of Jews and the large-scale movement of goods by illegal organizations. This was done to finance the move of Jewish displaced persons to Palestine. In addition, the Constabulary was constantly forced to intervene to check violence between the German population and the Jews.

One major Polish uprising occurred in May 1946 at the Altenstadt displaced persons camp that was run by the UNRRA. Thousands of Poles were poised to challenge the selection of their new leader, one appointed by UNRRA rather than one of their own choosing. The incident moved to a riot stage when the Constabulary was called in to extract the deputy camp leader, Josep Dutkiewicz, who was challenging UNRRA's selection in favor of himself. Historian Brian A. Libby, who made an extensive study of the incident, claimed Dutkiewicz was making bogus claims such as that the camp's hospital had been killing babies. The actions of the Constabulary to subdue the riots further inflamed the Poles, resulting in a call for reinforcements who eventually restored order in the camp. The camp then was placed under the control of the Constabulary until the rioters settled down. Libby gave credit to the firm action of the Constabulary troops, especially D Troop, 2nd Squadron, for their professionalism. Dutkiewicz and his followers were removed; no Poles were shot or injured.[57] What repercussions could have occurred if just one Constabulary trooper had become trigger-happy?

The Constabulary also had to deal with refugees from the Baltic countries who were adamant about not returning to their homelands because they had been overrun by Soviet military forces and communized. The most sensitive issue was the question of the Ukrainians and the anti-Soviet Russians, many of whom had joined the German military during the war. They presented a peculiar problem because they placed U.S. authorities in the position of harboring enemies of the Soviet Union, a wartime ally. They were aided by remnants of the Russian Orthodox clergy, who were active in finding emigration opportunities in other countries. Other people in Eastern European counties, such as Czechoslovakia, Hungary, and Romania, fled their homelands and refused repatriation because of Soviet domination there.[58]

By now, the Constabulary had moved from a police-security force to more of a law-enforcing agency. It was evident that the Circle C Cowboys were again influenced by international events. Change became a constant.

In mid-April, the 427th Counter Intelligence Corps (CIC) Detachment was assigned to the U.S. Constabulary. This relationship became critical to resolving the security issue. Throughout the zone, cases of *Frageboden* or questionnaire falsification continued to be a serious problem as concealed Nazi Party members attempted to avoid automatic arrest. Part of the job of the CIC, working with the Constabulary, was to interrogate and apprehend suspected National Socialist officials.

Shortly before the U.S. Constabulary became operational, General Harmon acted on the GYA program for groups of eight- to eighteen-year-olds. He sent out letters to his officers and enlisted men indicating that he wanted the Constabulary to organize an effective German youth organization program. Furthermore, he emphasized that "every officer and enlisted man be encouraged to assist appropriate youth activities; and that every commander take a personal and continuing interest in the activities of his unit in this program." By the summer all the brigades, regiments, and squadrons had in place some type of a GYA program.[59]

On July 1, 1946, the U.S. Constabulary finally became operational, and the former VI Corps Headquarters was phased out. General Harmon's command had a chief of staff and a G-1, G-2, G-3, and G-4 similar to the wartime division or corps. In Headquarters Troop, special troops were established consisting of a signal squadron, a band, a school squadron, a military police company, a car platoon, an intelligence detachment, a counterintelligence corps unit, and elements from the criminal investigation division. In addition, a special service company and an ordnance maintenance company were established in support of the Air Liaison Squadron. By now, planners had established the structure and stationing of units in three *Laender* capitals and nine administrative districts as proposed earlier. The 1st Constabulary Brigade commanded by General Prickett (later by Arthur G. Trudeau and then George W. Reed Jr.) was located in Wiesbaden, the capital of Greater Hesse. The 2nd Constabulary Brigade commanded by General Gay (later by Edmund B. Sebree and Bruce C. Clarke) was located at Munich, the capital of Bavaria. The 3rd Constabulary Brigade commanded by Brigadier General Thomas Harrold (later by Bruce C. Clarke) was located at Stuttgart, the capital of Baden-Wurttemberg. The units designated for inclusion in the Constabulary were as follows:[60]

1st Constabulary Regiment in Kassel (Hesse)
2nd Constabulary Regiment in Freising (Bavaria)
3rd Constabulary Regiment in Wetzlar (Hesse)
5th Constabulary Regiment in Gablingen (Bavaria)
6th Constabulary Regiment in Bayreuth (Bavaria)
10th Constabulary Regiment in Mohringen (Baden-Wurttemberg)
11th Constabulary Regiment in Weiden in der Oberpfalz (Bavaria)
14th Constabulary Regiment in Kitzingen (Bavaria)
15th Constabulary Regiment in Sonthofen (Bavaria)
Constabulary School Squadron in Sonthofen (Bavaria)

The 4th Constabulary Regiment was stationed in Linz, Austria, which was controlled by American zonal occupation forces. The areas they were responsible for were the provinces of Salzburg and Upper Austria, and parts of Vienna, the capital. Three constabulary squadrons were attached to each regiment.

An element of the 16th Constabulary Regiment patrolled the American Sector of Berlin. Horse troops and a Special Troops Signal Squadron were also established. The latter was to expedite communications necessary for mobile operations.

While the Constabulary brigades and regiments generally conformed to *Laender* boundaries, the squadrons, composed of six troops, conformed to boundaries that usually followed smaller *Landkreis,* or German counties. The squadrons, composed of Headquarters Troop, three mechanized troops, and two motorized troops, usually patrolled several counties. Once organized, Constabulary elements began what was called "swoop raids," meaning search and seizure where known or suspected illegal activities existed. After careful planning, the method of operation was surprise, meaning to hit the objective early in the morning while the suspects were still sleeping. In addition, Constabulary elements began to set up a network of speed traps to curb a major problem of speeding in the American Zone, which had caused numerous fatalities. Offenders were stopped and punishments were handed out by roving Summary Court officers. Surprise spot checks were established at crossroads to examine credentials of people passing through the American Zone. Railroads were subjected to the same spot checks, with baggage and passengers subject to inspection. Later, Constabulary stations were established, manned by six or eight men, to monitor troublesome areas and to secure intelligence. In addition, Constabulary Headquarters formed up plans establishing a policy of frequent rotation so troopers could keep active their training programs and prevent them from going stale during the extended troop deployment.[61] With the busy and difficult organizational months behind it, the Constabulary was now up and running and setting a new course in cavalry history.

In this same month of official operations, the Constabulary met another difficult challenge when a Jewish mob attacked German policemen. Apparently, according to historian Libby, the incident became inflamed because of an earlier rumor over German-made soap that was stamped "RIF." The Jews took this to mean *Reines-Judisches-Fett* or "pure Jewish fat," meaning made from the remains of concentration camp victims. In fact, the letters stood for Reichs-Industrie-Fettwaren, or National Fat Products Company. A few weeks later emotions again overtook reason when a crowd of Jews threatened German police who were carrying out routine black-market checks. As the crowd became more violent, the policemen opened fire, killing one Jewish displaced person and wounding another. At the nearby Fohrenwald Camp, the Jewish population became furious over the incident. To prevent violence, a Constabulary platoon and an infantry company were sent into the camp to restore order. During the funeral service, a rabbi added to the emotions with some inflammatory remarks. When the Constabulary and infantry troops started to move out of the camp, they were spat on and called "American Gestapo" and members of the SS.[62]

Libby, who made an interesting statistical analysis of the personnel instability affecting the Constabulary, noted that early in the operational month the Constabulary was made up of 494 officers and 31,730 enlisted men. By the end of the month, however, as a result of demobilization, redeployment, and the point system, 66 officers and 1,439 enlisted men had left the Constabulary. They were replaced by 127 new officers and 3,732 green replacements. Libby, like Ziemke earlier, offers another interesting analysis, concluding that the green replacements were "on a whole of low quality." What Harmon got from the United States, according to Libby's study, were typical draftees and practically untrained recruits. This, he maintains, was due to the War Department's efforts to keep up the army's authorized strength in Germany. His thesis questioned the idea that the Constabulary was the elite force conceived by the planners. Furthermore, he argued, draftees arriving in the Constabulary became a discipline problem. General Harmon made every effort to deal with this problem by encouraging duty before self-interest. However, this was not the case in many incidents. The crime rate was "slightly higher than that for other troops in the theater," as was venereal disease, according to Libby. About four hundred cases of VD were reported per month. Libby quoted one Constabulary officer who wrote that the German fräuleins were perceived "as wholesome farm girls." They "smelled like the delicious odor of new-mown hay"; however, "they are giving most Constabulary troopers the burn of their life."[63]

Statistical analysis, however, failed to reveal that in spite of the personnel problems, the young troopers inherited by the Constabulary were trained and carried out numerous difficult missions, many of which were widespread throughout the American Zone. Libby's analysis also failed to consider the operational side of the Constabulary, which planned and executed many difficult missions in spite of the high turnover rate. Credit was surely earned by the generals and colonels who commanded the brigades and regiments that Harmon recruited. Harmon claimed they were exceptional. The problem with statistical analysts, perhaps, is that they are too immersed in their quantifying analytic function to understand the nature of training and daily operations necessary to complete missions. Missions are usually fixed. However, they take on a life all their own due to the ongoing efforts to deal with the chafing and haze of operations that interpose, arise, and disappear as the combat situation develops. The Constabulary was continually confronted with this situation as international dynamics became more threatening. Gradually, constabulary troopers were becoming experienced cold war warriors.

In addition, during the first month the U.S. Constabulary became operational, General Harmon was inundated with the unending fräulein problem. He called Colonel Sam Williams, the commander of the 26th Infantry Regiment in the large communication center of Nurnberg, about female inspections routinely held to diagnose venereal diseases. Williams told Harmon the German women were treated with consideration. However, Williams also told Harmon that German civilians reluctantly accepted the fact that American soldiers were going out with the fräuleins. The civilians objected "to fraternization with colored troops, especially because the colored boys were picking up 12–13 year olds." Williams told Harmon he was controlling this situation "somewhat." The two also talked about the use of German women for clerical work in the communication center. Williams claimed that this new policy was working. Harmon advised Williams to be sure that appropriate screening of female applicants was performed because some of them may be going out with "high class Nazis."[64] What Harmon and Williams failed to realize was the possible infiltration of German communists loyal to the Soviet cause.

General Harmon's method of operation for monitoring discipline and morale was to make use of Göring's three-car train, which had kitchen and dining facilities, sleeping berths, and a conference compartment with communication equipment. When on inspection trips, he made sure that his complement and staff arrived at the exact time he did. Whenever he stepped from his train, he was impeccably dressed in his cavalry britches and boots. This performance became Harmon's hallmark. His image created much anxiety for troopers, especially those who had not met his expectations. Colonels greeted him with snappy salutes, which were returned in kind. This was all planned to impress the German population, many of whom at first called the Constabulary "Harmons's Gestapo," until they realized its importance in maintaining law and order and prosecuting the perpetrators.[65] In his memoirs, Harmon claimed the populace of Munich called his arrival "The Second Coming."[66]

Harmon's purpose for the inspections was not only to draw the attention of the civilians to the importance of occupation law and order but also to instill discipline and pride in the Constabulary and draw attention to the troopers with their distinctive uniforms, a point of pride that Professor Davis had not understood at the time. All the units and their troopers periodically were subjected to one of Harmon's zealous inspections. Pity the poor trooper or officer who did not meet the general's expectations.

General Ernest N. Harmon reviews and inspects troops from the 10th Constabulary, May 1946. Courtesy of the National Archives and Records Administration.

Within a few months, an active patrol system was finalized to take action against any disturbances of law and order, including riots, rebellions, and especially border violations. Border control at first was confined to key static posts recently occupied by infantry tactical units. It was soon determined that this system did not work. The problem was that many illegal crossers had no problem circumventing the static posts. To remedy this situation, border operations were reorganized by eliminating the manning of nonessential crossing points by Constabulary troops and replacing them with German border police. This enabled the Constabulary to move from an occupation force to more of a controlling force along the interzone borders and frontiers in the American Zone of occupation. Its mission-type orders were to prevent illegal travel along approximately fourteen hundred miles of the border. To control border violations a plan was implemented that made provisions for horse-mounted and foot patrols, as well as M8 and M20 armored cars and jeeps to operate parallel to and in a depth of ten miles behind the borders to apprehend violators who circumvented the border posts. Motorcycles and light aircraft supplemented these patrol operations. This change in tactics improved results by apprehending illegal crossers who had evaded the static posts at the border.[67] During the first six months of operation, twenty-eight hundred Constabulary troops spread out in 120 border posts turned back twenty-six thousand undocumented transients.

The utilization of and relationship with the German border police came about as a necessity due to the large-scale redeployment and rapid demobilization of American soldiers in Germany. As an expedient to resolve this situation, U.S. Forces European Theater envisioned a border police modeled after the Grenzpolizei, the Third Reich's Nazi national border police service. Though the service was abolished when American forces occupied Germany, Grenzpolizei facilities were used. Once thoroughly screened and devoid of all Nazi affiliations, a *Land* Border Police was established. By November 1945, recruiting was under way, and an eight-week course was begun. The border police were to wear dark blue–stained Wehrmacht uniforms and be armed with light weapons, such as pistols and American M1 carbines. The *Land* Border Police assumed duty on March 15, 1946, and began patrolling borders and manning checkpoints with American tactical units. On July 1, 1946, they came under Constabulary control and proved to be an asset, especially when dealing with German-speaking people.[68]

This was an important issue for Harmon, who had to build a trustworthy German police organization. This did not take long to put into effect, according to Harmon, who had the assistance of a German nobleman, Count von Henneberg. The Germans who were picked for police duty were those cleared of having connections with the National Socialist Party, its philosophy, and its institutions, such as the dreaded SS or Schutzstaffel. Once activated, German police worked with Constabulary sections. The establishment of a German police force had a positive effect on the civilian population as the force gained in prestige. As time went on, ties between the Constabulary and the German police became stronger. The *Land* Border Police began to take over routine policing of German civilians.[69]

With the use of German border police, Constabulary troops were removed from nonessential posts and subsequently employed on foot, horse, and wheeled mounted patrols to intercept the illegal crossers in the rear of the border posts. One of the most difficult problems now was dealing with displaced persons at the borders, especially along the Czechoslovakian and Austrian frontiers. The displaced persons were primarily Jews who came from Poland via Czechoslovakia and Austria expecting to find a refuge in the American Zone that provided opportunities for resettlement, especially to Palestine. By the end of the year, this became considerably less of a problem when the Jewish Rehabilitation Organization started to handle and process these immigrants.[70]

While General Harmon was taking stock of the first month of operations, Secretary of State James E. Byrnes at the July meeting in Paris of the Allied Council of Foreign Ministers set the stage for the creation of Bizonia, the merging of the American and British zones. The search for a future peace accord in Germany and Austria was carried on by the victorious powers, the United States, the United Kingdom, the Soviet Union, and then France. The mechanism to achieve this goal was the Allied Council of Foreign Ministers, which was originally established to produce a quadripartite solution for dealing with a postwar Europe, especially Germany. By now, however, Byrnes viewed the zone situation in Germany as sealed compartments that prevented, rather than facilitated, the exchange of communications. This arrangement, he claimed, also caused inflation and economic paralysis. Byrnes offered to economically merge the American Zone with the other zones. The Soviet Union and France rejected his proposal. The latter's reason was historical, preferring to see Germany emerge as a loose confederation of disunited states that politically benefited France before the unification of Germany in 1870. In addition, France's political future was in disarray. About 25 percent of the electorate was communist. Conversely, the Soviet Union favored a highly centralized one-party state. A few weeks later General McNarney repeated the Bizonia offer to the British, who promptly accepted. This made the exchange of native exports and imports between the two zones a joint effort and lessened the Constabulary's activities between the two zones.

In September 1946, while Harmon was still struggling with the manpower situation and trying to maintain law and order, and border control, Secretary of State Byrnes made his famous Stuttgart speech. The secretary indicated there would be a change in U.S. foreign policy regarding the German situation. Byrnes stated that the military role of the Allied powers in West Germany had changed from one of occupation and control. Now the goal of the Western Allies was to protect Germany as well. As a nudge to Stalin, he said that the removal of heavy industries should not be carried out if they were necessary for a peacetime German economy.[71] Now the United States and the United Kingdom had moved to establish Bizonia as an economic unit. After initial French resistance, a unified economic area was established to create some semblance of restoring German self-sufficiency. This decision went against the goals of Stalin, who still was in the process of removing heavy German equipment to the Soviet Union and demanding rights over determining the future of the industrial Ruhr. Meanwhile, the United States and the United Kingdom temporarily suspended dismantling operations and started to raise production to lessen the unfavorable balance in Germany's foreign trade. For all practical purposes, Germany no longer could be treated as a separate economic unit for each of the occupying powers. This development began to affect the original mission of the Constabulary.

During this period of political turmoil and economic transition, Harmon, for morale purposes, issued a six-month history of the Constabulary and had it distributed to all units. The history indicated that the downturn in illegal crossings was due to the effectiveness of Constabulary border enforcement. During the first month and five succeeding months the Constabulary made 168,000 patrols in jeeps, in M8 and M20 armored cars, in M24 light tanks, and on horseback and foot. In addition, the Constabulary turned back at border crossings 25,638 people and arrested an additional 17,388 who were turned over to the Military Government for trial. The latter remained constant monthly because this group contained people who were refused permission to leave their zone, those who wished to escape their zone, and those who did not wish to have their whereabouts known for clandestine operations. The number turned over for trial declined monthly because of a growing awareness of the Constabulary's aggressive efforts in monitoring border crossings. It was statistically quantified during this period that the Constabulary traveled more than five million miles, with jeeps becoming the primary vehicle used for patrolling. L-4 and L-5 light observation planes flew more than fourteen hundred hours on eleven thousand missions. All these operations were impressive, especially in view of the ongoing personnel turbulence.[72] They demonstrated the ability of the Constabulary leadership and the young troopers to deal with the ever-present change that presented itself.

After the war ended, efforts were made to put passenger railroads back on line, especially the Orient Express and the London–Switzerland train. As these international trains came back, they also became a problem. Many passengers were not properly documented to enter or leave Germany. The Constabulary solved this problem by stationing special select troopers on board the trains when they entered the zone. While en route, these troopers checked the documents of all persons attempting to board or detrain while in the American Zone.[73]

During the first six months, numerous black-market rings involving scarce merchandise were uncovered. These were operated by all classes and groups in the American Zone. By the end of December 1946, a total of 2,681 major black-market operations had been exposed. The problem leading to this situation was an overabundance of money and a scarcity of goods, especially food, which was being hoarded by many German farmers. Most of the goods being dealt with were jewelry, drugs, cigarettes, clothing, large amounts of money, and small quantities of food. Virtually all the raids were conducted in displaced persons camps and German houses. In November, the 91st Constabulary Squadron in Berra discovered the largest cache of black-market money to date. An alert sergeant found more than 2 million reichmarks hidden in a civilian sedan crossing over from the Soviet Zone into the American Zone. The following month a large cache of money and jewelry was uncovered in Kraiberg. Three Jewish displaced persons were detained at a roadblock. Discovered in their possession were watches, rings, and 700,000 reichmarks. It was reported that the Jews tried to bribe the Constabulary troopers with part of their illegal goods.[74] No luck!

Other than border control and black-market operations, the Constabulary had to deal with civil security and subversive activities. These included vicious attacks on American personnel and the unauthorized use of firearms. The rise in subversive activities manifested in various forms of fraternization. One example was a bürgermeister's wife, who gave a Nazi salute followed by the banned "Heil Hitler" as a Constabulary patrol passed through a small town. It was common in these cases for the German perpetrators to display Nazi feelings and emblems. It was also discovered that there were clubs organized by the Germans for subversive purposes. One of their main methods was placing obstructions and decapitating wires on and across roads frequented by U.S. Army vehicles and Constabulary patrols. The primary subversive activity occurred in October when three buildings and a Military Police Headquarters were bombed in Stuttgart by an ex-SS officer and fourteen of his men. The following month a CIC detachment and the German police finally arrested the perpetrators. It was also reported that a general "free-for-all" occurred in a number of cafés between U.S. troops, displaced persons, and German civilians. Most of the time, these incidents were instigated by those seeking revenge over fraternization.[75]

As the year ended, black-market activities were still a serious problem. One interesting incident occurred in November, when the Constabulary was called on to search, check contraband operations, and apprehend the offenders. Operation Camel was typical, revealing one of the many situations the Circle C Cowboys had to deal with. Members of the 27th Squadron surrounded the Polish Displaced Persons Camp at Ulana Kaserne, where black-market and other nefarious activities were reported. Assisted by military police, the Constabulary troopers from the 10th and 13th Squadrons entered the camp and began a room-to-room search in earnest. Eventually the troopers discovered a small, worn suitcase hidden on the roof of one of the buildings. Opening the suitcase, the troopers discovered morphine, codeine, and penicillin reportedly worth about $52,120. In addition, quantities of army clothing, rifle ammunition, knives, and a small amount of black powder were confiscated. More than a hundred Poles were appended for interrogation, resulting in the arrest of eighty-four. A total of 676 officers and men participated in the successful raid.[76]

It was also reported in November 1946 that American soldiers were a source of illegal firearms. Many were found in the hands of displaced persons and German civilians. A Constabulary investigation in the 1st Brigade reported that a disproportionate number of troop disorders were "directly attributed to colored troops." A provost marshal analysis stated that almost half of the serious disorders involved soldiers connected with varying degrees of assaults, including the use of firearms.[77] Colonel Bard's Judge Advocate General Section was bogged down with hundreds of legal matters, such as domestic affairs, divorces, intended marriages of American soldiers to foreign nationals, and the preparation of numerous legal documents. In addition, the JAG Section had more than a full slate of court-martial cases. By the end of the year, hundreds of summary and special courts-martial cases had been processed, many of them dealing with dereliction of duty by officers, who were punished under Article 104, Articles of War, that gave disciplinary powers to commanding officers. For serious violations general courts-martial were appointed by higher headquarters. Since this was a period of military occupation, a number of special courts-martial and general courts-martial also involved civilians. As was the normal procedure, these cases were forwarded to the War Department under Article 50.5, Articles of War, which required a review and rehearing of the accused.[78]

As usual, the Constabulary found itself understrength early in 1947 due to the all-familiar redeployment of enlisted men and lack of replacements from the Zone of Interior. In January, a heavy burden fell on General Harmon. Constabulary Headquarters were moved from Bamberg to Third Army Headquarters at Heidelberg and consequently took over the functions of the Third Army. This added more responsibilities at a time when the Constabulary was already undermanned. During this period of consolidation, certain administrative functions performed by base sections were reorganized under newly created tactical units, the military posts. Harmon was not exactly content with all these developments, especially those regarding his Constabulary. He was concerned that his personnel, especially married officers, would be placed on static commitment or temporary duty to assist the regular army tactical military posts. In a memorandum he said, "I have no intent whatever in the desire of married men to stay on post." Furthermore, all commanders, he said, "are to go out and visit all squadrons and detached troops assigned to their command." Next, he chastised commanders for allowing sloppy, unmilitary-looking soldiers in their commands. Harmon expected commanders to vigorously punish any infractions. He came down heavily on the tendency of brigade and regimental staff officers to take it easy while sitting in their command posts.[79]

The move to Heidelberg was also made to fit the needs of the G-3 function. When the move was completed, Constabulary G-3 was superimposed upon the G-3 Section of the Third Army. With this move, General Harmon was appointed commander of Third Army concurrent with his duties as commander of the Constabulary. By the end of the first quarter, the Constabulary was the only major ground force operating in the American Zone that was free of static commitments. However, the ongoing problem of personnel furor still plagued

Harmon when he transferred out of the Constabulary. Meanwhile, the light tank troops were inactivated and reassigned, which added to the reorganization.

Meanwhile, after reviewing hundreds of informal and twenty-seven formal investigations ordered by higher headquarters, the Inspector General's (IG) Section reported that at the end of the year the main difficulty encountered by the Constabulary was the large number of untrained enlisted personnel, including too many young, inexperienced troop commanders. This instability was caused by the ongoing redeployment problems. The IG Section was experiencing the same problem. Part of the solution recommended was that long-term enlistees be assigned to the Constabulary.[80]

Steps were finally taken in the first quarter to establish an important Historical Sub-Section to recount Constabulary operations for the Historical Program of the United States Forces in Europe.[81] The main outgrowth of Harmon's directive was the extensive history *The Establishment and Operations of the United States Constabulary, 3 October 1945–30 June 1947*, which was begun in the summer of 1946 by Major Nicholais Harathias and continued by Lieutenant Colonel Robert C. Ingals. These field artillery officers collected mountains of documents and made miscellaneous notes. These were gathered by Major James M. Snyder (Cavalry), who finalized the important history with the assistance of a War Department civilian, Warren Goldman. Snyder's work is one of the most extensive and thoroughly researched historical projects of the early Constabulary. It became a foremost contribution to the early history of the Constabulary and demonstrated the significant adjustment made by wartime cavalry units under Harmon's leadership.

In addition, during the first three months of 1947, all U.S. ground forces began to withdraw from large, overcrowded cities to former *Wehrmacht Kasernes,* or military installations. This move eventually consolidated all squadrons, with the exception of a few border detachments, into a concentrated area. This move to the *Kasernes* prompted many Germans to feel that the departure of the "yellow-scarved" troopers provided encouragement to criminal elements.[82]

There was also a concern because of a lack of intelligence when it came to preparing factual reports regarding crimes. More so, troopers had problems interpreting the basic laws of military government and military law. The G-1, who was in charge of personnel, blamed this situation on the character of these troops, many of whom had a record of conviction for crimes "involving moral turpitude." The solution was to establish a Theater Placement Board to eliminate the undesirables and place soldiers in suitable duty assignments. The Constabulary G-1 reported to the Historical Section the problems with undesirables and subcaliber soldiers. This, he noted, was evident early during the establishment of the Constabulary.[83] Regarding the subcaliber soldiers, it eventually changed.

The new replacements were trained and thus formed the legacy that made the Constabulary an innovative organization in spite of the personnel mayhem. As Al Irzyk of the 1st Constabulary Brigade noted, "It was a miracle." They did their job more professionally than history had given them credit for. They were instruments of change and innovation, and unlike the future professor Davis, they were proud of their accomplishments. James Deming, who served forty-three years in the military, recalled that of all those years none was more memorable than when he was with the Constabulary from 1946 to 1949. He did duty with A Troop, 53rd Squadron. Ken Driscoll recalled when he was assigned to the B Troop, 97th Signal Squadron, that it was a "nice tour of duty with great guys." Leonard A. "Honey" Huntington recalled that as a member of an ordnance field shop in Kassel, his unit had the "utmost most respect for the Constabulary" during those dangerous years. These were a few of the many young American soldiers on foreign soil meeting the demands of the cold war. Kiki Goodwin, wife of Brigadier General Samuel McC. Goodwin, who commanded the 16th Constabulary (Separate) in Berlin in 1946, pointed out another aspect of the difficulties that faced the Circle C Cowboys. She noted that her husband's command was distinctive not because of its physical distance but because logistical and command responsibilities were independent of the regular chain of command in the American Zone.[84] Nevertheless, it accomplished its mission, demonstrating that all echelons of the Constabulary performed its job in spite of postwar downsizing.

In the March 1947 issue of *Military Review*, Al Irzyk published "Mobility, Vigilance, Justice," the first article to explain the critical and important role the Constabulary was playing in the American Zone at that time. Irzyk, who helped create the Constabulary, saw the whole picture and experienced it until he departed for Fort Knox in September 1947. Irzyk's firsthand knowledge provided a favorable picture of the successful but difficult operations that the Constabulary had to deal with on a day-to-day basis. His article was written months before publication, when Irzyk was the brigades' operations officer. It was a period when things were occurring that made the Constabulary a legacy, and was not bleak as the G-1 and IG Section described later. Irzyk disagreed with the negative attitude, noting that the Constabulary was alive, functioning, and accomplishing its mission beyond all expectations. One could argue this was due to the cavalry spirit, leadership, and organizational model brought to the Constabulary. Upon leaving Germany, Irzyk recalled he was upbeat and proud of what the Constabulary was doing and had accomplished. Even though the scribes described how chaotic the American Zone was, Irzyk and his wife

General Fay B. Prickett bidding farewell to Colonel Albin F. Irzyk as he leaves the Constabulary in 1947. The two were the chief architects of transforming the 4th Armored Division from an armored division to the nucleus of the Constabulary. Courtesy of BG Albin F. Irzyk, USA (Ret.).

felt perfectly safe as they traveled throughout the country.[85] This was due to the actions of the Constabulary.

General Irzyk also recalled that General Harmon on his numerous inspections caused considerable excitement among the troopers. He noted, "They were upbeat and morale was high after they performed and passed the critical eye of Harmon and afterwards were praised by him."[86] These young Americans who once were called green, ill trained, and low scoring, he wrote, set a high standard.

By the end of the year, activity in GYA, which concentrated on athletics, had risen steadily. Colonel Harry J. Wheaton, the director of GYA, stated that the plan to make German youths help themselves and reduce delinquency was working. The Constabulary had more GYA programs in effect than any army unit in EUCOM, although its status was constantly in doubt because of a lack of supplies. Also, in isolated cases, Germans were suspicious of the GYA, commenting that "the programs were too active for the skimpy food and clothing rations now allowed German youth."[87] Nevertheless, General Clay gave credit to General McNarney and those involved in the youth program for one of the major contributions to restoring a democratic Germany. The Soviets, however, accused the Americans of teaching German youths baseball to keep alive traditional martial spirit.[88]

By now, Harmon was tired. Organizing the Constabulary had taken its toll. The difficult task of setting up a police-security force at a time when rapid demobilization of officers and enlisted men affected the European military force structure had been accomplished. As the first anniversary of the establishment of the U.S. Constabulary approached, Harmon requested a transfer. He had spent almost five years overseas. On May 1, a graduate of the Virginia Military Institute and an infantry division commander, Major General Withers A. Burress, replaced General Harmon. Before he departed, Harmon made sure that a new training program was put into effect that again placed heavy emphasis on military formations, discipline, and courtesy. This issue had irritated Harmon since he took over the Constabulary. It was back to the basics—all Constabulary schools were expected to teach the proper hand salute.[89]

No sooner had Burress taken command than EUCOM received a cut in troop strength. Again, the Constabulary was affected by this ongoing situation. At the same time, intelligence operations with CIC detachments assumed greater importance. To deal with the situation, a new table of organization and equipment was put into effect. The 3rd Constabulary Brigade was eliminated, four of the nine regiments were deactivated, and equipment turned in. Eleven of twenty-seven squadrons were also deactivated. The 1st Brigade now moved to control the states of Hesse and Wurttemberg-Baden, while the 2nd Brigade took control of Bavaria. However, the Constabulary security missions remained the same, with the exception of patrolling activities generally performed by the German police. Burress was concerned that the quarterly reports of Constabulary operations were far from adequate. Like Harmon's earlier directive, he directed that all units appoint a historian and that he expected all unit commanders to cooperate.[90]

To say the least, it was a quite an accomplishment to put together a new organization created mainly from armor and mechanized cavalry elements. All this occurred during a period of personnel and international instability. To those who organized and served faithfully in the Constabulary during the first two years, it was more than a demanding task. It again gave a new meaning to cavalry. As Colonel Biddle had stated earlier when addressing the 11th Constabulary Regiment, "Troubles are not new to the 11th Cavalry. For example, in 1901 [when the regiment was organized], the commander of the 1st Squadron telegraphed the War Department for more officers, saying, 'I have 400 horses that have never seen a soldier, 400 recruits that have never seen a horse, and four second lieutenants that have never seen either a soldier or a horse.'" Biddle added that the squadron got over that hump. He reminded the troopers that the 11th Cavalry motto is "Allons," in other words, "Let's go."[91]

General Irzyk, who was more involved and perhaps the last standing senior authoritative voice of the Constabulary, summed up the early achievements. He noted that the Constabulary became an elite force based on the mechanized cavalry organization in spite of the personnel turbulence of the early cold war. How it came into being, he maintained, was a marvel and a credit to postwar noncommissioned officer leadership and the young troopers who made things happen. They dealt with change and accepted new opportunities, making their mark during those early difficult years when two major powers began confronting each other and perhaps moving closer to another world war.

Now the Constabulary entered into a new era. During the regrouping period, it was reduced to 18,000 men from a table and organization strength of 30,185 in May 1947. A few weeks later Operation Drill began and revealed several weaknesses in alert plans during the reorganization phase. As international tensions began to mount in Germany, the Constabulary started to move from a military police–security force to a tactical force. A light tank troop was reactivated, and a recoilless rifle troop was added. Because of the reorganization, Constabulary troops gradually began focusing more on tactical training and normal military duties as they received additional weapons and more firepower, including artillery. Plans were also made to arm the Constabulary with mortars and rocket and grenade launchers.[92] The army faced many problems as it moved to convert the Constabulary to a tactical force at a time when the Soviets were carving up central and Eastern Europe.

An economy-minded, Republican-controlled Congress made the army's future even more uncertain. Senator Robert A. Taft, an influential Republican isolationist, challenged the country's postwar role in world affairs and was a proponent of limited government. The Ohio senator was not enthusiastic about committing U.S. ground forces in Europe. Instead, he supported the navy and a policy of reliance on airpower and nuclear weapons for national defense.[93]

Significantly, the army was steeply downsizing, and funds for armor research and development were being slashed as well. The cuts were so drastic that during General Eisenhower's tour as chief of staff, between November 1945 and February 1948, he remarked that implementing the rapid demobilization of the wartime army was more unpleasant than being head of the occupation forces in Germany. Adding to his frustrations were ongoing debates over the unification of the military services.[94]

Adjustments for the Circle C Cowboys began when Headquarters Army Ground Forces completed a study on reorganizing the Constabulary as a corps headquarters and armored division headquarters. The effect was the reorganization of all Constabulary regiments as armored cavalry regiments (light). A study by Constabulary Headquarters in December 1947 found this decision had a number of disadvantages mainly stemming from a lack of appropriate equipment. Overall tank requirements suffered from a lack of research and development that seriously compromised preparedness and thus the supply of modern warfighting equipment. The army in Europe had to depend on what was left over from World War II. Equipping the Constabulary as it reorganized with recoilless rifles was not seen as practical. The back blast of the gun made its use as a close support weapon difficult. In addition, the current prime mover 2½-ton truck was considered not realistic for tactical operations because of its high silhouette. Nevertheless, by July 1, 1948, the regrouping requested by EUCOM was finalized and plans were made for establishing a light tank M24 training school under the direction of the 2nd Constabulary Brigade. The school focused on operating the M24s as reconnaissance and security vehicles. At the same time, a number of Constabulary squadrons became inoperative and converted into carrier units for field artillery and antiaircraft battalions all under the organization of the 7732nd Field Artillery Group with headquarters at Sonthofen. Plans also were completed to organize the 2nd, 6th, and 14th Constabulary Regiments as armored cavalry regiments (light).[95]

By the beginning of 1948, the main security threats the scaled-back Constabulary had to deal with were the communists and their propaganda and espionage activities directed from the Soviet Union. There were still the issues of locating former Nazis, and the German population, which was being subjected to the continuous hammering of communist propaganda. In addition, the nefarious black-market activities of two divergent groups of displaced persons, the Jews and Poles, continued. The Military Government relied on the Constabulary for assistance in curbing black-market ventures and unrest, stopping riots, conducting raids, engaging in check-and-search operations, and preserving security.

The most noteworthy activity during the regrouping period was the Constabulary's G-2 function. One historian wrote in 1949 that the intelligence net established by Colonel A. R. Reeves, Constabulary G-2 Section, was "exemplary." This was because the Circle C Cowboys were spread throughout the American Zone and tasked with police-security responsibilities. Consequently, they were able to acquire pertinent intelligence. Colonel D. D. Yeuell from the Constabulary G-2 Section became the new chief of the Soviet Border Commission. Consequently, the intelligence-gathering team became of even greater value to G-2 EUCOM. Weekly classes were held on intelligence methods and how to cooperate with other intelligence agencies attached to Constabulary elements. Each trooper was constantly reminded to be "intelligence-conscious." The arrangement was important because it kept the intelligence network abreast of all activities that potentially threatened internal security, such as communist-driven strikes, instigating population unrest, propaganda activities, and undesirable political uprisings and riots. Also, because of the security situation and the deteriorating relationship with the Soviet authorities, the Military Government issued a directive to the liaison and security officers (LSOs) to no longer advise the German police. Reeves found that this directive hampered his section in gathering critical intelligence.[96]

The issue of intensive intelligence gathering arose back in mid-1947, when Constabulary operations began to place more emphasis on intelligence and to rely more on the Germans for policing. Squadrons, working primarily with the 470th and 970th CIC Detachments and Military Government personnel, started to pay considerably more attention to the political activities of the German Communist Party (KPD) and the Socialist Unity Party of Germany (SED) in the American Zone. The Military Government was concerned given that there was a shortage in such essentials as food, shelter, and clothing. This could lead to political agitation and a major German uprising directed against American supply areas and command posts. The Military Government made the Constabulary G-2 Section aware that interviews with various communists and Socialist Party members indicated their prevailing suspicious attitude of anything American and that their mood was becoming increasingly more antagonistic. The information provided to the G-2 noted that the transition from Allied cooperation during the war to the current period of suspicion and dislike was reflected in the communist publication *Das Neue Wort*. The publication mentioned that some communists in the American Zone were convinced that a final military conflict between communists and capitalists was inevitable. A few of the communists interviewed, however, admitted that in the Soviet Zone political freedom was restricted. Their excuse was that "it only affected the reactionaries who deserve to be restricted." It was becoming evident as the Soviet Union was consolidating its control over central and Eastern Europe that the authoritative one-party communist system could not tolerate dissenting opinions. It was also propagandized by KPD in the American Zone that the Soviet Zone was more "security-conscious" because it was supposedly under constant attack from the west.[97]

Regarding the delicate German prisoners of war issue, those interviewed supported Stalin's position that they should be retained to help rebuild a damaged Soviet Union. An August 1947 intelligence report from EUCOM to the Constabulary G-2 noted that in Joachimsthal, Czechoslovakia, the Soviets were operating three uranium mines with eight hundred German prisoners of war working at depths of more than seven hundred meters. One mine was reported to have the richest uranium vein, about ten inches in diameter. Once extracted, the ore was crushed, washed, rinsed, and boxed for shipment to the Soviet Union. Numerous Soviet army personnel dressed in civilian clothes not only heavily guarded the three mines but also engaged in counterespionage and countersabotage.[98]

One other problem Constabulary Headquarters was made aware of was the issue of dismantling and reparations. The communists, more than the other zonal powers, were determined to strip East Germany of heavy industries, plus acquire what they could from West Germany. The rationale was that the Soviet Union demanded Germany make good its war losses. Stalin was determined to transform East Germany not only by establishing a single-party communist political system but also by denuding it of the economic base for a unified Germany. Coal mining equipment, aluminum, locomotive engines, jet engines, and ball bearing plants, to mention just a few, were among wartime industries dismantled. I. G. Farben chemical facilities and the Krupp works located in the Eastern Zone met the fate of total dismantling. The hydrogenation Schwellwerk plant in Altenburg and the hydrogenation Bravack A. G. plant in Magdeburg, which were engaged in the production of synthetic gasoline, were also totally dismantled and sent to the Soviet Union. In some cases, it took the communists almost two years to dismantle a plant. In addition, there were numerous cases of plumbing, stoves, windows, and electrical appliances being shipped to the Soviet Union. Once the goods and fixtures were removed, the buildings were dynamited. The Soviets also had not hesitated to dismantle all plants engaged in manufacturing arms, ammunition, tank parts, and other military equipment. The dismantling was usually done by forced German laborers, who were guarded by detachments of Soviet soldiers under the watchful eyes of communist security elements and commissars.[99]

The KPD and SED members interviewed believed that a united socialist Germany likely would be allied with the Soviet Union. The report from the Office of the Military Government concluded after the interviews with KPD and SED members that, "As long as the policies and the relative strength of the occupation powers remain the same, unification of Germany is not possible."[100] The problem was that the relative military strength in relation to that of the Soviet Union was rapidly diminishing due to rapid redeployment and demobilization, hence putting more pressure on the Constabulary as a counteracting military force.

Toward the second half of 1947, it was apparent that dealing with Jewish displaced persons was still an ongoing problem for the Constabulary. There seemed to be no end to their black-market activities. In September, Reeves's intelligence team reported that the situation had become an even more serious security issue.

At the end of August, three to five thousand cartons of American cigarettes were discovered in a wrecked truck on the autobahn near Benzheim. Two Jews drove the truck. The trip tickets were altered to reroute the truck, supposedly carrying chemicals, to mask the movements of black-market items. Another cigarette incident occurred in Schwarzenfeld. After the confiscation and subsequent investigation it was determined that there was a New York City connection. The illegal import system originated with the American Levent Industrial Company in Brussels, Belgium, whose home office was on West Twenty-second Street, in New York City. Shortly after this discovery, German border police intercepted a truck containing one ton of flour, six hundred cases of meat products, and various other food items. An investigation by the 8th Constabulary Squadron and 970th CIC indicated the items originated in a house occupied by a Jewish displaced person. When the house was searched, 500 pounds of flour, 250 pounds of potatoes, and 40 pounds of nails were discovered buried under hay in the cellar. All the items found in the truck and house were of American origin. Two German truck drivers and four Jewish displaced persons were arrested. The Jew who occupied the house was never seen again.[101]

In areas that contained large German populations, increased anti-Semitism was also a constant problem. Constabulary personnel believed former Nazis who had not been arrested fostered this attitude. The racist attitude was further enhanced by the black-market operations, which generated funds. This situation made clandestine behavior more intense given that Jewish refugees were attempting to leave the American Zone for a homeland in Palestine.

In addition to the Jewish problem, the Poles were still strongly despised by the population for their incidences of violence. The Constabulary G-2 Section noted that if the opinions of the Germans were taken at face value, the Jews were responsible for 80 percent of the black-market activities, the Poles were responsible for 80 percent of the violence, and the remaining 20 percent of both was committed by U.S. personnel. Sarcastically, Reeves reported "that each true German should be provided with a shiny halo."[102]

Nevertheless, the check-and-search operations continued. Upon request by the Military Government, troops from the 25th Constabulary Squadron were called upon to search for an illegal distillery operation. Fifteen troopers under the direction of a Military Government liaison and security officer arrested five inmates of Jewish Hohenfels Displaced Persons Camp. The distillery, however, was not located. Instead, what was discovered was an illegal slaughtering operation, five hundred pounds of beef, several cowhides, and 250 gallons of beer. At the same time, the Constabulary carried out other operations. The 2nd Constabulary Squadron confiscated ammunition and seven pistols of foreign make, two automobiles, twelve head of cattle, and four thousand pounds of freshly slaughtered beef at the Jewish Leipheim camp. At the Vogelsherd camp, the 53rd Constabulary Squadron confiscated 1,500 pounds of raw tobacco, several cowhides, and miscellaneous items of U.S. Army clothing. Eleven arrests were made, nine of which were Turkish nationals.[103]

By October 1947, when the mission of the Constabulary began to change, EUCOM issued a requirement that troops be prepared to reinforce U.S. military forces in the Bremen Enclave, in Austria, and in Berlin. The requirement was part of EUCOM's protective security plan to deal with anticipated Soviet moves. Provisions were made for attaching other combat ground troops to the Constabulary. As tensions increased with the Soviet Union, the Constabulary moved farther away from its original police-security mission.

In November, an analysis of the employment of Constabulary personnel was requested by EUCOM. The report stated that 7 percent were employed as overhead, 32 percent in area operations, 4 percent in border patrol, 43 percent in training, and 14 percent in purely reserve status. These reports led again to a paring down of the Constabulary, as well as of the training programs at Sonthofen. The horse platoons' missions were changed to backup patrolling. They began to function more in ceremonial roles, as the *Laender* border police continued to take over more Constabulary operations. The Constabulary unit-training center at Kafertal continued to operate the Negro Personnel Center, offering academic instruction for colored soldiers integrated into the Constabulary. These developments in November and December indicated that a totally new mission for the Constabulary was moving forward.[104]

Disciplinary problems among Constabulary troops still remained a major concern. In December alone, 58 cases were tried by general courts-martial, 397 cases by special courts-martial, and 545 cases by summary courts-martial. All cases were for violations of various military codes listed in the Articles of War. Some of the cases tried by general courts-martial involved officers, but most involved enlisted men engaged in theft, manslaughter, murder, embezzlement, and being AWOL, with a smaller number of rape, carnal knowledge, assault, and sodomy cases. A few cases were dismissed, or the soldiers involved were acquitted. The majority were found guilty and received dishonorable discharges, total forfeiture of pay, and confinement to hard labor. To say the least, the Constabulary Judge Advocate General Staff was overloaded.[105]

The purpose for providing the Constabulary with intelligence information was that it was responsible for area security. With growing Soviet activities, it was important that the Constabulary be made aware of all potential threats that could create further instability in the American Zone. The mission of EUCOM was to make sure all intelligence agencies disseminated information in a timely fashion. The G-2 noted, "If the [intelligence] liaison should break down the mission of the Constabulary and the security of the U.S. Zone would definitely be adversely affected."[106]

Additional intelligence reports received by the Constabulary at the end of the year and into January 1948 indicated the continued intensity of Soviet activities in mining uranium ore throughout occupied eastern Europe, especially ore containing blue-green streaks. Some of the mineral was being shipped to Freiberg, which had previously contained a large research laboratory engaged in testing and analyzing uranium ore. The hydrogenation plant at Zietz employing six thousand workers was producing synthetic gasoline solely for Soviet military forces. In Soviet-controlled Saxony, a *Volkspolizei* was established, charged with political espionage against the civilian population. One German source stated that anyone making a derogatory statement about the Soviet-controlled government was subject to arrest. Intelligence also reported a marked increase in the movement of Soviet army units into the Eastern Zone. Noted were units of brigade size with heavy, medium, and light tanks, assault artillery, and mechanized assault troops including specialists such as engineer troops, sharpshooters, sappers, and mine-laying detachments.

The most noticeable effect on the Constabulary resulted from the breakdown of the Four Power London Protocol, especially over the sensitive issue of German reparations. This disagreement was one of the main issues that aborted joint peace efforts. As expected, Germany posed the most difficult problem since the war ended. During 1947, as Stalin's armies conquered and were thriving on his totalitarian goals and the miseries of occupied European nations, Truman, with a supportive Congress, believed international communism was bent on probing for European weaknesses. After the war ended, the Truman administration believed that the economic well-being of Europe was essential to the future security of the United States and its European allies. The events of 1947 now moved President Truman and Secretary of State George C. Marshall to search for a new national foreign policy. The conflicting ideological issue was over political self-determination and economic restoration versus a dictated one-party state transformation proposed by Stalin. By various measures enacted in 1947, such as the Truman Doctrine and the Marshall Plan, political and economic efforts were implemented to contain what was perceived as Soviet expansionism at the expense of weaker states. Under the Truman

Doctrine, economic and military assistance was provided to deal with the immediate communist menace in Greece and Turkey. The Marshall Plan was a cooperative recovery plan designed to deal with the economic plight of all of Europe, which was still recovering from the chaos of war and the demonic effects of hunger, poverty, desperation, and the harsh European winter of 1946–1947. Congress finally passed George C. Marshall's Economic Assistance Plan. Stalin, represented by his minister of foreign affairs, V. M. Molotov, rejected the plan, calling it a capitalistic scheme that meddled in the internal affairs of other countries. The Soviet Union, which now was gaining more control over occupied territories, made sure that those countries under its control did not partake because part of the plan required that participating nations have freely elected institutions.

When the Council of Foreign Ministers failed to come to an agreement over a future Germany, subversive activities became a principal concern in the American Zone. The Allied powers came to a deadlock over Soviet reparations, Ruhr industrial demands, German economic unity, and the establishment of a central government. As a result, the principal security threat facing the Military Government and the Constabulary now became communist subversive activities in the American Zone. For the Constabulary this threat replaced Jewish and Polish displaced persons as the primary problem. Nevertheless, the Jews' struggle for relocation and their black-market activities were still considered problems for the Constabulary. After the partition of Palestine was approved by the United Nations, the Constabulary expected illegal activities to continue for raising funds and equipment for the exodus. The Constabulary G-3 noted, "Jewish DPs are highly persistent in their emigration efforts and rather disrespectful of current regulations concerning border control."[107] On May 14, 1948, Israel declared its independence, and a home for many of the displaced persons was established.

At first, the largest party in East Germany was the German Socialist Party (SPD). It was pressed by the Soviet occupiers into merging with the KPD to form the Socialist Unity Party of Germany (SED). However, this was not a real merger. The SED became the one-party system for the Soviets to use in controlling the political process in East Germany. By mid-1947, it was evident that negotiations to establish a single economic unit in Germany had all but failed. That November, the Constabulary G-3 reported a marked increase in liaison between elements of the Communist Party in the American Zone and SED-KPD elements from the East. The main purpose was to clarify party politics and disrupt economic facilities, such as by initiating a wave of minor strikes. This, Reeves reported, was in response to General Lucius D. Clay's announced anticommunist campaign. Reeves was concerned that this volatile situation could become more intensified. He reported, the German population in the American Zone was and had been skeptical of a solution for unity since the formation of the Council of Foreign Ministers. Also noted, the attitude toward American occupation forces had not improved, nor had the respect for authority. There was also an increase in biting vitriolic accusations against U.S. policies by communist-controlled eastern newspapers. The Soviet government virtually closed its East German border. Consequently, crossings decreased noticeably, while at the same time Soviet Army violations of border control directives increased. The report by Reeves's intelligence team in November concluded, "The primary interest of this section [G-2] in political fields is concerned with these political moves which are of security threatening nature."[108]

After one and a half years of acting as a police force, the Constabulary was shifting to a purely military mission. It was finally determined by the War Department in April 1947 that Cavalry and Armored Forces should be consolidated into a single Armored Cavalry arm. The decision to move the Constabulary to a tactical force had been in the planning stages for months. The decision was finalized when General Clay responded to Lieutenant General Albert C. Wedemeyer, director of army plans and operations in the Pentagon, suggested the evacuation of dependents as plans were being developed for joint service operations in Europe. Clay told Wedemeyer that plans had been drawn up to round out U.S. forces in Europe and that he proposed to reorganize the Constabulary into armored cavalry regiments, with two additional artillery battalions in support and provisional recoilless rifle troops. In addition, Clay requested three armored force–type infantry battalions with equipment for use with armored cavalry regiments to be made available from the Zone of Interior.[109]

Accordingly, Army Ground Forces Headquarters completed plans for reorganizing the Constabulary as armored cavalry elements. Subsequently, all their regiments were redesignated as armored cavalry (light). In February 1948, Constabulary Headquarters was forced to move from Heidelberg to Stuttgart. The move was made to provide new headquarters for EUCOM, which left Frankfurt to the administrators of Bizonia.

The Constabulary, in anticipation of being reorganized, began training in the use of weapons not listed in its table of organization and equipment. More time was spent at Grafenwohr and by June the 2nd Constabulary Brigade had made plans for the operation of a tank school. During this period of anticipation and planning for reorganization, intelligence activities of the Constabulary continued.[110]

In the meantime, labor strikes in Bizonia were reported as still being orchestrated by communist elements. During the April 1948 local elections, however, the KPD was dealt a setback. Nevertheless, the propaganda continued to be geared to the working classes. What really concerned Constabulary Headquarters was a communist coup in Czechoslovakia on February 25. In typical Soviet fashion, the communists infiltrated most government services and trade unions. In March a flow of Czech Sudeten Germans streamed across the border. Adding to the Czech influx, *Volksdeutsh* Romanians, Hungarians, and others from the Balkans continued to cross the borders, especially at the Austrian frontier. For the next succeeding months, democratic elements in Czechoslovakia were purged, resulting in the country becoming a Soviet satellite.[111] The fallout of the coup started an intense intelligence operation in the American Zone, which typically involved the Constabulary and the 427th and 970th CIC Detachments and 7827th Military Intelligence Company all working together now on a much larger scale. The Czech coup d'état finally induced the recalcitrant French to join the Americans and British and to form Trizonia.

With the implementation of the Marshall Plan, communist propaganda became more intense, calling the economic assistance plan an effort to enslave the German population. The KPD weekly in Mannheim, the *Badisches Volksecho,* carried a proclamation in the April 28 issue by Wilheim Pieck and Otto Grotewohl, the two top communists in Germany. They proclaimed the Marshall Plan was America's "monocapitalism" means to enslave and split up Germany. Colonel Reeves's G-2 Section continued to monitor KPD activities, reporting on the increasingly vitriolic propaganda. Even General Heinz Guderian, the famous panzer leader of World War II, was accused of being the chief of the American espionage network.[112]

Problems with displaced persons persisted. A raid in March by the 10th and 22nd Squadrons on the Eschwege Jewish displaced persons camp yielded eleven weapons, quantities of ammunition, a large number of livestock, thousands of American cigarettes, hundreds of pounds of American coffee, and a large quantity of jewelry. It was still evident that these raids were just stopgap measures. Colonel Reeves reported "that in spite of these raids and check and search operations these activities are resumed shortly after the action by U.S. authorities to curb them." Black-market prices continued to escalate.[113] Meanwhile, Reeves's intelligence team reported that a so-called drafting organization called the *Sochut* ordered all Jews to report for possible service with the Jewish army in Palestine. Additional reports noted that military training was being carried out in the majority of Jewish camps in the American Zone.[114]

In May 1948, Major General Isaac D. White succeeded General Burress. An expert polo player and a graduate of Norwich University in Vermont, White was commissioned in the cavalry. His idol was J. E. B. Stuart, the dashing Confederate cavalryman from the Civil War who notoriously rode around Union lines. As a young lientenant, White spent time with the mechanized cavalry at Fort

Major General I. D. White and his aide, Captain Jonathan R. Burton, inspect training of 14th Cavalry Regiment (USCON) at Grafenwohr, Germany, 1950. General White was the last commander of the U.S. Constabulary. Note the M24 light tank in the background. Courtesy of the Patton Museum.

Knox and during World War II, serving under Harmon. He spent most of the war as an armor commander, and later commanded the 2nd Armored Division. Before moving again to Germany, White commanded the Cavalry School at Fort Riley. He left the post to take command and prepare to reorganize the Constabulary based on an Army General Board recommendation for a new table of organization and equipment for a new Armored Cavalry Regiment (Light). When he reported to General Huebner, commander U.S. Army Europe, he was told to start turning the Constabulary from a police-security force into a multitask, hard-hitting, maneuver-oriented armored cavalry fighting unit.

White's job was far from easy. Like Harmon earlier, he had the task of organizing and training a new military force predisposed by the changing tensions of the cold war. The tactical concept he instilled was a philosophy he maintained throughout his wartime career, namely, the use of armor as a fast-moving fighting force to penetrate and disrupt the enemy's communications and supply installations. However, he planned to combine several combat arms branches in one armored cavalry organization. The challenge for White was distressing, if not demanding. He had to mold armored cavalry into a self-contained combined organization similar to the regimental combat teams created in World War II. The goal was to provide a highly mobile armored force capable of fighting mounted. White provided the final blow to the traditional stealth "sneak and peek" reconnaissance mission. The tactical mission now was to fight fully mechanized and to provide an economy of force structure for use on a wide-ranging battlefield for offensive and defensive operations. White also had to train the Constabulary to prevent the enemy from obtaining information on friendly units for the division's support of the corps' main effort against hostile forces. This was his goal when he took over the new command.[115]

Again, as before, the Constabulary was confronted with the ever-present change that made it unique. The cavalry again had to struggle with doctrinal innovation and organizational adaptation. White began by establishing a noncommissioned officer academy for the Constabulary and, with Huebner's support, a tank training center at Vilseck. In addition to commanding a new tactical force with new missions, White was given another mission by the newly reorganized Department of the Army created under the National Defense Act of 1947. He was to organize and train an army equestrian team capable of competing in the 1948 Olympics in England. For a year or two, White recalled, his Constabulary equestrian team was made up of former cavalrymen and artillerymen, all with considerable experience with horses. This team competed in shows and "compiled a very magnificent record of wins" for their efforts, he recalled proudly.[116] Most important, General James Polk recalled, White took positive action by beginning to restore discipline and self-confidence and faith in his men.[117] It was a restoration of the traditional cavalry spirit.

In the meantime, another issue the Constabulary faced arose in June when the German crisis became acute over Soviet reaction to currency reform proposed by the three Western powers. In a series of conferences in London between February and June, representatives from England, France, the Benelux countries, and the United States decided to act unilaterally to resolve the issue of economic unity without Soviet participation. Because of these meetings, the Soviet delegates to the Allied Control Council in Berlin walked out. This signaled the end of the quadripartite administration of Germany and the Allied Control Council. Historically the Allied Control Council rarely acted in unanimity. The French, smarting from their exclusion from the Potsdam conference, usually resisted any suggestions made by American and British representatives regarding the centralization of German administrative departments.

Excessive printing of currency by the Soviets and sluggish German productivity brought about the necessity for the currency change. This devaluated the occupation mark and Germany's own currency. To resolve the financial dilemma and stimulate economic growth, the Western powers decided to issue a new currency. The other problem, which was a constant issue for the Constabulary, was the inflated deutsche mark that encouraged black-market operations and economic exchange. Cigarettes generally served as the accepted medium of exchange throughout the Constabulary era. Thus by causing deflation, currency reform undercut the thriving black-market operations, especially those operated by Jewish displaced persons.

On June 16, 1948, the Soviets withdrew their representatives from the quadripartite administration of Berlin. Days later, the Western powers officially announced currency reform for Trizonia. The next day the Soviets stopped all ground traffic between Berlin and Trizonia and announced their own currency reform for their occupied territories in Germany as well as Berlin. Thus began the infamous Berlin blockade.

Colonel Reeves's G-2 Section began to notice that the German population was fast losing its apathy. However, there was concern that the new deutsche mark was easy to counterfeit. Nevertheless, most Germans favored the reform, hoping it could stabilize the economy. All the agitation against reform came from the KPD members and what Reeves called "short-sighted persons." In addition, the G-2 Section noticed that immediately after the conversion, rationed and unrationed goods appeared in quantity throughout shops in the American Zone. Farmers were peddling vegetables in the streets. It was apparent to the G-2 Section that the Germans had hoarded a great quantity of goods during the preceding months. As expected, Jewish displaced persons camps continued to foster illegal activities. Riots and fights were reported among residents over petty matters. It was also reported that the Jewish recruiting program continued. During June, several ships left Germany for Palestine. Also, border crossings continued. Reeves reported an all-time high in illegal crossings from Czechoslovakia, while simultaneously the Soviets virtually closed their borders. For all practical purposes, liaison with Soviet border authorities had broken down.[118]

The following month, the Constabulary relinquished control of all border posts to the German police. Reeves was concerned that this move could create inaccurate intelligence reporting. Since the announcement of currency reform, the border areas had been critical for gathering intelligence. At the same time, the realignment of the Constabulary as armored cavalry was expected to hamper intelligence gathering as the Soviets continued to violate border regulations. Since first activated, the G-2 Section—working with the CIC detachments and a military intelligence platoon—accumulated much valuable intelligence. Now there would be a void because of a lack of accessibility. Meanwhile, only two

Constabulary Squadrons, the 22nd and 53rd, were given border operation missions. The 22nd Constabulary Squadron was assigned to the northern portion of the eastern border facing the Soviet Zone. The 52nd Constabulary squadron patrolled the southern part of the eastern border with Czechoslovakia. Both squadrons usually patrolled in the rear in support of the German border police.[119]

Understandably, General Clay was now concerned that any reduction in troop strength in Europe possibly could place the U.S. military in a precarious situation. He told General Huebner in February 1948 that the Constabulary strength at eighteen thousand was at a minimum for the security of Germany. Clay was disturbed that steps were being taken to again reduce the authorized strength of his command. He emphasized to Huebner that troop strength "is now much too low for the situation which exists today." He concluded, "We do regard the proposed cut as dangerous to the accomplishment of the mission for which we are responsible." Meanwhile, Clay had been urging—for years, in fact—the transfer of the Office of the Military Government from army control to the State Department. George Marshall finally announced in January that the State Department was ready to take over military government of Germany.[120]

In addition, in June 1948, the Constabulary school at Sonthofen was closed, and during the summer and fall all Constabulary regiments were finally reorganized as armored cavalry. Meanwhile, the school became the headquarters of the Field Artillery Group. The brigades assumed functions similar to those of armored divisions' combat commands. What was left of the Constabulary was now classified as Special Troops. In spite of the changes, the black markets continued to flourish unabated. The economic situation in the American Zone caused some serious concerns due to high prices and the general feeling that there might be another currency reform. The communists took advantage of this attitude by propagandizing that "communism is not dangerous to the workers, only to the capitalists." Colonel Reeves reported that apparently Soviet border personnel were given orders to "shoot on sight." At the end of the year the G-2 Section cautioned that with the Berlin blockade numerous border incidents and communist propaganda would be potential threats in the American Zone then and in the near future.[121]

In August, General White took steps to publish a training schedule to meet the demands of transition. His purpose was to maintain the bearing and appearance of enlisted personnel "at the high desired constabulary standard." Field exercises were planned. A twenty-four-hour training program was established for armored cavalry elements in September, and the following month a Constabulary M26 medium tank and unit training center was established. By December, the long preparation for the reorganization of the Constabulary regiments became a reality. The armored cavalry regiment organization was influenced by mechanized cavalry experiences during the war and the comprehensive and unbiased reports from wartime cavalry leaders, corps and army commanders. In the United States, elements the 3rd Cavalry Group based at Fort Meade, Maryland, became the first redesignated armored cavalry regiment. Subsequently in Germany, the 2nd, 6th, and 14th Regiments became armored cavalry regiments (light).[122]

The table of organization and equipment called for three squadrons, composed of a complement of 2,883 men, seventy-two Chaffee M24 light tanks, fifty-one Sherman M4A3 medium tanks, 105mm howitzer assault guns, and more than three hundred jeeps. No M8 armored cars were included in the new organization. The important additions were dismounted infantry and antitank capabilities, which made up for serious shortcomings experienced and noted in the *Report of the General Board: Tactics, Employment, Technique, Organization, and Equipment of Mechanized Cavalry Units* (1945). Unfortunately, army supplies of tanks had not met the requirements for armored cavalry regiments because of a national policy favoring funding strategic airpower and the atomic bomb and missiles. Not until the Army Reorganization Bill of 1950, followed shortly by the surprise invasion of South Korea spearheaded by North Korea's T34 tanks, did this policy change regarding the ground forces.

Nevertheless, by the end of 1948, the Constabulary redesignated as armored cavalry regiments stood as the only maneuver-oriented combat force available to deal with Soviet aggression. In the event of war, it would be no more than a speed bump because the army was far from being prepared to fight the first battle and win. It was apparent that the army's postwar doctrine on how to organize and fight its next war was not supported with the modern equipment necessary to execute its mission to blunt potential Soviet aggressions in western Germany.

Meanwhile, the 7766th Constabulary Horse Troop continued patrols along the American and Soviet borders. By the end of the last quarter of 1948, horse troops had engaged in 447 patrols that covered almost twelve thousand miles. The last month saw a reduction in patrols. At the end of the year, horse troop strength was at 142, with 40 having rotated to the States or transferred to other units in the European Command. The constant shortage and change of personnel, as with other Constabulary units, made the horse troop's mission difficult. The troop commander complained that this caused considerable pressure; however, he noticed improvements in performance and efficiency. Horsemanship continued to lead the list in training, although cavalrymen complained of a lack of spurs and horse covers. All this was indicative of the future. Within a year, the horse troop would be inactivated and its personnel assigned to tactical units.[123]

In May 1950, the 4th Constabulary Regiment in Austria was inactivated, followed in November by the 16th Constabulary Squadron in Berlin.

The final demise of the Constabulary as a police-security force came about because the United States, England, and France agreed on an Occupation Statute for Western Germany. This action assured West Germans of self-government, and at the same time removed dismantling provisions and industrial restrictions. This gave West Germany more economic freedom and opportunities. The next year, on May 8, 1949, the Basic Law was adopted; the Federal Republic of Germany was established weeks later, with Bonn as its capital. The North Atlantic Treaty Organization (NATO) was created in April and went into effect in August as a defensive counterbalance to Soviet aggressive overtures in Europe. For the first time in U.S. military history, a peacetime integrated armed force was established under a single command and employed to act as a deterrent supported by American nuclear power. There was no need for a U.S. Constabulary in this new international arena and a restored democratic West Germany.

General Clay retired to the United States with the satisfaction of seeing a transfer to civil authority. He wrote, "An occupying army had taken the lead in restoring self-responsibility to the German people and in encouraging the formation of democratic government. It was something of which the Army could be proud and which might well set a valuable historical precedent." Regarding the U.S. Constabulary, Clay said "it won the respect and admiration of all, including the German population." This was a ringing endorsement from one of the army's greatest leaders and administrators during the early cold war period.[124]

Four years after the war ended, the official army publications deleted cavalry as a separate branch and listed instead "armored cavalry." Armored cavalry was reorganized as "an arm of mobility, armor-protected firepower, and shock action." The new table of organization and equipment was structured for armored cavalry to engage in all types of combat in coordination with other combat arms and support services. This was a distinctive position for the U.S. Cavalry given that it had inherited a mechanized maneuver–oriented combined arms force as far back as the "ten lean years." The use of the term "armored cavalry" was a compromise between those who wanted a new armor branch and the diehards who were reluctant to part with cavalry as a branch. These supporters wanted to uphold the tradition of the horse. They were reluctant to accept that since World War I cavalry had changed because of mechanization and new ways of warfighting. The changes challenged the cavalry's traditional role as the arm of mobility. General Herr and his supporters continued to argue for a horse cavalry division. Their attitude was a major hurdle in cavalry mechanization. Neither group was satisfied with the term "armored cavalry." Within a year, Congress resolved the debate.

Meanwhile, on September 21, 1949, the Allied occupation statute came into effect, transferring the functions of OMGUS to the Allied High Commission. As part of the United States' commitment to NATO, President Truman announced in the fall of 1950 a buildup of American forces. All forces and personnel were absorbed into the newly created Seventh Army. On November 24 the Constabulary Headquarters was inactivated, and the last commander, General White, soon left to take command of the Armor School and Armor Center at Fort Knox. The 1st Constabulary Brigade was inactivated on August 15, 1951. On November 15, the 2nd Constabulary Brigade was inactivated. During the last few years, Constabulary personnel and units had been reorganized as armored cavalry regiments forming up as combat teams consisting of a battalion each of tanks, armored infantry, and armored field artillery. The Constabulary had moved from a police security force to a combined arms tactical force. The last remnants of the U.S. Constabulary, the 15th and 24th Squadrons, were inactivated in December 1952. Until then, the remaining Circle C Cowboys had been able to retain their distinctive insignia and uniforms.

There is no doubt that the U.S. Constabulary established an unusual history. Those faithful young troopers and their leaders who were able to accomplish their changing missions were a credit to the U.S. Army. In November 14, 1945, *Harper's* published an anonymous article entitled "Military Occupation Can't Succeed." The Constabulary proved that the author was wrong. By early 1947, the Constabulary had reached its peak strength of nearly 30,917 men, but then it started to decline. As General Harmon lamented, the Constabulary was constantly faced with personnel problems, which remained its foremost challenge. With this being the case, it became a major challenge to leadership. Leaders and troopers of the Constabulary had to improvise without any historical precedent. Nevertheless, as a major police-security force in the occupation of the American Zone of Germany, the U.S. Constabulary provided valuable aid in securing a future peacetime democratic Germany in spite of growing cold war antagonisms between the world's two major powers.

As Brigadier General Albin F. Irzyk stated in the January 2003 edition of *ARMY*, the Constabulary accomplished its mission and, in doing so, ensured the success of military occupation in the American Zone for almost six years. Irzyk quoted from the 1950 *Congressional Record*, Eighty-first Congress, which acknowledged the Constabulary as "probably the keenest, most vigilant eye we possess—ready to live up to [its] mission."[125]

The U.S. Constabulary was built on the cavalry organizational model and created in a tumultuous period in American military and political history. It was a period of unprecedented chaos when elements of the mechanized cavalry and supporting units were called upon to perform a unique mission under thorny international conditions. At home, the postwar rush to "get the boys home" became debilitating and stifled preparedness for the coming cold war. Yet, in spite of this, the Constabulary prevailed. Certainly, the Circle C Cowboys became unique and distinct. They subsequently filled the void caused by redeployment of World War II veterans eager to get home. The Circle C Cowboys met and adjusted to unparalleled challenges and changes, all during a period of extraordinary international complexities. They became a new breed—cold war warriors.

By 1950, the mission of the reconstituted armored cavalry regiments was to deploy as a light armored force "to engage in offensive or defensive combat, either mounted, dismounted, or a combination of both, primarily in execution of security and reconnaissance missions." The principle of economy of force was added to the field manuals, meaning high commanders now had the means to allocate minimum essential combat power with a minimum expenditure of sources. The armored cavalry regiments were to be assigned to field armies normally on the basis of one per corps or army. The regiments were to be retrained to develop the capability to deploy as light armored task forces in security and light combat missions when suitably reinforced. In addition, they were to act as screening and counterreconnaissance elements and, finally, deploy to conduct reconnaissance for higher echelons by independent action without reinforcements.[126] The problem was how to make it work with appropriate equipment. The new field manuals specified that the combat missions of offense, defense, and delays were the techniques to be used to execute reconnaissance and security. Those Constabulary troopers that stayed with the redesignated armored cavalry regiments continued a tradition that was upheld for decades and played itself out in the deep envelopment of the Iraqi army in 1991.

Conclusion

How can the embattled period from after World War II until the Army Reorganization Bill of 1950 be summed up? What effect did postwar American attitudes and the international turmoil of the early cold war period have on the future of the U.S. Cavalry and its leadership? In answering these questions it is important to discuss and understand the historical developments that shaped the future American military landscape.

It is said that when Henry Ford requested a loan in 1903, the banker told him "the horse is here to stay." Years later, after the Korean War, the loosely antic television series *M*A*S*H** depicted a mobile army surgical hospital treating casualties near the front during that war. With typical showmanship, the commanding officer and surgeon, Colonel Sherman J. Potter (as portrayed by Harry Morgan), was depicted as still rooted to his old branch, the horse cavalry. Potter arrived at his new command with a saddle and numerous pictures of famous horses, which he displayed proudly in his office. The colonel was overjoyed when he finally received a horse from his company clerk. Upon entering Potter's office the excited animal proceeded to void on the floor. Potter exclaimed cheerfully, "It was like tiptoeing through the tulips." One of his maxims during the series was that looking at life from the top of a horse was much more rewarding than looking at life from the back of a horse.

The Ford incident and the television rendition of a nostalgic cavalryman demonstrated the utility and misguided mystique of the horse. The banker and Potter represented an attitude that resisted change and upheld the tradition of "man's noblest companion." During World War II, mechanized cavalrymen had adjusted by integrating new tactical means to advance the role of their combat arm. "Man's noblest companion" was no longer an issue. No sooner had the war ended than mechanized cavalry elements received a new mission, at a time when the United States first became a dominant force in international affairs. Certainly, neither Ford's reluctant banker nor Potter would have understood the effect it had on the cavalry.

The recording of cavalry history in a professional journal began in 1888, and since then the army's mounted force journal has undergone numerous changes. Initially it was called the *Journal of the U.S. Cavalry Association* (1888–1921), then the *Cavalry Journal* until 1946, when the name was changed to the *Armored-Cavalry Journal*. In 1950, when the cavalry branch was absorbed by the new armor branch, the name was changed to *ARMOR*, with little opposition from traditionalists. Through all these years the purpose of the journal has remained unchanged, namely, discussion of doctrine, tactics, equipment, sports, and at times heated debates over the horse versus mechanization. After World War I, the mounted journal reflected the dogma of the chiefs of cavalry until World War II, when Johnny Herr was removed for his hidebound traditionalism and lack of loyalty to the army chief of staff, George C. Marshall.

With antagonistic international events becoming more intensified, the U.S. Constabulary had to struggle continuously with the deteriorating manpower issue, as Ernie Harmon noted. The United States' national defense policy was moving toward reliance on nuclear weapons delivered by strategic airpower with a current reduction in ground forces. After World War II, nuclear airpower thinking had a great impact on national military policy. As a result, conventional combined arms doctrine, so successfully demonstrated in western Europe in World War II, was downplayed in favor of an ongoing concentration on airpower and nuclear weapons.

In January 1946, the *Army Navy Register* reported, "Demonstrating mobs of soldiers in near mutinous actions [in] overseas theaters." This headline must have made discharged veterans who had served faithfully wonder what had happened to that large army that defeated Nazi Germany. News from Europe indicated the arrival from the United States of a large number of low-point replacements. The *Register* blamed a portion of the personnel problems on civilians, noting, "Here again the politicians had not only a finger, but hand and forearm, in the pie."[1] "Bring Back Daddy" clubs sprang up in the United States and had a staccato effect on Congress so that the legislative body would speed demobilization. Eisenhower strongly believed that his position as chief of staff during this period was a sorry place to be after having commanded a theater of war.[2]

Most American civilians were not concerned with who performed occupation duties in Germany and Japan, so long as it was not their fathers, sons, or husbands. The attitude on the home front also had an effect on those soldiers just deployed for occupation duty. Certainly, they felt the frustrations from home. Once the war was over and the large army demobilized in Europe, local civilians and politicians did not understand that it was necessary to slow down for two reasons. First was to assure pacification and control of Germany and

Austria; second was to dismantle and liquidate huge German war capabilities used to fight a global war. The wartime consensus of the government, the military, and the people had broken down. It was no wonder Eisenhower's tenure as chief of staff was full of frustrations.

The wartime army was falling apart rather than demobilizing while Eisenhower was busy coping with Congress over budgetary problems and the ongoing public outcry to "bring the boys home." Adding to this dilemma, troop discontent over inequities in demobilization almost turned into a mutiny. At the same time, Eisenhower struggled with the need to redeploy the army for occupation duty in Germany and Austria.[3] The burden of those duties fell more on the army than the other services, placing undue stress on the ground forces' ability to handle postwar turbulence in a country devastated by total war and the resulting millions of displaced persons. Added to this predicament was the threat of communist expansionism.

Speaking on national security at the Nebraska State Fair in Lincoln on August 31, 1947, General Jacob L. Devers, then the commanding general, Army Ground Forces, observed that army strength was not adequate if a crisis occurred. "It became evident that, with every reduction in the power at our disposal, there was a corresponding deterioration in the international situation," he said.[4] President Truman, meanwhile, continued to implement a defense policy based on deep cuts in military expenditures and primary reliance on U.S. nuclear power delivered by air.

As political imperatives developed, General Devers became critical of the nation's policymakers. He claimed they made no plans to raise the level of education so the public could have an informed opinion, cognizant of world problems. Regarding the future army, he was disappointed that the president and the War Department's plan for universal military training necessary to fill the ranks of the National Guard and Organized Reserves was negatively viewed by a recalcitrant Congress. Because the bulk of what was left of a demobilized regular army was on occupation duty and garrisoning U.S. territories, Devers continued, there could be a major manpower problem if a war occurred.[5] Two years later the army would be stretched even further because ground troops were required for the United States' contribution to the North Atlantic Treaty Organization, which along with the Truman Doctrine and Marshall Plan were part of the nation's strategy for containing Soviet expansionism.

When the economy-minded Republicans gained control of both houses of Congress in the 1946 elections, the army's future became even more uncertain. Perhaps the influential Republican senator and conservative from Ohio, Robert A. Taft, displayed the best example of congressional temperament during the postwar period. He was a proponent of limited government, challenging the country's postwar role in internationalism. The senator was a leading isolationist who viewed the United States' new commitment to Europe as being imperialistic. He was far from enthusiastic about committing U.S. ground forces in Europe. Taft even questioned sending the much-needed military equipment to Europe. Instead, he supported the navy and a policy of reliance on airpower and nuclear weapons for national defense.[6] Eisenhower came to consider Taft "as hot-tempered and demagogic."[7]

Throughout this period Congress tenaciously held the illusion of insular security despite growing Soviet intransigence and aggressiveness. By controlling the purse, Congress was able to influence national strategic policy, limiting military force levels and weapon systems development programs. In addition, Congress was unable to make provisions for integrating military budgets with a national military policy because of an archaic decentralized appropriations structure. The army suffered the most under the fiscal restraints of the legislative branch, having its appropriations, especially for research and development, cut each year until the war broke out in Korea. Before he left office in February 1948, General Eisenhower warned that the unbalanced budget situation had rendered the army increasingly unable to mobilize in a national emergency. The outgoing army chief of staff stated that the ground forces had in essence purchased no new equipment, including tanks, since World War II. Therefore, the army, he warned, was in no position to train and arm its soldiers adequately to meet the demands of emerging international threats. Consequently, the ground forces were not in a state of readiness to deal with contingencies and to carry out the defensive plans, which according to Eisenhower, were nothing but "mere scraps of paper."[8]

The resulting postwar military manpower turbulence produced a political and economic ambivalence about the future role of the nation's ground forces despite an emerging, turbulent, new world order. Postwar America was more concerned with bringing forth suburbia in the haze of rising prices and inflation, labor unrest, a crisis in education, housing shortages, and tax disputes. Meanwhile, military manpower continued to decline, not for a lack of volunteers but due to army budget cuts. As for the horse cavalry, James Forrestal, the secretary of the navy and soon to become the first secretary of defense under the 1947 National Defense Act, wrote in his diary the day President Harry Truman decided to abolish the cavalry as a combat arm. Forrestal quoted Truman in April 1946 recalling that General Pershing was a cavalry officer, and during the post–World War I period horse cavalrymen were spending too much time claiming they had won the war. Truman was adamant that this would not happen again. Incidentally, he added, to keep the cavalry as a combat arm, "I don't know just what bearing this has on unification." At the same cabinet meeting, the outspoken, unabashed liberal, Secretary of Commerce Henry A. Wallace, took a jab at the cavalry traditionalists. Commenting on the British cavalry of old, he recalled that one unit's motto read "Mount and ride away." When the unit was mechanized and its horses replaced, it became advisable to change the motto to "Screw and bolt."[9]

Meanwhile, the so-called unification National Defense Act of 1947 had separated the air force from the army, giving airpower equal status with the army and navy. The new Defense Department establishment under a civilian head with cabinet status was intended to improve wartime operations of the services. In fact, it politicized the process by making it difficult to establish centralized planning due to multiservice bickering and squabbling among the service chiefs, thus increasing the competition for military technology moneys during a period of budget constraints. With the technologically driven airpower proponents striving to achieve a greater nuclear delivery capability and the navy, traditionally the most expensive of the military services, fighting for its share, funds needed for armor research and development were virtually nonexistent. These events weakened the army's political situation, depriving the ground forces of the means to develop a proper relationship between doctrine and technology required for mechanized warfighting as envisioned by the War Department Equipment Board chaired by General Joseph W. Stilwell. This board recommended that the cavalry branch be discontinued as a separate arm and suggested that a combined arms force be established to conduct extended service tests of new weapons and equipment. Furthermore, the board recommended the army formulate a doctrine for employment, specifically aimed at providing a ready force quickly available for any military contingency.[10]

These lofty goals recommended by the board were not possible in light of rapid demobilization and the "bring the boys home" attitude. Congress, the president, and the American people were reluctant to intervene militarily against communist expansion, a new aggression that was becoming evident. Instead, the United States initially relied on political and economic endeavors to meet cold war threats. With the breakdown of cooperation within the Council of Foreign Ministers, it became evident the military reality of the cold war had set in. This meant that the ground forces in Germany, including the Constabulary, had to reorganize to deal with any possible contingency. The Constabulary moved from a police-security force to a tactical force and slowly organized to prepare for possible military action in Europe. Unfortunately, the lack of modern equipment, such as tanks, continued to plague the American forces in Europe for years.

Continually driven by domestic policies that focused more on postwar economy and social programs, President Truman was repeatedly adamant about defense cuts. During 1948, it was estimated that the army could expect an 80 percent reduction in equipment funds. Again, the army had to defer equipment modernization. In 1948, when the Joint Chiefs of Staff submitted a $30 billion defense budget based on their perceptions of national security needs, Truman forced a major reduction to the cap of $14.4 billion set in 1947 and progressively reduced it in succeeding fiscal years until January 1950, when it was down to $13.5 billion. Meanwhile, Congress reduced the authorized army end strength from 677,000 to 630,000 men. When the North Korean People's Army invaded South Korea, the U.S. Army's actual strength was about 591,000 men. Furthermore, six thousand serviceable tanks remained in 1950, compared with more than twenty-eight thousand tanks at the end of World War II.[11]

President Truman had blamed rapid post–World War II demobilization of America's mighty military force on the people, the press, and Congress. Nevertheless, he himself went to great lengths to hold down defense spending.[12] Truman's ambitious secretary of defense, Louis Johnson, whose economy drive on the eve of the Korean War again fell heavily on the army, best illustrated this. Johnson believed that the national defense policy rested on nuclear airpower. Unlike Johnson, Secretary of State Dean Acheson favored a more flexible policy based on deployable military power to augment American diplomacy. This policy found support in a recommendation made shortly before the invasion of South Korea in a secret National Security Council study (NSC-68), which finally called for a stronger ground force to deal with increasing challenges caused by the spread of communism worldwide.[13]

Adding to this dilemma was atomic bomb scientist and influential author Vannevar Bush. Bush was head of the Office of Scientific Research and Development during World War II and, beginning in September 1947, director of the Joint Research and Development Board created to resolve technological differences between the several departments and agencies in the military establishment. Earlier he had suggested to Congress that the military limit its work to improvements in existing equipment rather than pursuing technological development. This pronouncement was perceived by the services as an effort to inhibit their ability to prepare for war, adding to problems stemming from inadequate troop strength and turbulence in Europe. Later, and shortly before the war started in Korea, Bush wrote the army chief of staff, General Omar N. Bradley, that the day of the tank's dominance was fading. He argued that for the cost of one tank, one hundred antitank guns could be built and, using new ammunition, could fight and hold defensive lines in Europe against a preponderance of Soviet tanks.[14]

Early in 1949, an Advisory Panel on Armor reported that the U.S. Army had no tanks in production or in development capable of meeting types possessed by the country's potential enemies. The panel considered this situation critical. Unless the army's tank development situation was improved, the members reported, the United States would not have enough tanks to support a major ground war for at least two and a half years after the beginning of hostilities. This had an impact on reforming the Constabulary into armored cavalry regiments. One suggested solution, again as during World War II, was to take advantage of America's great industrial capabilities and the mechanical aptitudes of its people.[15]

The importance of the offensive role of armor was again emphasized in a 1949 field manual. The mounted arm's philosophy, it noted, was that the faster armor moves and the quicker it accomplishes its offensive mission of penetration and envelopment, the smaller the losses and more effective the gains. The cavalry tradition of exploitation was considered, meaning a continuation of penetration and envelopment. Tankers were expected to plan boldly and execute their missions with aggressiveness and violence, employing heavy firepower, mobility, and speed.[16] Armored Cavalry was expected to provide mobility, armor-protected firepower, and shock action. As the U.S. ground forces stood in Europe, they were lucky if they could just provide a speed bump if Stalin decided to move into West Germany with the Soviet horde.

By 1950, army doctrine had been revised in many ways; however, it was a refinement based on the experiences of World War II. It was Eurocentric, designed to fight a total war rather than contingency operations in present and future less-than-total war situations around the world.[17] Virtually all ten army divisions had been skeletonized as a result of congressional and White House actions. Furthermore, peacetime training had become less rigorous. The army was in no shape to fight and win its first battle.

In March 1950, the Hodge Board, chaired by Lieutenant General John R. Hodge, reported on the status of doctrine and equipment, noting that armor was more effective when employed as part of a combined arms team of tank, infantry, artillery, combat engineers, and tactical airpower. Armor's mission in the combined arms team was destruction of enemy forces using firepower, mobility, and shock action. The report added that attacking toward deep objectives in pursuit and exploitation over considerable distances was the role for armor at the operational level. In the design of tanks, the report stated, firepower, maneuverability, and mobility were more important than armor protection, although armor remained important. Like the War Department Equipment Board, which recommended cavalry be discontinued as a separate arm, the Hodge Board concurred and recommended tanks be made organic to infantry regiments and divisions, and that three types of functional tanks be developed: light, medium, and heavy. Dishearteningly, the Hodge Report noted that army research and development had been curtailed and would likely be further reduced.[18]

While the Hodge Board was dealing with doctrine and equipment for the ground forces, in the spring of 1950 Congress held hearings on an Army Organization Act to place the army on a sound statutory basis. A new legal base was needed for the Department of the Army, especially after the passage of the 1947 National Defense Act. In addition, the army had to deal with rapid demobilization of its huge wartime army, and at the same time had to reorganize to take care of changing global commitments and military occupation during an uneasy peace. The previous statute that governed the army had become outmoded as the military forces underwent legislative changes based on emergency authority during wartime expansion. The army was still governed by regulations authorized by the president under the War Powers Act that went back to an organizational pattern prescribed by the National Defense Act of 1916 and modified by the National Defense Act of 1920. The 1920 act gave legal substance to the bureau system during the postwar period. With expiration of the antiquated War Powers Act, the army would revert to the pre–World War II organization.[19] This meant that compartmentalization of the combat arms branch chiefs would return, as would their autonomy.

One of the issues was the status of cavalry as a combat branch. Since World War II, there had been a movement in army circles and the executive and legislative branches to eliminate the cavalry. Appearing before the Senate Committee on Armed Services, Major General Verne D. Mudge, who commanded the 1st Cavalry Division in World War II, added his regrets. However, he agreed with the army chief of staff, General J. Lawton Collins, that the only horses now in "the army were used in a small demonstration pack team maintained by the Quartermaster General to keep alive the technique of supply in mountainous terrain." Senator Virgil Chapman, Democrat from Kentucky, remarked that he, too, regretted the passing of the horse cavalry. For decades, the horse had been so much a part of military life, and "he still had difficulty in reconciling himself to it."[20]

In June, the new law went into effect. The Offices of the Chief of Infantry, Chief of Cavalry, Chief of Field Artillery, and Chief of Coast Artillery were abolished. This congressional action was the legal confirmation of the de facto changes that taken place in 1942. The Armored Forces now had a legal branch, which replaced the cavalry branch. The long-running branch debate between the two had ended. Cavalry was absorbed by armor, as were armored cavalry and mechanized cavalry elements. The act was the coup de grâce to traditional cavalrymen who held fast to past exploits.[21] As Johnny Herr lamented, "One basic and immutable truth stands out through all our wars. Sometimes our commanders have had to learn it the hard way: There is no substitute for cavalry."[22]

The popular military writer and analyst Hanson W. Baldwin wrote the cavalry's obituary in the *New York Times*. "No more of the glint of sunlight on the saber, the sweet music of the creak of saddle harness, the champ of bits. The sound of 'Boots and Saddles' sings no more across the Great Plains; the horse has retired from the field of battle."[23]

Thus, the long history of U.S. Cavalry as a separate branch had ended. Looking to the past, there were great cavalry innovators who adjusted to the ever-changing political, doctrinal, organizational, and technological realities of warfare. As the Civil War (1861–1865) progressed, cavalry units began to function more often as mounted infantry. In 1863, Brigadier General John Buford's Union cavalry used dismounted tactics that set the tone for the Battle of Gettysburg. Lieutenant General Nathan Bedford Forrest used his "Critter Cavalry" frequently in dismounted tactics to defeat Union infantry and cut their supply lines. Major General Philip H. Sheridan carried dismounted tactics even further in 1864 when, armed with rapid-firing carbines, his troopers got the better of General Robert E. Lee's infantry. One of the most innovative of all cavalrymen, Major General James H. Wilson, was referred to by one historian as the "Yankee Blitzkrieger" for his exploits when he carried out spectacular raids through Alabama and Georgia.[24] These cavalrymen are examples of past innovators who demonstrated the ability to adjust to meet the evolving demands of warfare. They had the keen sense to adapt and then change. They understood the past but were not captives of it. Many cavalry units rode the Great Plains to subdue the native Indians. Yet history cannot forget Colonel George A. Custer's decision at the Little Bighorn, where he placed his cavalrymen in a compromising position, resulting in an overwhelming number of Indians employing tactics he had never seen before. Surrounded, Custer and all his "yellowlegs" were killed. By 1916, much of the tradition of mounted cavalry was a cherished part of American history. However, change was apparent when General John J. Pershing led his punitive expedition into Mexico. Pershing's cavalry and infantry were accompanied by primitive, motorized elements. On the western front in World War I it was evident to a few that the "yellowlegs" and their horses had no place on the modern battlefield. Frank Parker tried to create awareness of the necessity to adopt and change to meet the conditions of mechanized warfare. His discussion with J. F. C. Fuller was illuminating, as was his influence on Chaffee.

Throughout the 1920s, the U.S. Cavalry tried to find its reason to be. Would it be the horse, mechanization, or horse, motor, and mechanized? Once the Mechanized Force was disbanded in 1931, the cavalry was divided. Fort Knox became the home of mechanization and mounted warfare, while at Fort Riley cavalrymen clung tenaciously to the horse and then tried to marry the horse with the machine. Adding to this divisive situation was a policy that compartmentalized each combat branch, thus degrading the cross-cooperation of a combined arms concept. The autonomous combat arms branch chiefs were more concerned with their traditional defined roles and specialized branch equipment, a situation that carried over until the time the United States entered the war against Germany. By then, the cavalry, along with the field artillery and infantry, had to rely on equipment not suitable for dealing with German arms, especially armor, artillery, and supportive airpower.

The one who was most resistant to change and innovation, causing the immense debate between cavalrymen, was the last chief of cavalry, Johnny Herr. As Bob Grow noted, Herr lost it all at a critical time when modernity clashed with tradition. There would be no horse and no mechanized cavalry divisions. It was the armored divisions of World War II that took over the cavalry role of pursuit and exploitation. Meanwhile, all remaining horse-motorized and horse-mechanized units were totally mechanized and initially trained for "sneak and peek" missions. As U.S. forces invaded occupied Europe, the concept of the old cavalry was forced to change. As during the Civil War, cavalry units were again fighting dismounted and rarely engaging in reconnaissance. By the end of the war, cavalry once again took on a new meaning with the establishment of the U.S. Constabulary. It was almost like the frontier cavalrymen of the second half of the nineteenth century. Armor proponents dominated postwar thinking, and as a result of the communist threat in Europe, the Constabulary was tactically rearmed to form armored cavalry regiments. This then gave way to the cavalry being absorbed into the newly created armored branch.

Looking back at the history of U.S. Cavalry, there always emerges agents of change that challenged the traditional ways of the past, understood the present, and then envisioned a future cavalry. Frank Parker, Daniel Van Voorhis, Adna Chaffee, Robert Grow, Bruce Palmer, Charles Scott, H. H. D. Heiberg, Truman Boudinot, Ernest Harmon, John Shirley Wood, Willis Crittenberger, I. D. White, plus many other cavalry leaders of World War II, such as Edward M. Fickett, Jimmy Polk, Charles H. Reed, Albin Irzyk, and even Cyrus Dolf, an infantryman, all demonstrated their ability to adjust to the demands of preparedness and innovation and to deal with mechanized warfare during a critical period in U.S. Cavalry history. Polk became commander of the 3rd Armored Cavalry Regiment and eventually the commander in chief of United States Army Europe (USAREUR), retiring as a four-star general in 1971. He and Creighton W. Abrams were the last senior officers on active duty who had begun their military careers as horse soldiers. It was Patton, however, who finally redirected cavalry thinking in the modern age when he moved to armor in 1940 and eventually drove his Third Army to an operational level of warfare. This notion was nurtured during the interwar period by the marketeers of mechanization at Fort Knox but was never moved to a broader tempo of time and space beyond tactics.

In the postwar period, cavalry elements and supporting units were again given a new mission, the U.S. Constabulary. Once more cavalry leadership prevailed. General Harmon and his staff, Fay Prickett and Al Irzyk, William Biddle, Henry Newton, A.R. Reeves, and many brigade, regiment, and squadron commanders, trained and led young and inexperienced troopers to become an elite force of cold war warriors. The Constabulary became an important legacy in American military history, but one rarely recognized for its contribution during that critical cold war period.

Regarding the postwar period and the establishment of the U.S. Constabulary, it is difficult to assess their impact in its formative years due to personnel turbulence and international events. General Harmon stated that the personnel situation was the most difficult ongoing problem in fielding a suitable force. However, it happened. During the police-security stage the Constabulary persevered and executed an almost impossible mission under the restraints of declining manpower and the loss of key personnel. All this occurred during the breakdown of the wartime Grand Alliance that defeated Nazi Germany. The Constabulary's legacy was that it was a unique army organization given an almost unattainable police-security mission considering the big-power hostility over the future restoration or transformation of Germany. These international debates over the political and economic unity of Germany had an enormous impact in the American Zone and on Constabulary operations.

During the Constabulary period, the troopers had a nearly improbable mission of dealing with a listless German population and at the same time dealing with the hostilities in Polish and Jewish camps, rampant black-market operations, riots, refugees and border crossings, returning German prisoners of war, *Volksdeuche* migration, and Soviet border violations. They had to manage and control millions of people who entered the American Zone, which was already densely populated and struggling with the chaos of total defeat. Unfortunately, the size of the Constabulary declined from nearly 35,000 to 18,000 men in just a few years, affecting its ability to systematically carry out all missions. Many were nearly impossible.

One of their most important achievements was their ability to gather intelligence. The G-2 Section, especially while under the direction of Colonel Reeves, had a positive effect because of its ability to work with the Counter Intelligence Corps to gain critical information on communist activities, espionage, and information, which was then forwarded to higher headquarters. This was critical because the Office of Military Government in Germany was considering moving the Constabulary from a police-security force to armored cavalry regiments to contain anticipated Soviet aggression in Germany. Again, a new manifestation of cavalry reemerged—from a police-security and intelligence force to a tactical force.

Through the last days of the Constabulary, it had to deal with a German population that the American high commissioner, John J. McCloy, in early 1950 chastised in an address, delivered in Stuttgart, for "failure to assume duties imposed upon them by their new position in Europe." He added that Germany's "future was more than occupation and economic disorganization."[25] It was a credit to the Constabulary that it was able to sustain some semblance of order and control. A new West Germany was finally brought into being, embracing the democratic process the high commissioner had called for. The Constabulary's presence contributed to a free West Germany and, along with the Military Government, prevented the American Zone from having to impose a one-party state based on Stalin's dictatorship.

For ten years, the Constabulary was promised a place of honor at the Fort Riley Cavalry Museum. Numerous administrations did little to expedite a special place to display its history. Finally, it happened through the efforts of Constabulary veteran Albin Iryzk, a noted horseman in his early military career who became an armor commander during World War II, commanded the 14th Cavalry Regiment during the cold war, and served in Vietnam. He was assisted by Master Sergeant William M. Tevington (Ret.) and a number of Constabulary veterans. Through their efforts, a section was established in memory of all those troopers who faithfully served during a most trying time in American history.

The Patton Museum of Cavalry and Armor at Fort Knox has on display an M8 armored car and a jeep that were used during the Constabulary period. This small but important historical display has the unique vehicle markings that identified the Circle C Cowboys, which was so visible as the troopers covered millions of miles throughout the American Zone from 1946 to 1952.

It can be argued that General Herr and those who supported his traditional position left behind a legacy of unwillingness to change. Herr, who ignored the accomplishments of the Constabulary, was too embedded in the past. He was not an example of the innovative thinking necessary to maintain preparedness. His adamant attitude changed the course of U.S. Cavalry and caused the "Great Debate" that was only settled with the creation of the Armored Force in 1940, and the subsequent absorption of the cavalry into the armor branch finally created by law in 1950. Nevertheless, in his speeches and writings Herr left behind a heritage of the mounted horsemen whose spirit and tradition soldiers live by. In fairness to Johnny Herr, this spirit is a potent force, even to this day. Tradition and spirit were understood and provided sustenance as cavalrymen adapted to change and innovation.

Recently a renovated Boudinot Hall, the home of the Command and Staff Departments of the Armor School at Fort Knox, was rededicated in honor of Brigadier General Truman E. "Boudie" Boudinot died in December 1945 of respiratory failure before he could take command of Camp Hood, Texas. On one of the walls are numerous pictures of armor and cavalry soldiers and an outline of their leadership and historical accomplishments during the cold war era. These soldiers are the recipients of the Gold Medallion of the Order of Saint George. The wall remains as one of the historical memorials at Fort Knox to those significant leaders who made cavalry and armor history.

The Patton Museum of Cavalry and Armor is in the process of expanding its facilities, displaying equipment cavalrymen have used ever since Fort Knox was established as the home of the mechanized cavalry. Its repository of textual documents and photos chronicles a transformation from a cultural and intellectual conservatism that caused hesitation in the cavalry community. Adjacent to the museum are over a hundred memorials to individual cavalry and armored units that were deployed in World War II. In addition, at the entrance of the museum are hundreds of bricks donated by loved ones to commemorate armor and cavalry veterans. These facilities and memorials at Fort Knox are a reminder of all the cavalrymen who struggled with doctrinal innovation and organizational adaptation to prevail on the modern battlefield. They were the marketeers for change who did not plan to fight the next war in terms of the last war.

Of the two pioneers of the mechanized cavalry regiments, Bruce Palmer retired on June 30, 1942, at sixty-four years of age. His obvious persistent bitterness over not being selected chief of cavalry in 1938 had a negative effect on Adna Chaffee and the army chief of staff, General Marshall.[26] Considered not a team player, Palmer was not given a major command in the Armored Force. Charles Scott was the first commander of the 2nd Armored Division and during Chaffee's illness was acting chief of the Armored Force. He was an observer in the North African campaigns in 1942 and then became commander of the Armored Training Center at Fort Knox, where he was known for his dynamic training speeches: "To survive in battle against a highly trained and physically fit enemy who revels in killing, our men must learn to endure great hardships, hunger, thirst, loss of sleep, and still remain physically fit and mentally alert." His maxim became legendary: "Kill first or be killed."[27] The other marketeers for change, General I. D. White and Lieutenant General Willis D. Crittenberger, retired after highly successful military careers. After an impressive combat career in the 3rd Armored Division during the war, Brigadier General Truman E. Boudinot died before he could take command of Camp Hood. Colonel H. H. D. Heiberg, who was one of the first troop commanders to deal with the Christie combat car and then became an aide to General Chaffee, eventually retired and returned to the Fort Knox area after a successful career.

Of all the cavalry leaders and pioneers discussed in this story, only Bob Grow, who felt the strong resentment of Herr, finished his career on a disappointing note.[28] He became a victim of unlawful command influence managed by Lieutenant General Maxwell D. Taylor, who in 1952 was deputy chief of staff for administration and later became army chief of staff, assistant for military affairs under President John F. Kennedy, chairman of the Joint Chiefs of Staff, and ambassador to South Vietnam. He was one of President Lyndon B. Johnson's closest advisors in decisions to escalate the war in Vietnam.[29] While the Constabulary was engaged in police-security operations, Grow was assigned to head the American Military Mission to Iran and after two years took brief command of Fort Devens in Massachusetts. Soon he was reassigned again. In 1950, he became the senior U.S. military attaché in Moscow. In 1952, he was the first general officer since Billy Mitchell to undergo a general court-martial for using questionable judgment in recording impolitic statements in his personal diary. However, it was common throughout American military history for high commanders, such as Generals Dwight Eisenhower, Omar Bradley, and Mark Clark, to keep personal diaries and extensive notes. Bradley and Clark were still on active duty when their books were published.[30]

In spite of the miscarriage of military justice, Grow's career followed that of many cavalry officers who later became well-known armored leaders. Like many of his peers, he had a strong sense of history. Throughout the 1930s, he played a considerable role in the development of mechanization and its application to cavalry and then took this experience to the Armored Force and became one of Patton's top armored division commanders. As Grow said, "I have always been a cavalryman and not a tankman nor a horseman." The tank, he believed, offered one means for the soldier to fight mounted. However, that was not enough, "since cavalrymen must fight both mounted and dismounted." Hence, "There must be means for cavalry that would permit the arm to carry out all its missions. This demanded a variety of 'horses' of which the tank is one, an essential one but not the sole one." Grow believed: "There is not, nor ever should have been an independent Tank Corps (consisting of tanks); infantry (dismounted soldiers); and cavalry (mounted soldiers). Both require tanks, however, the latter in greater proportion."[31]

He held that the conduct of war was an art rather than a science and that "the principle of the objective, often referred to as the mission, alone remains fixed," while "the other principles of war varies [*sic*] with each situation and even during the development of an action." He was saying that "secondary objectives interpose themselves in each operation, arise and disappear as the situation develops, but taking any one operation, large or small, the basic objective does not change."[32] One of his organizational methods was to make sure his staff was well trained and active at all levels to avoid confusion and assure that each element was in the proper place at the proper time with an understanding of the intended mission. He expected each soldier in the division to know his job and that of his immediate superior. Key to this philosophy was that all soldiers in the division were trained and imbued with the highest degree of confidence, comradeship, and the will to win. Unlike some armored and division commanders, Grow had the ability to amplify his relationship with higher commanders.

Grow remained a dedicated cavalryman and prominent marketeer until he died on November 3, 1985. J. Caleb Boggs, former U.S. senator from Delaware and Grow's wartime G-4, said, "We all lost a dedicated friend. He was a great combat commander and a good man and continued [an] interest in and [a] concern for all of us since the war." Grow seldom missed a Sixth Armored Division Association annual meeting or local task force events. Colonel Jim Moncrief said of his wartime commander, he "was a man, who, with people, appreciated and cared; and who, in combat, dominated and dared." Colonel Clyde J. Burk, assistant G-3, summed up the tribute: "I, like so many others who served under General Grow, received a legacy which we will never lose—the spirit and the will to fight as a team for what we know is right."[33] Grow was the only armored division commander who stayed with his division from the time it landed on Omaha Beach in July 1944 until the war in Europe ended. As General Patton wrote his wife during the Battle of Bastogne, "To day has been a slugging match, but I got Bob Grow (6th Armored Division) in and things are better."[34]

The question one often hears today is what ever happened to the four combat car T1s built for the mechanized cavalry by the cantankerous J. Walter Christie? Fort Knox veterans claim they ended up on tank firing ranges and their pieces are scattered who knows where over the Kentucky landscape. The Aberdeen Proving Ground Ordnance Museum has the only Christie tank (or combat car?) that was part of the seven vehicles contracted for with Christie. Christie died on January 11, 1944, in Falls Church, Virginia, destitute, facing eviction from his home.[35] After the war ended in Europe, Christie's estate filed a brief with the chairman of the House Committee on Claims requesting reimbursement for development costs to the extent of $250,000.[36] The claim was never settled. Yet even today considerable curiosity exists regarding his troubled life and his contribution to U.S. Cavalry and Infantry tank history, not to mention the impact his vehicles had on the Red Army and later the British army.

Notes

Introductory Essay

1. Baron de Jomini, *The Art of War* (Lippincott, 1863), 304. This observation is from one of the two remarkable observers of the campaigns of Napoleon. The other, and possibly better known, is of course Carl von Clausewitz. Interestingly, Clausewitz wrote in German, Jomini in French. However, Jomini's accounts of Napoleon's wars were standard fare at West Point, where French was the second language through most of the nineteenth century, and from which school were commissioned most U.S. Army officers in the years between the institution's founding in 1802 and the American Civil War. Clausewitz, apparently not translated into English until well after the Civil War, became famous for having opined that war was merely the extension of politics by other means. Jomini, on the other hand, wrote that the proper military objective was ever the enemy army in the field, a fact of which President Abraham Lincoln reminded his General Hooker, then commanding the Union army, in an 1863 letter with this admonition: "I think Lee's army and not Richmond is your true objective point." Eliot Cohen makes this point quite eloquently in *Supreme Command: Soldiers, Statesmen and Leadership in Wartime* (New York: Free Press, 2002).

2. General Creighton W. Abrams Jr., U.S. Army chief of staff, 1972–1974, when he tragically died in office. Of his generation he was clearly the outstanding, tough, enigmatic, demanding, yet charismatic leader and commander. The quotation is in the form of a comment about things not being what they once were, usually made by a subordinate or friend, and his response to that comment, both heard many times by those who served with and for him. Also characteristic of Abrams was a response to the effect that "no it is not the same as it was, and it's a damned good thing it's not."

3. George F. Hofmann and Donn A. Starry, eds., *Camp Colt to Desert Storm: The History of U.S. Armored Forces* (Lexington: University Press of Kentucky, 1999).

4. V. Vuksic and Z. Grbasic, *Cavalry: The History of a Fighting Elite 650 B.C.–A.D. 1914* (London: Cassell, 1993), 37.

5. Department of the Army, *Order of Battle of the United States Land Forces in the World War (1917–1919), Zone of the Interior, Directory of Troops* (Washington, D.C.: GPO, 1949).

6. See John S. Gray, *Custer's Last Campaign: Mitch Boyer and the Little Big Horn Reconsidered* (Lincoln: University of Nebraska Press, 1991), 198, 199; Charles K. Hofling, *Custer and the Little Big Horn* (Detroit: Wayne State University Press, 1981), 24, 29–30; James Willert, *Little Big Horn Diary* (Duarte, Calif.: Published by author, 1982), 203.

7. Jerome A. Greene, *Nez Perce Summer, 1877: The U.S. Army and the Nee Me Poo Crisis* (Helena: Montana Historical Society Press, 2000), 272–79.

8. Francis B. Heitman, *Historical Register and Dictionary of the United States Army from Its Organization. September 29, 1789 to March 2, 1903* (Washington, D.C.: GPO, 1903), 487. Also see Greene, *Nez Perce Summer,* 250.

9. Perhaps most illustrative of the obstinate character of people and institutions is the story of Major General James Wolfe Ripley and the repeating rifle. Commissioned in the corps of artillery out of West Point in 1813, Ripley would transfer to the Ordnance Corps and become chief of ordnance of the army at the onset of the Civil War. Here he stubbornly rejected procurement and issue of the Sharps repeating rifle to the Union army, favoring instead manufacture and issue of more muzzle-loading rifles, then standard issue of the day. His reasons were twofold—there were not enough muzzleloaders to equip the rapidly expanding army anyway, and there was insufficient ammunition manufacturing capacity to support repeating rifles in the hands of troops, who would inevitably fire more ammunition if they had repeaters. It was a decision that would affect the U.S. Army until the Spanish-American War in 1898, after which the army switched to Krag-Jorgenson rifles—repeaters using smokeless powder. So the army fought the Indian wars with single-shot black powder carbines, successors to the muzzleloaders, against Indians generally armed with repeating rifles bought with money the U.S. government had given them to buy food. How many soldiers lay dead or wounded on the western battlefields because they were outgunned by better-armed Indians would be difficult to estimate, but probably frightening to know. See William H. Hallahan, *Misfire: The History of How America's Small Arms Have Failed Our Military* (New York: Scribner's, 1994), 132–38. Ripley's curriculum vitae can be found in Heitman, *Historical Register and Dictionary of the United States Army,* 832.

10. U.S. Army, *FM 100-5 Operations,* (Washington, D.C.: Department of the Army, 1982), 7–2, 7–3.

11. David T. Zabecki, *Steel Wind: Colonel Georg Bruchmuller and the Birth of Modern Artillery* (Westport, Conn.: Praeger, 1986), 14–18.

12. Donn A. Starry, *To Change an Army* (Fort Leavenworth, Kans.: Military Review, March 1983). See also Donn A. Starry, "Changing an Army: An American Perspective," in Michael Evans, ed., *Changing the Army* (Brisbane: Australian Army Land Warfare Studies Centre, 2000), 5–21; and Michael Evans, *Cultural Implications of Hardening and Networking the Army* (Brisbane: Land Warfare Studies Centre. Australian Army Chief of Army Exercise 2004: 2004).

13. Harold R. Winton and David R. Mets, eds., *The Challenge of Change: Military Institutions and New Realities, 1918–1941* (Lincoln: University of Nebraska Press, 2000), xv.

14. Jacob W. Kipp, "Military Reform and the Red Army, 1918–1941" in Winton and Mets, *Challenge of Change,* 108–9.

15. George F. Hofmann, "Doctrine, Tank Technology, and Execution: I. A. Khalepskii

and the Red Army's Fulfillment of Deep Offensive Operations," *Journal of Slavic Military Studies* 9:2 (June 1996), 283–334.

16. Victor Alexandrov, *The Tukachevsky Affair* (Englewood Cliffs, N.J.: Prentice-Hall, 1963), 167. With their chief of staff on that tragic night in June 1937 there died, at the hands of Stalin's secret police, without trial, Generals Yakir, Uborevitch, Kork, Eidemann, Primakov, Feldmann, and Putna, all proponents of a military doctrine that saw independent armored forces in deep attack into enemy force deployments as the essential means to operational-level success. The "Young Red Commanders" (Stalin's favorites) were, on the other hand, proponents of a doctrine that saw armor attacking with and among infantry, at the infantry rate of advance. In true "people's army" fashion, the peasant masses would yet win, this time with the help of armored vehicles. The purge extended downward through the ranks, even into the design bureaus themselves, taking in all several thousand lives. The remaining deep attack proponent, General Triandafillov, escaped execution with this group, having died in an aircraft accident in 1931.

17. Trevor N. Dupuy, *A Genius for War: The German Army and the General Staff 1807–1945* (Englewood Cliffs, N.J.: Prentice-Hall, 1977).

18. Roger Edwards, *Panzer: A Revolution in Warfare 1939–1945* (London: Brockhampton Press, 2003).

19. Will Fowler, *Barbarossa the First 7 Days. Nazi Germany's 1941 Invasion of the Soviet Union* (London: Amber Books, 2004), 188–89. See also Klaus Christian Richter, *Cavalry of the Wehrmacht 1941–1945* (Westchester, Penn.: Schiffer Military History, 1995).

20. Bruce Condell and David T. Zabecki, eds., *Truppenfuhrung: On the German Art of War* (London: Lynne Reiner, 2003).

21. Ibid.

22. Shimon Naveh, *In Pursuit of Military Excellence: The Evolution of Operational Theory* (London: Frank Cass, 1997), 116.

23. Wolfgang Fleischer, *The Wehrmacht Weapons Testing Ground at Kummersdorf* (Westchester, Penn.: Schiffer Military History, 1997). At the fall of the Third Reich in 1945, Kummersdorf fell into the Soviet Zone of occupation, later the territory of East Germany—the German Democratic Republic. Under Soviet occupation, as happened in many places, the records and machinery of the Kummersdorf Testing Ground were removed to the Soviet Union proper, and once there became difficult, if not impossible, to recover. There is no record in Fleischer's book of the Guderian demonstration for Hitler. Kenneth Macksey, in *Guderian, Creator of the Blitzkrieg* (New York: Stein & Day, 1976), 58, records the Guderian-Hitler meeting as having occurred in "early 1934" and includes photographs of the meeting, with Blomberg, head of the Truppenamt, and Goring in the party accompanying Hitler. In his own book *Panzer Leader,* Guderian makes no mention of the Kummersdorf meeting but on page 35 cites a demonstration at the Munster-Lager Training Area, in June 1935, of an experimental panzer division. General Lutz, commanding the Motorized Troops Command, sponsored the exercise; the division itself was commanded by General Freiherr von Weichs, with Generals Blomberg and Freiherr von Fritsch from the high command, but not Hitler himself, in attendance. The evidence suggests that there was indeed such a demonstration at Kummersdorf, but that it was apparently staged totally without armored vehicles, since

none were available at the time, using as surrogates small trucks and motorcycles to demonstrate force mobility and tactical agility. Frequently cited is Guderian's persuasive excellence as a briefer, and at the end of the demonstration Hitler's exclamation to the effect that "That's what I want," referring to the panzer force and its operational concept of employment so eloquently described by Guderian.

24. Richter, *Cavalry of the Wehrmacht.*

25. Winton and Mets, *Challenge of Change.*

26. Field Marshal Lord Carver, *The Apostles of Mobility: The Theory and Practice of Armoured Warfare* (New York: Holmes and Meier, 1979), 52.

27. J. P. Harris, *Men, Ideas and Tanks: British Military Thought and Armoured Forces 1903–1939* (Manchester, England: Manchester University Press, 1995)

28. Roman Johann Jarymowycz, *Tank Tactics from Normandy to Lorraine* (Boulder, Colo.: Lynne Reiner, 2001).

29. G. McLeod Ross, *The Business of Tanks 1933–1945* (Devon, England: Arthur S. Stockwell, 1976). The quotation is from the foreword to Ross's book, written by General Sir James Marshall-Cornwall in 1974.

30. See Kenneth Macksey, *The Tank Pioneers* (London: Janes, 1981).

31. Ibid.

32. See Major General Sir Ernest D. Swinton, *Eyewitness* (London: Hodder and Stoughton Ltd., 1932). In *Men, Ideas, and Tanks,* Harris debunks Swinton's claim as inventor of the original British tank, remarking that there would have been no tanks at all had it not been for the ubiquitous Winston Churchill and his Admiralty Landships Committee, a comment somewhat beside the point of who was the inventor of the machine in the first place. Harris does, however, acknowledge that it was Swinton who first tried his hand at doctrine—tactics, techniques, and procedures for tank employment—even before the actual machinery had taken form. Blame, according to Harris, lies at the feet of the Royal Tank Corps, whose leaders failed several times over to fashion a doctrine acceptable to enough of the armor community—users of the tanks—to have the doctrine embraced by most, if not all, tank men and their corps. On the other hand, Ross, in *The Business of Tanks,* points an accusing finger at the technical side of the British armor community, in all categories from insufficient mobility and agility—speed made good—to inadequate armor protection, and most especially to an unwillingness to demand fielding of the British 17-pounder (76mm) cannon, a gun whose proponents claimed it was the only such weapon superior to late-model Wehrmacht armor and gun lethality at range. The truth of the latter assertion is of course subject to considerable question—the German 88mm dual-purpose gun and the armor design of later model German tanks were, by any standards, the best in the field. However one sides in this (yet unresolved) debate, there is sufficient blame for all those involved in Britain.

33. B. H. Liddell Hart, *The Tanks* (New York: Praeger, 1959).

34. See Harris, *Men, Ideas, and Tanks.*

35. Ibid., 317.

36. Harold R. Winton, *To Change an Army: Sir John Burnett-Stuart and British Armored Doctrine, 1927–1938* (Lawrence: University Press of Kansas, 1988).

37. Carver, *The Apostles of Mobility,* 100.

38. Alistair Horne, *To Lose a Battle: France 1940* (Boston: Little, Brown, 1969), 33–42.

39. Alistair Horne, *The French Army and Politics, 1870–1970* (New York: Macmillan, 1984). This book contains three lectures by Alistair Horne from the Lees Knowles Lecture Series. The first three, "Flags in the Wind: The Commune to Dreyfus, 1870–1900," "The Union Sacrée, 1919–1940," and "Collapse 1940," are without question the best available accounts of the complex relationships between France's political leaders and the French army from the end of the Franco-Prussian War to the collapse of France under the onslaught of the Wehrmacht in 1940.

40. See Horne, *To Lose a Battle*, 577. See also Donn A. Starry, *La Guerre Revolutionaire* (Fort Leavenworth, Kans.: Military Review, U.S., Army Command and General Staff College, February 1967), 61–70; and Horne, *The French Army and Politics*, 67–92, which is the fourth and last of the Alistair Horne Lees Knowles Lectures on this subject.

41. Horne, *To Lose a Battle.*

42. See Robert Allan Doughty, *The Seeds of Disaster: The Development of French Army Doctrine 1919–1939* (Hamden, Conn.: Archon Books, 1985), 91.

43. Ibid., 176–77.

44. J. F. C Fuller, *Tanks in the Great War 1914–1918* (London: John Murray, 1920). Chapter 25, "The French Tank Corps," even though written shortly after the 1914–1918 war, is likely the most comprehensive account of this subject in any language.

45. Martin Blumenson, *The Patton Papers, 1885–1940*, vol. 1 (Boston: Houghton Mifflin, 1972). In chapter 29, "San Mihiel," and again in chapter 30, "Meuse Argonne," we find Patton diary entries reflecting vividly the basic problems of controlling and coordinating the operation of armored vehicles. There was first the problem of communication inside the vehicles. The smaller tanks (French FT17s) had no intercommunication system. Crews communicated by various means—physical contact, to direct changes in speed, direction, or other commands relevant to moving and fighting the vehicle. Second, as between tanks, there were no radios or other means of signaling what groups of tanks should do—to support one another, to support infantry, to coordinate a change of mission, command an attack, to withdraw, and engage in defense. (Then) Colonel Patton solved this problem by moving about the battlefield afoot among the tanks, in the process of which he was wounded by small arms or machine gun fire. The third problem of this set was that of communicating with higher headquarters regarding where the tanks were on the battlefield and what progress was being made toward accomplishing the mission. The practice of the day was to extend telephone wire as far forward as possible, normally to regimental level, where a "wire head" would be established, then communicate forward of the "wire head" by runner. Obvious risks to runners for that most important function on the most dangerous part of the battlefield had led to the use of carrier pigeons—among them the famous "Cher Ami." The pigeons, with cryptic messages attached, would fly from the front back to some higher level, perhaps corps or army level (Tank Corps Headquarters in this case) and deliver their precious messages. So important was this link that at one juncture we find Colonel Patton directing a tank crew whose vehicle has overturned in a shell crater to get back into their vehicle and rescue their cage of precious pigeons so that he could inform General Rockenback

(commanding the Tank Corps) of his whereabouts and about how the battle was going. These problems had been encountered and reported on by the British after the first tank attack at Cambrai in 1917, but obviously no acceptable solution had been found by a year later when U.S. Army tanks went forward, first at Saint-Mihiel, and again in the Meuse-Argonne. See Robert Wollcombe, *The First Tank Battle: Cambrai 1917* (London: Arthur Barker, 1967), for an excellent description of the encumbrances to armored operations represented by these three problems.

46. Charles de Gaulle, *Vers l'Armée de Métier* (Paris, 1935)

47. Horne, *To Lose a Battle*, 554.

48. See Eugenia C. Kiesling, "Resting Comfortably on Its Laurels: The Army of Interwar France," in Winton and Mets, *Challenge of Change*, 1–34.

49. John Tupper Cole, Brigadier General, U.S.A., conversations with his aide-de-camp, circa 1953–1954. After graduating from West Point in April 1917, he quite soon became Captain Cole, commanding a troop of the 2nd U.S. Cavalry, deployed to France with the AEF. He frequently described reporting a-horse to General Pershing, with his troop mounted on horses borrowed from the French Cavalry School at Saumur. But he frequently also described the frustrations of horse cavalrymen of the AEF over the inability to employ horse cavalry in battles on the western front. In the years following that war, he would become one of the army's most distinguished dressage riders and a member of the U.S. Olympic equestrian team. He would, on the eve of the 1939–1945 war, command the horse-mounted 2nd Cavalry Regiment, and was the last commander of that regiment as it stabled its mounts and mechanized. He himself would command Combat Command B of the 5th Armored Division in operations in the European theater with considerable distinction. Following the 1939–1945 war, he would command the 3rd Armored Division during its service as the training division at Fort Knox, Kentucky. A remarkably accomplished trainer of soldiers and armor units, he nonetheless never lost his love for the horse cavalry and equestrian activities, serving as chef d'equire of the U.S. Olympic equestrian team for many years after retirement from active army service. I served as his aide-de-camp during his last years of active military duty.

50. See John C. Sparrow, *History of Personnel Demobilization in the United States Army: Department of the Army Pamphlet 20-210* (Washington, D.C.: Department of the Army, July 1952), 11–19.

51. David E. Johnson, "From Frontier Constabulary to Modern Army," in Winton and Mets, *Challenge of Change*, 168.

52. Emory Upton, *The Military Policy of the United States* (1904) (New York: Greenwood Press, 1968 reprint).

53. Raphael P. Thian, *Legislative History of the General Staff of the Army of the United States from 1775 to 1901* (Washington, D.C.: GPO, 1901).

54. Russell Weigley, *History of the United States Army* (New York: Macmillan, 1967), 305.

55. Spenser Wilkinson, *The Brain of an Army: A Popular Account of the German General Staff* (Westminster: Archibald Constable, 1895; reprint, Carlisle Barracks, Pa.: U.S. Army War College, 1983).

56. Upton, *The Military Policy of the United States*, iii.

57. See Major General Otto L. Nelson Jr., *National Security and the General Staff*

(Washington, D.C.: Infantry Journal Press, 1946). Also James E. Hewes Jr., *From Root to McNamara: Army Organization and Administration, 1900–1963* (Washington, D.C.: GPO, 1975), esp. 3–50. For a somewhat milder treatment of General Ainsworth, see Siert F. Riepma, "Portrait of an Adjutant General: The Career of Major General Fred C. Ainsworth," *Journal of the American Military History Foundation* II (Spring 1938): 26–35.

58. Weigley, *History of the United States Army,* 348–50.

59. See Hewes, *From Root to McNamara,* 298–385, 366–74, for an excellent summary.

60. War Department Document No. 1120, *Field Service Regulations: United States Army 1923* (Washington, D.C.: GPO, 1923), for example, p. 11, para. 44 (hereafter cited as *FSR 1923*).

61. War Department *Field Service Regulation 1939* (tentative) (Washington, D.C.: GPO, 1939), 7 (hereafter cited as *FSR 1939*).

62. Obituary of Samuel McClure Goodwin, *Assembly* (Association of Graduates, United States Military Academy), January–February 2001.

63. Don M. Fox, *Patton's Vanguard: The United States Army Fourth Armored Division* (Jefferson, N.C.: McFarland, 2003).

64. Creighton W. Abrams Jr., Lieutenant Colonel, U.S.A., conversations with his officers, 63rd Heavy Tank Battalion, 1st Infantry Division, U.S. Army Europe, circa 1949–1950. He had enjoyed a remarkable wartime record for battle success. The battalion task force he commanded (37th Tank Battalion, 4th Armored Division) had made itself (and him) famous in the relief of Bastogne in the winter of 1944–1945, as already noted. Later he would serve as a combat commander in the 4th Armored Division and become the division chief of staff. After the war, and reverted in rank, he found himself back in command of a tank battalion, where he applied the same techniques that had made his command so famous during the war. He was forever asking of his officers fundamental questions about tactics, techniques, procedures, organization—starting with how many tanks there should be in a tank platoon, a company, a battalion, command control, soldiers, sergeants, maintenance—all details of good unit performance. He frequently commented that he had learned his leadership under the tutelage of Major General John Shirley Wood, legendary commander of the 4th Armored Division in the war just ended in Europe. He had a way of causing his subordinates to think their way through problems the question asker probably had already resolved to his own satisfaction, but which Wood wanted his people to solve for themselves in their own way. As time wore on and his subordinates themselves rose in rank, we all found ourselves using the techniques he taught us, and more and more emulating his methods of teaching by asking questions.

65. William Eugene Depuy, General, U.S. Army. The first commander of the Training and Doctrine Command (1973–1977), he believed sound doctrine to be the engine of change, that doctrine drives requirements for everything—from equipment capabilities, to force structure and organization, to training and education of soldiers, leaders, and units. Even so, he reminded those of us who struggled mightily under his tutelage to find the right doctrine for the post-Vietnam U.S. Army that what we were postulating was no better than the willingness of those who would execute the doctrine to follow

what was written and what they had been taught. "Doctrine," he would say, "is what 51 percent of the people are willing to do on a given day under whatever circumstances they find themselves in." Thus, it would be imperative that the doctrine developer also be a consensus builder—a marketeer, in more modern terms. For once what was written stretched beyond the comprehension of those who would live, and perhaps die, in its execution, they would promptly ignore it—set it aside and devise their own doctrine in situ.

1. Frank Parker: Early Mechanized Cavalry Theorist

1. Elbridge Colby, "Our Arm of the Service," *Infantry Journal* (Jan. 1924): 94–103. For an interesting discussion on the issue of open warfare and firepower, see James W. Rainey, "Ambivalent Warfare: The Tactical Doctrine of the AEF in World War I," *Parameters,* journal of the U.S. Army War College (Sept. 1983): 34–40.

2. Frank Parker Papers, Southern Historical Collection, Manuscript Department, University of North Carolina Library at Chapel Hill, Collection No. 1516, p. IV. This collection contains numerous letters and other papers, official, semiofficial, and personal, as well as pictures. It also contains correspondence with important military and political figures. In addition, a collection of scrapbooks at the Army War College Circular Files, U.S. Army Military History Institute (USAMHI), Carlisle Barracks, Pennsylvania, contains newspaper clippings of Parker's postwar activities. The U.S. Army Military History Institute also has a number of Parker studies and lectures on mechanization, including a dossier. This is the best source for viewing his concepts on doctrine (hereafter cited as MHI). See also Parker Collection at the First Division at Cantigny, Illinois.

3. C.T. Lanham with Edwin F. Harding, *Infantry in Battle* (Washington, D.C.: Infantry Journal, 1939, reprinted Oct. 1986 by Marine Corps Association, Second Edition), ix.

4. Reynold M. Wik, *Benjamin Holt and Caterpillar Tracks and Combines* (St. Joseph, Mo.: American Society of Agricultural Engineers, 1984), 89–96.

5. John R. M. Taylor, "Cavalry and the Aeroplane," *Journal of the United States Infantry Association* (July 1909): 84–88. Also see Edward M. Coffman, *The Regulars: The American Army 1898–1941* (Cambridge, Mass.: Belknap Press of Harvard University Press, 2004), 197–99.

6. LeRoy Hodges, *Post-war Ordnance* (Richmond, Va.: Richmond Press, 1923), 105–6; W. N. Dickinson, *History of Anti-aircraft Guns* (Washington, D.C.: GPO, 1920), pamphlet 1941; and Dr. Peter Kindsvatter to author, Aug. 10, 2000, Subject: SP Artillery, U.S. Army Ordnance Museum, Aberdeen Proving Ground, Maryland.

7. Carl Boyd, Subject: Armored Automobiles for Cavalry, Feb. 24, 1916, War College Division Correspondences 1903–19, 4767-11, Records of the War Department General and Special Staffs, Record Group (RG) 165, National Archives (NA), 1–6.

8. Parker, Memorandum: To Accompany "Report on Armored Automobiles for Cavalry," Feb. 25, 1916, 4767-11, RG 165, NA, 1–2, and interview, 1–3.

9. W. A. Castle, Subject: British armored motor cars, Sept. 16, 1916, 8959-3, RG 165, NA, 1–2: "The Mysterious Tanks," *London Times*, Sept. 16, 1916, 1; and Editorial, "The New British Attack: Our Latest Military Weapon," *Iron Age* (Sept. 16, 1916): 1.

10. W. E. Freeland, "Automatic Land Cruisers: Part Played by American Company in Development of 'Tanks' of French Battlefields," *Iron Age* (Sept. 28, 1916): 694–95.

11. L. T. Hillman, Subject: Tanks, Oct. 19, Nov. 24, and Dec. 17, 1916, 8957-7, RG 165, NA, passim.

12. J. W. Barker, Subject: Notes on French "Tanks," or Armored Land Ships, Dec. 1, 1916, 4767-14, RG 165, NA, 1–2; [French source] Tanks-Tactical Considerations, Apr. 6, 4767-14, 1917, RG 165, NA, 1–6; and St. Chamond, 4768-14, RG 165, NA, 1–5.

13. American Military Mission, Subject: Construction of Tanks, Report No. 95, Insert No. 1, Apr. 6, 1917, 4767-17, RG 165, NA, 1-6.

14. M. Churchill, Subject: French Tanks, Notes based on Allied Offensive of Apr. 1917, May 9, 1917, 4767-27, RG 165, NA, 1; and Subject: The Tactical Employment of Tanks, May 21, 1917, AEF G-3 Report 1468, War Department Historical Files, RG 120, NA, 1 and attachments. Parker supplied this information.

15. American Military Mission, Subject: Tanks, Apr. 12, 1917, 4767-26, RG 165, NA, 1–2.

16. American Military Mission, Subject: Tanks, Apr. 27, 1917, 4767-26, RG 165, NA, 1–2; and Training Note No. 16, "Tank Tactics."

17. I. B. Holley Jr., *General John M. Palmer: Citizen Soldiers, and the Army of Democracy* (Westport, Conn.: Greenwood Press, 1982), 312.

18. Maurer Maurer, ed., "Parker: The Role of Aviation 2 Jul. 1917," in Maurer Maurer, ed., *The U.S. Air Service in World War I*, vol. 2 (Washington, D.C.: GPO, 1978), 118–20.

19. Conner, Subject: Tanks, July 19, 1917, AEF G-3 Report 1468, RG 120, NA, 1-2, and attached proposals; LeRoy Eltinge, "Tank Organization," Nov. 2, 1917, AEF G-3 Report 1468, Records of the American Expeditionary Forces (World War I), 1917–1923, RG 120, NA, 1.

20. General Rene Omnes, Duree de recherché, Paris, France: *"Chars FT de la lere guerre mondiale,"* Aug. 12, 1997; and Pierre Zigmant, Direction de la Communication, Billancourt, photos, *"L. Renault à la tête de la délégation americaine dans une des rues de l'usine,"* June 9, 1998.

21. Parker, Conner, Margetts, and Williams, "Reports on Tanks," Memorandum: To: Chief of Operations Section, AEF-GHQ, AEF G-3 Report 1468, Sept. 1, 1917, RG 120, NA, 1. Parker prepared this report. Also see John J. Pershing, Memorandum: "Tanks" to Adjutant General AEF-GHQ, Sept. 23–24, 1917, AEF G-3 Report 1472, RG 120, NA, 1–4, and attachments.

22. Fuller, "The Tactical Employment of Tanks in 1918," in, "Reports on Tanks," Appendix 6, 1–23.

23. J. F. C. Fuller, *Memoirs of an Unconventional Soldier* (London: Ivor Nicholson and Watson, 1936), 157–58.

24. J. F. C. Fuller, *Tanks in the Great War 1914–1918* (London: John Murray, 1920), 277–78; and Fuller, *Memoirs of an Unconventional Soldier*, 157–58.

25. J. F. C. Fuller, *Decisive Battles of the U.S.A.* (New York: Thomas Yoseloff, 1942), 398 n. 6.

26. "Polish Theater: Blitzkrieger," *Time* (Sept. 25, 1939): 25–27.

27. Kirby Smith, Subject: Organization for Tanks, Memorandum for Chief of Staff, 1–2, John J. Pershing, Cablegram to Chief of Staff: The Organization of Material and

Personnel Necessary for Tanks, Sept. 14, 1917, 1; and LeRoy Eltinge, "Tank Organization," Nov. 2, 1917, AEF G-3 Report 1459, RG 120, NA, 1. For an excellent study on the history of the U.S. Army's Tank Corps, see Dale E. Wilson, *Treat 'Em Rough! The Birth of American Armor, 1917–1920* (Novato, Calif.: Presidio Press, 1989).

28. Redding F. Perry, "The 2nd Cavalry in France," *Cavalry Journal* (Jan. 1928): 28.

29. Ibid., 32.

30. Ibid., 27; and E. N. Harmon, "Second Cavalry in the Meuse-Argonne," *Cavalry Journal* (Jan. 1923): 10–18.

31. E. N. Harmon with Milton MacKaye and William Ross MacKaye, *Combat Commander: Autobiography of a Soldier* (Englewood Cliffs, N.J.: Prentice-Hall, 1970), 17–45; and Edward M. Coffman, *The War to End All Wars: The American Military Experience in World War I* (Lexington: University Press of Kentucky, 1998), 281–82.

32. Harmon, *Combat Commander*, 35.

33. Martin Blumenson, *The Patton Papers, 1885–1940*, vol. 1 (Boston: Houghton Mifflin, 1972), 527–28.

34. George F. Hofmann, "Army Doctrine and the Christie Tank," in George F. Hofmann and Donn A. Starry, eds., *Camp Colt to Desert Storm: The History of U.S. Armored Forces* (Lexington: University Press of Kentucky, 1999), 96–97.

35. H. B. Fiske, A.C. of S, G-5, Memorandum for the Chief of Staff, Subject: Training, France, July 1918, *United States Army in the World War 1917–1919: Reports of the Commander-in-Chief, Staff Sections and Services*, vol. 14 (Washington, D.C.: U.S. Army Center of Military History, 1991), 303–4.

36. On this issue, see David F. Trask, *The AEF and Coalition Warmaking, 1917–1918* (Lawrence: University Press of Kansas, 1993), 176–77.

37. Benedict Crowell and Robert Forrest Wilson, *The Armies of Industry: Our Nation's Manufacture of Munitions for a World in Arms, 1917–1918*, vol. 1 (New Haven, Conn.: Yale University Press, 1921), 199.

38. Charles B. Cory Jr., "Mechanical Cavalry—Henry Ford's Tiny Tanks," *Illustrated World*, July 1918, 648–51; and Robert L. Collins, "Report on the Development of the Tank Corps," in *United States Army in the World War 1917–1919: Reports of the Commander-in-Chief, Staff Sections and Services*, vol. 15 (Washington, D.C.: U.S. Army Center of Military History, 1991), 222.

39. General Headquarters, AEF, France, "Tanks: Organization and Tactics," December 1918, AEF G-3 Report 1459, RG 120, NA, 5; and S. D. Rockenbach, "The Role of Tanks in Modern Warfare," GHQ, AEF, France, Folder 42-3 1918, Oct. 20, 1918, MHI, 1–7.

40. "Notes on the Relation between Tanks and Cavalry," Weekly Tank Note No. 8, in Raymond E. Carlson, "Papers on Development of Tanks," Mar. 16, 1921, Ordnance Technical Intelligence Files (OKD) 451.25/56.1, Records of the Chief of Ordnance, RG 156, NA, 1–2.

41. Rockenbach, "The Role of Tanks in Modern Warfare," 1–2.

42. Peyton C. March, *The Nation at War* (Garden City, N.Y.: Doubleday, Doran, 1932), 274–76; and Hunter Liggett, *Commanding an American Army: Recollections of the World War* (New York: Houghton Mifflin, 1925), 115–16.

43. March, *The Nation at War*, 275.

44. Coffman, *The War to End All Wars*, 352–53; and Donald Smythe, "A.E.F. Snafu at Sedan," *Prologue* (Sept. 5, 1973): 134–49.

45. John J. Pershing, *My Experiences in the World War* (London: Hodder and Stoughton, 1931), 686.

46. Ltr., Summerall to Rockenbach, HQ, Fifth Army Corps, AEF, France, Jan. 13, 1919, MHI, 1–2.

47. Charles P. Summerall, "Leadership," *Cavalry Journal* (Jan. 1921): 4.

48. "First Had Valiant Part in the War: Was Praised by Pershing," *New York Times*, Sept.. 1, 1919, 20.

49. John J. Pershing, "A Message to the Cavalry," *Cavalry Journal* (Apr. 1920): 5–6.

50. Perry, "The 2nd Cavalry in France," 40–41.

51. John B. Wilson, *Maneuver and Firepower: The Evolution of Divisions and Separate Brigades* (Washington, D.C.: GPO, Center of Military History, 1998), 79–81.

52. *Howitzer* (1906): 46.

53. Stephen H. Sherrill, "The Experiences of the First American Troop of Cavalry to Get into Action in the World War," *Cavalry Journal* (Apr. 1923): 158.

54. Christian A. Bach and Henry Noble Hall, *The Fourth Division: Its Service and Achievements in the World War* (Garden City, N.Y.: Issued by the Division, 1920), 221–22.

55. Sherrill, "The Experiences of the First American Troop of Cavalry to Get into Action in the World War," 158–59; and *American Occupation of Germany*, "Operations Reports," Third Army, vol. 11, disc. 2, Dec. 9 to 11, 1918, U.S. Army Center of Military History, 1988, CD-ROM, 67, 69, 73; *Order of Battle of the United States Land Forces in the World War*, "Records of Events," Third Army, vol. 1, disc. 3, U.S. Army Center of Military History, 1988, CD-ROM, 179 (hereafter cited as Center of Military History, 1988).

56. *American Occupation of Germany*, From: Chief of Staff, To: Commanding General, III Corps, IV Corps, VII Corps, Army Staff, vol. 11, disc. 2, Jan. 8, 1919, Center of Military History, 1988, 98; and *Order of Battle of the United States Land Forces in the World War. Zone of the Interior: Directory of Troops*, vol. 3, disc. 3, Center of Military History, 1988, 1270–72.

57. *American Occupation of Germany*, Operations Report, Third Army, vol. 11, disc. 2, June 18, 1919, Center of Military History, 1988, 123.

58. Edward Mandell House and Charles Seymour, *What Really Happened at Paris: The Story of the Peace Conference, 1918–1919* (New York: Scribner's, 1921), 49, 52.

59. *United States Army in World War I*, "The Rhenish Separatist Movement," in "American Military Government of Occupied Germany," 1918–1919, vol. 11, disc. 2, Center of Military History, 1988, 241–42; and Paul-Marie de la Gorce, *The French Army: A Military-Political History* (New York: George Braziller, 1963), 159–71.

60. "After the Armistice," in *American Armies and Battlefields in Europe*, chap. 13, May 22, 1919, disc. 3, Center of Military History, 1988, 487–93.

61. *Order of Battle of the United States Land Forces in the World War*, "Records of Events," Third Army, vol. 1, disc. 3, Center of Military History, 1988, 404.

62. Donald Smythe, *Pershing: General of the Armies* (Bloomington: Indiana University Press, 1986), 244.

63. Robert G. L. Waite, *Vanguard of Nazism: The Free Corps Movement in Postwar Germany 1918–1923* (New York: Norton, 1952), passim.

64. *American Occupation of Germany*, Intelligence Summary, Third Army, vol. 11, disc. 2, Nov. 20, 1918, Center of Military History, 1988, 15–17.

65. Coffman, *The Regulars*, 6.

66. Henry, "Mobility," *Cavalry Journal* (Apr. 1920): 23–27.

2. A Reason to Be?

1. Rockenbach, "The Role of Tanks in Modern Warfare," GHQ Tank Corps, AEF, France, Folder 42-3, 1918, Army War College Curricular Files, MHI, 11–7.

2. "Change in Tactics Owing to Use of Tanks," *Neue Freie Presse*, Sept. 10, 1918, in R. E. Carlson, Paper on Development of Tanks, Mar. 16, 1921, Ordnance Technical Files, Office of the Chief of Ordnance, OKD Files 451.25/56.1, RG 156, NA.

3. Bradford Grethen Chynoweth, *Memoirs: Bellamy Park* (Hicksville, N.Y.: Exposition Press, 1975), 84; and Report of Board of Officers, War Department, 1918, Ordnance Intelligence Files 334.3/1.17, RG 156, NA, 24.

4. William Westervelt, "A Challenge to American Engineers," *Army Ordnance* 1 (1920): 60.

5. John B. Wilson, *Maneuver and Firepower: The Evolution of Divisions and Separate Brigades* (Washington, D.C.: GPO, Center of Military History, 1998), 84–86.

6. Cavalry: General Considerations, Report of the Superior Board on Organization and Tactics 1919, General Headquarters, AEF, Apr. 19, 1919, Command and General Staff College Library, Fort Leavenworth, Kansas, 64–68.

7. Tactical Lessons—Tanks, 29–31; and Pershing's conclusions, ibid., 7.

8. Liggett extracts, "Lessons to be gained from the experience of the present war in so far as they affect the tactics and organization of the infantry," Proceedings of the Lewis Board 1919, General Headquarters, AEF, Apr. 18, 1919, Command and General Staff College Library, 14.

9. Parker extracts, ibid., 15.

10. Summerall extracts, ibid., 11.

11. Chaffee extracts, ibid., 12A.

12. Fuqua extracts, ibid., 4.

13. "Tactical Lessons," ibid., 3–5.

14. Mary Lee Stubbs and Stanley Russell Conner, *Army Lineage Series: Armor-Cavalry: Part I: Regular Army and Army Reserve* (Washington, D.C.: GPO, 1969), 52.

15. Constance McLaughlin Green, Harry C. Thomson, and Peter C. Roots, *The Technical Services: The Ordnance Department: Planning Munitions for War* (Washington, D.C.: GPO, Office of the Chief of Military History, 1955), 33–34.

16. Clarence C. Williams, "Automotive Ordnance Equipment," *Society of Automotive Engineers* (S.A.E.) *Journal* (Oct. 1919): 304.

17. Carlson, "Memorandum on Development of Tanks," Aug. 1, 1919, Ordnance Technical Files, OKD File 451.25/56.1, RG 156, NA, 1–2.

18. "Paper on the Development of Tanks," Mar. 21, 1921, Ordnance Technical Files, OKD File 451.25/56.1, RG 156, NA, passim.

19. "Tanks: Functions in Relation to Design," Samuel D. Rockenbach Papers, Box 4, Virginia Military Institute (VMI) Archives, Lexington, Virginia, 1–7 (hereafter cited as Rockenbach Papers).

20. "Development Record Medium Tank," Ordnance Office (OO), 451/1018, Jan. 1921, 1, and "History of Medium Tank Development," Ordnance Committee Minutes 7814, Aug. 25, 1929, RG 156, NA, 45.

21. Ltr., Summerall to Rockenbach, Hdq., V Corps, AEF, Jan. 13, 1919, Folder 42-3, 1918, MHI, 1–3.

22. Memorandum for the Chief of Staff: Subject: Tank Development, Ordnance Office, OO 451.25/1017, RG 156, NA, 1.

23. Rockenbach, "Remarks at Conference of Department and Division Commanders," Jan. 12–19, 1920, Washington, D.C., Box 4, Rockenbach Papers, 1–22.

24. Rockenbach, "Tanks and Their Cooperation with Other Arms," *Infantry Journal* (Jan. 1920): 541, 543, and ibid. (Feb. 1920): 662, 673.

25. "Army Reorganization Bill," *Army and Navy Register*, Aug. 30, 1919, 303.

26. Quoted in Edward M. Coffman, *The Hilt of the Sword: The Career of Peyton C. March* (Madison: University of Wisconsin Press, 1966), 188.

27. "House Committee Hearings on the Army," *Army and Navy Register*, Sept. 6, 1919, 303.

28. Ibid.

29. "House Committee Hearings on the Army," *Army and Navy Register*, Sept. 13, 1919, 51.

30. "Senate Hearings on the Army," *Army and Navy Register*, Sept. 20, 1919, 365.

31. "Senate Hearings on the Army," *Army and Navy Journal*, Sept. 27, 1919, 114.

32. "Confusion in Military Legislation," *Army and Navy Journal*, Sept. 27, 1919, 115.

33. "Senate Military Hearings," *Army and Navy Register*, Oct. 25, 1919, 526.

34. *Historical Documents Relating to the Reorganization Plans of the War Department and to the Present National Defense Act*, pt. 1 (Washington, D.C.: GPO, 1927), 365–66, 404–5.

35. James E. Hewes Jr., *From Root to McNamara: Army Organization and Administration, 1900–1963* (Washington, D.C.: GPO, 1975), 50–51; and Ray S. Cline, *Washington Command Post: The Operations Divisions* (Washington, D.C.: GPO, Office of the Chief of Military History, 1951), 19–20.

36. George S. Patton Jr., "Tanks in Future Wars," (Apr. 1920): 958–62; and Eisenhower, "A Tank Discussion," (Nov. 1920): 53–58.

37. Dwight D. Eisenhower, *At Ease: Stories I Tell to Friends* (Garden City, N.Y.: Doubleday, 1967), 173.

38. Eisenhower, "Tanks with Infantry" [May 1921], in Daniel D. Holt and James W. Leyerzapf, eds., *Eisenhower: The Prewar Diaries and Selected Papers, 1905–1941* (Baltimore: Johns Hopkins University Press, 1998), 35–42.

39. Martin Blumenson, "The Many Faces of George S. Patton, Jr.," The Harmon Memorial Lectures in Military History, No. 14, U.S. Air Force Academy, Colorado, Mar. 16, 1972, 11–12.

40. John L. Daley, "Patton versus the 'Motor Maniacs': An Interwar Defense of Horse Cavalry," *ARMOR* (Mar.–Apr. 1997): 12–15.

41. Ltr., Chynoweth to author, June 7, 1971; and ltr., Chynoweth to Brig. Gen. S. R. Hinds, Jan. 4, 1972, author's collection. All correspondence from Chynoweth and Hinds to author is also on file at the MHI, Carlisle Barracks, Pennsylvania, and Special Collections and Archives Division, U.S. Military Academy, West Point, New York.

42. Bradford Grethen Chynoweth, "Cavalry Tanks," *Cavalry Journal* (May 1921): 247–51; and Chynoweth, "Tank Infantry," *Infantry Journal* (May 1921): 504–7.

43. Bradford Grethen Chynoweth, "Military Conservatism," *ARMOR* (Mar.–Apr. 1974): 20–23; and ltr., Brig. Gen. S. R. Hinds to Lt. Gen. W. D. Crittenberger, Re: "Summary on the Development of Armor," May 30, 1971, author's collection, 2.

44. "Editorial Comment," *Cavalry Journal* (Oct. 1921): 420.

45. R. M. P. Preston, *The Desert Mounted Corps* (London: Constable and Company, 1921), passim.

46. G. T. Denison, *A History of Cavalry: From the Earliest Times, with Lessons for the Future*, 2nd ed. (London: Macmillan, 1913), 447. Denison, a Canadian, won Czar Alexander II's award in 1877 for producing the best book at the time on the history of cavalry.

47. For an analysis of the pamphlet, see C. Burnett, "Cavalry in the World War," Guy V. Henry Papers, MHI, 1–11 (hereafter cited as Henry Papers).

48. "Military Value of Cavalry," Testimony Given by Colonel Glasgow before the Appropriations Committee, House of Representatives in 1924, George S. Patton Jr. Papers, Library of Congress, 1–5.

49. 1921 Annual Report of the Chief of Cavalry, quoted in Vincent J. Tedesco III, "Greasy Automations and Horsey Set: The U.S. Cavalry and Mechanization, 1928–1940" (master's thesis, Pennsylvania State University, 1995), 1.

50. Adna R. Chaffee Jr. , "The Maneuvers of the First Cavalry Division, September–October 1923," *Cavalry Journal* (Apr. 1924): 133–62.

51. George S. Patton Jr., "Armored Cars with Cavalry," *Cavalry Journal* (Jan. 1924): 6–8.

52. R. P. Hunnicutt, *Armored Cars: A History of Wheeled Combat Vehicles* (Novato, Calif.: Presidio Press, 2002), 21–22.

53. Edward M. Fickett, "A Study of the Relationship between the Cavalry and the Air Service in Reconnaissance," *Cavalry Journal* (Oct. 1923): 412–22.

54. John T. Broom, "It Isn't Doctrine until They All Understand It: *FM 100-5 Field Service Regulations, 1941*, the Armored Force, and the Conduct of the 1944 European Campaign" (paper presented at the U.S. Armor Conference, Fort Knox, Kentucky, May 1993), 9.

55. *Field Service Regulations United States Army 1923* (Washington, D.C.: GPO, 1924), 11. For an excellent study on the influence of *FSR 1923* on army doctrine, see William O. Odom, *After the Trenches: The Transformation of U.S. Army Doctrine, 1918–1939* (College Station: Texas A&M University Press, 1999).

56. *FSR 1923*, 17–18.

57. Air Corps–Cavalry, Air Service (Misc. Div.), AG 321.91 (6-4-25), 5th Ind., RG 407, NA, 2, and Subject: Observation Squadron, Air Service, To: Adjutant General, June 6, 1925, Office of the Chief of Cavalry, File No. 322.09, RG 407, NA, 1.

58. Parker, "On the Conduct of Operations," lecture, Army War College, Sept. 11, 1923; "Organization of Mobilized Forces," Nov. 15, 1923; and "Organization of Mobilized Forces: Combat Organization," Dec. 10, 1923, MHI, passim.

59. On the Morrow Board, see Charles F. Downs II, *Calvin Coolidge, Dwight Morrow, and the Air Commerce Act of 1926*, www.Calvin-Coolidge, org/pages/history/research/, 1–2. For an extended study, see David E. Johnson, *Fast Tanks and Heavy Bombers: Innovation in the U.S. Army, 1917–1945* (Ithaca, N.Y.: Cornell University Press, 1998), 81–103.

60. Judge Advocate General's Corps, *The Army Lawyer: A History of the Judge Advocate General's Corps, 1775–1975* (Washington, D.C.: GPO, 1976), 143–46. Also, see Alfred F. Hurley, *Billy Mitchell: Crusader for Air Power* (Bloomington: Indiana University Press, 1964), 101.

61. Testimony of General Parker before the Aviation Board, Oct. 15, 1925, Parker Papers, University of North Carolina at Chapel Hill, 1–10.

62. Fuller, "Tactics and Mechanization," *Infantry Journal* (May 1927): 457–65.

63. Ibid.

64. "Discussion," *Infantry Journal* (May 1927): 465–76.

65. Ibid., 465–68.

66. Ibid., 476.

67. Allen, "A Resume of Tank Development," lecture, Army War College, Oct. 27, 1927, MHI, 1–5; and Timothy K. Nenninger, "The Experimental Mechanized Forces," *ARMOR* (May–June 1969): 33.

68. Charles Summerall, *Report of the Secretary of War to the President, Annual Reports, War Department, Fiscal Year Ended June 30, 1929* (Washington, D.C.: GPO, 1928), 81.

69. Crosby, "Our New Cavalry Organization," lecture, Army War College, Sept. 19, 1928, MHI, 1–16; and George Dillman, "1st Cavalry Division Maneuvers," *Cavalry Journal* (Jan. 1928): 47.

70. H. H. Arnold, "The Cavalry–Air Corps Team," *Cavalry Journal* (Jan. 1928): 70–76.

71. Memorandum for the Assistant Chief of Staff, G-3, Subject: A Mechanized Force, Mar. 19, 1928, Office of the Adjutant General, AG 537.3 (3-20-28), RG 407, NA, 1.

72. Memorandum for the Chief of Staff, Subject: A Mechanized Force, Mar. 20, 1928, AG 537.39 (3-20-28), RG 407, NA, 1–23.

73. Adna R. Chaffee Jr., "James Harrison Wilson," *Cavalry Journal* (July 1925): 289.

74. Stephen E. Ambrose, *Upton and the Army* (Baton Rouge: Louisiana State University Press, 1964), 43–53; Chaffee, "James Harrison Wilson," 200.

75. John T. See Broom, "The Commander's Vision in Blue and Gray: The Roles of Adna R. Chaffee, Jr., James H. Wilson and the American Civil War in the Development of American Armor Doctrine" (Ph.D. diss., Union Institute, 1993); and ltr., Broom to author, Aug. 21, 1993, author's collection. Also see James Pickett Jones, *Yankee Blitzkrieger: Wilson's Raid through Alabama and Georgia* (Lexington: University Press of Kentucky, 1976), 187–88.

76. Denison, *History of Cavalry*, 447; and interview, Grow with author, May 14, 1971, Patton Museum of Cavalry and Armor, Fort Knox, Kentucky (hereafter cited as Patton Museum).

77. Memorandum: To the Chief of Infantry, Subject: An Analysis of the Chronological History of Progress toward the Formation of a Mechanized Force in the United States Army, Mar. 5, 1931, Records of the Chief of Arms, CI 537.3/7884-B, RG 177, NA, 1–12.

78. Wiley Howell for the Chief of Infantry, Memorandum for the Chief of Staff, G-3, Mar. 26, 1928, CI 537.3/7884-B, RG 407, NA, 1–3; Parker, "A Mechanized Unit," Apr. 2, 1928, AG 537.3 (3-20-28), RG 407, NA, 3–4; and ltr., Parker to Brig. Gen. James C. Dozier, AG, State of South Carolina, Jan. 9, 1929, Parker Papers, University of North Carolina Library at Chapel Hill, 1.

79. Special Orders, No. 110, War Department, May 10, 1928, AG 537.3 (5-8-28), RG 407, NA, 1.

80. Nenninger, "The Experimental Mechanized Forces," 33–34; Lewis Bernstein, "Army Experimental Formations and Their Possible Influence on the Establishment of the Force XXI Experimental Force," CAC History Branch, Research Division, Center for Army Lessons Learned, TRADOC History Conference/Workshop, Oct. 23, 1996, Part 1, 3–4; and ltr., Brig. Gen. S. R. Hinds to Lt. Gen. W. D. Crittenberger, Re: "Summary on the Development of Armor," May 30, 1971, author's collection, 6.

81. Memorandum for: The Chief of Infantry, Subject: Chronological History of Progress toward the Formation of a Mechanized Force in the United States Army, Feb. 20, 1931, CI 537.3/7884-B, RG 177, NA, 16.

82. From: Sub-committee on Tanks, To: The Ordnance Committee, Technical Staff, Subject: Light Tanks T1E2, July 16, 1929, OCM Item 7786, RG. 156, NA, 1–36.

83. Bernstein, "Army Experimental Formations and Their Possible Influence on the Establishment of the Force XXI Experimental Force," 4.

84. Subject: A Mechanized Force, To: The Adjutant General, Proceedings of a Board of Officers… Oct. 1, 1928, CI 537.3/7884-B, RG 177, NA, 1–18.

85. Memorandum for the Chief of Staff, Subject: A Mechanized Force, Report of Board of Officers Convened by S.O. 110, W.D. 1928, Oct. 31, 1928, CI 537.3/7884-B, RG 177, NA, 1–2.

86. 1st Cavalry Division Board, Fort Bliss, Texas, Nov. 20, 1928, and Subject: Ordnance Report on Armored Cars, To: President, 1st Cavalry Division Board, Fort Bliss, Texas, Nov. 27, 1929, OO 451.24/170, RG 156, 1–3.

87. Levin H. Campbell Jr., "Special Automotive Equipment of the Army," *S.A.E. Journal* (Sept. 1928): 294, 299.

88. John K. Christmas, "The Mechanization of Armies," *Military Engineer* (Sept.–Oct. 1929): 455.

89. Frank Parker, "The G-3 Division of the War Department General Staff and Its Problems," lecture, Army War College, Sept. 11, 1928, MHI, 6–8.

90. Parker, to Major General W. D. Conner, War Department General Staff, G-3, Dec. 17, 1928, AG 537.3 (12-17-28), RG 407, NA, 1.

91. Ltr., Brig. Gen. S. R. Hinds to Brig. Gen. Brad Chynoweth, Jan. 7, 1972; ltr., Brig. Gen. S. R. Hinds to Lt. Gen. W. D. Crittenberger, Re: "Summary on the Development of Armor," May 30, 1971; and ltr., Brig. Gen. S. R. Hinds to author, Feb. 13, 1972, author's collection. On the issue over the expanding torrent, see Martin Blumenson, *The Patton Papers, 1885–1940* (Boston: Houghton Mifflin, 1972), 735; and Basil Liddell Hart, *The Liddell Hart Memoirs 1895–1938*, vol. 1 (New York: Putnam's, 1965), 48–49.

92. Chaffee, "The Status of Mechanized Combat Organization and the Desired Trend in the Future," lecture, Army War College, Sept. 19, 1929, MHI, 2–10.

93. For a more in-depth examination of the Christie affair, see Hofmann, "Army Doctrine and the Christie Tank," in George F. Hofmann and Donn A. Starry, eds., *Camp Colt to Desert Storm: The History of U.S. Armored Forces* (Lexington: University Press of Kentucky, 1999), 92–143.

94. From: Sub-committee on Tanks, To: Ordnance Committee, Technical Staff, Subject: Procurement of Christie Armored Car, Mar. 5, 1929, Office of the Chief of Ordnance, Item 7522, RG 156, NA, 14.

95. War Department, Ordnance Office, To: The Adjutant General, Washington, D.C., July 17, 1929, Office of the Chief of Ordnance, OO 451.24/122, RG 156, NA, 8.

96. Procurement of the Christie Armored Car, To: The Chief of Ordnance, From: Office of the Chief of Cavalry, Feb. 26, 1929, Item 7522, Mar. 5, 1929, RG 156, NA, 2.

97. Ltr., Brig. Gen. Hinds to Lt. Gen. Willis Crittenberger, May 30, 1971, author's collection, 9.

98. Ltr., Icks to author, Jan. 25, 1970, author's collection.

99. Maj. George S. Patton Jr. to Commandant, the Tank School, Oct. 28, 1929, Subject: Christie Tank, TS 470 (Christie) RG 177, NA, 1–6; and Col. H. L. Cooper to Chief of Cavalry, Nov. 22, 1929, Subject: Comments on Tank School Report to the Chief of Infantry, Office of the Chief of Cavalry, TS 470, RG 177, NA, 1–3. Also see Blumenson, *Patton Papers, 1885–1940*, 841; and "Special Characteristics of Track and Wheel-Type Vehicle (Christie)," Patton Papers, Library of Congress, 1.

100. Patton, "Tactical Employment of Armored Cars (Experimental)," Patton Papers, Library of Congress, 1; and Martin Blumenson, *Patton: The Man behind the Legend, 1885–1945* (New York: Morrow, 1985), 121; and interview, George S. Patton with author, May 14, 1971, Fort Knox, Kentucky.

101. Subject: Procurement of Christie Armored Car, From: Sub-committee on Tanks, To: The Ordnance Committee, Technical Staff, Mar. 5, 1929, Item 7522, 13–14; and Feb. 7, 1930, Item 8095, RG 156, NA, 1–2; *War Department Appropriations Bill for 1931*, Military Activities, Hearings before the Subcommittee of House Committee on Appropriations, Dec. 11, 1929, 71st Cong., 3rd sess., pt. 1 (Washington, D.C.: GPO, 1929), 538–49; and ltr., Brig. Gen. S. R. Hinds to Lt. Gen. W. D. Crittenberger, Re: "Summary on the Development of Armor," May 30, 1971, author's collection, 7.

102. C. C. Benson, "Mechanization Aloft and Allow," *Cavalry Journal* (Jan. 1929): 154.

103. Ibid.; and C. C. Benson, "Ships on the Battlefield," *Infantry Journal* (Mar.–Apr. 1929): 239.

104. Benson, "The New Christie, Model 1940: An Estimate of the Tank Armored Car Combined," *Infantry Journal* (Sept.–Oct. 1929): 255–61; and Benson, *Army Ordnance* (Sept.–Oct. 1929): 34–41.

105. War Department Appropriations Bill for 1931, Military Activities, Hearings before the Subcommittee of House Committee on Appropriations, 538–49; and Chronology of Christie Tanks Procurement, Office of the Chief of Ordnance, Apr. 1, 1931, OO 451.25/3693, RG 156, NA, 1–2.

106. Affidavit, J. Walter Christie, Dec. 21, 1935, author's collection, 11; and War Department Appropriations Bill for 1931, 1–2.

107. "Convening of Board of Officers to Review the Manual for Commanders of Large Units," Washington, Mar. 25, 1929, AG 062.11 (3-14-29), RG 407, NA, 1–5. Chaffee recorded the proceedings. Also see Gary L. Bounds, *Larger Units: Theater Army–Army Group–Field Army*, chap. 1, CSI Report No. 6, Jan. 1985, Combat Studies Institute, U.S. Command and Staff College, Fort Leavenworth, Kansas, 6.

3. The Struggle for an Innovative Doctrine and a Combat Car

1. "National Defense and World Disarmament," in *The Memoirs of Herbert Hoover: The Cabinet and the Presidency 1920–1933* (New York: Macmillan, 1952), 338–39, 355. For an excellent discussion of the economic impact on the army, see John W. Killigrew, "The Impact of the Great Depression on the Army, 1929–1936" (Ph.D. diss., Indiana University, 1960), Fort Knox Library. Also see John R. M. Wilson, "The Quaker and the Sword: Herbert Hoover's Relations with the Military," *Military Affairs: The Journal of Military History, Including Theory and Technology* (Apr. 1974): 41–47.

2. Charles Summerall, *Report of the Secretary of War to the President, Annual Reports, War Department, Fiscal Year Ended June 30, 1929* (Washington: GPO, 1929), 108.

3. "Statement of Maj. Gen. Adna R. Chaffee, the Commanding General of the Armored Force, United States Army," *Military Establishment Appropriation Bill, 1942* (Washington, D.C.: GPO, 1942), 557.

4. "Prelude to Armor," in *The Armored Force Command and Center*, Study No. 27, 1946, Historical Section, Army Ground Forces, RG 337, NA, 2.

5. Charles P. Summerall, "New Developments in Mechanization," *Infantry Journal* (Feb. 1931): 91–93.

6. Ibid., 92.

7. Subject: Mechanized Forces, To: The Adjutant General, Washington, D.C., Fort George Meade, Maryland, Apr. 17, 1930, AG 573.3, RG 407, NA, 1–10; and "Prelude to Armor," RG 337, NA, 6. Also see Timothy K. Nenninger's excellent study, "Organizational Milestones in the Development of American Armor, 1920–40," in George F. Hofmann and Donn A. Starry, eds., *Camp Colt to Desert Storm: The History of the U.S. Armored Forces* (Lexington: University Press of Kentucky, 1999), 43.

8. Memorandum for the Chief of Cavalry, Subject: Study of Mechanized Forces by Colonel James Kelly Parsons, Office of the Chief of Cavalry, May 19, 1930, AG 573.3, Office of the Adjutant General, RG 407, NA, 1–7.

9. Ibid.

10. Summerall, *Report of Secretary of War to the President*, 124–26; and "Prelude to Armor," RG 337, NA, 6.

11. Ltr., Brig. Gen. S. R. Hinds to Lt. Gen. W. D. Crittenberger, Re: "Summary on the Development of Armor," May 30, 1971, author's collection, 2.

12. Robert W. Grow, "The Ten Lean Years: From the Mechanized Force (1930) to the Armored Force (1940)," ms., Feb. 1969, author's collection, 2–12. His "Ten Lean Years" was edited and published in serial format in four issues of *ARMOR* in 1987 (hereafter cited as "Ten Lean Years").

13. Ibid., 13–14.

14. For a personal observation of the force's organization, see Arthur R. Wilson, "The Mechanized Force: Its Organization and Present Equipment," *Infantry Journal* (May–June 1931): 252–55.

15. Van Voorhis, "Mechanization" lecture, Army War College, Oct. 13, 1937, MHI, 2–3.

16. Memorandum For: General Henry, Office of the Chief of Cavalry, Air Corps Cavalry, File No. 322.09 (8-30-21/2-21-42), Records of the Chief of Arms, RG 177, NA, 2.

17. Memorandum For: General Mosley, Subject: Organization—Mechanized Force, Office of the Chief of Infantry, Apr. 27, 1931, CI 537.3/7884-B, RG 177, NA, 2.

18. Grow, "Ten Lean Years," 16.

19. George F. Hofmann, "Doctrine, Tank Technology, and Execution: I. A. Khalepskii and the Red Army's Fulfillment of Deep Offensive Operations," *Journal of Slavic Military Studies* vol. 9:2 (June 1996): 289–90, 327. Also see Jacob W. Kipp, "Soviet Military Doctrine and the Origins of Operational Art, 1917–1936," in Willard C. Frank Jr. and Philip S. Gillette, eds., *Soviet Military Doctrine from Lenin to Gorbachev* (Westport, Conn.: Greenwood Press, 1992), 114–19.

20. Reasons why General Hof opposed the initial purchase of six Christies, Office of the Chief of Ordnance, June 1, 1931, OO 160/5435, Records of the Chief of Ordnance, RG 156, NA, 1.

21. Christmas comment on meeting with Christie, June 18, 1930, Office of the Chief of Ordnance, OO 451.25/3387, RG 156, NA, 1.

22. War Department, Ordnance Office, To: The Adjutant General, Washington, D.C., July 17, 1929, Office of the Chief of Ordnance, OO 451.24/122, RG 156, NA, 3–4.

23. "Chronology of Christie Tank Procurement," compiled in the Ordnance Office from original data, personal knowledge, and data prepared by John K. Christmas, Apr. 1, 1931, OO 451.25/3693, RG 156, NA, 9.

24. "And Still No Tanks," *Chicago Tribune*, Feb. 11, 1931, 14.

25. Re: Tanks and Newspaper Clipping, Office of the Chief of Ordnance, Feb. 11, 1931, OO 451.25/3626, RG 156, NA, 1–2.

26. Correspondence from Henry E. Barbour, House of Representatives: To: Chief of Staff, War Department, Adjutant General's Office, Mar. 7, 1931, OO 451.25/3666, RG 156, NA, 1–2.

27. War Department Appropriations Bill for 1932, Military Affairs, Hearings before the Subcommittee of House Committee on Appropriations, 71st Cong. 3rd sess., pt. 1, Dec. 5, 1930 (Washington, D.C.: GPO, 1930), 551; and "Christie Tanks," *Army and Navy Journal* (June 10, 1931): 436.

28. War Department Appropriations Bill for 1932, Military Affairs, Hearings before the Subcommittee of House Committee on Appropriations, 71st Cong. 3rd sess., pt. 1, Dec. 5, 1930 (Washington, D.C.: GPO, 1930), 542–629.

29. Correspondence from Henry E. Barbour, 1–2.

30. General MacArthur's reply to Re Barbour, Office of the Chief of Staff, Mar. 9, 1931, OO 451.25/3667, RG 156, NA, 1–2.

31. Taped interview, author with J. Edward Christie, June 6, 1971, Miami, Florida, author's collection.

32. Ltr., Christie to Honorable Patrick J. Hurley, Mar. 11, 1931, OO 451.24/290, RG 156, NA, 1.

33. Ltr., Robert J. Icks to author, Jan. 7, 1970, author's collection, 1.

34. "Chronology of Christie Tank Procurement," 14.

35. Subject: Report of a Board of Officers Appointed by Letter, Dated October 4, 1930 on the Acceptance Tests of the Christie Tank, and the Adjutant General through the Chief of Ordnance, Mar. 16, 1931, OO 451.25/3676, RG 156, NA, 4.

36. "Chronology of Christie Tank Procurement," 15–16.

37. Ltr., Christie to Maj. Gen. L. H. Campbell Jr., Chief of Ordnance, War Department, Washington, D.C., Oct. 19, 1943, author's collection, 1–2; and House of Representatives, Inventive Contributions Award Board, H.R. 7316, 82d Cong., 2nd sess., May 14 and 24, 1952 (Washington, D.C.: GPO, 1955), 66.

38. David Fletcher, *Mechanized Force: British Tanks between the Wars* (London: HMSO Publication Center, 1991), 122.

39. Memorandum for Executive Officer, Subject: Comments on Chronological History Mechanized Force Attached Hereto, Office of Chief of Infantry, Mar. 11, 1931, CI 537.3/7884-B, RG 177, NA, 2; and Subject: An Analysis of the Chronological History of Progress toward the Formation of a Mechanized Force in the United States Army and the Purpose of This Organization, Office of Chief of Infantry, Mar. 24, 1931, CI 537.3/7884-B, RG 177, NA, 1–6.

40. Memorandum For: General Mosley, Subject: Organization—Mechanized Force, Office of the Chief of Infantry, Apr. 27, 1931, CI 537.3/7884-B, RG 177, NA, 1–3.

41. Grow, "Ten Lean Years," 15.

42. Ibid., 16.

43. Killigrew, "The Impact of the Great Depression on the Army," chap. 4, p.15; and Wilson, "The Quaker and the Sword," 41–47.

44. "Chief of Staff Sets Forth General Principles to Govern the Extension of Mechanization and Motorization of the Army," War Department, May 5, 1931, Henry Papers, MHI, 1–3.

45. "A Brief Narrative of the Life of Guy V. Henry, Jr.," An Autobiography, Henry Papers, MHI, 65.

46. Arthur R. Wilson, "With the Mechanized Force on Maneuvers," *Infantry Journal* (July–Aug. 1931): 331–35.

47. Grow, "Ten Lean Years," 18; and Thomas H. Greer, *The Development of Air Doctrine in the Army Air Arm 1917–1941* (Washington, D.C.: GPO, 1985), 66.

48. Report of Mechanized Force 1931 Headquarters, Mechanized Force, Fort Eustis, Virginia, July 1, 1931, author's collection, 1, 22–23. General Grow provided this document.

49. Ibid., 5–11.

50. Correspondence dealing with von Schell's visit to the United States came from the Military Intelligence Division (MID) Files 2257-B-78, July 1930 to June 1931, War Department General and Special Staff, RG 165, NA, passim.

51. Heinz Guderian, *Panzer Leader* (London: Michael Joseph, 1952), 316; and Truman Smith, Memo to the Chief of the G-2 Historical Section, Department of the Army, Subject: Activities of the Office of Military Attaché, American Embassy, Berlin, Germany—August 1935 to March 1939, Truman Smith Papers, Herbert Hoover Presidential Library, West Branch, Iowa, 14.

52. Adolf von Schell, *Battle Leadership: Some Personal Experiences of a Junior Officer*

of the German Army with Observations on Battle Tactics and the Psychological Reactions of Troops in Campaign (Fort Benning, Ga.: Benning Herald, 1933), 90–95. Von Schell's book used for this history was the original copy available at the Fort Knox Armor School Library. Since then the Marine Corps Association has reprinted it in September 2004. In addition, the Marine Corps Association reprinted a second edition of *Infantry in Battle* (Infantry Journal, 1934) in October 1986, which contains a number of von Schell's lectures. The Military History and Publications Section of the Infantry School prepared the first edition in 1934 under the direction of Colonel Marshall. See 18–19, 154–55, 167, 349, 358.

53. Ibid., 17, 63; and Adolf von Schell, "A Typical Fight in Open Warfare," *Infantry Journal* (July –Aug. 1931): 345–47.

54. Steve E. Dietrich, "The Professional Reading of General George S. Patton, Jr." (paper presented at the annual conference of the American Military Institute, Lexington, Virginia, Apr. 14–15, 1989), 37.

55. George S. Patton Jr., "Motorization and Mechanization in the Cavalry," *Cavalry Journal* (July 1930): 331–48. Also see Randy Steffen, *The Horse Soldier, 1776–1943*, vol. 4 (Norman: University of Oklahoma Press, 1979), 130–32.

56. Hofmann, "Doctrine, Tank Technology, and Execution" 300–301; and Ralph E. Jones, George H. Rarey, and Robert J. Icks, *The Fighting Tanks from 1916 to 1933* (Old Greenwich, Conn.: WE, 1933), 168–70.

57. Hamilton S. Hawkins, "Cavalry and Mechanized Force," *Cavalry Journal* (Sept.–Oct. 1931): 19–25.

58. Douglas O. MacArthur, *Report of Secretary of War to the President, Annual Reports War Department Fiscal Year Ended June 30, 1931* (Washington, D.C.: GPO, 1931), 42–43; U.S. Congress, Subcommittee of the Committee on Appropriations, *War Department Appropriations Bill for 1933 Military Activities*, Hearings, 72nd Cong., 1st sess., pt. 1 (Washington, D.C.: GPO, 1932), 17–18; and D. Clayton James, *The Years of MacArthur, 1880–1941*, vol. 1 (Boston: Houghton Mifflin 1970), 354–55.

59. Subject: "Methods for the Tactical Employment and for the Training of a Mechanized Cavalry Regiment," To: The Chief of Cavalry, Office of the Chief of Cavalry, Oct. 30, 1931, Document File 1923–1942, RG 177, NA, 1–3.

60. Ibid.

61. George S. Patton Jr. and C. C. Benson, "Mechanization and Cavalry," *Cavalry Journal* (Apr. 1930): 234.

62. Alexander M. Miller, "Professional Notes and Discussions: Mechanized Cavalry," *Cavalry Journal* (Nov.–Dec. 1931): 54.

63. Alexander M. Miller, "Professional Notes and Discussions: Mechanized Cavalry," *Cavalry Journal* (Nov.–Dec. 1931): 54.

64. John B. Smith, "The Effects of Mechanization Upon Cavalry," *Cavalry Journal* (Nov.–Dec. 1931): 24.

4. Cavalrymen Looking for New Mounts and a Tactical Doctrine

1. "Prelude to Armor," in *The Armored Force Command and Center*, Study No. 27, Historical Section, Army Ground Forces, 1946 RG 337, NA, 1.

2. Research Report, "Operation of Cav. Rcn. Sq. Integral to Armd. Div.," The Armored School, 1949–1950, Fort Knox Armor School Research Library, May 1950, Fort Knox, Kentucky, 9–10.

3. Ibid., ii.

4. Scott Davenport, interview with H. H. D. Heiberg, Colonel, USA (Ret.), Concerning Adna R. Chaffee Jr., "Doe Run," Brandenberg, Kentucky, July to Sept. 1986, Chaffee Papers, Patton Museum, 2–3. Also see *Senior Officers Debriefing Program*, Conversation between General Isaac D. White and Colonel Charles S. Stodter, MHI, 58. Also see Grow: "Ten Lean Years," 28–29, 51.

5. Grow, "Ten Lean Years," 29, 34.

6. Edward M. Coffman, *The Regulars: The American Army 1898–1941* (Cambridge, Mass.: Belknap Press of Harvard University Press, 2004), 270.

7. Scott Davenport, interview with H. H. D. Heiberg, Colonel, USA (Ret.), Concerning Adna R. Chaffee Jr., Chaffee Papers, Patton Museum, 21.

8. George F. Hofmann, "A Yankee Inventor and the Military Establishment: The Christie Tank Controversy," *Military Affairs* (Feb. 1975): 13–16, reprinted in *ARMOR* (Mar.–Apr. 1976): 13–17, 50–52; and Grow, "Ten Lean Years," 34–41. For more on the Christie controversy, see Hofmann, "The Troubled History of the Christie Tank," *ARMY* (May 1986): 54–65.

9. Ltr., Icks to author, Jan. 7, 1970; interview, Icks with author, Mar. 25, 1970, Elmhurst, Illinois; and Garret to author, Oct. 1, 1970, and Sept. 2, 1971. Colonel Icks is still considered one of the foremost chroniclers of tank history. He passed away on Feb. 19, 1985, at the age of eighty-four.

10. Ltr., Campbell to author, July 9, 1973, author's collections; and interview, author with Campbell, Annapolis, Maryland, Aug. 18, 1973.

11. Interview, Sidney R. Hinds with author, May 14, 1971, Patton Museum.

12. Heiberg, "Organize a Mechanized Force," n.d., ms., Heiberg Collection, Patton Museum, 8. Heiberg's manuscript was published in *ARMOR* (Sept.–Oct. 1976): see 8–11, 48–51. Also interviews, author with Heiberg and Grow, May 14, 1971, Patton Museum.

13. Armored Cars, Hdq. Det., 1st Cav. (Mech), Fort Knox, Kentucky, Oct. 10, 1932, File No. 451/A.C. M1, Records of the Chief of Arms, RG 177, NA, 1–3.

14. Smith, "Memorandum for the Assistant Commandant, Subject: Present and Proposed Cavalry Divisional Organization," Course at the Army War College, 1931–1932, G-3, Feb. 15, 1932, MHI, 1–4.

15. Command Information Branch, "The Boudinot Story," Public Affairs Office, Mar. 1979, Fort Knox, Kentucky, 4.

16. U.S. Department of State, *Foreign Relations of the United States, 1932*, vol. 1 (Washington, D.C.: GPO, 1948), 65–67, 70, 180–82; Information Section, *The League from Year to Year (1931–1932)*, League of Nations, Geneva, Switzerland, 28–46; "Suggest 'Police' Needs as Basis for Arms Cut," *Army and Navy Journal* (June 18, 1932): 985–86; and statement of General Douglas MacArthur, Chief of Staff, United States Army, before War Policies Commission, May 13, 1932, Office of the Chief of Cavalry, File No. 334.8-337, RG 177, NA, 1–24. Also see Subject: Abolition or Limitation of Tanks, To: The Assistant Chief of Staff, G-3, From: Maj. Gen. Guy V. Henry, Office of the Chief of Cavalry, Sept. 14, 1932, File No. 334.8-337, RG 177, NA, 1–2.

17. Charles L. Scott, "Early History of Mechanization," ms., Charles L. Scott Papers, Ac. No. 10,842, Manuscript Division, Library of Congress, Washington, D.C., 1.

18. Lucian K. Truscott Jr., *The Twilight of the U.S. Cavalry: Life in the Old Army 1917–1942*, ed. Lucian K. Truscott III (Lawrence: University Press of Kansas, 1989), 123–26.

19. "The Organization and Employment of TANKS," prepared under the direction of Major General Frank Parker, n.d., MHI, 1–5. Also see Parker, "Detailed Critique of Corps and Division Orders," Second Army Tactical Exercises, Sixth Corps Area, Chicago, Illinois, Apr. 6, 1933, Manuscript Volumes, Parker Papers, Manuscript Department, University of North Carolina Library at Chapel Hill, Collection No. 1516.

20. "The Black Hawks Bid Farewell to the Horse," Hdg. 1st Cavalry (Mech), Unit Box in 1st AD, Box 1, Dec. 14, 1932–Jan. 27, 1933, Patton Museum, 1–6; and address by Colonel William A. Austin, Hdq., 1st Cavalry Regiment, Fort D. A. Russell, Marfa, Texas, Dec. 12, 1932, 1–2, Operations Report, Movement of 1st Cavalry from D. A. Russell to Fort Knox, Kentucky, Dec. 17, 1932–Jan. 16, 1933, Hdq. 1st Cavalry (Mech), Unit Box, 1st Armored Division, Box 1, Patton Museum.

21. Grow, "Ten Lean Years," 45–46.

22. Directive for the Future Development of Combat Cars and Tanks, Apr. 29, 1933, Ordnance Office, OO. 451.24/622, Office of the Chief of Ordnance, RG 156, NA, 1.

23. Ltr., Grow to author, Nov. 16, 1973, author's collection, 1. This letter reflected Grow's daily diary entries for the year 1933.

24. Grow, "Ten Lean Years," 55–56; and ltr., Brig. Gen. R. W. Strong to author, Nov. 13, 1971, author's collection, 1. Strong referred to a knowledgeable German officer called Luttwitz. Apparently, it was Oswald Lutz, who played a key role in developing principles of armor deployment. For clarification, see Robert M. Citino, *The Path to Blitzkrieg: Doctrine and Training in the German Army, 1920–1939* (Boulder, Colo.: Lynne Rienner, 1999), 202–4.

25. Subject: Annual Report for the Command and General Staff School, 1931–1933, To: Adjutant General, War Department, Washington, D.C., Aug. 14, 1933, Command and General Staff College Library, 7–8. Von Greiffenberg ended up as a lieutenant general of infantry, with combat experience in the Balkans and on the eastern front. He became somewhat of a controversial figure when he blamed the Moscow campaign in 1941 more on leadership misjudgment and a lack of intelligence than on the weather.

26. Ltr., Grow to author, Nov. 16, 1973, author's collection, 1. Also see Grow, "Ten Lean Years," 55–56; and Van Voorhis, "Prelude to Armor," in *The Armored Force Command and Center*, 5. Also see Mildred Hanson Gillie, *Forging the Thunderbolt: A History of the Development of the Armored Force* (Harrisburg, Pa.: Military Service Publishing, 1947), 85–89. For an excellent assessment of the German maneuvers, see Citino, *Path to Blitzkrieg*, 190–95, 212–16.

27. George F. Hofmann, "Tactics vs Technology: The U.S. Cavalry Experience," *ARMOR* (Sept.–Oct. 1973): 10–14. Most of this information came from Grow's daily diary. Also see Grow, "The Role of Armor," *ARMOR* (Sept.–Oct. 1961): 30.

28. For an excellent analysis of the situation Parker faced in the Philippines, see Brian McAllister Linn, *Guardians of the Empire: The U.S. Army and the Pacific, 1902–1940* (Chapel Hill: University of North Carolina Press, 1997), 229–30.

29. Parker to the Adjutant General, Digest of Information, July 1934, Aug. 1, 1934, Manila Islands, AG 1920–39, RG 407, NA, 4, 13, 17.

30. Parker, "Scrapbook 1934," MHI, passim. This scrapbook is replete with his interest in aviation.

31. George F. Hofmann, "The Marine Corps's First Experience with an Amphibious Tank," in George F. Hofmann and Donn A. Starry, eds., *Camp Colt to Desert Storm: The History of U.S. Armored Forces* (Lexington: University Press of Kentucky, 1999), 67–84.

32. Ltr., Grow to author, Apr. 15, 1973, author's collection, 1–4; and Grow, "The Role of Armor," 30–31.

33. George S. Patton Jr., "Mechanized Forces," *Cavalry Journal* (Sept.–Oct. 1933): 5–8.

34. "Major General Guy V. Henry," *Cavalry Journal* (Mar.–Apr. 1934): 6; and Gillie, *Forging the Thunderbolt*, 64–65.

35. "A Brief Narrative of the Life of Guy V. Henry, Jr.," an autobiography, Guy V. Henry Jr. Papers, MHI, 65.

36. George Grunert, "Cavalry," lecture, Army War College, Apr. 2, 1934, Library, Field Artillery School, Fort Sill, Oklahoma, 3, 11, 25.

37. "The Cavalry Maneuvers, Fort Riley, Kansas, May 1934," author's collection, 1–2. Grow prepared this report and supplied his copy to the author. Scott Davenport, interview with H. H. D. Heiberg, Colonel, USA (Ret.), Concerning Adna R. Chaffee Jr., Chaffee Papers, Patton Museum, 15. Also see Gillie, *Forging the Thunderbolt*, 62–64.

38. "The Cavalry Maneuvers, Fort Riley, Kansas, May 1934," 1–2.

39. Chaffee, "Report of Field Exercises and Maneuvers at Fort Riley, Kansas, May 1934," Headquarters 1st Cavalry (Mech), Fort Riley, Kansas, June 10, 1934, author's collection, 1–19.

40. Edward Croft, "Developments in Organization, Armament and Equipment of Infantry," lecture, Army War College, Sept. 18, 1934, 1–8, and "Notes of Discussion Following Lecture," Circular Archives, MHI, 1–5.

41. "Report of Maneuvers: The Cavalry School, Apr.–May 1934," Fort Riley, Kansas, May 1934, AG 353 (9-25-34), RG 407, NA, 406–19.

42. Kromer, "The Cavalry Maneuvers, Fort Riley, Kansas, May 1934," lecture, Army War College, June 22, 1934, MHI, 1–27. The lecture was prepared in the Office of the Chief of Cavalry and included observations brought out in reports and letters from General Henry, Colonel Bruce Palmer, and Lieutenant Colonels Chaffee and Patton.

43. Jesse A. Ladd, "Report of Observations of Fort Riley Maneuvers," May 14 to 26, 1934, author's collection, 1–10.

44. Ltr., Grow to author, Sept. 9, 1971, author's collection, 1–2.

45. Gillie, *Forging the Thunderbolt*, 68.

46. Report of Technical Committee, Mar. 24, and Proceedings of Board of Officers, Mar. 25, 1934, HQ, 1st Cavalry (Mech), OO 451.24/1789, RG 156, NA, 1–5 and 1–4; and Daily Log of Combat Cars T4 and T5, during Test at Fort Riley, May 8–12, 1934, author's collection. Grow was a member of the technical committee at Fort Knox that recommended the combat car be declared standard and procured. Also see H. H. D. Heiberg, "Organize a Mechanized Force," 13–15; and "Mechanization in the Army," lecture: Soci-

ety of Mechanical Engineers, Pittsburgh, Pennsylvania, Apr. 23, 1940, Heiberg Collection, Patton Museum, 11–13.

47. For an excellent developmental history of light tanks, see R. P. Hunnicutt, *Stuart: A History of the American Light Tank* (Novato, Calif.: Presidio Press, 1992).

48. Ltr., Grow to Hofmann, Sept. 23, 1971, author's collection, 1–2. Most of this information came from Grow's daily diary.

49. Heiberg, "Organize a Mechanized Force," 13–15; and ltr., Icks to author, Jan. 13, 1970, author's collection.

50. On the evolution of the volute suspension system, see Daniel Chase, "Combat Car," in "The Development Record in Combat Vehicles," in vol. 2, "Research and Development," Icks Collection, Patton Museum, 18–21; Memorandum: for the Chief of Staff, Subject: Volute Suspension for the Light Tank, T2, Apr. 25, 1934; Subject: Light Tank T2—Application of Volute Spring Type Suspension, Sub-committee on Automotive Equipment to Ordnance Committee Technical Staff, May 1, 1934; and Subject: Light Tank T2, To: Adjutant General, Item 11450, May 4, 1934, RG 156, NA, 1. Also interview, S. R. Hinds with author, May 14, 1971, Patton Museum. Hinds was a young officer assigned to the Mark XIII Liberty tanks, which were still used by infantry in the early 1930s.

51. "German Expert Finds U.S. Tanks Would Not Stand Test of War," *New York Times*, Nov. 21, 1936, 1–2.

52. Heinz Guderian, *Achtung-Panzer! The Development of Armoured Forces, Their Tactics and Operational Potential* (London: Arms and Armour Press, 1993), 153.

53. "Army Denies Tanks Are Second Rate," *New York Times*, Nov. 22, 1936, 7; Omar N. Bradley, *A Soldier's Story* (New York: Henry Holt, 1951), 40–41. The German observation soon proved correct. When U.S. Army tanks were first employed in Tunisia in February 1943, they lacked sufficient armor and armaments to engage German tanks. This disparity was never corrected until after the Battle of the Bulge, according to General Omar N. Bradley, the commander of the 12th Army Group during its assault on Fortress Europe. See Bradley, *Soldier's Story*, 40–41.

54. R. P. Hunnicutt, *Armored Cars: A History of American Wheeled Combat Vehicles* (Novato, Calif.: Presidio Press, 2002), 34–42; and Larry Jesensky, "Scout Cars: White Scout Car Story," pt. 1, *Army Motors* vol. 96 (2001): 17–24.

55. Ltr., Grow to author, Feb. 19, 1972, author's collection. Also see Williamson Murray, "Transformation Lessons Learned between World Wars" (paper presented at the winter meeting of the Association of the United States Army, Fort Lauderdale, Florida, Feb. 27, 2003).

56. Combined Arms Research Library, *Conversations with General J. Lawton Collins*, Combat Studies Institute Report No. 5, May 1983, U.S. Command and General Staff College, 4–5.

57. *Senior Officers Debriefing Program*, Conversation between General Isaac D. White and Colonel Charles S. Stodter, MHI, 58–60.

58. "Major General Guy V. Henry," 6; and Coffman, *The Regulars*, 270.

59. Bruce Palmer Jr., "Oral History," Interviews: Lieut. Colonels James E. Shelton and Edward P. Schmidt, Dec. 5, 1975, MHI, 12–13; and Larry I. Bland, ed., and Sharon Riten-

our Stevens, assoc. ed., *The Papers of George Catlett Marshall: "The Soldierly Spirit" December 1880–June 1939*, vol. 1 (Baltimore: Johns Hopkins University Press, 1981), 501 n. 3.

60. Kromer, "Cavalry," lecture, Army War College, Sept. 17, 1934, MHI, 1–11.

61. Bradford Grethen Chynoweth, *Memoirs: Bellamy Park* (Hicksville, N.Y.: Exposition Press, 1975), 142–43; ltr., Chynoweth to author, Mar. 8, 1972, MHI, 1. Also see Chynoweth, "Military Conservation," *ARMOR* (Mar.–Apr. 1974): 20–23.

62. Douglas O. MacArthur, *Report of Secretary of War to the President, Annual Reports War Department Fiscal Year Ended June 30, 1933* (Washington, D.C.: GPO, 1933), 29–31; MacArthur, *Report of Secretary of War to the President, Annual Reports War Department Fiscal Year Ended June 30, 1934* (Washington, D.C.: GPO, 1934), 45; and MacArthur, *Report of Secretary of War to the President, Annual Reports War Department Fiscal Year Ended June 30, 1935* (Washington, D.C.: GPO, 1935), 61. Also see Thomas H. Greer, *The Development of Air Doctrine in the Army Air Arm 1917–1941* (Washington, D.C.: GPO, 1985), 71–75.

63. William C. Lee, "Fast Tanks as Leading Tanks and Exploiting Tanks," lecture, Tank School, Jan. 12, 1932, Tank Notes, MHI, 14.

64. Chynoweth, *Memoirs*, 184.

65. Alexander D. Surles, "The Cavalry and Mechanization, 1936," *Cavalry Journal* (Jan.–Feb. 1936): 6–7.

66. Palmer interview, "Horse and Motors," *Argus*, as reported in (Springfield) *Illinois State Journal*, *Cavalry Journal* (Mar.–Apr. 1936): 105; and Bruce Palmer Jr., "Oral History," Interviews: Lieut. Colonels James E. Shelton and Edward P. Schmidt, Dec. 5, 1975, MHI, 12.

67. "Organization and Activities," *Cavalry Journal* (May–June 1935): 310; and ibid. (July–Aug. 1936): 348. Also see "History of the 68th Field Artillery," in *First Armored Division* (Nashville, Tenn.: Army and Navy Publishing Company, 1941), 57–58.

68. Willis D. Crittenberger, "Mechanized Cavalry," Hdq, 1st Cavalry (Mech), Fort Knox, Kentucky, July 23, 1936, Willis D. Crittenberger Papers, MHI, 5 (hereafter cited as Crittenberger Papers, MHI).

69. Bruce Palmer, "Mechanized Cavalry in the Second Army Maneuvers," *Cavalry Journal* (Nov.–Dec. 1936): 461–69, 476.

70. Andrews, "The GHQ Air Force," lecture, Army War College, Oct. 15, 1936, MHI, 1–23.

71. Ibid., 470–78; and Grow, "Ten Lean Years," 82–84.

72. *Senior Officers Debriefing Program*, Conversation between General Isaac D. White and Colonel Charles S. Stodter, MHI, 68.

73. Forrest C. Pogue, *George C. Marshall: Education of a General, 1880–1939* (New York: Viking Press, 1963), 290–92, 402–3 nn. 23, 24.

74. Ltr., Marshall to Major General Frank R. McCoy, Aug. 16, 1936, in Bland and Stevens, *Papers of George Catlett Marshall: "The Soldierly Spirit,"* 499–500.

75. Second Army Maneuvers 1936, Report of the Mechanized Force, vol. 1, sec. I, George C. Marshall Papers, Marshall Foundation, VMI, Lexington, Virginia, 1–18. All the Second Army maneuvers reports were apparently microfilmed in RG 94, which no longer exists as AG Bulky File 353 (10-3-35) (1) Sec. 2 C. Also see Palmer, "Mechanized Cavalry in the Second Army Maneuvers," 460.

76. *Senior Officers Debriefing Program*, Conversation between General Isaac D. White and Colonel Charles S. Stodter, MHI, 71.

77. Charles L. Scott, "Progress in Cavalry Mechanization: Scout Car Developments," *Cavalry Journal* (July–Aug. 1936): 281–84.

78. Hunnicutt, *Armored Cars*, 34–58.

79. "The Employment of Tanks in Combat," in Organization and Conduct of Field Forces. The Application of Strategy by Tactics and Combat, 1–10; and ltr., Parker to Major General John H. Hughes, War Department General Staff, G-3, Sept. 10, 1936, MHI, 1.

80. Christopher R. Gabel, *The U.S. Army GHQ Maneuvers of 1941* (Washington, D.C.: Center of Military History, 1991), 6.

81. Ronald Spector, "The Military Effectiveness of the U.S. Armored Forces, 1919–1939," in Allan R. Millett and Williamson Murray, eds., *Military Effectiveness* (Boston: Allen and Unwin, 1988), 70–93.

82. Robert W. Grow, "Military Characteristics of Combat Vehicles," *Cavalry Journal* (Nov.–Dec. 1936): 508–11; and Grow, "Ten Lean Years," 83.

83. "Military Science Review" (the official publication of the German War Ministry), Subject: Motorized Combat Troops in America (A German Estimate), MID Report No. 14,504, Berlin, Feb. 1, 1936, War Department and Special Staff, RG 165, NA, 1.

84. Guderian, "Armored Troops and Their Cooperation with Other Arms," MID Report No. 14,994, Berlin, Dec. 15, 1936, RG 165, NA, 1–25.

85. *Tables of Organization Mechanized Division (Tentative)* (Fort Leavenworth, Kans.: Command and General Staff School Press, 1936), 3–24; and Grow, "Ten Lean Years," 69.

86. Kingman, Memorandum for the Assistant Commandant, Subject: "Mechanized Units," course at the Army War College, 1933–1934, G-3, May 5, 1934, 1–3; and Annex No. 2, "Detailed Discussion of Facts Bearing on Study," MHI, 1–6.

87. "Prelude to Armor," 4; and *Tactical Employment of the Mechanized Division (Tentative)* (Fort Leavenworth, Kans.: Command and General Staff School Press, 1937), 3–4, 6, 23–24, 31.

88. *The Tactical Employment of Cavalry (Tentative), Part One–Horse and Part Two–Mechanized* (Fort Leavenworth, Kans.: Command and General Staff School Press, 1937), 8, 79.

89. Ibid., 8–10, 19–22, 53, 80, 124–25.

90. *Tactical Employment of Infantry (Tentative)* (Fort Leavenworth, Kans.: Command and General Staff School Press, 1937), 64.

5. The "Great Cavalry Debate" over New Opportunities

1. Memorandum: Brigadier General Daniel Van Voorhis to the Chief of Cavalry, Revised Tables of Organization, Dec. 19, 1936, Office of Chief of Cavalry, 322.02, 1–2; and Memorandum: Major Robert W. Grow to Colonel Kent, Cavalry Regiment (Mech), Dec. 1936, Office of the Chief of Cavalry, 322.02, RG 177, NA, 1–2.

2. Russell F. Weigley, *History of the United States Army* (New York: Macmillan, 1967), 416.

3. "Cavalry Combat," *Cavalry Journal* (Sept.–Oct. 1937): 456–57. The review of "Cavalry Combat" appeared in the *Infantry Journal* in the Sept.–Oct. issue.

4. "The Cavalry School Visits Fort Knox," *Cavalry Journal* (July–Aug. 1937): 336–41.

5. Ibid., 340–41.

6. Ltr., Smith to Marshall, Jan. 2, 1936; and ltr., Smith to Marshall, Nov. 20, 1938, 670 n. 1, in Larry I. Bland, ed., and Sharon Ritenour Stevens, assoc. ed., *The Papers of George Catlett Marshall: "The Soldierly Spirit" December 1880–June 1939* (Baltimore: Johns Hopkins University Press, 1981), 484; and Subject: Autumn 1938 Changes in German High Command, G-2 Report, from: MA Berlin, MID Report 16,261, Nov. 18, 1938, RG 165, NA, 3. Von Schell's contribution while he was an instructor at the Kriegsakademie was a book on combat and armored cars entitled *Kampf gegen Panzerwagen* (Berlin: Gerhard Stalling, 1936).

7. The bases were as follows: Fort George G. Meade, Maryland; Fort Knox, Kentucky; Jefferson Barracks, Missouri; Fort Douglas, Utah; Fort Francis E. Warren, Wyoming; Fort Lewis, Washington; Presidio of San Francisco, California; Presidio of Monterey, California; Fort Bliss, Texas; Fort Sam Houston, Texas; Fort Riley, Kansas; and Fort Benning, Georgia.

8. Truman Smith; Memo to the Chief of the G-2 Historical Section, Department of the Army, Subject: Activities of the Office of Military Attaché, American Embassy, Berlin, Germany—August 1935 to March 1939, Truman Smith Papers, Herbert Hoover Presidential Library, West Branch, Iowa, 1–16; and ltr., Smith to Marshall, Jan. 2, 1936, in Bland and Stevens, *Papers of George Catlett Marshall: "The Soldierly Spirit,"* 483. The cordial relationship began to end when an increasing antagonism developed between the Nazi Party and the German army in 1937 and into 1938, and in October 1938 when Hitler committed a series of purges of the Jews throughout Germany. See Smith, Memo to the Chief of the G-2 Historical Section, 7, 14. Also see Harold C. Deutsch, *Hitler and His Generals: The Hidden Crisis January–June 1938* (Minneapolis: University of Minnesota Press, 1974).

9. Memorandum for the Assistant Chief of Staff, G-2, Subject: Visit of German Army Officer to Army Posts in the United States to see Latest Type Light Tanks and Combat Cars Used in the United States Army, MID Report 343-W-97, June 16, 1937, RG 165, NA, 1–2.

10. Subject: Visit of Lt. Colonel von Schell to Fort Meade, To: Commanding General, HQ 66th Infantry (Light), MID Report 343-W-97, June 25, 1937, RG 165, NA, 1, 3.

11. Ibid.

12. Subject: Visit of Lt. Colonel von Schell, German Army, To: The Commandant, the Infantry School, July 21, 1937, 1–3; Memorandum For: The Commandant, July 24, 1937, 1–2; and notes regarding visit of Foreign Officer (Col. von Schell, German Army) and certain views expressed by him, MID Report No. 343-W-97, RG 165, NA, 1–2.

13. Memorandum for: The Commandant, the Infantry School, Office of the Assistant Commandant, July 24, 1937, MID Report No. 343-W-97, RG 165, NA, 1–2.

14. "Nazi Officer Views Knox: Troop Inspection Is Set for Military Tourist," *Louisville Times*, June 29, 1937; and "German Army Colonel Sees Demonstration at Fort Knox," *Louisville Courier-Journal*, June 30, 1937.

15. Subject: Inspection of Light Tanks or Combat Cars by Lieutenant Colonel von Schell, German Army, To: The Adjutant General. Thru: the Chief of Cavalry, Washington, D.C., July 26, 1937, MID Report No. 343-W-97, RG 165, NA, 2; Memorandum: To: The Commanding General, Fort Knox, Kentucky, Hdq., Seventh Cavalry Brigade, July 30, 1937, G-2 and Military Attaché Reports, Crittenberger Papers, MHI, 1–4; and Subject: Visit of Lt. Colonel von Schell, German Army, to Fort Knox, Kentucky, To: Adjutant General of the Army, Washington, D.C., MID Report No. 343-W-97, July 23, 1937, RG 165, NA, 1–4.

16. Ibid.

17. Ibid.

18. Van Voorhis, "Mechanization," lecture, Army War College, Oct. 13, 1937, MHI, 14–15.

19. Subject: Visit of Lt. Colonel von Schell, German Army, to Fort Knox, 4.

20. Ltr., Crittenberger to Senator H. C. Lodge, June 17, 1937, Correspondences, Crittenberger Papers, MHI, 1–2.

21. Grow, "Ten Lean Years," 38.

22. Charles L. Scott, "Early History of Mechanization," ms., Charles L. Scott Papers, Ac. No. 10,842, Manuscript Division, Library of Congress, Washington, D.C., 3; "Prelude to Armor," in *The Armored Force Command and Center*, Study No. 27, Historical Section, Army Ground Forces, 1946, RG 337, NA, 5; and Van Voorhis, "Mechanization," MHI, 14.

23. Ltr., Marshall to von Schell, July 7, 1937, in Bland and Stevens, *Papers of George Catlett Marshall: "The Soldierly Spirit,"* 552.

24. Ibid.

25. James S. Corum, "The Spanish Civil War: Lessons Learned and Not Learned by the Great Powers," *Journal of Military History* 62 (Apr. 1998): 325–27; and Jonathan M. House, *Combined Arms Warfare in the Twentieth Century* (Lawrence: University Press of Kansas, 2001), 82–83.

26. "Observation Squadron to Fort Knox," *Cavalry Journal* (May–June 1937): 258; and Press Release: Observation Squadron to Fort Knox, June 17, 1937, Air Corps–Cavalry, File No. 322.09, RG 407, NA, 1.

27. War Department, *Field Service Regulations United States Army 1923: Operations* (Washington, D.C.: GPO, 1924), 21–23; William Mitchell, *Winged Defense: The Development and Possibilities of Modern Air Power—Economic and Military* (New York: Putnam's, 1925), 188–89; and Thomas H. Greer, *The Development of Air Doctrine in the Army Air Arm 1917–1941* (Washington, D.C.: GPO, Office of Air Force History, 1985), 12, 66–67.

28. Albert C. Wedemeyer, *Wedemeyer Reports!* (New York: Holt, 1958), 61–62.

29. Heinz Guderian, *Panzer Leader* (London: Michael Joseph, 1952), 316–17.

30. Subject: The Panzer Division Germany (Combat) MID Report No. 15,596, Berlin, Nov. 24, 1937, RG 165, NA, 4; Subject: Reform of the German Automobile Industry (An Industrial Mobilization Measure), MID Report No. 16,301, Berlin, Dec. 1, 1938, RG 165, NA, 1–5; and Smith, Memo to the Chief of the G-2 Historical Section, 13.

31. Synopsis, Colonel Bradford's visit to Fort Knox, July 1, 1937, 1; The Cavalry School, Fort Riley, Kansas, To: Chief of Cavalry, Washington, D.C., July 30, 1937, 1–3; and War

Department, Office of Chief of Cavalry, To: The Adjunct General, Nov. 16, 1937, 1–3, Document File, 1923–1942, Office of the Chief of Cavalry, RG 177, NA, 1–3.

32. Memorandum for the Chief of Cavalry: Subject: Relationship between the Chief of Cavalry and Mechanized Units, June 25, 1937, 1–3; and Army Regulations No. 80-10, Cavalry Board, War Department, Washington, D.C., Document File, 1923–1942, Office of the Chief of Cavalry, RG 177, NA, 1–2.

33. Command Information Branch, "The Boudinot Story," Public Affairs Office, AG-3748-0-Army-Knox-Mar. 1979-SM, Fort Knox, Kentucky, 4.

34. Burton S. Boudinot, "The Way I Remember It," ms. 1989–1992, Radcliff, Kentucky, 14. Lieutenant Colonel Boudinot (Ret.) wrote this manuscript for interested friends and as a personal reminiscence of his career in the army.

35. Ibid.

36. For a more expanded study on Fuqua's influence and military intelligence reports, see George F. Hofmann, "The Tactical and Strategic Use of Attaché Intelligence: The Spanish Civil War and the U.S. Army's Misguided Quest for a Modern Tank Doctrine," *Journal of Military Affairs* 62 (Jan. 1998): 101–33.

37. Hamilton S. Hawkins, "We Must Have Cavalry," *Cavalry Journal* (Sept.–Oct. 1937): 405–9.

38. Van Voorhis, "Mechanization," Oct. 13, 1937, MHI, 1–16.

39. Committee No. 6, "Mechanization," pt. 1, Synopsis of Report, G-3 Committee, Subject: Course at the Army War College, 1937–1938, Oct. 21, 1937, MHI, 1–2.

40. David Fletcher, *Mechanized Force: British Tanks between the Wars* (London: HMSO Publication Center, 1991), 122; and Giffard Martel, *An Outspoken Soldier: His Views and Memoirs* (London, Sifton Praed, 1949), 127–28, 137.

41. Subject: Christie Convertible Chassis, Characteristics of Christie Convertible Chassis, and press releases, To: The Chief of Ordnance, War Department, Washington D.C., Dec. 15, 1937, Ordnance Office (OO) 451.25/7384, RG 156, NA, 1–4; and Subject: Demonstration—Christie Tank, To: Assistant Chief of Staff, G-4, Dec. 16, 1937, OO 451.25/7385, RG 156, NA, 1–3.

42. Fletcher, *Mechanized Force*, 122.

43. Headquarters, Seventh Cavalry Brigade, Fort Knox, Kentucky, To: Chief of Cavalry, War Department, Combat Cars, June 7, 1939, OO 451.24/3429, RG 156, NA, 1.

44. Grow, "Ten Lean Years," 57

45. Committee No. 2, "The Spanish Revolution," "Tanks," Report, G-2 Committee, Subject: Course at the Army War College, 1937–1938, Jan. 24, 1938, MHI, 10–11.

46. "Lessons from the Spanish Civil War," Command and General Staff School, Fort Leavenworth, Kansas, Dec. 1, 1937, MID 2607-S-144, RG 165, NA, 3, 13.

47. Notice For: Instructors, Subject: The Panzer Division, the Command and General Staff School, Fort Leavenworth, Kansas, Jan. 24, 1938, G-2 and Military Attaché Reports, Crittenberger Papers, MHI, 1–12.

48. Statement by Brigadier General Daniel Van Voorhis on the Subject of the American Doctrine for Mechanization, Hdq., 7th Cavalry Brigade, Jan. 28, 1938, Crittenberger Papers 1931–1941, MHI, 1. This statement was read by Van Voorhis to the Brigade Officers' Tactical School.

49. "Government Officials at Fort Knox," *Army and Navy Journal* (Jan. 29, 1938): 457; and "Roosevelt Is Sending McIntyre as Emissary to Barkley Dinner," *Louisville Courier-Journal*, Jan. 22, 1938, 1.

50. "Tour of Fort Knox," *Louisville Courier-Journal*, Jan. 22, 1938, 3; and "Mechanized Division Seen for Fort Knox," *Louisville Courier-Journal*, Jan. 25, 1938, 3.

51. Grow, "Ten Lean Years," 38.

52. Marvin A. Kreidberg and Merton G. Henry, *History of Military Mobilization in the United States Army 1775–1945*, Department of the Army Pamphlet No. 20-212, Nov. 1955, 542–43.

53. Press release, Harry S. Truman Papers: Papers as U.S. Senator and Vice President: Speech Files, Mar. 7, 1938, Truman Library, Independence, Missouri, NA, 1–4.

54. Memorandum for Colonel Chaffee, Office of the Deputy Chief of Staff, Subject: Tank Casualties in Spanish Civil War, Jan. 27, 1938, MID 2657-S-144, RG 165, NA, 1–2.

55. Mildred Hanson Gillie, *Forging the Thunderbolt: A History of the Development of the Armored Force* (Harrisburg, Pa.: Military Service Publishing, 1947), 106.

56. Grow, "Ten Lean Years," 86; and George F. Hofmann, "Tactics vs Technology: The U.S. Cavalry Experience," *ARMOR* (Sept.–Oct. 1973): 13–14.

57. Grow, "Ten Lean Years," 86.

58. "Prelude to Armor," RG 337, NA, 4; and "Policies Governing Mechanization, and the Tactical Employment of Mechanized Units," Apr. 6, 1938, AG 537.3 (3-25-38), RG 407, NA, 1–4; and "Chief of Staff before Subcommittee," *Army and Navy Register* (Mar. 26, 1938): 4.

59. For example, see "Lessons from the Spanish Civil War."

60. Malin Craig, "Mechanization and Tanks," special statement in "General Craig's Hearing," *Army and Navy Register* (Mar. 26, 1938): 4.

61. General Van Voorhis's Comments on War Department Directive Defining Mechanization, dated Apr. 6, 1938, Hdq., 7th Cavalry Brigade, Correspondences 1935–1938, Crittenberger Papers, MHI, 1–3. The comments were presented to the Brigade Officers' Tactical School.

62. Ltr., Patton to Crittenberger, Apr. 27, 1938, Correspondences, Crittenberger Papers, MHI, 1–2.

63. Memorandum: Grow to Crittenberger, Aug. 8, 1938, Correspondences, Crittenberger Papers, MHI.

64. "Prelude to Armor," RG 337, NA, 3.

65. Gillie, *Forging the Thunderbolt*, 90–94.

66. "7th Cavalry Ready for 700-Mile March: Leaves Friday for Trek in 3 States," *Louisville Courier-Journal*, May 16, 1938.

67. Martin Blumenson, *The Patton Papers, 1940–1945* (Boston: Houghton Mifflin, 1974), 20.

68. Field Orders: Number 1, Hdq., 7th Cavalry Brigade, Fort Knox, Kentucky, May 19, 1938, Crittenberger Papers 1931–1941, MHI, 1–3; and "The Mechanized Cavalry Takes the Field," *Cavalry Journal* (July–Aug. 1938): 291–300.

69. "The Mechanized Cavalry Takes the Field," 291–300.

70. Ibid., 296–300.

71. "Gasoline for the Hill Buggies," *Shell Progress*, Nov.–Dec. 1938, Miscellaneous, Crittenberger Papers, MHI, 13.

72. Van Voorhis, "Mechanization," lecture, Army War College, Sept. 29, 1938, MHI, 25–26.

73. Scott Davenport, interview with J. R. Pritchard, Colonel, USA (Ret.), Concerning Adna R. Chaffee Jr., Elizabethtown, Kentucky, Nov. 23, 1986, Chaffee Papers, Patton Museum, 2–3.

74. Oral presentation, "Mechanization and Defense against Aviation," G-3 Course, AWC, 1938–1939, Report of Committee No. 5, Oct. 7, 1938, MHI, 11.

75. Subject: Policies Governing Mechanization, and Tactical Employment of Mechanized Units, Apr. 6, 1938, AG 537.3 (4-6-38), RG 407, NA, 1–4.

76. For a study of the cavalry's most prominent adherent to the horse, see Alexander M. Bielakowski, "The Last Chief of Cavalry: Major General John K. Herr," *Journal of America's Military Past* (Fall 2001): 67–82. Regarding Herr's own argument, see J. K. Herr, "Editorial Comment," *Cavalry Journal* (May–June 1946): 35–40.

77. Christopher R. Gabel, *The U.S. Army GHQ Maneuvers of 1941* (Washington, D.C.: GPO, 1991), 29.

78. "Prelude to Armor," RG 337, NA, 5.

79. J. K. Herr, Remarks: Re: Conference with Assistant Chief of Staff, G-3, Oct. 10, 1938, Office of the Chief of Cavalry, Doc. File, 1923–1942, RG 177, NA, 1–2.

80. Ibid., 1; and Hofmann, "Tactics vs Technology," 14.

81. John T. Broom, "It Isn't Doctrine until They All Understand It: *FM 100-5 Field Service Regulations, 1941*, the Armored Force, and the Conduct of the 1944 European Campaign" (presentation, 1993 U.S. Armor Conference, Fort Knox, Kentucky), author's collection, 11.

82. Robert W. Grow, "Mechanized Cavalry," *Cavalry Journal* (Jan.–Feb. 1938): 30–31.

83. Grow, "Mechanized Cavalry," *Cavalry Journal* (May–June 1938): 150–51; and Grow, "One Cavalry," *Cavalry Journal* (May–June 1938): 230.

84. Roland D. Tausch, Interview: General James H. Polk Conversations, Section 1, Senior Officers Debriefing Program, 28–29, and Summary of the Debriefing of General James M. Polk, May, 15, 1972, MHI, 2.

85. Ibid.

86. Ltr., Marshall to Chaffee, Oct. 3, 1938, Chaffee Papers, Patton Museum, 1.

87. Ltr., Chaffee to Marshall, Nov. 17, 1938, Chaffee Papers, Patton Museum, 1–2.

88. Crittenberger, "Recommendation," To: Chief of Cavalry, Correspondences, Crittenberger Papers, MHI, 1–2.

89. Subject: Letters of Transmittal, Re: Proceedings of Board, July 27, 1938; Chief of Cavalry to Chief of Field Artillery, Oct. 6, 1938; Chief of Field Artillery to Chief of Cavalry, Dec. 17, 1938, and Chief of Cavalry to Chief of Field Artillery, Jan. 5, 1939, OO 472/3476, RG 156, NA, 1–6; and Upton Birnie Jr., "Obsolescence of Horse-Drawn Artillery," *Army and Navy Register* (May 16, 1937): 11. Also see Boyd L. Dastrup, *King of Battle: A Branch History of the U.S. Army's Field Artillery* (Fort Monroe, Va.: U.S. Army Training and Doctrine Command, 1992), 192–93.

90. Robert M. Danford, "Horses and Motors for Artillery," Armor at Fort Knox 1934–1941, Crittenberger Papers, MHI, 1.

91. Subject: Self-Propelled Mount for a 75mm Howitzer, 8th Endorsement, Apr. 17,

1939, OO 472/3496, RG 156, NA, 13; and R. P. Hunnicutt, *Stuart: A History of the American Light Tank*, (Novato, Calif.: Presidio Press, 1992), 317.

92. Report of Inspection of Christie Tank at Hempstead, Long Island, Oct. 4, 1938, OO 451.25/8209, RG 156, NA, 1–3.

93. Conference on Christie Tank, Office of the Chief of Ordnance, Mar. 15, 1939, OO 451.25/8674, RG 156, NA, 1–2.

94. George F. Hofmann, "Christie's Last Hurrah: In 1941, the Army Reappraised the Christie Suspension for Use on Tank Destroyers," *ARMOR* (Nov.–Dec. 1991): 14–19.

95. Christie interview, *Elizabeth Daily Journal*, June 10, 1940, 3.

96. Robert Lee Bullard, "Give Us a Modern Army," *Liberty Magazine* (Feb. 11, 1939): 24–25.

97. Herr to Chaffee, June 1, 1939, Office of the Chief of Cavalry, 322.02, RG 177, NA, 1–2.

98. Robert W. Grow, "New Developments in the Organization and Equipment of Cavalry," *Cavalry Journal* (Mar.–Apr. 1939): 204–7.

99. Adna R. Chaffee Jr., "The Seventh Cavalry Brigade and the First Army Maneuvers," *Cavalry Journal* (Nov.–Dec. 1939): 450–61; Chaffee, "Mechanized Cavalry," lecture, Army War College, Sept. 29, 1939, MHI, 22–27; and H. H. D. Heiberg, "Organize a Mechanized Force," n.d., ms., Heiberg Collection, Patton Museum, 24–28.

100. "Prelude to Armor," RG 337, NA, 4.

101. Memorandum for the Assistant Chief of Staff, G-3, Subject: Report of First Army Maneuvers, Jan. 5, 1940, Office of the Chief of Cavalry, 322.02, RG 177, NA, 1.

102. Herr, "The Cavalry," lecture, Army War College, Sept. 19, 1939, MHI, 13.

103. Henry J. Reilly, "Horsed Cavalry and the Gas Engine's Children," *Cavalry Journal* (Jan.–Feb. 1940): 2–9; and Reilly, "Cavalry and Mechanized Force," *Cavalry Journal* (Jan.–Feb. 1940): 19–25.

104. Herr to the Chief of Staff, Subject: Expansion of Mechanized Cavalry, Oct. 3, 1939, Office of the Chief of Cavalry, 322.2, RG 177, NA, 1–3.

105. Gillie, *Forging the Thunderbolt*, 134–35; and Chaffee, "Mechanized Cavalry," 21.

106. Scott Davenport, interview with H. H. D. Heiberg, Colonel, USA (Ret.), Concerning Adna R. Chaffee Jr., "Doe Run," Brandenberg, Kentucky, July to Sept. 1986, Chaffee Papers, Patton Museum, 12.

107. Memorandum for the File, Federal Bureau of Investigation, U.S. Department of Justice, Washington, D.C., Sept. 8, 1939, 1. Edwin Krampitz Jr. supplied this memorandum to the author as a result of his execution of the Freedom of Information Act.

108. Scott Davenport, interview with H. H. D. Heiberg, Chaffee Papers, Patton Museum, 5.

109. J. K. Herr, "What of the Future?" *Cavalry Journal* (Jan.–Feb. 1939): 3–6.

110. Ltr., Crittenberger to Van Voorhis, Office of Chief of Cavalry, June 6, 1939, Crittenberger Papers, MHI, 1–3. For an excellent dissertation that describes Crittenberger during this period, see Matthew Darlington Morton's chapter "Changing Priorities, Drumbeats of War," in "Men on 'Iron Ponies': The Death and Rebirth of Modern U.S. Cavalry" (Ph.D. diss., Florida State University, 2004).

111. Notes for the Chief of Staff, Reserve Corps Affairs, Dec. 8, 1938, in Bland and Stevens, *The Papers of George Catlett Marshall: "The Soldierly Spirit,"* 662.

112. Gary C. Cox, "Beyond the Battle Line: U.S. Air Attack Theory and Doctrine, 1919–1941" (thesis, School of Advanced Airpower Studies, Air University, Maxwell Air Force Base, Alabama, 1995), 39–49.

113. Corum, "The Spanish Civil War," 318–22.

114. Charles Noble, "Defense against Air Attack," *Cavalry Journal* (July–Aug. 1938): 332–34.

115. See David N. Spires, *Patton's Air Force: Forging a Legendary Air-Ground Team* (Washington, D.C.: Smithsonian Press, 2002), passim. Also see Kenneth A. Steadman, "A Comparative Look at Air-Ground Support Doctrine and Practice in World War II," Fort Leavenworth, Kans.: Combat Studies Institutes, U.S. Army Command and Staff College, Sept. 1, 1982, 1–2, 8–10.

116. George F. Hofmann, *The Super Sixth: A History of the Sixth Armored Division in World War II* (1975; Nashville, Tenn.: Battery Press, 2000), 160–63, 286, 446–47.

117. Ibid., 437–38.

118. Van Voorhis, "Mechanization," lecture, Army War College, Oct. 13, 1937, 1–14.

119 Crittenbeger, "Efficiency Report," June 2, 1938, Correspondences, Crittenberger Papers, MHI, 1.

120. Van Voorhis, "Mechanization," lecture, Army War College, Oct. 13, 1937, 1–14.

121. Scott Davenport, interview with J. R. Prichard, Colonel, Chaffee Papers, Patton Museum, 2–3.

122. Ltr., Grow to author, Feb. 2, 1972, author's collection, 1.

123. Interview: Charles S. Stodter Conversations with Isaac D. White, *Senior Officers Debriefing Program*, Oct. 1977, MHI, 61.

124. Bruce Palmer Jr., "Oral History," Interviews: Lt. Colonels James E. Shelton and Edward P. Schmidt, Dec. 5, 1975, MHI, 14.

125. Ibid., 40; and Ltr., Chaffee to Marshall (Personal), July 26, 1940, Hqr., Armored Force, Fort Knox, Kentucky, Chaffee Papers, Patton Museum, 1–3.

126. William M. Hoge, "Engineer Memoirs," Web site: redstone.army.mil/history, 108.

127. Scott, "Early History of Mechanization," 4.

128. Ltr., Chaffee to Marshall (Personal), July 26, 1940, Hqr., Armored Force, Fort Knox, Kentucky, Chaffee Papers, Patton Museum, 1.

129. Ibid., 1. The autonomy of the combat branch system provided no organization to coordinate combat functions like the post–World War II Combat Development Command and today's Army Training and Doctrine Command (TRADOC). Since July 1973, TRADOC's mission has been to synchronize doctrine, training, leadership development, organizational structure, and material readiness to prepare the army for today's wars and guide its future.

130. Broom interview with Colonel Pritchard, Dec. 29, 1991, Fort Knox, Kentucky, author's collection, 23. However, there is a Van Voorhis Collection at the Patton Museum that contains a few pictures and numerous newspaper articles of interest.

131. Van Voorhis, quoted in "Prelude to Armor," Army Ground Forces, RG 337, NA, 5–6.

132. Ltr., Brig. Gen. Hinds to Lt. Gen. Willis Crittenberger, May 30, 1971, author's collection, 4.

133. *Lebenslauf*, Hdq., EUCON, Office of the Chief Historian, Operational History

(German) Branch, BIOG File, MHI, and *Erganzende Angaben zum militarischen Werdegang des Adolf von Schell, Militargeschichtliches Forschungsamt, Potsdam, Germany.* According to the editor of the Marshall papers, Larry I. Bland, Marshall interceded on behalf of von Schell, who was wanted by the Communist regime in Poland. Conversation, author with Bland, June 8, 2005.

134. Bland and Stevens, *The Papers of George Catlett Marshall: "The Soldierly Spirit,"* 671 nn2–3.

6. "So He Lost It All?"

1. George F. Hofmann, "The Tactical and Strategic Use of Attaché Intelligence: The Spanish Civil War and the U.S. Army's Misguided Quest for a Modern Tank Doctrine," *Journal of Military History* 62 (Jan. 1998): 101–33.

2. William O. Odom, *After the Trenches: The Transformation of U.S. Army Doctrine, 1918–1939* (College Station: Texas A&M University Press, 1999), 36.

3. "Gaps in U. S. Preparedness," *Newsweek* (Aug. 28, 1939): 11.

4. Scott Davenport, interview with Bruce C. Clarke, General, USA (Ret.), Concerning Adna R. Chaffee Jr., McLean, Va., Dec. 5, 1986, Chaffee Papers, Patton Museum, 2.

5. Quoted in Russell F. Weigley, *History of the United States Army* (New York: Macmillan, 1967), 428.

6. Christopher R. Gabel, *The U.S. Army GHQ Maneuvers of 1941* (Washington, D.C.: Center of Military History, 1991), 31–32; Edward Croft, "Developments in Organization, Armament and Equipment of Infantry," lecture, Army War College, Sept. 18, 1934, MHI, 3; and L. J. McNair, *Antimechanized Defense (Tentative)*, The Command and General Staff School, Fort Leavenworth, Kans., passim.

7. "A-Tk Defense: Conflicting French and German Views," "Foreign Military News-Digest." *Army and Navy Journal* (May 21, 1938): 853.

8. Forrest C. Pogue, *George C. Marshall: Ordeal and Hope, 1939–1942* (New York: Viking Press, 1966), 84.

9. Memorandum for the Chief of Cavalry, Office of the Chief of Cavalry, June 28, 1940, G-2 and Military Attaché Reports (Germany) 1940, Crittenberger Papers, MHI, 1–4.

10. Henry C. Lodge Jr., "Condition of the Army and Formulation of a Military Policy," Proceedings and Debates, 76th Cong., 3rd sess., *Congressional Record* (Washington, D.C.: GPO, May 13, 1940), 1.

11. William M. Grimes, "The 1940 Spring Maneuvers," *Cavalry Journal* (Mar.–Apr. 1940): 98–114.

12. "Prelude to Armor," in *The Armored Force Command and Center*, Study No. 27, Historical Section, Army Ground Forces, 1946, RG337, NA, 7.

13. Christopher R. Gabel, "The 1940 Maneuvers: Prelude to Mobilization: The Louisiana Maneuvers," Louisiana Center of Tourism, 3; and Mildred Hanson Gillie, *Forging the Thunderbolt: A History of the Development of the Armored Force* (Harrisburg, Pa.: Military Service Publishing, 1947), 162–63.

14. "Prelude to Armor," 7.

15. Ibid.

16. Ltr., Colonel Patton to General Joyce, Office of the Commanding Officer, Third Cavalry, Fort Meyer, Feb. 6, 1940, 1–2. Major Robert Bateman, Military Fellow, Center for Strategic and International Studies, Washington, D.C., provided this letter.

17. J. K. Herr's official communications in "Editorial Comment," *Cavalry Journal* (May–June 1946): 34–39.

18. Ibid.

19. Scott Davenport, interview with Robert Merrill Lee, General, USAF (Ret.), Concerning Adna R. Chaffee Jr., Colorado Springs, Colorado, Apr. 25, 1988, Chaffee Papers, Patton Museum, 12–13, 18.

20. "Initial Structure," in *The Armored Force Command and Center*, 9; and Memorandum for the Chief of Staff, G-3, Subject: Organization of the Armored Force, G-3/41665, AG 320.2 (10-2040), Nov. 19, 1940, Dec. 7, 1940, Officer of the Chief of Cavalry, RG 177, NA, 5–9; and Herr, "Editorial Comment," 38.

21. "Prelude to Armor," 8.

22. Charles L. Scott, "Early History of Mechanization," ms., n.d., Papers of Major General Charles L. Scott, Folder 10, Library of Congress, 3.

23. Percy G. Black, "The U.S. Armored Force: The Possibilities Fit Strategical and Tactical Employment," *Cavalry Journal* (Sept.–Oct. 1940): 400.

24. Unrevised Copy of Lecture Given by Major General Adna R. Chaffee to the Officers of the Armored Force, Sept. 13, 1940, Chaffee Papers, Patton Museum, 1, 7, 8, 29.

25. "Training: The Armored Force School," in *The Armored Force Command and Center*, 61.

26. Scott Davenport, interview with Harvey Alness, Lieutenant General, USAF (Ret), Concerning Adna R. Chaffee Jr., San Antonio, Texas, Dec. 8, 1988, Chaffee Papers, Patton Museum, 5–6.

27. "Organization and Tactics," in *The Armored Force Command and Center*, 13–17.

28. Ibid.

29. Interview, Grow with author, May 14, 1971, Patton Museum.

30. Bruce Palmer, "Turn 'Em Over!" *Cavalry Journal* (Nov.–Dec. 1940): 539–42.

31. Ibid.; memorandum for the Chief of Staff, G-3, Subject: Organization of the Armored Force, Your G-3/41665, AG 320.2 (10-2040), Nov. 19, 1940, Dec. 7, 1940, Officer of the Chief of Cavalry, RG 177, NA, 5–9; and Herr, "Editorial Comment," 38.

32. "Redesignation of the Armored Force," in *The Armored Force Command and Center*, 108.

33. Ibid.

34. Ltr., Crittenberger to Lt. Gen. Daniel Van Voorhis, Hdq., Panama Canal Department, Dec. 2, 1940, MIH, 1–5; and Ltr., Crittenberger to Col. Carl J. Rohsenberger, Panama Canal Department, Nov. 25, 1940, Crittenberger Papers, MHI, 1.

35. Ltr., Crittenberger to Senator H. C. Lodge, Jan. 20, 1941, Crittenberger Papers, MHI, 1–2.

36. John T. Broom, "Developing Doctrine When Nobody Cares: Adna R. Chaffee and Mechanization in the U. S. Army, 1925–1941" (paper presented at the 1996 annual conference of the Society for Military History, Birmingham, Alabama, author's collection, 1–35; and U.S. Army, *Field Service Regulations 100-5: Operations, 1941* (Washington, D.C.: War Department, 1941), 6–7, 13–14, 50–51, 97–103, 275–78.

37. However, not until the successes of Patton's Third Army in 1944–1945 was there an indication of conducting war at a maneuver-oriented level of operational warfare. Understanding operational level of war was for the army a struggle to conceptualize and was never defined until after the Vietnam War. It finally entered the U.S. military lexicon in the 1980s.

38. Marvin A. Kreidberg and Merton G. Henry, *History of Military Mobilization in the United States Army 1775–1945*, Department of the Army Pamphlet No. 20-212, Nov. 1955, 630.

39. Quoted in ibid.

40. "Redesignation of the Armored Force," 109.

41. "The Cavalry: How Men on Horseback Team Up with Machines," *Life*, Apr. 21, 1941, 86–92.

42. Dan Janquitto on *Jeeps: The Big Rigs of Combat*, the History Channel, New Video Group, New York, 1996.

43. "The Cavalry," 86–92.

44. James W. Cocke, "Cavalry in World War II," Command and Staff Department, U.S. Armor School, July 1964, Armor School Library, Fort Knox, Kentucky, 2–3.

45. Memorandum for the Chief of Staff, Subject: Evaluation of Modern Battle Forces, in Robert R. Palmer, "Organization and Training of New Ground Combat Elements," in *The Organization of Ground Combat Troops* (Washington, D.C.: Historical Division, U.S. Army, 1947), 308–9, 388–89; and Jonathan M. House, *Combined Arms Warfare in the Twentieth Century* (Lawrence: University Press of Kansas, 2001), 97, 136–38.

46. "Statement of Maj. Gen. Adna R. Chaffee," *Hearings Military Establishment Appropriations Bill, 1941,* House of Representatives, Subcommittee on Appropriations, 77th Cong., 1st sess., May 14, 1941 (Washington, D.C.: GPO, 1941), 552–53.

47. For an excellent study on the maneuvers, see Gabel, *U.S. Army GHQ Maneuvers of 1941.*

48. "Big Maneuvers Test U.S. Army," *Life,* Oct, 6, 1941; and Gable, "The 1940 Maneuvers," Louisiana Center of Tourism, 1.

49. Gabel, *U.S. Army GHQ Maneuvers of 1941*, 118.

50. Bruce Palmer Jr., "Oral History," Interviews: Lt. Colonels James E. Shelton and Edward P. Schmidt, Dec. 5, 1975, MHI, 26–27.

51. John K. Herr and Edward S. Wallace, *The Story of the U.S. Cavalry, 1775–1942* (Boston: Little, Brown, 1953), 251.

52. Gabel, *U.S. Army GHQ Maneuvers of 1941*, 189.

53. Rickey Robertson, "More Stories of the Great Louisiana Maneuvers," "The Louisiana Maneuvers," Louisiana Center of Tourism, 1–5.

54. Charles E. Kirkpatrick, "Lesley J. McNair: Training Philosophy for a New Army," *Army History* (Apr. 1990): 11–17.

55. General Krueger's letter to Patton quoted in Martin Blumenson, *The Patton Papers, 1940–1945* (Boston: Houghton Mifflin, 1974), 42.

56. "Big Maneuvers Test U.S. Army," 5–6.

57. Christopher R. Gabel, *Seek, Strike, and Destroy: U.S. Army Tank Destroyer Doctrine in World War II*, Leavenworth Paper No. 12 (Washington, D.C.: GPO, Sept. 1985),

192–94. For another excellent study on the flawed doctrine, see Charles M. Baily, *Faint Praise: American Tanks and Tank Destroyers during World War II* (Hamden, Conn.: Archon Books, 1983).

58. Ltr., S. Hinds to author, Jan. 20, 1972; and ltr., S. Hinds to Brad Chynoweth, Jan. 7, 1972, author's collection.

59. George F. Hofmann, "Christie's Last Hurrah: In 1941, the Army Reappraised the Christie Suspension for Use on Tank Destroyers," *ARMOR* (Nov.–Dec. 1991): 14–19.

60. Ltr., Churchill to Marshall, Harvard Medical School–Massachusetts General Hospital, Dec. 12, 1940, Chaffee Papers, Patton Museum, 1.

61. Donald E. Houston, *Hell on Wheels: The 2nd Armored Division* (Novato, Calif., Presidio Press, 1977), 59–101.

62. "Statement," *Hearings Military Establishment Appropriations Bill, 1941*, 552–53.

63. Gillie, *Forging the Thunderbolt*, 205.

64. John Cranston, "Chaffee Instrumental in Bringing about Change," *Inside the Turret Armor Conference Issue*, Apr. 29, 1994, 37.

65. Ltr., Grow to Colonel Barrows, editor, *Military Review*, Fort Leavenworth, Kansas, Sept. 12, 1945, author's collection, 1–2.

66. Gillie, *Forging the Thunderbolt*, 208.

67. "The Armored Force: Commanders and Principles," in *The Armored Force Command and Center*, 16.

68. "Antitank Conference," July 14–17, 1941, Andrew D. Bruce Papers, MHI, 81.

69. U.S. Army, *FM 2-15 Cavalry Field Manual: Employment of Cavalry* (Washington, D.C.: GPO, 1941), 1–5.

70. Larry I. Bland, ed., and Sharon Ritenour Stevens, assoc. ed., *The Papers of George Catlett Marshall, "The Right Man for the Job" December 7, 1941–May 31, 1943*, vol. 3 (Baltimore: Johns Hopkins University Press, 1991), 129–30.

71. "Statement," *Hearings Military Establishment Appropriations Bill, 1941*, 561.

72. James E. Hewes Jr., *From Root to McNamara: Army Organization and Administration, 1900–1963* (Washington, D.C.: GPO, Center of Military History, U.S. Army, 1975), 68–69.

73. Herr and Wallace, *Story of the U.S. Cavalry*, 258–61; and "Statement," *Hearings Military Establishment Appropriations Bill, 1941*, 248–61.

74. Edward M. Coffman, *The Hilt of the Sword: The Career of Peyton C. March* (Madison: University of Wisconsin Press, 1966), 245.

75. Fred Glueckstein, "The Last Mounted Cavalry Charge: Luzon 1942," *ARMY* (July 2005), 47–50.

76. Herr and Wallace, *The Story of the U.S. Cavalry*, vii–vii, 257. Also see Edwin Price Ramsey and Stephen J. Rivele, eds., *From Horse Soldier to Guerrilla Commander* (Washington, D.C.: Brassey's, 1996).

77. Shelby L. Stanton, *World War II Order of Battle* (New York: Galahad Books, 1984), 22.

78. Ibid., 18–19.

79. John T. Broom, "Developing Doctrine When Nobody Cares: Adna R. Chaffee and Mechanization in the U.S. Army, 1925–1941," author's collection, 22–23.

80. Broom, "Armor's Four Horsemen," unpublished ms., 1998, author's collection, 24–25. There is a small collection at the Patton Museum, called the Chaffee Papers, that contains scrapbooks, comments on tank casualities during the Spanish civil war, a lecture given by Chaffee to officers at the Armored Force, and letters between Chaffee and Marshall and between Marshall and Mrs. Chaffee after her husband's death. Most important in this collection are five interviews conducted by Scott Davenport during the 1980s with individuals who were closely associated with Chaffee and Van Voorhis during the 1930s.

81. "Organization and Training of New Ground Combat Elements," *Organization of Ground Combat Troops*, 388–91; and "Armored Force: Commanders and Principles," in *The Armored Force Command and Center*, 18–19. Also see Baily, *Faint Praise*, 28–29.

82. Baily, *Faint Praise*, 58–59.

83. Grow, "Ten Lean Years," 116.

84. Kreidberg and Henry, *History of Military Mobilization in the United States Army 1775–1945*, 597.

85. E. N. Harmon with Milton MacKaye and William Ross MacKaye, *Combat Commander: Autobiography of a Soldier* (Englewood Cliffs, N.J.: Prentice-Hall, 1970), 14–16, 57.

86. Lucian K. Truscott Jr., *The Twilight of the U.S. Cavalry: Life in the Old Army, 1917–1942,* ed. Lucian K. Truscott III (Lawrence: University Press of Kansas, 1989), viii, 155, 157.

87. Roland D. Tausch, Interview: General James H. Polk Conversations, Section 1, Senior Officers Debriefing Program, 28–29; and "Summary of the Debriefing of General James M. Polk," May 15, 1972, MHI, 2.

88. John Goolrick's interview, "Super Sixth: Importance of Armored Force Recalled by Patton Favorite," May 4, 1985, *Free Lance-Star*, Fredericksburg, Virginia, author' collection.

89. Ltr., Grow to author, Feb. 2, 1972, author's collection, 1.

90. Robert W. Grow, "Mounted Combat: Lesson from the European Theater," *Cavalry Journal* (Nov.–Dec. 1945): 35–36; and Herr, "Editorial Comments," 35–40.

91. Interview, Grow with author, May 14, 1971, Patton Museum. The interview was documented in my "Tactic vs Technology: The U.S. Cavalry Experience," *ARMOR* (Sept.–Oct. 1973), 14. Much of the interview was based on Grow's daily diary, which he made available to me for this article.

92. Discussions on doctrine with Starry and author, May 23 and 28, 2005.

7. "Sneak and Peek" or Fight

1. H. H. D. Heiberg, "The Reconnaissance Battalion," Officers School, Conference No. 7, 1st Armored Division, Apr. 1, 1941, Fort Knox, Kentucky, Heiberg Papers, Patton Museum, 1–8.

2. C. L. Scott, "Organization and Operation of Mechanized Reconnaissance," ms.,

Charles L. Scott Papers, Ac. No. 10,842, Library of Congress, Washington, D.C., and published as "Armored Reconnaissance," *Cavalry Journal* (Nov.–Dec. 1942): 20–22.

3. Martin Blumenson, *The Patton Papers, 1940–1945* (Boston: Houghton Mifflin, 1974), 133, 138, 150–51.

4. "Convening of Board of Officers to Review the Manual for Commanders of Large Units," Washington, D.C., Mar. 25, 1929, AG 062.11 (3-14-29), Office of the Adjunct General, RG 407, NA, 1–5. Chaffee recorded the proceedings.

5. U.S. Army, *FM 100-15, Field Service Regulations: Larger Units, "The Corps,"* (Washington, D.C.: GPO, 1942), 56–58.

6. Charles J. Hoy, "Reconnaissance Lessons from Tunisia," *Cavalry Journal* (Nov.–Dec. 1943): 16–20.

7. For a more detailed examination of the 81st combat engagements, see George F. Howe, *The Battle History of the 1st Armored Division* (Washington, D.C.: Combat Forces Press, 1954); and Louis D. DiMarco, "The U.S. Army's Mechanized Cavalry Doctrine in World War II" (master's thesis, Command and General Staff College, 1995), 35–74.

8. For a brief examination of command failures, see Steven L. Ossad, "Lesson Learned from Lloyd R. Fredendall," *ARMY* (Mar. 2003): 45–52.

9. Blumenson, *Patton Papers, 1940–1945*, 173–74.

10. Stephen E. Ambrose, *The Supreme Commander: The War Years of General Dwight D. Eisenhower* (Garden City, N.Y.: Doubleday, 1969), 167–69; and Alfred D. Chandler Jr., ed., *The Papers of Dwight David Eisenhower: The War Years*, vol. 2 (Baltimore: Johns Hopkins University Press, 1970), 1007.

11. E. N. Harmon with Milton MacKaye and William Ross MacKaye, *Combat Commander: Autobiography of a Soldier* (Englewood Cliffs, N.J.: Prentice-Hall, 1970), 119–20.

12. Blumenson, *Patton Papers, 1940–1945*, 177.

13. Chandler, *Papers of Dwight David Eisenhower*, 1007. Also see Ambrose, *Supreme Commander*, 167–69.

14. Hamilton H. Howze, *A Cavalryman's Story: Memoirs of a Twentieth-Century Army General* (Washington, D.C.: Smithsonian Institution Press, 1996), 54–59.

15. Harmon, *Combat Commander*, 120, 127.

16. Hoy, "Reconnaissance Lessons from Tunisia," 20.

17. The movie was based on Ladislas Farago, *Patton: Ordeal and Triumph* (New York: Ivan Obolensky, 1964). General Bradley was a consultant to the movie.

18. Chandler, *Papers of Dwight David Eisenhower*, 1101.

19. Harmon, *Combat Commander*, 134.

20. Omar N. Bradley, *A Soldier's Story* (New York: Henry Holt, 1951), 40.

21. Earl Alexander, *The Alexander Memoirs 1940–1945* (London: Cassell, 1962), 44.

22. David G. Chandler, *The Campaigns of Napoleon* (New York: Macmillan, 1966), 351–52. Also see Roger H. Nye, *The Patton Mind: The Professional Development of an Extraordinary Leader* (Garden City Park, N.Y.: Avery, 1993), 30, 141.

23. Harry W. Candler, "91st Reconnaissance Squadron in Tunisia," *Cavalry Journal* (Mar.–Apr. 1944): 14–22.

24. Perrin H. Long, "Patton…and the Slap Heard round the World," *Medical Times* (Nov. 1970): 244–54. At the time, Dr. Long was deeply involved in investigating the

two slapping incidents, and the article, which first appeared in the August 1963 issue of *Medical Times*, was based on his personal diary.

25. Roy R. Grinker and John Spiegel, "The Management of Neuropsychiatric Casualties in the Combat Zone," in *Manual of Military Neuropsychiatry* (Philadelphia: Sanders, 1945), 128–45, 515–40. Also see William S. Mullins and Albert J. Glass, *Neuropsychiatry in World War II: Overseas Theaters*, vol. 2 (Washington, D.C.: GPO, 1973), 303.

26. Adolf von Schell, *Battle Leadership: Some Personal Experiences of a Junior Officer of the German Army with Observations on Battle Tactics and the Psychological Reactions of Troops in Campaign* (Fort Benning, Ga.: Benning Herald, 1933), 12.

27. Robert L. Wagner, *The Texas Army: A History of the 36th Division in the Italian Campaign* (Austin: self-published, 1972), 47. This book takes Clark to task for sacrificing the division in a frontal assault against tremendous odds.

28. Roland D. Tausch, interviewer, General James H. Polk Conversations, Section 1, Senior Officers Debriefing Program, 28–29; and "Summary of the Debriefing of General James M. Polk, May 15, 1972, MHI, 2.

29. L. K. Truscott Jr., *Command Mission: A Personal Story* (New York: Dutton, 1954), 543–44; and D. Eldred, "Are Horses Essential in Modern War?" *Cavalry Journal* (May–June 1944): 4.

30. Subject: Transmittal of Operations in Italy for December 1943, and June 1944, After Action Reports, 36th Cavalry Reconnaissance Troop, 36-CAV-0.3, RG 407, NA.

31. Mark W. Clark, *Calculated Risk* (New York: Harper and Brothers, 1950), 2.

32. For Walker's comments on the crossings, see 36th Infantry Division Association, "General Walker's Story of the Rapido Crossing" in *Memories Never Forgotten* (1946) manuscripts of 36th Infantry Division Association, 1–11.

33. Interview with Field Marshal Kesselring, "How Hitler Could Have Won," *U.S. News and World Report*, Sept. 2, 1955, 65.

34. "General Walker's Story of the Rapido Crossing," 1–11. For an extended history, see "Cassino: The Rapido River Crossing," in Wagner, *Texas Army*, 92–127.

35. Quoted from Walker's diary, in Felix R. McKnight, "T-Patchers' Exploits Won't Be Forgotten," *Dallas Morning News*, Sept. 6, 1996, 9.

36. Frido von Senger und Etterlin, *Neither Fear nor Hope* (London: Macdonald, 1963), 190–92.

37. Clark, *Calculated Risk*, 263–82.

38. Howze, *Cavalryman's Story*, 90–91.

39. Truscott, *Command Mission*, 548.

40. Special Investigation and Interrogation Report, Subject: Operation "Lighting," "The Rapido Offensive and the Allied Landing Operation at Anzio (15 Jan. 1944 to 30 Apr. 1944)," U.S. Army Military Service in Austria, Mar. 15, 1947, Office of Center of Military History, Washington, D.C., 1947, MHI, 57–71.

41. Siegfried Knappe and Ted Brusaw, *Soldat: Reflections of a German Soldier, 1936–1949* (London: Airlift, 1993), 243.

42. Howe, *Battle History of the 1st Armored Division*, 306–7.

43. Ibid., 308.

44. Anthony J. Abati, "Cisterna di Littoria," *Army History* (Fall 1991): 13–21.

45. Interview: General Mark W. Clark by Sidney T. Matthews, May 18, 1948, MHI, 49–51, 60.

46. Truscott, *Command Mission*, 547.

47. Ibid.; and interview: Lt. Gen. Lucian K. Truscott with Sidney T. Mathews, "The Change in the Axis of the Attack," Apr. 3, 1948, MHI, 1–10.

48. Subject: Transmittal of Operations in Italy for June 1944, After Action Reports, 36th Cavalry Reconnaissance Troop, 36-CAV-0.3, RG 407, NA, 1–4; and Wagner, *Texas Army*, 180.

49. Research Report, "Dismounted Patrolling," in "Operation of Cav. Rcn. Sq. Integral to the Armd. Div.," Armor School 1949–1950, Fort Knox, Kentucky, 34. This study relied primarily on the 81st ARB/CRS after-action reports.

50. Ibid., 33–46.

51. Ibid., 40, 46.

52. Clark, *Calculated Risk*, 3.

53. Interview: Clark by Matthews, May 18, 1948, MHI, 49–50.

54. Alexander, *Alexander Memoirs*, 42, 146–53.

55. Dan S. Ray, "The Mission," 36th Reconnaissance Unit, 36th Infantry Division, *Memories Never Forgotten*, manuscript of 36th Infantry Division Association, 1–3.

56. Subject: Observer's Report No. 112 (Cavalry Operations), To: Chief, Combat Lesson Branch, G-3, Mar. 7, 1945, AGF, RG 337, NA, 1–2; and "Lesson Learned in Combat," Subject: Observer's Report No. 112 (Cavalry Operations), To: Chief, Combat Lesson Branch, G-3, Mar. 7, 1945, AGF, RG 337, NA, 5.

57. "Lesson Learned in Combat," RG 337, NA, 1–5.

58. Ltr., M. Carl Ramga to author, Apr. 12, 2004; and "The 91st Recon Troop Remembered," *Powder River Journal*, 1995, 1–3 (*PRJ* is a publication for the 91st Infantry Division).

59. See DiMarco, "The U.S. Army's Mechanized Cavalry Doctrine in World War II," 1–6.

60. See Hoy, "Reconnaissance Lessons from Tunisia"; Candler, "91st Reconnaissance Squadron in Tunisia"; and Scott, "Armored Reconnaissance."

61. Bruce Palmer Jr., "New Battle Lessons on Reconnaissance," *Cavalry Journal* (Sept.–Oct. 1943): 36–37. Hoy is quoted in this article.

62. Ibid. Also see Charles J. Hoy, "The Last Days of Tunisia," *Cavalry Journal* (Jan.–Feb. 1944): 8–13.

63. B. H. Liddell Hart, *The Tanks*, vol. 2 (New York: Praeger, 1959), 39, 41.

64. G. L. Verney, *The Desert Rats: The History of the 7th Armoured Division 1938–1945* (London: Hutchinson, 1954), 17.

8. Mechanized Cavalry from Normandy to End of the War

1. Research Report, "Introduction to the Problem," in "Operation of Cav. Rcn. Sq. Integral to the Armd. Div.," Armor School Research Library, Fort Knox, Kentucky, 1949–1950, 1–2

2. U.S. Army, *Field Service Regulations 100-5: Operations, 1944* (Washington, D.C.: GPO, 1944), 9–10.

3. Ibid., 8–9.

4. Research Report, "Forward" and "Introduction to the Problem," in "Operation of Cav. Rcn. Sq. Integral to the Armd. Div.," ii, 1–3.

5. Shelby L. Stanton, *World War II Order of Battle* (New York: Galahad Books, 1984), 23.

6. Operations Memorandum No. 2, Operations of Corps Cavalry on Terrain of Normandy, Headquarters V Corps, June 25, 1944, Records of Review of Tactical Doctrine, Organization, and Doctrine Branch, G3, Hdq., Army Ground Forces, Aug. 28, 1947, RG 337, NA, 1–2. Also see "Organization and Tactics," available at http://mars.wnec.edu/~dwilliams/history/106cav.html.

7. Martin Blumenson, *Breakout and Pursuit* (Washington, D.C.: GPO, Office of the Chief of Military History, 1961), 206. Cavalry group historian Louis D. DiMarco noted that a number of 102nd Cavalry Group veterans he interviewed claimed the poor performance at Normandy was due to the spinelessness of the 38th Cavalry Squadron commander and the New Jersey National Guard commander of the group. Ltr., DiMarco to author, Feb. 16, 2005.

8. Stanton, *World War II Order of Battle*, 308–11.

9. *History of the 6th U.S. Cavalry*, Veterans Association, 6th U.S. Cavalry, 1947, MHI, 3–9.

10. George S. Patton Jr., *War as I Knew It* (Boston: Houghton Mifflin, 1947), 413–14.

11. Subject: Directive, To: Commanding Officer, 6th Cavalry Group, APO 436, U.S. Army, May 16, 1944, AG 312.3, RG 407, NA, 1; and Subject: Army Information Service, Through: Commanding General—Corps, To: Commanding General—Division. The cavalry groups executed all types of offensive actions, May 16, 1944, AG 320.2, RG 407, NA, 1–2.

12. Oscar W. Koch with Robert G. Hays, *G-2: Intelligence for Patton* (Philadelphia: Whitemore, 1971), 58.

13. Letters and Notes of James H. Polk in World War II (1944–1945), Polk Papers, MHI, 51.

14. Robert M. Batley, "Mechanized Cavalry Group in Defense," *Cavalry Journal* (Sept.–Oct. 1945): 10–12; and Charles B. MacDonald, *The Siegfried Line Campaign* (Washington, D.C.: GPO, Office of the Chief of Military History, 1963), 101–15.

15. Christopher N. Prigge, "Crossed Sabers in Steel: The Evolution of World War II Mechanized Cavalry Doctrine" (paper presented at the Society of Military History conference, Apr. 28–30, 2000, Marine Corps University, Quantico, Virginia), 1.

16. James W. Cocke, "Cavalry in World War II," June 1964, Command and Staff Department, U.S. Army Armor School, Fort Knox, Kentucky, 16–17. This history was provided by Lt. Col. Louis D. DiMarco, Command and General Staff College, Fort Leavenworth, Kansas. A copy is now on file at the Armor School Research Library.

17. Omar N. Bradley, *A Soldier's Story* (New York: Henry Holt, 1951), 416–18.

18. History of the Army Information Service, Oct. 1944, 6th Cavalry Group, Army Ground Forces (AGF), RG 407, NA, 1–6.

19. Training Memorandum No. 17, Operations of a Mobile Force, General, Part I, Hdq., 6th Cav. Group (Mech) (Reinf), Nov. 22, 1944, AGF, RG 407, NA, 1.

20. Ibid.

21. Training Memorandum No. 18, Training Program: Preparation of a Mobile Force, Hdq., 6th Cav. Group (Mech) (Reinf), RG 407, NA, 1.

22. Training Memorandum No. 17, Operations of a Mobile Force, Outline of the Preparation of the Mobile Force, Part III, Hdq., 6th Cav. Group (Mech) (Reinf), 5.

23. Training Memorandum No. 17, Operations of a Mobile Force, General, 7.

24. Fickett, Observations and Recommendations, Section IV, After Action Reports, Apr. 1945, Hdq., 6th Cav. Group (Mech) (Reinf), AGF, RG 407, NA, 13–26.

25. "Mechanized Cavalry Group-Offensive, 6th Cavalry Group: Attack to seize L'Hopital and Clear Karlsbrunn Forest," Appendix 6, Dec. 2–5, 1944, in "Examples of Opinions of Officers Who Commanded Mechanized Cavalry Units," in *Report of the General Board, United States Forces, European Theater*, APO 408, Armored Section, Study Number 49 (1945), Biddle Papers, MHI, 1. Most of the cavalry units' actions described by numerous commanders at all levels of command in the *Report of the General Board* were taken from combat and after-action reports now located in National Archives II, College Park, Maryland (hereafter cited as the Biddle Papers, MHI).

26. For a brief history covering the period discussed, see Organization Day 1945, General Orders No. 11, Hdq., 6th Cavalry Group (Mech), May 4, 1945, AGF, RG 407, NA, 1–7.

27. "Early History," in Research Report, "Cavalry Group as an Economy of Force" (4th Cav. Gp., 19–30 Dec. 44), 1949–1950, Armor School Research Library, Fort Knox, Kentucky, 13–15.

28. Research Report, "Cavalry Group as an Economy of Force," 1.

29. E. N. Harmon with Milton MacKaye and William Ross MacKaye, *Combat Commander: Autobiography of a Soldier* (Englewood Cliffs, N.J.: Prentice-Hall, 1970), 236.

30. See Research Report, "Cavalry Group as an Economy of Force," passim.

31. Charles B. MacDonald, *The Mighty Endeavor: American Armed Forces in the European Theater in World War II* (New York: Oxford University Press, 1969), 354–55.

32. Russell F. Weigley, *Eisenhower's Lieutenants: The Campaigns of France and Germany, 1944–1945* (Bloomington: Indiana University Press, 1981), 432–33.

33. E-mail: correspondence, William O. Hickok, Subject: 28th Reconnaissance Troop, 28th Infantry—Hurtgen Forest, June 16, 2004; and Charles B. MacDonald and Sidney T. Matthews, *Three Battles: Arnaville, Altuzzo, and Schmidt* (Washington, D.C.: Office of the Chief of Military History, GPO, 1952), 254. Mr. Hickok is the 28th Infantry Division historian.

34. "Mechanized Cavalry Group-Security," Appendix 6, Report on Mechanized Cavalry, Biddle Papers, MHI, 8.

35. Research Report, "Cavalry Group as an Economy of Force," 76.

36. Subject: Observer's Report No. 172 (Cavalry Operations), To: Chief, Combat Lessons Branch, G-3, Feb. 2, 1945, AGF, RG 337, NA, 1.

37. Subject: Observer's Report No. 28 (Cavalry Operations), To: Chief, Combat Lessons Branch, G-3, Apr. 24, 1945, AGF, RG 337, NA, 3.

38. Letters and Notes of James H. Polk (1944–1945), Polk Papers, MHI, 13.

39. Ibid., June 30, 1944, 69.

40. Ibid., 73.

41. Martin Blumenson, *The Patton Papers, 1940–1945* (Boston: Houghton Mifflin, 1974), 485.

42. Hugh M. Cole, *The Lorraine Campaign* (Washington, D.C.: GPO, 1950), 120; and Drury obituary, *Assembly*, Association of Graduates, U.S. Military Academy, West Point, Jan. 1990.

43. Letters and Notes of James H. Polk, Sept. 11, 1944, 71.

44. Ibid., Sept. 22, 1944, 75–76.

45. Ibid., Sept. 22, 1944, 76, 83.

46. Cocke, "Cavalry in World War II," 5–6.

47. Letters and Notes of James H. Polk, Nov. 23, 1944, 110, 152.

48. Robert W. Grow, "Mobility Unused," *Military Review* (Feb. 1953): 18–24. The article was in response to Hugh M. Cole's *Lorraine Campaign*.

49. Ibid. The other principles of war that shape military operations are the offensive, security, unity of command, mass, maneuver, surprise, and simplicity. The principles were published in 1923 in *Field Service Regulations United States Army*. They vary with military operations, providing a link between theory and application. However, without a definite objective, there cannot be a definite plan, and thus military actions cannot be coordinated.

50. F. W. von Mellenthin, *Panzer Battles: A Study of the Employment of Armor in the Second World War* (Norman: University of Oklahoma Press, 1955), 334.

51. Personal Papers of Col. Robert S. Allen, Box 1, Handwritten Diaries, Jan. 31, 1943–Feb. 15, 1945, Patton Museum. Also see Roger H. Nye, *The Patton Mind: The Professional Development of an Extraordinary Leader* (Garden City Park, N.Y.: Avery, 1993), 154.

52. Extract: Tactical Operations, Battalion Headquarters and Headquarters Company, IX TAX D 253; and George Ruhlen, "9th Air Force Battle of the Bulge Warning," *Battalion Channel*, 14th Tank Battalion Association, 9th Armored Division, Sept. 2000, 3–4.

53. Blumenson, *Patton Papers, 1940–1945*, 605.

54. Personal Papers of Col. Robert S. Allen, Patton Museum.

55. Letters and Notes of James H. Polk, Mar. 19 and 28, 1945, MHI, 155, 163–64.

56. "Historical Example of Employment of 316th Provisional Mechanized Cavalry Brigade in Combat," Appendix 19, Report on Mechanized Cavalry, Biddle Papers, MHI, 1–28.

57. Letters and Notes of James H. Polk, Mar. 28, 1945, MHI, 164.

58. Carlo D'Este, *Patton: A Genius for War* (New York: HarperCollins, 1995), 723; and Letters and Notes of James H. Polk, Apr. 11, 1945, MHI, 165.

59. Letters and Notes of James H. Polk, May 28, 1945, MHI, 182.

60. Cocke, "Cavalry in World War II," 5–6.

61. Letters and Notes of James H. Polk, May 28, 1945, MHI, 182–83.

62. General Walker's letter quoted in Letters and Notes of James H. Polk, May 22, 1945, MHI, 184–85.

63. Patton, *War as I Knew It*, 330; and Letters and Notes of James H. Polk, May 23, 1945, MHI, 187.

64. History, Customs and Traditions, *3rd Armored Cavalry Regiment, "The Regiment of Mounted Riflemen," 1846–1967*, pt. 2, MHI, 14.

65. Subject: Summary of the Debriefing of General James H. Polk, Thru: Major James J. Steinbach, To: Colonel George S. Pappas, Director, U.S. Army Military History Research Collection, May 15, 1972, MHI, 3.

66. "Examples of Opinions of Officers Who Commanded Mechanized Cavalry Units in Combat Concerning the General Mission," Biddle Papers, MHI, 1.

67. Ibid., 15.

68. Cocke, "Cavalry in World War II," 4.

69. Cole, *Lorraine Campaign*, 220–21.

70. Cocke, "Cavalry in World War II," 4–5.

71. Subject: AGF Report No. 686, Resume of the 106th Cavalry Group (Mech) From Aug. 5, 1944 to Jan. 28, 1945, Submitted by: Col. George W. Coolidge, Cavalry, Hdq., ETO, APO 887, Feb. 28, 1945, Fort Leavenworth Library, Command and General Staff College, Kansas, 1–3; and James W. Cocker, "A Cavalry Group in Delaying Action," *Cavalry Journal* (Nov.–Dec. 1945): 39–44.

72. Introduction: "Tactics, Employment Organization of Mechanized Cavalry Units," Biddle Papers, MHI, 7.

73. For an excellent history of the days before the 86th shipped to Europe and its postwar association, see Charlie Barbour, "Mount Up! A History of the 86th Cavalry Reconnaissance Squadron Mechanized in World War II," Barbour Web site: www.86reconmindspring.com.

74. Patton, *War as I Knew It*, 96–97; and Blumenson, *Patton Papers, 1940–1945*, 485, 491.

75. George F. Hofmann, *The Super Sixth: History of the Sixth Armored Division in World War II* (1975; Nashville, Tenn.: Battery Press, 2000), 61.

76. "Route Reconnaissance," chap. 7 in "Operation of Cav. Rcn. Sq. Integral to the Armd. Div.," 118–28.

77. "John Hughes Remembers August 1944," *The Super Sixer*, Sixth Armored Division Association (Aug. 1998): 24.

78. Kenneth Koyen, *The Fourth Armored Division from the Beach to Bavaria* (1946; Nashville, Tenn.: Battery Press, 2000), 58.

79. Quoted in Roger Crillo *Ardennes-Alsace*, The U.S. Army Campaigns of World War II, Center of Military History Publication 72–26 (Washington, D.C.: GPO, 1995), 7.

80. "Combat Reconnaissance," chap. 5 in "Operation of Cav. Rcn. Sq. Integral to the Armd. Div.," 81–92; and Ltr, Brig. Gen. Albin F. Irzyk to author, Aug. 7, 2004, 1–2. General Irzyk closely worked with each troop and got to know them intimately.

81. Blumenson, *Patton Papers, 1940–1945*, 539.

82. "Reinforced Squadron in Offensive Action," chap. 8 in "Operation of Cav. Rcn. Sq. Integral to the Armd. Div.," 132–53.

83. "Mechanized Cavalry Reconnaissance Squadron Armored Division-Offensive," Appendix 6, Report on Mechanized Cavalry, Biddle Papers, MHI, 12.

84. "Defense of a Roadblock," chap. 6 in "Operation of Cav. Rcn. Sq. Integral to the Armd. Div.," 101–17.

85. Ibid., 115–16.

86. "Mechanized Cavalry Reconnaissance Squadron, Armored Division-Defense," Appendix 6, Biddle Papers, MHI, 13.

87. Robert W. Grow, "The Capture of Muhlhausen," *Military Review* (Nov. 1946): 3–7.

88. "Mechanized Cavalry Reconnaissance Squadron, Armored Division-Security," Appendix 6, Biddle Papers, MHI, 15.

89. Ibid., 14.

90. "General Mission, Tactical Doctrine and Technique, and Future Role of Mechanized Units," Section 1, Combat Analysis, Biddle Papers, MHI, 7–8.

91. Hofmann, *Super Sixth*, 367–68.

92. "Mechanized Cavalry Reconnaissance Troop, Infantry Division-Reconnaissance," Appendix 6, Biddle Papers, MHI, 17.

93. Bradley, *Soldier's Story*, 110.

94. Bradley's directive No. 20 quoted in J. Lawton Collins, *Lightning Joe: An Autobiography* (Baton Rouge: Louisiana State University Press, 1979), 319, 320.

95. "Mechanized Cavalry Reconnaissance Troop, Infantry Division-Offensive," Appendix 6, Biddle Papers, MHI, 16; and Collins, *Lightning Joe*, 321.

96. For a complete discussion, see Report of the General Board, United States Forces, European Theater, "Tactics, Employment, Technique, Organization, and Equipment of Mechanized Cavalry," Armor Section, ETO, Study No. 49, dated Nov. 20, 1945, MHI, passim. Generally, it is known as the Biddle Report.

97. Subject: Post War Equipment Review, To: Commanding Officer, 6th Cavalry Group (Mech), June 8, 1945, AGF, RG 407, NA, 1–2.

98. Fickett, Observations and Recommendations, sec. III, 6th Cavalry Group (Mech), June 8, 1945, AGF, RG 407, NA, 6–7.

99. Ltr., Louis D. DiMarco to author, Feb. 16, 2005; and D'Este, *Patton*, 742–73.

100. "Group Commanders," available at www.louisdimarco.com.

101. 101st Cavalry Regiment: World War II, History, New York State Military Museum and Veterans Research Center, New York State Division of Military and Naval Affairs.

102. Vennard Wilson, address to members of 106th Cavalry Group, Memorial Day, May 30, 1945, Hdq., 106th Cavalry Group (Mech), St. Wolfgang, Austria.

103. Fickett, Observations and Recommendations, Sec. III, 6th Cavalry Group (Mech), 1–2.

104. "Examples of Opinions of Officers Who Commanded Mechanized Cavalry Units in Combat Concerning the General Mission, Tactical Doctrine and Techniques, and Future Role of Mechanized Cavalry," Appendix 11, Biddle Papers, MHI, 1–2.

105. Ibid., 2.

106. "Examples of Opinions of Senior Officers Concerning General Mission, Tactical Doctrine and Technique and Future Role of Mechanized Cavalry Units," Appendix 15, Biddle Papers, MHI, 1–2.

107. "Summaries of Mechanized Cavalry Organizations Proposed by Headquarters First, Third, and Seventh Armies," Appendix 17, Biddle Papers, MHI, 1–2.

108. "Examples of Opinions of Senior Officers Concerning Organization and Equipment of Mechanized Cavalry," Appendix 15, Biddle Papers, MHI, 1–5; and "Examples of Opinions of Officers Who Commanded Mechanized Cavalry Units in Combat Concerning the Organization and Equipment of Mechanized Cavalry," Appendix 16, Biddle Papers, MHI, 1–6.

109. Ibid.

110. *Reconnaissance Battalion, Armored Division* (Washington, D.C.: GPO, 1951), 3.

111. *Stilwell Board Report* (W. D. Equipment Board), Section II, Recommendations, 1945, Patton Museum, 9.

9. The Terrible Turmoil of Postwar Germany and the U.S. Constabulary

1. Albin F. Irzyk, "Mobility, Vigilance, Justice," *ARMY* (Jan. 2003): 44; and Ltr., Irzyk to author, Aug. 7, 2004. William M. Tevington's book *The United States Constabulary: A History* (Paducah, Ky.: Turner, 1998) provides suitable pictures, biographies, general accounts, and personal stories both then and now. It is no longer in print but does provide an interesting personal history of the pride these Constabulary troopers felt as they executed their difficult missions during the early period of the cold war.

2. William E. Stacy, "Early Postwar Border Operations 1945–1946," in *U.S. Army Border Operations in Germany: 1945–1983*, Hdq., U.S. Army, Europe and 7th Army, Military History Office, GSM 5-1-1984, U.S. Army Center of Military History, 5–6.

3. Winston Churchill, *The Second World War: Triumph and Tragedy*, vol. 6 (Boston: Houghton Mifflin, 1953), 573.

4. Oliver J. Frederiksen, *The American Military Occupation of Germany 1945–1953* (Darnstadt: Historical Division, Headquarters, U.S. Army, Europe, 1953), 66.

5. Alfred D. Chandler Jr., ed., *The Papers of Dwight David Eisenhower: The War Years*, vol. 4 (Baltimore: Johns Hopkins University Press, 1970), 2580.

6. Frederiksen, *American Military Occupation of Germany 1945–1953*, 66.

7. *The First Year of the Occupation*, Occupation Forces in Europe Series, 1945–46, Office of the Chief Historian, European Command, Frankfurt-am-Main, Germany, 1947, MHI, 18–19.

8. Frederiksen, *American Military Occupation of Germany 1945–1953*, 66.

9. *The First Year of the Occupation*, MHI, 19–20.

10. Allen W. Dulles, "The Present Situation in Germany," digest of an off-the-record speech at a meeting of the Council on Foreign Relations, Dec. 3, 1945, in "That Was Then: Allen W. Dulles on the Occupation of Germany," *Foreign Affairs* (Nov.–Dec. 2003).

11. Earl Ziemke, *The U.S. Army in the Occupation of Germany 1944–1946* (Washington, D.C.: GPO, Center of Military History, 1975), 421–24.

12. Malcolm Mackinstosh, *Juggernaut: A History of the Soviet Armed Forces* (New York: Macmillan, 1967), 271–72.

13. Telephone interview, Tevington with author, Dec. 9, 2004; and David Coley, "A Little History of the Constabulary and Photo of Uniform," Web site: usConstabulary/history.

14. *The First Year of the Occupation*, MHI, 20.

15. Harry L. Coles and Albert K. Weinberg, *Civil Affairs: Soldiers Become Governors* (Washington, D.C.: GPO, 1964), x; Lucius D. Clay, *Decision in Germany* (New York: Doubleday, 1950), 55–56; and Dwight D. Eisenhower, *Crusade in Europe* (New York: Doubleday, 1952), 434–35.

16. E. N. Harmon, with Milton MacKaye and William R. MacKaye, *Combat Commander: Autobiography of a Soldier* (Englewood Cliffs, N.J.: Prentice-Hall, 1970), 280.

17. Ibid.

18. Untitled typescript by Harmon, Re: U.S. Constabulary, n.d., E. N. Harmon Papers, MHI, 1.

19. Memorandum: To: Commanding General, Third Army U.S. Army, Subject: "Organization of the Zone Constabulary," Jan. 20, 1946, Biddle Papers, MHI, 2–4.

20. Ibid., 1.

21. Biddle, "The U.S. Zone Constabulary," ms., n.d., Biddle Papers, MHI, 1–13.

22. Ibid.

23. Ltr., Irzyk to author, Aug. 7, 2004.

24. Stacy, "Early Postwar Border Operations," 1–2.

25. Ibid., 29–30; and Ray E. Williams, "U.S. Constabulary Horse Cavalry," *Cavalry Journal* (May–June 1948): 2–6.

26. "History of the United States Constabulary," For the Biennial Report of the Chief of Staff, U.S. Army, Harmon Papers, MHI, 2.

27. Quoted in Michael A. Rauer, "Order out of Chaos: The United States Constabulary in Capital Germany," *Army History* (Summer 1998): 26. Rauer's article is one of the most succinct and well-researched short histories of the Constabulary.

28. Memorandum, To: G-1, Subject: Army Assistance to German Youth Activities, Hdq., U.S. Constabulary APO 46, Jan. 6, 1947, in General Historical and Operational Reports, 97-USF8-O.1, Office of the Adjunct General, RG 407, NA, 1–2.

29. According to David Brion Davis, "World War II and Memory," *Journal of American History*, vol. 77 (Sept. 1990). At that time, "colored" was the preferred term for African Americans. The term "black" was considered a derogatory term. See page 584.

30. Clay, *Decision in Germany*, 227–31.

31. Memorandum: Headquarters, MTOUSA, To: Chief of Staff, War Department, Washington, D.C., Aug. 13, 1945, Truman Presidential Museum and Library, Independence, Missouri, 1.

32. "Cavalry-Armor Merger," *Army Navy Register* (Apr. 6, 1946): 11; and "Merger Certain Branches," *Army Navy Register* (Apr. 6, 1946), 1.

33. Dee W. Pettigrew, Subject: Report of Operations, To: The Commanding General, U.S. Constabulary, U.S. Army, July 10, 1946, "Background of the Constabulary School," sec. II, Henry C. Newton Papers, MHI, 1–3.

34. "The Constabulary School," vol. 3, Early History of the Constabulary School, 1946, in General Historical and Operational Reports, 97-USF8-O.1, RG 407, NA, 5.

35. "History of the United States Constabulary," For the Biennial Report of the Chief of Staff, MHI, 5.

36. James. M. Snyder, *The Establishment and Operations of the United States Constabulary, 3 October 1945–30 June 1947*, Historical Sub-Section C-3, U.S. Constabulary 1947, Headquarters U.S. Constabulary, MHI, 51.

37. Irzyk, "Mobility, Vigilance, Justice," 20–21.

38. Snyder, *Establishment and Operations of the United States Constabulary*, 51.

39. Pettigrew, Subject: Report of Operations, Newton Papers, MHI, 5–15.

40. Ibid.; "History of the United States Constabulary," For the Biennial Report of the Chief of Staff, MHI, 25; and Snyder, *Establishment and Operations of the United States Constabulary*, 95–96.

41. Untitled typescript by Harmon, Re: U.S. Constabulary, n.d., Harmon Papers, MHI, 4.

42. Harmon, *Combat Commander,* 286–87. For General White's experience, see U.S. Constabulary Train, Box 7, The Isaac D. White Collection, Norwich University.

43. "Historical Narrative," Headquarters, U.S. Constabulary (VI Corps), Apr. 1946, APO 46, U.S. Army, WWII Operations Reports 1940–48, European Theater, in General Historical and Operational Reports, 97-USF8-0.1, RG 407, NA, 1–14.

44. Untitled typescript by Harmon, Re: U.S. Constabulary, n.d., Harmon Papers, MHI, 2.

45. Memorandum for G-3, Subject: G-4 Historical Notes for the Month of Apr., 1946, MHI, 1–2.

46. Biddle Speech, "Officers and Men of the 11th Constabulary Regiment," Hdq., 11th Constabulary Regiment, Apr. 15, 1946, Biddle Papers, MHI, 1–6.

47. "Historical Narrative," Headquarters, U.S. Constabulary (VI Corps), Apr. 1946, RG 407, NA, 11–14.

48. Ltr., Brig. Gen. Hobart R. Gay to Col. W. S. Biddle, Oct. 16, 1946, Biddle Papers, MHI, 1.

49. Ibid., 14.

50. Ltr., Irzyk to author, Oct. 17, 2004.

51. Ltr., Col. W. S. Biddle to Lt. Col. Harry J. Candler, Apr. 17, 1946; and Ltr., Irzyk to author, Oct. 17, 2004, 1.

52. Davis, "World War II and Memory," 580–87.

53. Ibid.

54. Clay, *Decision in Germany*, 227–31.

55. Irzyk, "Mobility, Vigilance, Justice," 43, 47.

56. Snyder, *Establishment and Operations of the United States Constabulary*, 248–49.

57. Brian A. Libby, "The United States Constabulary in Germany," *Indiana Military History Journal* vol. 13:3 (Oct. 1988): 7. Libby's article was based on his doctorial dissertation, "Policing Germany: The United States Constabulary, 1946–1952" (Purdue University, 1977).

58. Snyder, *Establishment and Operations of the United States Constabulary*, 248.

59. Memorandum, To: G-1, Subject: Army Assistance to German Youth Activities, RG 407, NA, 2.

60. The following unit redesignation to the Constabulary was from James M. Snyder, *Establishment and Operations of the United States Constabulary*, 54–57.

The 1st Constabulary Regiment received 11th Armored Infantry Battalion, 1st Armored Division, which became the 2nd Constabulary Squadron; the 6th Armored Infantry Battalion, 1st Armored Division, became the 12th Constabulary Squadron; and the 91st Armored Field Artillery, 1st Armored Division, became the 91st Constabulary Squadron.

The 2nd Constabulary Regiment received the 2nd Mechanized Cavalry Squadron (Separate), 2nd Cavalry Group (Mech), which became the 2nd Constabulary Squadron; the 42nd Mechanized Cavalry (Separate), became the 42nd Constabulary Squadron; and the 66th Armored Field Artillery Battalion, 4th Armored Division, became the 66th Constabulary Squadron

The 3rd Constabulary Regiment received the 37th Tank Battalion, 4th Armored Division, which became the 37th Constabulary Squadron; the 68th Armored Field Artillery Battalion, 1st Armored Division, became the 68th Constabulary Squadron; and the 81st Cavalry Reconnaissance Squadron (Mech), 1st Armored Division, became the 81st Constabulary Squadron.

The 5th Constabulary Regiment received the 8th Tank Battalion, 4th Armored Division, which became the 8th Constabulary Squadron; the 35th Tank Battalion, 4th Armored Division, became the 35th Constabulary Squadron; and the self-propelled 474th Antiaircraft Automatic Weapons Battalion became the 74th Constabulary Squadron.

The 6th Constabulary Regiment received the 6th Mechanized Cavalry Squadron (Separate), 6th Cavalry Group (Mech), which became the 6th Constabulary Squadron; the 53rd Armored Infantry Battalion, 4th Armored Division, became the 53rd Constabulary Squadron.

The 10th Constabulary Regiment received the 13th Tank Battalion, 1st Armored Division, which became the 13th Constabulary Squadron; the 4th Tank Battalion, 1st Armored Division, became the 72nd Constabulary Squadron; and the 771st Tank Battalion became the 71st Constabulary Squadron.

The 11th Constabulary Regiment received the 25th Mechanized Cavalry Reconnaissance Squadron, 4th Armored Division, which became the 25th Constabulary Squadron; the 94th Armored Field Artillery Battalion, 4th Armored Division, became the 94th Constabulary Squadron; and the 51st Armored Infantry Battalion, 4th Armored Division, became the 51st Constabulary Squadron.

The 14th Constabulary Regiment received the 10th Armored Infantry Battalion, 4th Armored Division, which became the 10th Constabulary Squadron; the 22nd Armored Field Artillery Battalion, 4th Armored Division, became the 22nd Constabulary Squadron; and the 27th Armored Field Artillery Battalion, 1st Armored Division, became the 27th Constabulary Squadron.

The 15th Constabulary Regiment received the 15th Cavalry Reconnaissance (Mech), 15th Cavalry Group, which became the 15th Constabulary Squadron; the 1st Tank Battalion, 1st Armored Division, became the 1st Constabulary Squadron; and the 14th Armored Infantry Battalion, 1st Armored Division, became the 14th Constabulary Squadron.

Also see Rauer, "Order Out of Chaos," 25.

61. Irzyk, "Mobility, Vigilance, Justice," 16–18.

62. Libby, "The United States Constabulary in Germany," 8.

63. Ibid.

64. Telephone conversation, Colonel Sam Williams with Major General Ernest Harmon, 26th Inf. Enclave, Jul. 18, 1946, 97-USF8-7, RG 407, NA, 1.

65. Irzyk, "Mobility, Vigilance, Justice," 17.

66. Harmon, *Combat Commander*, 287.

67. Irzyk, "Mobility, Vigilance, Justice," 17.

68. Stacy, "Early Postwar Border Operations 1945–1946," 11–15.

69. Ibid., 289; and "Cooperation with the German Police," sec. 2, 1946, in Staff Sections, 97-USF8-7, RG 407, NA, Extract.

70. Border Control, Quarterly Reports of Operations, July to Dec. 1946, in General Historical and Operational Reports, 97-USF8-0.3, RG 407, NA, 1–2.

71. James F. Byrnes, *Speaking Frankly* (New York: Harper and Brothers, 1947), 188–89.

72. Ibid., 1–2; and Subject: The History of the United States Constabulary, To: See Distribution, Hdq., U.S. Constabulary, APO 46, Jan. 25, 1947, Charles E. Rousek Papers, MHI, 7.

73. Border Control, Quarterly Reports of Operations, July to Dec. 1946, 2.

74. Black Market, Quarterly Reports of Operations, July to Dec. 1946, 1–2.

75. Weekly Intelligence Summary No. 24, for the Period 160001 Nov. to 222400 Nov. 1946, pt. 2, Civil Security, 97-USF8-2.6, in G-2 Section, RG 407, NA, 1–2.

76. Snyder, *Establishment and Operations of the United States Constabulary*, 151, 152.

77. Weekly Intelligence Summary No. 24, 1.

78. Subject: Report of Operations for the year 1946, From: Office of the Staff Judge Advocate, To: Commanding General, U.S. Constabulary, APO 46 (Attention G-1), Jan. 13, 1947, History (Establishment and Operations), 97-USF8-0.1, in General Historical and Operational Reports, RG 407, NA, 1–4.

79. Subject: General Comments, To: Commanding Generals, I, II, II Constabulary Brigades, Mar. 11, 1946, 1947, 97-USF8-0.1, in General Historical and Operational Reports, RG 407, NA, 1–3.

80. Subject: Report of Operations, To: Headquarters, U.S. Constabulary, APO 46, From: Office of the Inspector General, Jan. 6, 1947, History (Establishment and Operations), 97-USF8-0.1, in General Historical and Operational Reports, RG 407, NA, 1–2.

81. The Quarter in Retrospect, chap. 1, Quarterly Report of Operations, July 46–Dec. 48, 97-USF8-0.3, in General Historical and Operational Reports, RG 407, NA, 3–5.

82. H. P. Rand, "A Progress Report on the United States Constabulary," *Military Review* (Oct. 1947): 32.

83. Subject: G-1 History, U.S. Constabulary, To: G-3, Historical Section, Hdq., U.S. Constabulary, Jan. 6, 1947, 97-USF8-0.1, in General Historical and Operational Reports, RG 407, NA, 1.

84. *First Hand Information from Constabulary Troopers and Their Families*, Web site: usConstabulary/history. For a short and interesting history of the 16th Constabulary Squadron (Separate), see Al Solosky, "A Brief History of the 16th Constabulary Squadron (Separate), Berlin," in Tevington, ed., *United States Constabulary*, 35–36. This article takes note of the extensive looting of Berlin by the Red Army.

85. Ltr., Irzyk to author, Oct. 17, 2004.

86. Ibid.

87. Memorandum, To: G-1, Subject: Army Assistance to German Youth Activities, RG 407, NA, 1–2.

88. Clay, *Decision in Germany*, 64, 228.

89. The Month in Retrospect, Apr. 1947, 97-USF8-3, G-3 Reports of Operations in G-3 Section, RG 407, NA, 1.

90. The Month in Retrospect, May 1947, 97-USF8-3, G-3 Reports of Operations in G-3 Section, RG 407, NA, 1–2.

91. Biddle Speech, "Officers and Men of the 11th Constabulary Regiment," Biddle Papers, MHI, 1–6.

92. The Month in Retrospect, July 1947, 97-USF8-3, G-3 Reports of Operations in G-3

Section, Office of the Adjutant General, RG 407, NA, 1–2; and Rand, "Progress Report on the United States Constabulary," 37, 31.

93. Thomas D. Boettcher, *First Call: The Making of the Modern U.S. Military, 1945–1953* (Boston: Little, Brown, 1992), 116–17.

94. Dwight D. Eisenhower, *At Ease: Stories I Tell to Friends* (Garden City, N.Y.: Doubleday, 1967), 316–20. For a provocative account of mobilization and military unpreparedness, see Michael Kendall, "An Inflexible Response: United States Army Manpower Mobilization Policies, 1945–1950" (Ph.D. diss., Duke University, 1982).

95. The Month in Retrospect, Jan. and July 1948, 97-USF8-3, G-3 Reports of Operations in G-3 Section, RG 407, NA, 1.

96. Rand, "A Progress Report on the United States Constabulary," 37; and Subject: Intelligence Reports for the Month of Feb. 1948, EUCOM, Frankfurt, APO 757, Hdq., U.S. Constabulary, 97-USF8-2.6, Intelligence Reports in G-2 Section, RG 407, NA, 1–3.

97. KPD Attitudes to Current Problems, Annex No. 5, To: G-2 Weekly Intelligence, Summary No. 68, Hdq., U.S. Constabulary, Aug. 1947, 97-USF8-2.6, Weekly Intelligence Summaries in G-2 Section, RG 407, NA, 1–3.

98. Subject: Intelligence Report No. 29, To: Deputy Director of Intelligence, Headquarters, EUCOM, Frankfurt, APO 757, Hdq., U.S. Constabulary, Aug. 30, 1947, 97-USF8-2.6, Intelligence Reports in G-2 Section, RG 407, NA, 14.

99. Subject: Intelligence Reports, Reports 1947, To: Deputy Director of Intelligence, Headquarters, EUCOM, Frankfurt, APO 757, Hdq., U.S. Constabulary, End of 1947, 97-USF8-2.6, Intelligence Reports in G-2 Section, RG 407, NA, all.

100. "KPD Attitudes to Current Problems," Annex 5 to G-2, Weekly Intelligence, Summary 68, Hdq., U.S. Constabulary, Aug. 22, 1947, 97-USF8-2.6, Intelligence Reports in G-2 Section, RG 407, NA, 3.

101. Weekly Intelligence Summary, Period 25001 Aug. to 012400 Sept. 1947, 97-USF8-2.6, Intelligence Reports in G-2 Section, RG 407, NA, 1–6.

102. Ibid., 3.

103. The Month in Retrospect, Oct. 1947, 97-USF8-3, G-3 Reports of Operations in G-3 Section, RG 407, NA, 7.

104. The Month in Retrospect, Nov. and Dec. 1947, 97-USF8-3, G-3 Reports of Operations in G-3 Section, RG 407, NA, 1–2.

105. Report of Operations, Judge Advocate Section, Dec. 1947, 97-USF8-5 in G-5 Section, RG 407, NA, 1–3.

106. Subject: Intelligence Reports No. 40 to 50, End of 1947 and Jan. 1948, 13–16; and The Month in Retrospect, To: Deputy Director of Intelligence, Headquarters, EUCOM, Frankfurt, APO 757, and U.S. Constabulary, Nov. 1947, 97-USF8-2.6, Intelligence Reports in G-2 Section, RG 407, NA, 1.

107. The Month in Retrospect, Nov. 1947, 97-USF8-3, G-3 Reports of Operations in G-3 Section, RG 407, NA, 3–4.

108. Ibid., 3–6.

109. Jean Edward Smith, ed., *The Papers of General Lucius D. Clay, Germany 1945–1949*, vol. 2 (Bloomington: Indiana University Press 1974), 640.

110. The Month in Retrospect, Jan. to June 1948; Smith, ed., *The Papers of General Lucius D. Clay,* passim.

111. The Month in Retrospect, Mar. 1948, 97-USF8-3, G-3 Reports of Operations in G-3 Section, RG 407, NA, 1–10.

112. Weekly Intelligence Summary, Period 27001 April to 03240 May 1948, 97-USF8-2.6, Intelligence Reports in G-2 Section, RG 407, NA, 1–16.

113. The Month in Retrospect, Mar. 1948, 97-USF8-3, G-2 Reports of Operations in G-2 Section, RG 407, NA, 9.

114. The Month in Retrospect, Mar. and June 1948, 97-USF8-3, G-2 Reports of Operations in G-2 Section, RG 407, NA, 6, 9.

115. A Pre-interview with: General Isaac Davis White, Ret., Conducted by: Charles Pittsburgh Roe, Oct. 27, 1977, MHI, 19–20.

116. Isaac D. White, conversation with Charles S. Stodter, Senior Officers Debriefing Program, Fall 1977, I. D. White Papers, MHI, 68.

117. Interview, General James H. Polk with Lt. Col. Roland D. Taush, Feb. 1, 1972, Polk Papers, MHI, 1.

118. The Month in Retrospect, Mar. 1948, 97-USF8-3, G-2 Reports of Operations in G-2 Section, RG 407, NA, 5.

119. Stacy, "The Constabulary Takes Charge," in *U.S. Army Border Operations in Germany: 1945–1983*, 38.

120. Ibid., 562; and Clay, *Decision in Germany*, 239.

121. The Month in Retrospect, Aug. 1948, G-2 Reports of Operations in G-2 Section, 1; and Weekly Intelligence Summary No. 132, General Summary, 070001 to 132400 Dec. 1948, in G-2 Section, 97-USF8-3, RG 407, NA, 1.

122. The Month in Retrospect, Aug. to Dec. 1948, 97-USF8-3, G-3 Reports of Operations in G-3 Section, RG 407, NA, 1.

123. Quarterly Historical and Operational Report, 7766th Horse Troop, U.S. Constabulary, 1 Oct. 1949 to 31 Dec. 1949, Special Troops, Headquarters, U.S. Constabulary, 97-USF8 in Headquarters Troops, RG 407, NA, 1–4.

124. Clay, *Decision in Germany*, 65, 438.

125. Irzyk, "Mobility, Vigilance, Justice," 43–47.

126. *Table of Organization and Equipment, FM 17-51, Armored Cavalry Regiment (Light)* (Washington, D.C.: GPO, Department of the Army, Oct. 7, 1948), passim; and *Field Manual 17-35: Reconnaissance Battalion, Armored Division* (Washington, D.C.: GPO, 1951), 3.

Conclusion

1. "Demobilization Pressure," *Army Navy Register* (Jan. 12, 1946): 8.

2. Louis Galambos, ed., *The Papers of Dwight David Eisenhower: The Chief of Staff,* vol. 7 (Baltimore: Johns Hopkins University Press, 1978), entry numbers 528, 645, 657.

3. Dwight D. Eisenhower, *At Ease: Stories I Tell to Friends* (Garden City, N.Y.: Doubleday, 1967), 316–20; and R. Alton Lee, "The Army 'Mutiny' of 1946," *Journal of American History*, vol. 53:3 (Dec. 1966): 555–71. For a provocative account of mo-

bilization and military unreadiness, see Michael Kendall, "An Inflexible Response: United States Army Manpower Mobilization Polices, 1945–1950" (Ph.D. diss., Duke University, 1982).

4. Address by General Jacob L. Devers at Veterans Day Observance, Aug. 31, 1947, Lieutenant General George W. Read Jr., Papers, author's collection, 2.

5. Ibid., 2, 4.

6. James T. Patterson, *Mr. Republican: A Biography of Robert A. Taft* (Boston: Houghton Mifflin, 1972), 436–37; and Thomas D. Boettcher, *First Call: The Making of the Modern U.S. Military, 1945–1953* (Boston: Little, Brown, 1992), 116–17.

7. David Flick, "The Late Sen. Taft 'Demagogic,' Eisenhower Wrote," *Cincinnati Post*, July 31, 1978, 15.

8. Louis Galambos, ed., *The Papers of Dwight David Eisenhower: The Chief of Staff*, vol. 9 (Baltimore: Johns Hopkins University Press, 1978), 2254–56. For an excellent study on Congress and its ability to influence the military through power of the purse, see Edward A. Kolodzeij, *The Uncommon Defense and Congress, 1945–1963* (Columbus: Ohio State University Press, 1966).

9. Walter Millis, ed., with E. S. Duffield, *The Forrestal Diaries* (New York: Viking Press, 1951), 151; and Cabinet meeting, Friday, Apr. 5, 1946, Matthew J. Connelly Papers: Notes on Cabinet Meeting I, Truman Library, Independence, Missouri, NA.

10. Report of War Department Equipment Board, Jan. 19, 1946, Falkovich Collection, Patton Museum, 8–9, 42–43.

11. William W. Epley, *America's First Cold War Army 1945–1950* (Arlington, Va.: Institute of Land Warfare, Association of the U.S. Army, 1999), 6, 11.

12. Harry S. Truman, *Memoirs by Harry A. Truman: Years of Trial and Hope 1946–1952* (Garden City, N.Y.: Doubleday, 1956), 345.

13. Steven L. Rearden, *History of the Office of the Secretary of Defense: The Formative Years 1947–1950* (Washington, D.C.: Historical Office, Office of the Secretary of Defense, 1984), 406–10. For brief discussions of NSC-68, see Maurice A. Mallin, *Tanks, Fighters, and Ships: U.S. Conventional Force Planning since WWII* (Washington, D.C.: Brassey's, 1990), 41–62; and John Edward Wilz, "Korea and the United States, 1945–1950," in Stanley Sander, ed., *The Korean War: An Encyclopedia* (New York: Garland, 1995), 176–77.

14. S. Everett Gleason and Fredrick Aandahl, gen. eds., *Foreign Relations of the United States 1950: National Security Affairs; Foreign Economic Policy*, vol. 1 (Washington, D.C.: GPO, 1977), 231–33. Also see Nathan Reingold, "Vannevar Bush," in John A. Garraty and Mark C. Carnes, eds., *American National Biography*, vol. 4 (New York: Oxford University Press, 1999), 80.

15. Report of Army Field Forces Advisory Panel on Armor, vol. 1, Feb. 18, 1949, Falkovich Collection, Patton Museum, 5–7.

16. Department of the Army, *FM 17-100 Armored Division and Combat Command*, Dec. 1949, Patton Museum, 85–87.

17. Robert A. Doughty, *The Evolution of U.S. Army Tactical Doctrine, 1946–76*, Leavenworth Papers (Fort Leavenworth, Kans.: Combat Studies Institute, U.S. Army Command and General Staff College, August 1979), 1.

18. Report of the Army Equipment Board 1950, Fort Monroe, Va., Mar. 8, 1950, Falkovich Collection, Patton Museum, 27–29.

19. J. Lawton Collins, *Lightning Joe: An Autobiography* (Baton Rouge: Louisiana State University Press, 1979), 350–51.

20. Report No. 1776, Senate Discussion, *Army Organization Act of 1950,* Eighty-first Congress, 2nd sess., Mar. to June 1950, 1–2; and "New Army Set-Up Will Go to Senate—End of Cavalry Lamented," *New York Times,* May 27, 1950, 6.

21. *Congressional Record,* Eighty-first Congress, 2nd sess., vol. 96, pt. 6, May 17, 1950, to June 13, 1950, 8258–59.

22. John K. Herr and Edward S. Wallace, *The Story of the U. S. Cavalry, 1775–1942* (Boston: Little, Brown, 1953), 261.

23. Hanson W. Baldwin, "The Cavalry Charges On: It's Dead in Fact and Act of Congress," *New York Times,* July 2, 1950, 4.

24. See James Pickett Jones, *Yankee Blitzkrieger: Wilson's Raid through Alabama and Georgia* (Lexington: University Press of Kentucky, 1976).

25. John J. McCloy. "The Future of Germany," *Army Navy Register* (Feb. 11, 1950): 6.

26. Ltr., Chaffee to Marshall (Personal), Hdq., Armored Force, July 26, 1940, Chaffee Papers, Patton Museum, 1–3.

27. "Training," in *The Armored Force Command and Center,* Study No. 27, Historical Section, Army Ground Forces, 1946, RG 337, NA, 75.

28. On this issue, see George F. Hofmann, *Cold War Casualty: The Court-Martial of Major General Robert W. Grow* (Kent, Ohio: Kent State University Press, 1993).

29. On Taylor and the Vietnam issue, see H. R. McMaster, *Dereliction of Duty: Lyndon Johnson, Robert McNamara, the Joint Chiefs of Staff, and the Lies That Led to Vietnam* (New York: HarperCollins, 1997).

30. Portions of Grow's diary were photocopied by an alleged Soviet agent in Frankfurt, sensationalized by communist propagandists, and then by the Western news media led by the *Washington Post.* The incident and subsequent court-martial became a cause célèbre during the tensions of the cold war and the political intolerance of McCarthyism. The contents of Grow's diary that were exploited by the communists and Western press were falsified or taken out of context. Only later, after the court-martial, was this made clear. Even before the pretrial investigation, the State Department and the Central Intelligence Agency recommended refuting the implications of portions of the published diary. Taylor disagreed. He decided to move his management team toward establishing a court-martial.

Years later Taylor stated, "I was not involved in the court-martial of General Grow… never got into it because of command influence.…I have no side and had no personnel involvement." Brigadier General Fay B. Prickett, who was so influential in organizing the Constabulary and at the time of the court-martial assigned to the Inspector General's Office, admitted later that some irregularities took place at the court-martial and that under oath he himself did not tell the truth. Prickett believed he should have changed his testimony. An examination of hundreds of records from the National Archives in Washington, D.C., indicated Taylor's heavy involvement in managing the trial.

These documents and others were only made available after the extensive use of the Freedom of Information Act and a number of legal appeals that took years to resolve. Grow was never made aware of the extent of Taylor's involvement.

Perhaps Colonel Jim Moncrief Jr., Grow's G-1 during the war, can throw some light on Taylor's actions. He recalled that serious friction occurred between the two during the Battle of the Bulge when Taylor's 101st Airborne Division failed to provide flank support when the 6th Armored Division was ordered by Patton to break out of the Bastogne pocket. Grow at the time called Taylor a "boy general," believing that Taylor was not with his troops during the critical stages of the encirclement. Certainly, Grow's name did not make it to Taylor's little black book, where he listed names of promising officers.

Grow was found guilty and sentenced to be reprimanded and suspended from command for a period of six months. Shortly, he retired as a major general, having fallen victim to Taylor and his management group, which was assisted by Judge Advocate General Ernest M. Brannon, and especially the assistant chief of staff, G-2 (Intelligence), Major General Alexander R. Bolling. He had the opportunity to counter the diary propaganda but then failed to testify that there was no compromise of national security. Bolling had taken possession of the diary in question. Surely, he viewed the contents and was aware that communist propagandists had forged the diaries or had taken them out of context. Actually, Bolling was on the list as a witness for the defense. However, for unexplained reasons he remained indisposed during the trial. Colonel Robert C. Bard, who was head of General Harmon's Judge Advocate General's Section during the early period of the Constabulary, was the nominal accuser and drafted a set of specifications. In spite of all the pretrial and trial irregularities and Taylor's deceitful, unlawful command influence, no action was taken under the Uniform Code of Military Justice to consider the violation of Article 37 (Unlawful Influencing the Action of the Court) and Article 46 (Opportunity to Obtain Witness and Other Evidence). In July 1957, President Eisenhower, under Article 71, reluctantly took action to resolve the Grow case, that "the findings…be approved but that the sentence be remitted." See Hofmann, *Cold War Casualty*, passim.

31. Interview, author with Grow, May 14, 1971, Fort Knox, Kentucky; and Ltr., Grow to author, Apr. 8, 1974, author's collection, 1–2.

32. Robert W. Grow, "Mobility Unused," *Military Review* (Feb. 1953): 18.

33. *The Super Sixer*, Sixth Armored Division Association (Jan. 1986), 4–5.

34. Martin Blumenson, *The Patton Papers, 1940–1945* (Boston: Houghton Mifflin, 1974), 610.

35. *Elizabeth (N.J.) Daily Journal*, Jan. 8, 1944, 1; and *New York Times*, Jan. 12, 1944, 23.

36. J. Walter Christie vs United States, *House Resolution 1332* (Washington, D.C.: GPO, June 1946), 3.

Select Bibliography

The following selected sources were consulted over many years. The list is far from conclusive; however, it provides the reader a working knowledge. The chapter notes contain an extensive use of primary sources, many for the first time. I have not listed them all in detail but advise the reader to review those of interest. Extensions of the sources from National Archives Record Groups are detailed in the notes. The author collection mentioned in the notes consists of years of accumulation of oral interviews, letters, and manuscripts that in most cases were forwarded over the years to the U.S. Army Military History Institute, Patton Museum of Cavalry and Armor, and Special Collections at the U.S. Military Academy.

Selective Use of the Following Records

First Division Museum at Cantigny, Illinois
Frank Parker Collection

Fort Knox Armor School Research Library, Fort Knox, Kentucky
Fort Knox Cavalry and Armor Circular Files
Research Reports on Cavalry and Armor, Armor School

Herbert Hoover Presidential Library, West Branch, Iowa
Truman Smith Papers

Library of Congress, Washington, D.C.
George S. Patton Jr. Papers
Charles L. Scott Papers

National Archives II, College Park, Maryland
Record Group 120: Records of the War Department Historical Files
Record Group 156: Records of the Chief of Ordnance and Technical Files
Record Group 165: Records of the War Department and Special Staff

Record Group 177: Records of the Chief of Arms
Record Group 337: Records of the Army Ground Forces
Record Group 407: Records of the Adjutant General

Norwich University, Northfield, Vermont

Ernst N. Harmon Papers
I. D. White Collection

Patton Museum of Cavalry and Armor, Fort Knox, Kentucky

Fort Knox History Circular Files
Robert S. Allen Papers
Mark Falkovich Collection
Robert W. Grow Collection
Robert J. Icks Papers and Photo Collection
Richard P. Hunnicutt Photo and Manuals Collection
First Armored Division History Collection
Daniel Van Voorhis Collection
Ardna R. Chaffee Jr. Papers and Scott Davenport Interviews
Mechanized Force Photo Collection
H. H. D. Heiberg Papers and Interviews
Stephen Henry Photo Collection

Truman Presidential Museum and Library, Independence, Missouri

Visits to Fort Knox Photos
Oral History Interview with Bruce C. Clarke

United States Army Military History Institute, Carlisle Barracks, Pennsylvania
Army War College Circular Files

Senior Officers Oral History De-briefing Program
Bruce Palmer Jr.
William S. Biddle Papers
Bradford G. Chynoweth Papers
Willis D. Crittenberger Papers
Robert W. Grow Papers
Earnest N. Harmon Papers
Guy V. Henry Papers
Hamilton H. Howze Papers
Leon B. Kromer Papers
George A. Lynch Papers
Henry C. Newton Papers
Frank Parker Collection
James H. Polk Papers
Charles E. Rousek Papers
Kenyon A. Joyce Papers

University of North Carolina at Chapel Hill
Frank Parker Papers

University of North Dakota, Grand Forks
George S. Patton Jr. Papers (Sereno E. Brett Collection)

U.S. Military Academy, West Point, New York
George S. Patton Jr. Book Collection
John K. Herr Collection

Virginia Military Institute, Lexington, Virginia
Samuel D. Rockenbach Papers
George C. Marshall Papers (Marshall Foundation)

Dissertations and Theses

Broom, John T. "The Commander's Vision in Blue and Gray: The Roles of Adna R. Chaffee, Jr., James H. Wilson and the American Civil War in the Development of American Armor Doctrine." Ph.D. diss., Union Institute, 1993.

Cameron, Robert S. "Americanization of the Tank: U.S. Army Administration and Mechanized Development within the Army, 1917–1943." Ph.D. diss., Temple University, 1994.

Cox, Gary C. "Beyond the Battle Line: U.S. Air Attack Theory and Doctrine, 1919–1941." Thesis, School of Advanced Airpower Studies, Air University, 1995.

Daly, John L. "From Theory to Practice: Tanks, Doctrine, and the U.S. Army, 1916–1940." Ph.D. diss., Kent State University, 1993.

DiMarcio, Louis A. "The U.S. Army's Mechanized Cavalry Doctrine in World War II." Master's thesis, Army Command and General Staff College, 1995.

Hofmann, George F. "Rejection of Christie's Armored Fighting Vehicles." Master's thesis, University of Cincinnati, 1970.

Johnson, Dave E. "Fast Tanks and Heavy Bombers: The United States Army and the Development of Armor and Aviation Doctrines and Technologies, 1917 to 1945." Ph.D. diss., Duke University, 1990.

Kendall, Michael. "An Inflexible Response: United States Army Manpower Mobilization Policies, 1945–1950." Ph.D. diss., Duke University, 1982.

Killigrew, John W. "The Impact of the Great Depression on the Army, 1929–1936." Ph.D. diss., Indiana University, 1960.

Libby, Brian A. "Policing Germany: The United States Constabulary, 1946–1952." Ph.D. diss., Purdue University, Indiana, 1977.

Morton, Matthew D. "Men on 'Iron Ponies': The Death and Rebirth of the Modern U.S. Cavalry." Ph.D. diss., Florida State University, 2004.

Nenninger, Timothy K. "The Development of American Armor, 1917–1940." Master's thesis, University of Wisconsin, 1968.

Tedesco, Vincent J. "Greasy Automations and the Horsey Set: The U.S. Cavalry and Mechanization, 1928–1940." Master's thesis, Pennsylvania State University, 1995.

Selected Studies and Manuscripts

The Armored Force Command and Center. Study No. 27. Washington, D.C.: Government Printing Office, 1946.

Bielakowski, Alexander M. "Crossed Sabers in Steel: Mechanized Cavalry after the 1940s." Manuscript presentation, Society for Military History Conference, Marine Corps University, Quantico, Virginia, April 2000.

Boudinot, Burton S. "The Way I Remember It." Manuscript, Radcliff, Kentucky, 1992.

Bounds, Gary L. *Larger Units: Theater Army–Army Group–Field Army.* Chap. l, CSI Report No. 6, Jan. 1985, Combat Studies Institute, U.S. Command and Staff College, Fort Leavenworth, Kansas.

Broom, John T. "It Isn't Doctrine until They All Understand It: *FM 100–5 Field Service Regulations, 1941,* the Armored Force, and the Conduct of the 1944 European Campaign." Manuscript, Fort Knox, Kentucky 1993.

Cocke, James W. *Cavalry in World War II.* Command and Staff Department, U.S. Armor School, Fort Knox, Kentucky, July 1964.

Grow, Robert W. "The Ten Lean Years: From the Mechanized Force (1930) to the Armored Force (1940)." Manuscript, Falls Church, Virginia, 1969.

Prigge, Christopher N. "Crossed Sabers in Steel: The Evolution of World War II Mechanized Cavalry Doctrine." Manuscript, Society for Military History Conference, Marine Corps University, Quantico, Virginia, April 2000.

Report of Special Armored Vehicle Board (Palmer Board). *Comparative Test of Armor Cars and Related Vehicles.* Washington, D.C.: War Department, December 5, 1942.

Report of the General Board, United States Forces, European Theater. *Tactics, Employment, Technique, Organization, and Equipment of Mechanized Cavalry.* Armor Section, Study No. 49. Washington, D.C.: Government Printing Office, November 20, 1945.

Research Report. *Critical Analysis of the History of Armor in World War II.* Armored School, Fort Knox, Kentucky, April 1953.

Research Report. *Operation of Cav RCN Sq Integral to the Armd. Div.* Armor School, Fort Knox, Kentucky, May 1950.

Severs, Hugh G. "The Controversy behind the Air Corps Tactical School's Strategic Bombardment Theory: An Analysis of the Bombardment versus Pursuit Aviation Data between 1930–1939." Research paper, Air Command and Staff College, Maxwell Air Force Base, Alabama, 1997.

Snyder, James M. "United States Constabulary: The Establishment and Operations of the United States Constabulary 3 October 1945–30 June 1947." Manuscript, Historical Sub-section G-3, Hdq., U.S. Constabulary, West Germany, 1947.

Stacy William E. "U.S. Border Operations in Germany." Manuscript, Military History Office, Hdq., U.S. Army, Europe and 7th Army, 1984.

Steadman, Kenneth A. "The Evolution of the Tank in the U.S. Army." Paper, Combat

Studies Institute, U.S. Army Command and General Staff College, Fort Leavenworth, Kansas, Government Printing Office, September 1, 1982.
Stilwell Board Report (W. D. Equipment Board). Section II. *Recommendations.* Patton Museum of Cavalry and Armor, Fort Knox, Kentucky, 1945.
Walker, Fred L. "General Walker's Story of the Rapido Crossing." Manuscript, 36th Infantry Division Association, 1946.

Selected Books

Alexander, Earl. *The Alexander Memoirs 1940–1945*. London: Cassell, 1962.
Allen, Robert S. *Lucky Forward, The History of Patton's Third U.S. Army.* New York: Vanguard Press, 1947.
Allen, Terry. *Combat Communication for Regiments and Smaller Units of Horse Cavalry.* Harrisburg, Pa.: Military Service Publishing, 1939.
Ambrose, Stephen E. *The Supreme Commander: The War Years of General Dwight D. Eisenhower.* Garden City, N.Y.: Doubleday, 1969.
———. *Upton and the Army.* Baton Rouge: Louisiana State University Press, 1964.
Bach, Christian A., and Henry Noble Hall. *The Fourth Division: Its Service and Achievements in the World War.* Issued by the Division, 1920.
Baily, Charles M. *Faint Praise: American Tanks and Tank Destroyers during World War II.* Hamden, Conn.: Archon Books, 1983.
Bland, Larry I., ed., and Sharon Ritenour Stevens, assoc. ed. *The Papers of George Catlett Marshall, "The Right Man for the Job" December 7, 1941–May 31, 1943*. Vol. 3. Baltimore: Johns Hopkins University Press, 1991.
———. *The Papers of George Catlett Marshall, "The Soldierly Spirit" December 1880–June 1939*. Vol. 1. Baltimore: Johns Hopkins University Press, 1981.
Blumenson, Martin. *Breakout and Pursuit.* Washington, D.C.: Government Printing Office, 1984.
———. *Patton: The Man behind the Legend, 1885–1945*. New York: Morrow 1985.
———. *The Patton Papers, 1885–1940*. Vol. 1. Boston: Houghton Mifflin, 1972.
———. *The Patton Papers, 1940–1945*. Vol. 2. Boston: Houghton Mifflin, 1974.
———. *Salerno to Cassino.* Washington, D.C.: Government Printing Office, Center of Military History, 1969.
Boettcher, Thomas D. *First Call: The Making of the Modern U.S. Military, 1945–1953.* Boston: Little, Brown, 1992.
Bradley, Omar N. *A Soldier's Story.* New York: Henry Holt, 1951.
Byrnes, James F. *Speaking Frankly.* New York: Harper and Brothers, 1947.
Chandler, Alfred D., Jr., ed. *The Papers of Dwight David Eisenhower: The War Years*. Vol. 2. Baltimore: Johns Hopkins University Press, 1970
Chandler, David G. *The Campaigns of Napoleon.* New York: Macmillan, 1966.
Churchill, Winston. *The Second World War: Triumph and Tragedy,* Vol. 6. Boston: Houghton Mifflin, 1953.
Chynoweth, Bradford Grethen. *Memoirs: Bellamy Park.* Hicksville, N.Y.: Exposition Press, 1975.

Citino, Robert M. *The Evolution of Blitzkrieg Tactics: Germany Defends Herself against Poland, 1918–1933*. Westport, Conn.: Greenwood Press, 1987.

———. *The Path to Blitzkrieg: Doctrine and Training in the German Army, 1920–1939*. Boulder, Colo.: Lynne Rienner, 1999.

Clark, Mark W. *Calculated Risk*. New York: Harper and Brothers, 1950.

Clay, Lucius D. *Decision in Germany*. New York: Doubleday, 1950.

Cline, Ray S. *Washington Command Post: The Operations Divisions*. Washington, D.C.: Government Printing Office, Office of the Chief of Military History, 1951.

Coffman, Edward M. *The Hilt of the Sword: The Career of Peyton C. March*. Madison: University of Wisconsin Press, 1966.

———. *The Regulars: The American Army 1898–1941*. Cambridge, Mass.: Belknap Press of Harvard University Press, 2004.

———. *The War to End All Wars: The American Military Experience in World War I*. Lexington: University Press of Kentucky, 1998.

Cole, Hugh M. *The Ardennes: Battle of the Bulge*. U.S. Army in World War II, ETO. Washington, D.C.: Government Printing Office, 1965.

———. *The Lorraine Campaign*. U.S. Army in World War II, ETO. Washington, D.C.: U.S. Government Printing Office, 1950.

Coles, Harry L., and Albert K. Weinberg. *Civil Affairs: Soldiers Become Governors*. Washington, D.C.: Government Printing Office, Office of the Chief of Military History, 1964.

Collins, J. Lawton. *Lightning Joe: An Autobiography*. Baton Rouge: Louisiana State University Press, 1979.

Collins, Robert L. "Report on the Development of the Tank Corps." In *United States Army in the World War 1917–1919: Reports of the Commander-in-Chief, Staff Sections and Services*. Vol. 15. Washington, D.C.: U.S. Army Center of Military History, 1991.

Corum, James S. *The Roots of Blitzkrieg: Hans von Seeckt and German Military Reform*. Lawrence: University Press of Kansas, 1992.

Crismon, Fred W. *U.S. Wheeled Vehicles*. Minneapolis, Minn.: Victory WW2 Publishing, 2001.

Crowell, Benedict, and Robert Forrest Wilson. *The Armies of Industry: Our Nation's Manufacture of Munitions for a World in Arms 1917–1918*. Vol. l. New Haven, Conn.: Yale University Press, 1921.

Dastrup, Boyd L. *King of Battle: A Branch History of the U.S. Army's Field Artillery*. Fort Monroe, Va.: U.S. Army Training and Doctrine Command, 1992.

De la Gorce, Paul-Marie. *The French Army: A Military-Political History*. New York: George Braziller, 1963.

Denison, G. T. *A History of Cavalry: From the Earliest Times with Lessons for the Future*. 2nd ed. London: Macmillan, 1913.

D'Este, Carlo. *Fatal Decision: Anzio and the Battle for Rome*. New York: HarperCollins, 1991.

———. *Patton: A Genius for War*. New York: HarperCollins, 1995.

Deutsch, Harold C. *Hitler and His Generals: The Hidden Crisis January–June 1938*. Minneapolis: University of Minnesota Press, 1974.

Dickinson, W. N. *History of Anti-aircraft Guns*. Washington, D.C.: Government Printing Office, 1920.

DiNardo, R. L. *Germany's Panzer Arm*. Westport, Conn.: Greenwood Press, 1997.

Doughty, Robert A. *The Evolution of U.S. Army Tactical Doctrine, 1946–1976*, Leavenworth Papers. Fort Leavenworth, Kans.: Combat Studies Institute, U.S. Army Command and General Staff College, August 1979.

Eisenhower, Dwight D. *At Ease: Stories I Tell to Friends*. Garden City, N.Y.: Doubleday, 1967.

———. *Crusade in Europe*. New York: Doubleday, 1952.

———. "Tanks with Infantry" [May 1921]. In Daniel D. Holt and James W. Leyerzapf, eds., *Eisenhower: The Prewar Diaries and Selected Papers, 1905–1941*. Baltimore: Johns Hopkins University Press, 1998.

Epley, William W. *American's First Cold War Army 1945–1950*. Arlington, Va.: Institute of Land Warfare, Association of the United States Army, 1999.

Farago, Ladislas. *Patton: Ordeal and Triumph*. New York: Ivan Obolensky, 1964.

Fletcher, David. *Mechanized Force: British Tanks between the Wars*. London: HMSO Publication Center, 1991.

Frank, Willard C., Jr., and Philip S. Gillette, eds. *Soviet Military Doctrine from Lenin to Gorbachev*. Westport, Conn.: Greenwood Press, 1992.

Fuller, J. F. C. *The Army in My Time*. London: Rich and Cowan, 1935.

———. *Decisive Battles of the U.S.A.* New York: Tomas Yoseloff, 1942.

———. *Memoirs of an Unconventional Soldier*. London: Ivor Nicholson and Watson, 1936.

———. *Tanks in the Great War 1914–1918*. London: John Murray, 1920.

Gabel, Christopher R. *The 4th Armored Division in the Encirclement of Nancy*. Fort Leavenworth, Kans.: Combat Studies Institute, 1986.

———. *Seek, Strike, and Destroy: U.S. Army Tank Destroyer Doctrine in World War II*. Leavenworth Paper No. 12. Washington, D.C.: Government Printing Office, 1985.

———. *The U.S. Army GHQ Maneuvers of 1941*. Washington, D.C.: Center of Military History, 1991.

Galambos, Louis, ed. *The Papers of Dwight David Eisenhower: The Chief of Staff*. Vols. 2 and 9. Baltimore: Johns Hopkins University Press, 1978.

Garland, Albert N., and Howard McGaw Smyth, assisted by Martin Blumenson. *Sicily and the Surrender of Italy*. The United States Army in World War II, Mediterranean Theater of Operations. Washington, D.C.: Center of Military History, 2002.

Garraty, John A., and Mark C. Carnes, eds. *American National Biography*. Vol. 4. New York: Oxford University Press, 1999.

Gillie, Mildred Hanson. *Forging the Thunderbolt: A History of the Development of the Armored Force*. Harrisburg, Pa.: Military Service Publishing, 1947.

Gleason, S. Everett, and Fredrick Aandahl, gen. eds. *Foreign Relations of the United States 1950: National Security Affairs; Foreign Economic Policy*. Vol. 1. Washington, D.C.: Government Printing Office, 1977.

Green, Constance McLaughlin, Harry C. Thomson, and Peter C. Roots. *The Technical Services: The Ordnance Department: Planning Munitions for War.* Washington, D.C.: Government Printing Office, Office of the Chief of Military History, 1955.

Greene, Joseph I., ed. *The Infantry Journal Reader.* Garden City, N.Y.: Doubleday, 1944.

Greenfield, Kent Roberts, Robert R. Palmer, and Bill I. Wiley. *The Army Ground Forces: The Organization of Ground Combat Troops.* Washington, D.C.: Department of the Army, Historical Division, 1947.

Greer, Thomas H. *The Development of Air Doctrine in the Army Air Arm 1917–1941.* Washington, D.C.: Government Printing Office, Office of Air Force History, 1985.

Grinker, Roy R., and John Spiegel. "The Management of Neuropsychiatric Casualties in the Combat Zone." In *Manual of Military Neuropsychiatry.* Philadelphia: Sanders, 1945.

Guderian, Heinz. *Achtung-Panzer! The Development of Armoured Forces, Their Tactics and Operational Potential.* London: Arms and Armour Press, 1993.

———. *Panzer Leader.* London: Michael Joseph, 1952.

Harmon, E. N., with Milton MacKaye and William Ross MacKaye. *Combat Commander: Autobiography of a Soldier.* Englewood Cliffs, N.J.: Prentice-Hall, 1970.

Herr, John K., and Edward S. Wallace. *The Story of the U.S. Cavalry, 1775–1942.* Boston: Little, Brown, 1953.

Hewes, James E., Jr. *From Root to McNamara: Army Organization and Administration, 1900–1963.* Washington, D.C.: Center of Military History, U.S. Army, 1975.

Historical Documents Relating to the Reorganization Plans of the War Department and to the Present National Defense Act Pt. 1. Washington, D.C.: Government Printing Office, 1927.

"History of the 68th Field Artillery." In *First Armored Division.* Nashville, Tenn.: Army and Navy Publishing, 1941.

Hodges, LeRoy. *Post-war Ordnance.* Richmond, Va.: Richmond Press, 1923.

Hofmann, George F. *Cold War Casualty: The Court-Martial of Major General Robert W. Grow.* Kent, Ohio: Kent State University Press, 1993.

———. *The Super Sixth: A History of the Sixth Armored Division in World War II.* Nashville, Tenn.: Battery Press, 2000.

Hofmann, George F., and Donn A. Starry, eds. *Camp Colt to Desert Storm: The History of the U.S. Armored Forces.* Lexington: University Press of Kentucky, 1999.

Holley, I. B., Jr. *General John M. Palmer, Citizen Soldiers, and the Army of Democracy.* Westport, Conn.: Greenwood Press, 1982.

Holt, Daniel D., and Leyerzapf, James W., eds. *Eisenhower: The Prewar Diaries and Selected Papers, 1905–1941.* Baltimore: Johns Hopkins University Press, 1998.

Hoover, Herbert. *The Memoirs of Herbert Hoover: The Cabinet and the Presidency 1920–1933.* New York: Macmillan, 1952.

House, Edward Mandell, and Charles Seymour. *What Really Happened at Paris: The Story of the Peace Conference, 1918–1919.* New York: Scribner's, 1921.

House, Jonathan M. *Combined Arms Warfare in the Twentieth Century.* Lawrence: University Press of Kansas, 2001.

Houston, Donald E. *Hell on Wheels: The 2nd Armored Division.* Novato, Calif.: Presidio Press, 1977.

Howe, George F. *The Battle History of the 1st Armored Division.* Washington, D.C.: Combat Forces Press, 1954.

Howze, Hamilton H. *A Cavalryman's Story: Memoirs of a Twentieth-Century Army General.* Washington, D.C.: Smithsonian Institution Press, 1996.

Hunnicutt, R. P. *Armored Cars: A History of American Wheeled Combat Vehicles.* Novato, Calif.: Presidio Press, 2002.

———. *Stuart: A History of the American Light Tank.* Novato, Calif.: Presidio Press, 1992.

Hurley, Alfred F. *Billy Mitchell: Crusader for Air Power.* Bloomington: Indiana University Press, 1964.

Infantry Journal. *Infantry in Battle.* Second reprint of 1939 edition. Quantico, Va.: Marine Corps Association.

James, D. Clayton. *The Years of MacArthur, 1880–1941.* Vol. 1. Boston: Houghton Mifflin, 1970.

Johnson, David E. *Fast Tanks* and *Heavy Bombers: Innovation in the U.S. Army, 1917–1945.* Ithaca, N.Y.: Cornell University Press, 1998.

Jones, James Pickett. *Yankee Blitzkrieger: Wilson's Raid through Alabama and Georgia.* Lexington: University Press of Kentucky, 1976.

Jones, Ralph E., George H. Rarey, and Robert J. Icks. *The Fighting Tanks from 1916 to 1933.* Old Greenwich, Conn.: WE, 1933.

Judge Advocate General's Corps. *The Army Lawyer: A History of the Judge Advocate General's Corps, 1775–1975.* Washington, D.C.: Government Printing Office, 1976.

Killigrew, John W. *Impact of the Great Depression on the Army.* New York: Garland, 1979.

Kipp, Jacob W. "Soviet Military Doctrine and the Origins of Operational Art, 1917–1936." In Willard C. Frank Jr. and Philip S. Gillette, eds., *Soviet Military Doctrine from Lenin to Gorbachev.* Westport, Conn.: Greenwood Press, 1992.

Knappe, Siegfried, and Ted Brusaw. *Soldat: Reflections of a German Soldier, 1936–1949.* London: Airlift Publishing, 1993.

Koch, Oscar W., with Robert G. Hays. *G-2 Intelligence for Patton.* Philadelphia: Whitemore, 1971.

Kolodzeij, Edward A. *The Uncommon Defense and Congress, 1945–1963.* Columbus: Ohio State University Press, 1966.

Koyen, Kenneth. *The Fourth Armored Division from the Beach to Bavaria.* Nashville, Tenn.: Battery Press, 2000.

Kreidberg, Marvin A., and Merton G. Henry. *History of Military Mobilization in the United States Army 1775–1945.* Department of the Army Pamphlet No. 20–212, November 1955.

Liddell Hart, Basil. *The Liddell Hart Memoirs 1895–1938.* Vol. 1 New York: Putnam's, 1965.

———. *The Tanks.* Vol. 2. New York: Praeger, 1959.

Liggett, Hunter. *Commanding an American Army: Recollections of the World War.* New York: Houghton Mifflin, 1925.

Linn, Brian McAllister. *Guardians of the Empire: The U.S. Army and the Pacific, 1902–1940.* Chapel Hill: University of North Carolina Press, 1997.

Lodge, Henry C., Jr. "Condition of the Army and Formulation of a Military Policy." Proceedings and Debates, 76th Cong., 3rd sess. *Congressional Record.* Washington, D.C.: Government Printing Office, May 13, 1940.

MacArthur, Douglas O. *Report of Secretary of War to the President, Annual Reports War Department Fiscal Year Ended June 30, 1931.* Washington, D.C.: Government Publications Office, 1931.

———. *Report of Secretary of War to the President, Annual Reports War Department Fiscal Year Ended June 30, 1933.* Washington, D.C.: Government Printing Office, 1933.

———. *Report of Secretary of War to the President, Annual Report War Department Fiscal Year Ended June 30, 1934.* Washington, D.C.: Government Printing Office, 1934.

———. *Report of Secretary of War to the President, Annual Report War Department Fiscal Year ended June 30, 1935.* Washington, D.C.: Government Printing Office, 1935.

MacDonald, Charles B. *The Mighty Endeavor: American Armed Forces in the European Theater in World War II.* New York: Oxford University Press, 1969.

———. *The Siegfried Line Campaign.* Washington, D.C.: Government Printing Office, Office of the Chief of Military History, 1963.

MacDonald, Charles B., and Sidney T. Matthews. *Three Battles: Arnaville, Altuzzo, and Schmidt.* Washington, D.C.: Office of the Chief of Military History, Government Printing Office, 1952.

Mackinstosh, Malcolm. *Juggernaut: A History of the Soviet Armed Forces.* New York: Macmillan, 1967.

Mallin, Maurice A. *Tanks, Fighters, and Ships: U.S. Conventional Force Planning since WWII.* Washington, D.C.: Brassey's, 1990.

March, Peyton C. *The Nation at War.* Garden City, N.Y,: Doubleday, Doran, 1932.

Martel, Giffard. *An Outspoken Soldier: His Views and Memoirs.* London: Sifton Praed, 1949.

Maurer, Maurer, ed. *The U.S. Air Service in World War I.* Vol. 2. Washington, D.C.: Government Printing Office, 1978.

McMaster, H. R. *Dereliction of Duty: Lyndon Johnson, Robert McNamara, the Joint Chiefs of Staff, and the Lies That Led to Vietnam.* New York: HarperCollins, 1997.

Millett, Allan R., and Williamson Murray, eds. *Military Effectiveness.* Vol. 2, *The Interwar Period.* Boston: Allen and Unwin, 1988.

Millis, Walter, ed., with E. S. Duffield. *The Forrestal Diaries.* New York: Viking Press, 1951.

Mitchell, William (Billy). *Winged Defense: The Development and Possibilities of Modern Air Power—Economic and Military.* New York: Putnam's, 1925.

Modern Reconnaissance: A Collection of Articles from The Cavalry Journal. Harrisburg, Pa.: Military Service Publishing, 1944.

Mullins, William S., and Albert J. Glass. *Neuropsychiatry in World War II: Overseas Theaters.* Vol. 2. Washington, D.C.: Government Printing Office, 1973.

Nye, Roger H. *The Patton Mind: The Professional Development of an Extraordinary Leader.* Garden City Park, N.Y.: Avery, 1993.

Odom, William O. *After the Trenches: The Transformation of U.S. Army Doctrine, 1918–1939*. College Station: Texas A&M University Press, 1999.

Palmer, Robert R. "Organization and Training of New Ground Combat Elements." *The Organization of Ground Combat Troops.* Washington, D.C.: Historical Division, U.S. Army, 1947.

Palmer, Robert R., Bill I. Wiley, and William R. Keast. *Procurement and Training of Ground Combat Troops: The United States Army in World War II.* Washington, D.C.: Historical Division, United States Army, 1948.

Patterson, James T. *Mr. Republican: A Biography of Robert A. Taft.* Boston: Houghton Mifflin, 1972.

Patton, George S., Jr. *War as I Knew It.* Boston: Houghton Mifflin, 1947.

Pershing, John J. *My Experiences in the World War.* London: Hodder and Stoughton, 1931.

Pogue, Forrest C. *George C. Marshall: Education of a General, 1880–1939*. New York: Viking Press, 1963.

———. *George C. Marshall: Ordeal and Hope, 1938–1942*. New York: Viking Press, 1966.

Preston, R. M. P. *The Desert Mounted Corps.* London: Constable, 1921.

Ramsey, Edwin Price, and Stephen J. Rivele, eds. *From Horse Soldier to Guerrilla Commander.* Washington, D.C.: Brassey's, 1996.

Ray, Dan S. "The Mission." In *Memories Never Forgotten.* 36th Infantry Division, 1946.

Rearden, Steven L. *History of the Office of the Secretary of Defense: The Formative Years 1947–1950*. Washington, D.C.: Historical Office, Office of the Secretary of Defense, 1984.

Reingold, Nathan, "Vannevar Bush." In John A. Garraty and Mark C. Carnes, eds., *American National Biography.* Vol. 4. New York: Oxford University Press, 1999.

Sander, Stanley, ed. *The Korean War: An Encyclopedia.* New York: Garland, 1995.

Smith, Jean Edward, ed. *The Papers of General Lucius D. Clay, Germany 1945–1949*. Vol. 2. Bloomington: Indiana University Press, 1974.

Smythe, Donald. *Pershing: General of the Armies.* Bloomington: Indiana University Press, 1986.

Solosky, Al. "A Brief History of the 16th Constabulary Squadron (Separate), Berlin." In William M. Tevington, *The United States Constabulary: A History.* Paducah, Ky.: Turner, 1998.

Spector, Ronald. "The Military Effectiveness of the U.S. Armored Forces, 1919–1936." In Allan R. Millett and Williamson Murray, eds., *Military Effectiveness.* Boston: Allen and Unwin, 1988.

Spires, David N. *Patton's Air Force: Forging a Legendary Air-Ground Team.* Washington, D.C.: Smithsonian Press, 2002.

Stanton, Shelby L. *World War II Order of Battle.* New York: Galahad Books, 1984.

Steffen, Randy. *The Horse Soldier, 1776–1943*. Vol. 4. Norman: University of Oklahoma Press, 1979.

Stubbs, Mary Lee, and Stanley Russell Connor. *Armor-Cavalry: Regular Army and Reserve.* Army Lineage Series. Pt. 1. Washington, D.C.: Government Printing Office,

Office of the Chief of Military History, 1969.
Summerall, Charles. *Report of the Secretary of War to the President, Annual Reports, War Department, Fiscal Year Ended June 30,1929*. Washington, D.C.: Government Printing Office, 1929.
Tevington, William M. *The United States Constabulary: A History*. Paducah, Ky.: Turner, 1998.
TO&E 17–51, Armored Cavalry Regiment (Light). Washington, D.C.: Government Printing Office, Department of the Army, October 7, 1948.
Trask, David F. *The AEF and Coalition Warmaking, 1917–1918*. Lawrence: University Press of Kansas, 1993.
Truman, Harry S. *Memoirs by Harry S. Truman: Years of Trial and Hope 1946–1952*. Garden City, N.Y.: Doubleday, 1956.
Truscott, Lucien K., Jr. *Command Mission: A Personal Story*. New York: Dutton, 1954
———. *The Twilight of the U.S. Cavalry: Life in the Old Army 1917–1942*. Ed. Lucian K. Truscott III. Lawrence: University Press of Kansas, 1989.
United States Army in the World War 1917–1919: Reports of the Commander-in-Chief, Staff Sections and Services. Vols. 14–15. Washington, D.C.: Government Printing Office, U.S. Army Center of Military History, 1991.
U.S. Army. *Field Service Regulations United States Army 1923*. Washington, D.C.: Government Printing Office, 1924.
———. *Field Service Regulations 100-5: Operations, 1939* (tentative). Washington, D.C.: Government Printing Office, 1939.
———. *Field Service Regulations 100-5: Operations, 1941*. Washington, D.C.: Government Printing Office, 1941.
———. *Field Service Regulations 100-5: Operations, 1944*. Washington, D.C.: Government Printing Office, 1944.
———. *FM 2–15 Cavalry Field: Employment of Cavalry*. Washington, D.C.: Government Printing Office, 1941.
———. *FM 17–35: Reconnaissance Battalion, Armored Division*. Washington, D.C.: War Department, Government Printing Office, 1950.
———. *FM 100–5: Operations, 1941*. Washington, D.C.: War Department, Government Printing Office, 1941.
———. *FM 100–15 Field Service Regulations: Larger Units, "The Corps."* Washington, D.C.: War Department, Government Printing Office, 1942.
———. *Tables of Organization Mechanized Division (Tentative)*. Fort Leavenworth, Kans.: Command and General Staff School Press, 1936.
———. *The Tactical Employment of Cavalry (Tentative), Part One–Horse and Part Two–Mechanized*. Fort Leavenworth, Kans.: Command and General Staff School Press, 1937.
———. *The Tactical Employment of Infantry (Tentative)*. Fort Leavenworth, Kans.: Command and General Staff School Press, 1937.
———. *The Tactical Employment of the Mechanized Division (Tentative)*. Fort Leavenworth, Kans.: Command and General Staff School Press, 1937.

U.S. Army Cavalry School. *Cavalry Combat.* Fort Riley, Kans.: U.S. Cavalry Association, 1937.

U.S. Cavalry School. *FM 2-10 Cavalry Field Manual.* Vol. 2, *Mechanized Cavalry.* Fort Riley, Kans.: 1938.

U.S. Congress. *Military Establishment Appropriation Bill, 1942.* Washington, D.C.: Government Printing Office, 1942.

U.S. Congress, Subcommittee of the House Committee on Appropriations. *War Department Appropriations Bill for 1931 Military Activities,* Hearings, Part I, 71st Congress 3rd Session. Washington, D.C.: Government Printing Office, 1928.

———. *War Department Appropriations Bill for 1932, Military Affairs.* Hearings, pt. 1. 71st Cong., 3rd sess., December 5, 1930. Washington, D.C.: Government Printing Office, 1930.

———. *War Department Appropriations Bill for 1933 Military Activities.* Hearings, pt. 1, 72nd Cong., 1st sess. Washington, D.C.: Government Printing Office, 1932.

U.S. Department of State. *Foreign Relations of the United States, 1932.* Washington, D.C.: Government Printing Office, 1948.

Verney, G. L. *The Desert Rats: The History of the 7th Armoured Division 1938–1945.* London: Hutchinson, 1954.

Veterans Association, 6th U.S. Cavalry. *History of the 6th U.S. Cavalry.* Washington, D.C.: U.S. Army Military Institute, 1947.

Von Mellenthin, F. W. *Panzer Battles: A Study of the Employment of Armor in the Second World War.* Norman: University of Oklahoma Press, 1955.

Von Schell, Adolf. *Battle Leadership: Some Personal Experiences of a Junior Officer of the German Army with Observations on Battle Tactics and the Psychological Reactions of Troops in Campaign.* Fort Benning, Ga.: Benning Herald, 1933.

———. *Kampf gegen Panzerwagen.* Berlin: Gerhard Stalling, 1936.

Von Senger, Frido und Etterlin. *Neither Fear nor Hope.* London: Macdonald, 1963.

Wagner, Robert L. *The Texas Army: A History of the 36th Division in the Italian Campaign.* Austin, Tex.: self-published 1972.

Waite, Robert G. L. *Vanguard of Nazism: The Free Corps Movement in Postwar Germany 1918–1923.* New York: Norton, 1952.

Walker, Fred L. "General Walker's Story of the Rapido Crossing." In *Memories Never Forgotten.* 36th Infantry Division Association, 1946.

War Department. *Field Service Regulations United States Army 1923: Operations.* Washington, D.C.: Government Printing Office, 1924.

Wedemeyer, Albert C. *Wedemeyer Reports!* New York: Holt, 1958.

Weigley, Russell F. . *Eisenhower's Lieutenants: The Campaigns of France and Germany, 1944–1945.* Bloomington: Indiana University Press, 1981.

———. *History of the United States Army.* New York: Macmillan, 1967.

Wik, Reynold M. *Benjamin Holt and Caterpillar Tracks and Combines.* St. Joseph, Mo.: American Society of Agricultural Engineers, 1984.

Wilson, Dale E. *Treat 'Em Rough! The Birth of American Armor, 1917–1920.* Novato, Calif.: Presidio Press, 1989.

Wilson, John B. *Maneuver and Firepower: The Evolution of Divisions and Separate Brigades.* Washington, D.C.: Government Printing Office, Center of Military History, 1998.

Wilz, John Edward. "Korea and the United States, 1945–1950." In Stanley Sander, ed., *The Korean War: An Encyclopedia.* New York: Garland, 1995.

Winton, Harold R., and David R. Mets, eds. *The Challenge of Change: Military Institutions and New Realities, 1918–1941.* Lincoln: University of Nebraska Press, 2000.

Yale, Wesley W., Isaac D. White, and Hasso E. von Manteufel. *Alternative to Armageddon: The Peace Potential of Lightning War.* New Brunswick, N.J.: Rutgers University Press, 1970.

Ziemke, Earl. *The U.S. Army in the Occupation of Germany 1944–1946.* Washington, D.C.: Government Printing Office, Center of Military History, 1975.

Military and Professional Journals

ARMOR

Chynoweth, Bradford Grethen. "Military Conservatism." March–April 1974.

Daley, John L. "Patton versus the 'Motor Maniacs': An Interwar Defense of Horse Cavalry." March–April 1997.

Grow, Robert W. "The Role of Armor." September–October 1961.

———. "Ten Lean Years." January–February, March–April, May–June, July–August 1987.

Heiberg, H. H. D. "Organize a Mechanized Force." September–October 1976.

Hofmann, George F. "Christie's Last Hurrah: In 1941, the Army Reappraised the Christie Suspension for Use on Tank Destroyers." November–December 1991.

———. "Tactics vs Technology: The U.S. Cavalry Experience." September–October 1973.

———. "A Yankee Inventor and the Military Establishment: The Christie Tank Controversy." March–April 1976.

Nenninger, Timothy K. "The Experimental Mechanized Forces." May–June 1969.

———. "A Revised Mechanization Policy." September–October 1969.

———. "The Tank Corps Reorganized." March–April 1969.

———. "The World War I Experience." January–February 1969.

ARMY

Glueckstein, Fred. "The Last Mounted Cavalry Charge: Luzon 1942." July 2005.

Hofmann, George F. "The Troubled History of the Christie Tank." May 1986.

Irzyk, Albin F. "Mobility, Vigilance, Justice." January 2003

Ossad, Steven L. "Lesson Learned from Lloyd R. Fredendall." March 2003.

Army and Navy Journal

"A-Tk Defense: Conflicting French and German Views." "Foreign Military News-Digest." May 21, 1938.

"Christie Tanks." June 10, 1931.

"Confusion in Military Legislation." September 27, 1919.

"Government Officials at Fort Knox." January 29, 1938.

"Senate Hearings on the Army." September 27, 1919.

"Suggest 'Police' Needs as Basis for Arms Cuts." June 18, 1932.

Army and Navy Register

"Army Reorganization Bill." August 30, 1919.

Birnie, Upton, Jr. "Obsolescence of Horse-Drawn Artillery." May 16, 1937.

"Calvary-Armor Merger." April 6, 1946.

"Chief of Staff before Subcommittee." March 26, 1938.

Craig, Malin. "Mechanization and Tanks." March 26, 1938.

"Demobilization Pressure." January 12, 1946.

"House Committee Hearings on the Army." September 6, 1919

"House Committee Hearings on the Army." September 13, 1919.

McCloy, John J. "The Future of Germany." February 11, 1950.

"Merger Certain Branches." April 6, 1946.

"Senate Hearings on the Army." September 20, 1919.

"Senate Military Hearings." October 25, 1919.

Army History

Abati, Anthony J. "Cisterna di Littoria." Fall 1991.

Bielakowski, Alexander M. "The Role of the Horse in Modern Warfare as Viewed in the Interwar *U.S. Cavalry Journal*." Summer–Fall 2000.

Kirkpatrick, Charles E. "Lesley J. McNair: Training Philosophy for a New Army." April 1990.

Rauer, Michael A. "Order Out of Chaos: The United States Constabulary in Capital Germany." Summer 1998.

Army Motors #96

Jesensky, Larry. "Scout Cars: White Scout Car Story." Parts 1 and 2. 2001

Army Ordnance

Benson, C. C. "The New Christie, Model 1940: An Estimate of the Tank Armored Car Combined." September–October 1929.

Westervelt, William. "A Challenge to American Engineers." V1 1920.

Cavalry Journal

Arnold, H. H. "The Cavalry-Air Corps Team." January 1928.

Batley, Robert M. "Mechanized Cavalry Group in Defense." September–October 1945.

Benson, C. C. "Mechanization Aloft and Allow." January 1929.
Black, Percy G. "The U.S. Armored Force: The Possibilities Fits Strategical and Tactical Employment." September–October 1940.
Candler, Harry W. "91st Reconnaissance Squadron in Tunisia." March–April 1944.
"Cavalry Combat." September–October 1937.
"The Cavalry School Visits Fort Knox." July–August 1937.
Chaffee, Adna R., Jr. "James Harrison Wilson." July 1925.
———. "The Maneuvers of the First Cavalry Division, September–October 1923." April 1924.
———. "The Seventh Cavalry Brigade and the First Army Maneuvers." November–December 1939.
Chynoweth, Bradford Grethen. "Cavalry Tanks." May 1921.
Cocker, James W. "A Cavalry Group in Delaying Action." November–December 1945.
Crittenberger, Willis D. "Cavalry Maneuvers at Fort Knox." September–October 1937.
Crockett, Cary I. "Five Days at Fort Knox." November–December 1937.
Dillman, George. "1st Cavalry Division Maneuvers." January 1928.
Edmunds, K. B. "Tactics of a Mechanized Force: A Prophecy." July 1930.
Eldred, D. "Are Horses Essential in Modern War?" May–June 1944.
Eliot, George Fielding. "The Future of American Cavalry." May–June 1939.
Fickett, Edward M. "A Study of the Relationship between the Cavalry and the Air Service in Reconnaissance." October 1923.
Grimes, William M. "The 1940 Spring Maneuvers." March–April 1940.
Grow, Robert W. "Mechanized Cavalry." January–February 1938.
———. "Mechanized Cavalry." May–June 1938.
———. "Military Characteristics of Combat Vehicles." November–December 1936.
———. "Mounted Combat: Lesson from the European Theater." November–December 1945.
———. "New Developments in the Organization and Equipment of Cavalry." March–April 1939.
———. "One Cavalry." May–June 1938.
Harmon, E. N. "Second Cavalry in the Meuse-Argonne." January 1923.
Hawkins, Hamilton S. "Cavalry and Mechanized Force." September–October 1931.
———. "We Must Have Cavalry." September–October 1937.
Henry, Guy V. "Mobility." April 1920.
Herr, J. K. "Editorial Comment." May–June 1946.
———. "What of the Future?" January–February 1939.
"Horse and Motors." *Argus*, as reported in (Springfield) *Illinois State Journal*. March–April 1936.
Hoy, Charles J. "The Last Days of Tunisia." January–February 1944.
———. "Reconnaissance Lessons from Tunisia." November–December 1943.
"Major General Guy V. Henry." March–April 1934.
"The Mechanized Cavalry Takes the Field." July–August 1938.
Miller, Alexander M. "Professional Notes and Discussions: Mechanized Cavalry." November–December 1931.

Noble, Charles. "Defense against Air Attack." July–August 1938.
"Observation Squadron to Fort Knox." May–June 1937.
"Organization and Activities." May–June 1935.
Palmer, Bruce. "Mechanized Cavalry in the Second Army Maneuvers." November–December 1936.
Palmer, Bruce, Jr. "New Battle Lessons on Reconnaissance." September–October 1943.
———. "Turn 'Em Over!" November–December 1940.
Parker, James. "The Mechanized Cavalry." September–October 1931.
Patton, George S., Jr. "Armored Cars with Cavalry." January 1924.
———. "Mechanized Forces." September–October 1933.
Patton, George S., Jr., and C. C. Benson. "Motorization and Mechanization in the Cavalry." July 1930.
Perry, Redding F. "The 2nd Cavalry In France." January 1928.
Pershing, John J. "A Message to the Cavalry." April 1920.
Reilly, Henry J. "Cavalry and Mechanized Force." January–February 1940.
———. "Horsed Cavalry and the Gas Engine's Children." January–February 1940.
Scott, Charles L. "Armor for Cavalry Reconnaissance Vehicles Is Essential." September–October 1938.
———. "Armored Reconnaissance." November–December 1942.
———. "Progress in Cavalry Mechanization: Scout Car Developments." July–August 1936.
Sherrill, Stephen H. "The Experiences of the First American Troop of Cavalry to Get into Action in the World War." April 1923.
Smith, John B. "The Effects of Mechanization upon Cavalry." November–December 1931.
Summerall, Charles P. "Leadership." January 1921.
Surles, Alexander D. "The Cavalry and Mechanization, 1936." January–February 1936.
Williams, Ray E. "U.S. Constabulary Horse Cavalry." May–June 1948.
Wilson, Arthur. "The Mechanized Force: Its Organization and Present Equipment." May–June 1931.

Infantry Journal

Benson, C. C. "The New Christie, Model 1940: An Estimate of the Tank Armored Car Combined." September–October 1929.
———. "Ships on the Battlefield." March–April 1929.
"Cavalry Combat." Review. September–October 1937.
Chynoweth, Bradford Grethen. "Tank Infantry." May 1921.
Colby, Elbridge. "Our Arm of the Service." January 1924.
"Discussion." May 1927.
"The New Christie, Model 1940: An Estimate of the Tank Armored Car Combined." September–October 1929.
Rockenbach, Samuel D. "Tanks and Their Cooperation with Other Arms." January–February 1920.

Summerall, Charles P. "New Developments in Mechanization." February 1931.
Von Schell, Adolf. "A Typical Fight in Open Warfare." July–August 1931.
Wilson, Arthur R. "The Mechanized Force: Its Organization and Present Equipment." May–June 1931.
———. "With the Mechanized Force on Maneuvers." July–August 1931.

Journal of Military History

Corum, James S. "The Spanish Civil War: Lessons Learned and Not Learned." April 1998.
Hofmann, George F. "The Tactical and Strategic Use of Attaché Intelligence: The Spanish Civil War and the U.S. Army's Misguided Quest for a Modern Tank Doctrine." January 1998.
Nenninger, Timothy K. "Leavenworth and Its Critics: The U.S. Army Command and General Staff School, 1920–1940." April 1994.

The Journal of Slavic Military Studies

Hofmann, George F. "Doctrine, Tank Technology, and Execution: I. A. Khalepskii and the Red Army's Fulfillment of Deep Offensive Operations." June 1996.

Journal of the United States Infantry Association

Taylor, John R. M. "Cavalry and the Aeroplane." July 1909.

Life

"Big Maneuvers Test U.S. Army." October 6, 1941.
"The Cavalry: How Men on Horseback Team Up with Machines" April 21, 1941.

Military Affairs, the Journal of Military History, Including Theory and Technology

Hofmann, George F. "The Demise of the U.S. Tank Corps and Medium Tank Development Program." February 1973.
Wilson, John R. "The Quaker and the Sword: Herbert Hoover's Relations with the Military." April 1974.

The Military Engineer

Christmas, John K. "The Mechanization of Armies." September–October 1929.

Military Review

Grow, Robert W. "The Capture of Muhlhausen." November 1946.
———. Letter to Colonel Barrows, Editor. September 12, 1945.
———. "Mobility Unused." February 1953.
Irzyk, A.F. "Mobility, Vigilance, Justice." March 1947.
Rand, H. P. "A Progress Report on the United States Constabulary." October 1947.

Parameters

Rainey, James W. "Ambivalent Warfare: The Tactical Doctrine of the AEF in World War I." September 1983.

Prologue

Smythe. Donald. "A.E.F. Snafu at Sedan." September 5, 1973.

Society of Automotive Engineers* (S.A.E.) *Journal

Campbell, Levin H., Jr. "Special Automotive Equipment of the Army." September 1928

Williams, Clarence C. "Automotive Ordinance Equipment." October 1919.

Other

Argus
Assembly
Battalion Channel
Chicago Tribune
Cincinnati Post
Dallas Morning News
Elizabeth (N.J.) *Daily Journal*
Foreign Affairs
Free Lance-Star
Howitzer
Illustrated World
Indiana Military History Journal
Inside the Turret Armor Conference Issue
Iron Age
Journal of American History
Journal of America's Military Past
Journal of the Society of Automotive Engineers
Liberty Magazine
London Times
Louisiana Center of Tourism
Louisville Courier-Journal
Louisville Times
Medical Times
Military Science Review
New York Times
Newsweek
Powder River Journal
Shell Progress
The Super Sixer
Time
U.S. News and World Report
Washington Post

Index